COLLISIONAL EFFECTS
ON MOLECULAR SPECTRA

COLLISIONAL EFFECTS ON MOLECULAR SPECTRA

Laboratory Experiments and Models, Consequences for Applications

Jean-Michel HARTMANN
Directeur de Recherche au CNRS
Laboratoire Inter-universitaire des Systèmes Atmosphériques
Universités Paris VII, Paris XII et CNRS
Faculté des Sciences et Technologie
61 avenue du Général de Gaulle
94010 Créteil Cedex, France

Christian BOULET
Professeur des Universités
Laboratoire de PhotoPhysique Moléculaire
Université Paris XI et CNRS
Campus d'Orsay (bâtiment 350)
91405 Orsay Cedex, France

Daniel ROBERT
Professeur émérite des Universités
Institut UTINAM
Université de Franche-Comté et CNRS
UFR Sciences & Techniques
16 route de Gray
25030 Besançon Cedex, France

ELSEVIER

Amsterdam • Boston • Heidelberg • London • New York • Oxford
Paris • San Diego • San Francisco • Singapore • Sydney • Tokyo

Elsevier
Radarweg 29, PO Box 211, 1000 AE Amsterdam, The Netherlands
Linacre House, Jordan Hill, Oxford OX2 8DP, UK

First edition 2008

Copyright © 2008 Elsevier B.V. All rights reserved

No part of this publication may be reproduced, stored in a retrieval system
or transmitted in any form or by any means electronic, mechanical, photocopying,
recording or otherwise without the prior written permission of the publisher

Permissions may be sought directly from Elsevier's Science & Technology Rights
Department in Oxford, UK: phone (+44) (0) 1865 843830; fax (+44) (0) 1865 853333;
email: permissions@elsevier.com. Alternatively you can submit your request online by
visiting the Elsevier web site at http://www.elsevier.com/locate/permissions, and selecting
Obtaining permission to use Elsevier material

Notice
No responsibility is assumed by the publisher for any injury and/or damage to persons
or property as a matter of products liability, negligence or otherwise, or from any use
or operation of any methods, products, instructions or ideas contained in the material
herein. Because of rapid advances in the medical sciences, in particular, independent
verification of diagnoses and drug dosages should be made

Library of Congress Cataloging-in-Publication Data
A catalog record for this book is available from the Library of Congress

British Library Cataloguing in Publication Data
A catalogue record for this book is available from the British Library

ISBN: 978-0-444-52017-3

For information on all Elsevier publications
visit our website at elsevierdirect.com

Printed and bound in Hungary

08 09 10 11 12   10 9 8 7 6 5 4 3 2 1

**Working together to grow
libraries in developing countries**

www.elsevier.com | www.bookaid.org | www.sabre.org

ELSEVIER   BOOK AID International   Sabre Foundation

This book was co-written by his father
in memory of Yann Robert

CONTENTS

FOREWORD		xiii
ACKNOWLEDGMENTS		xv
I	**INTRODUCTION**	**1**
II	**GENERAL EQUATIONS**	**9**
II.1	INTRODUCTION	9
II.2	DIPOLE AUTOCORRELATION FUNCTION	10
	II.2.1 General formalism	10
	II.2.2 The Hamiltonian of the molecular system	13
II.3	TOWARD "CONVENTIONAL" IMPACT THEORIES	16
	II.3.1 General properties of the correlation function	16
	II.3.2 The binary collision approximation	17
	II.3.3 Initial statistical correlations	19
	II.3.4 The impact approximation	20
II.4	BEYOND THE IMPACT APPROXIMATION	23
II.5	EFFECTS OF THE RADIATOR TRANSLATIONAL MOTION	25
II.6	COLLISION-INDUCED SPECTRA	28
II.7	CONCLUSION	33
	APPENDICES	
	II.A Spectral and time domain profiles in various spectroscopies	33
	1. Absorption, emission, and dispersion	33
	2. Rayleigh and spontaneous Raman scatterings	35
	3. Nonlinear Raman spectroscopies	38
	4. Time-resolved Raman spectroscopies	42
	II.B Some criteria for the approximations	44
	1. The large number of perturbers	44
	2. The local thermodynamic equilibrium	45
	3. The binary collisions	47
	4. The (full) impact assumption	49
	II.C The impact relaxation matrix	50
	1. Analysis through the time dependence	50
	2. Analysis through the frequency dependence	54
	II.D The Liouville space	55

	II.E	The resolvent approach		58
		1. Spectral-shape expression		58
		2. Rotational invariance		60
		3. Detailed balance		61

III ISOLATED LINES   63
III.1 INTRODUCTION   63
III.2 DOPPLER BROADENING AND DICKE NARROWING   73
     III.2.1 The Doppler broadening   74
     III.2.2 The Dicke narrowing   75
III.3 BASIC MODELS FOR SPECTRAL LINE SHAPES   77
     III.3.1 The Lorentz profile   77
     III.3.2 The Dicke profile   78
     III.3.3 The Voigt profile   79
     III.3.4 The Galatry profile   80
     III.3.5 The Nelkin-Ghatak profile   81
     III.3.6 Correlated profiles   83
     III.3.7 Characteristics of the basic profiles   85
III.4 SPEED-DEPENDENT LINE-SHAPE MODELS   90
     III.4.1 Observation of speed-dependent inhomogeneous profiles   90
     III.4.2 Basic speed-dependent profiles   98
     III.4.3 The Rautian-Sobelman model   104
     III.4.4 The Keilson-Storer memory model   114
III.5 AB INITIO APPROACHES OF THE LINE SHAPE   126
     III.5.1 The Waldmann-Snider kinetic equation   126
     III.5.2 The generalized Hess method   128
     III.5.3 Collision kernel method   130
     III.5.4 Approaches from a simplified Waldmann-Snider equation   133
III.6 CONCLUSION   139
APPENDIX
     III.A Computational aspects   140
          1. Algorithms for the Voigt and Galatry profiles   140
          2. Computation of speed-dependent profiles   142

IV COLLISIONAL LINE MIXING (WITHIN CLUSTERS OF LINES)   147
IV.1 INTRODUCTION   147
IV.2 THE SPECTRAL SHAPE   154
     IV.2.1 Approximations and general expressions   154
     IV.2.2 Asymptotic expansions   158
     IV.2.3 Computational aspects and recommendations   169

IV.3 CONSTRUCTING THE IMPACT RELAXATION MATRIX 173
- IV.3.1 Simple empirical (classical) approaches 174
- IV.3.2 Statistically based energy gap fitting laws 181
- IV.3.3 Dynamically based scaling laws 188
- IV.3.4 Semi-classical models 199
- IV.3.5 Quantum models 211

IV.4 DETERMINING LINE-MIXING PARAMETERS FROM EXPERIMENTS 218
- IV.4.1 Introduction 218
- IV.4.2 Relaxation matrix elements 222
- IV.4.3 First-order line-coupling coefficients 224
- IV.4.4 Mixed theoretical model and measured spectra fitting approaches 227

IV.5 LITERATURE REVIEW 227
- IV.5.1 Available line-mixing data 228
- IV.5.2 Comparisons between predictions and laboratory measurements 229
- IV.5.3 Comparisons between predictions and atmospheric measurements 232

IV.6 CONCLUSION 232
APPENDICES
- IV.A Vibrational dephasing 233
- IV.B Perturbed wave functions 237
- IV.C Resonance broadening 238

V THE FAR WINGS (BEYOND THE IMPACT APPROXIMATION) 241
V.1 INTRODUCTION 241
V.2 EMPIRICAL MODELS 243
- V.2.1 The χ factor approach 243
- V.2.2 The tabulated continua 246
- V.2.3 Other approaches 248

V.3 FAR WINGS CALCULATIONS: THE QUASISTATIC APPROACH 248
- V.3.1 General expressions 249
- V.3.2 Practical implementation and typical results 252
- V.3.3 The band average line shape: back to the χ factors 255

V.4 FROM RESONANCE TO THE FAR WING: A PERTURBATIVE TREATMENT 257
- V.4.1 General expressions 257
- V.4.2 Illustrative results 259

	V.5	**FROM RESONANCE TO THE FAR WING: A NON-PERTURBATIVE TREATMENT**	**261**
		V.5.1 General expression	261
		V.5.2 Illustrative results	263
	V.6	**CONCLUSION**	**265**
		APPENDIX	
		V.A The water vapor continuum	266
		1. Definition, properties and semi-empirical modeling of the H_2O continuum	268
		2. On the origin of the water vapor continua	269
		3. The self- and N_2-broadened continua within the ν_2 band	271
		4. Conclusion	272
VI	**COLLISION-INDUCED ABSORPTION AND LIGHT SCATTERING**		**275**
	VI.1	**INTRODUCTION**	**275**
	VI.2	**COLLISION-INDUCED DIPOLES AND POLARIZABILITIES FOR DIATOMIC MOLECULES**	**276**
	VI.3	**COLLISION-INDUCED SPECTRA IN THE ISOTROPIC APPROXIMATION**	**277**
		VI.3.1 Two illustrative examples: H_2 and N_2	277
		VI.3.2 Modeling of the line shape	281
	VI.4	**EFFECTS OF THE ANISOTROPY OF THE INTERACTION POTENTIAL**	**284**
	VI.5	**THE IMPORTANCE OF BOUND AND QUASIBOUND STATES IN CIA SPECTRA**	**290**
	VI.6	**INTERFERENCE BETWEEN PERMANENT AND INDUCED DIPOLES (CIA) OR POLARIZABILITIES (CILS)**	**293**
		VI.6.1 Depolarized light scattering spectra of H_2 and N_2	294
		VI.6.2 The HD problem	296
		VI.6.3 Intercollisional dips	300
	VI.7	**CONCLUSION**	**301**
VII	**CONSEQUENCES FOR APPLICATIONS**		**303**
	VII.1	**INTRODUCTION**	**303**
	VII.2	**BASIC EQUATIONS**	**304**
		VII.2.1 Radiative heat transfer	304
		VII.2.2 Remote sensing	307

	VII.3	ISOLATED LINES	311
		VII.3.1 The basic Lorentz and Voigt profiles	311
		VII.3.2 More refined isolated line profiles	314
	VII.4	LINE MIXING WITHIN CLUSTERS OF LINES	318
	VII.5	ALLOWED BAND WINGS AND CIA	325
		VII.5.1 Allowed band wings	325
		VII.5.2 Collision-induced absorption	331
	VII.6	CONCLUSION	333
VIII	**TOWARD FUTURE RESEARCHES**		**335**
	VIII.1	INTRODUCTION	335
	VIII.2	DICKE NARROWING IN SPEED-DEPENDENT LINE-MIXING PROFILES	335
		VIII.2.1 Models of profiles in the hard collision frame	335
		VIII.2.2 Experimental tests in multiplet spectra	339
	VIII.3	FROM RESONANCES TO THE FAR WINGS	343
		VIII.3.1 Semi-classical approach	344
		VIII.3.2 Generalized scaling approach	348
	VIII.4	TOMORROW'S SPECTROSCOPIC DATABASES	348
		VIII.4.1 Isolated lines	349
		VIII.4.2 Line mixing	351
		VIII.4.3 Far-wings and collision-induced absorption	352
	VIII.5	CONCLUSION	354

APPENDIX	**357**
ABBREVIATIONS AND ACRONYMS	357
SYMBOLS	360
UNITS AND CONVERSIONS	362

REFERENCES	**365**

SUBJECT INDEX	**409**

FOREWORD

Remote sensing is based entirely on spectroscopy. For astronomy or atmospheric science the physical and chemical properties of a distant object are all derived from the interpretation of spectra. The simulation of atomic and molecular spectra requires at a minimum line positions, line intensities and line shapes, regardless of the state of aggregation or spectral resolution. The position and intensity of a line or band are relatively straightforward to measure or even calculate with modern quantum mechanical methods. In contrast, the line shape or band shape is very difficult to model in part because it depends in a very complex way on the particular environment.

The study of line shapes is a venerable subject that predates quantum mechanics and yet is still a vigorous area of experimental and theoretical activity. Over the years a variety of sophisticated theoretical models and approaches have been developed to model spectral shapes of gases, but the subject has been relatively obscure. This has all changed in recent years because of the tremendous advances in spectroscopic instrumentation. For example, with tunable single frequency lasers the intrinsic line shape can be measured with a signal-to-noise ratio of several thousand in the laboratory and with Fourier transform spectrometers signal-to-noise ratios of several hundred are routinely obtained for atmospheric remote sensing. With such high spectral resolution and high quality spectra, none of the simple line shape functions reproduce the observations except at very low pressures.

The traditional favorite general-purpose line profile is the Voigt function, which includes the Gaussian and Lorentzian functions as special limiting cases. Modern high quality spectra are then said to need "non-Voigt" line shape functions in order to be modeled correctly. The physical source of this non-Voigt behavior may originate from many complex effects including collision-induced absorption, collisional line mixing, line shape failure in the far wings (weak collisions), and a variety of other speed-dependent collisional effects (*eg* Dicke narrowing). Collisions are hard to model correctly!

What has happened in the last few years is that the study of spectral shapes has changed from being a specialized research subject of a small group of physicists to being a necessary tool in a much wider community. Astronomers, atmospheric scientists, combustion scientists and chemists who deal with modern high quality spectra all encounter non-Voigt line shapes. It is therefore very timely to have this monograph by J.-M. Hartmann, Ch. Boulet, and D. Robert: it should find wide acceptance and utility.

Peter BERNATH
Department of Chemistry
University of York
Heslington, York
United Kingdom YO10 5DD

ACKNOWLEDGMENTS

The present authors are three successive links of a researchers chain which was initiated by Louis Galatry. They take the opportunity to express to him their deep gratitude for opening a path which has led, many years later, to the writing of this monograph. They are pleased that a new link has been added to this chain, through the recent nomination of Ha Tran as permanent scientist. More generally, the authors thank all the research workers with whom they collaborated in the past years and who, directly or indirectly, contributed to the results presented thereafter.

They express their gratitude to their colleagues who helped completing this book by providing data and figures, but also by their careful readings and suggested improvements. Among these, special thoughts to Jeanine Bonamy, Lionel Bonamy, Jean-Pierre Bouanich, Roman Ciurylo, Athéna Coustenis, Robert Gamache, Claude Girardet, Pierre Joubert, Bruno Lavorel, Massimo Moraldi, Jean-Pierre Perchard, François Rohart, Franck Thibault, Richard Tipping, Ha Tran, and Jean-Marie Vigoureux, for reviewing parts of this book. Gratitude, also, to Peter Bernath for his writing of the foreword and to Martine Bresson-Rosenmann for typing parts of this book.

Jean-Michel Hartmann is grateful to his wife Laurence and to his children Sarah, Léna and Raphaël, who kindly supported him while he was writing this book. He also thanks his parents who helped him all along the long studies which led him to a permanent research position.

Christian Boulet thanks all those close to him, and particularly Marie-José, Valérie and Martial for their encouragements, support, and forbearance during the considerable gestation period of this book, *ie* the last twenty years.

Daniel Robert wishes to express his hearty gratitude to his wife Suzanne for constant encouragements all along this work. He also thanks Franck Thibault for his kind and helpful hospitality during many short stays in Rennes University.

Finally, the authors thank the French research system (*Ministère de l'Enseignement Supérieur et de la Recherche* and *Centre National de la Recherche Scientifique*) for providing salaries and freedom of mind.

I. INTRODUCTION

Gas phase molecular spectroscopy is a field of chemical physics that was born about one hundred and fifty years ago. In spite of this long history, it remains an active discipline to which many international conferences and journals are devoted, fed by a large community of research workers. The reasons for the vitality of this field are many among which one finds, in the last two decades, the constant improvement of experimental devices and the growing concern about the influence of human activities on the evolution of the Earth atmosphere. Modern laboratory measurements, thanks to their considerable sensitivity, enable the study of smaller and smaller details of molecule-radiation interactions, thus requiring a constant improvement of the understanding and modeling of the processes involved. The very high spectral resolution and signal-to-noise ratio achieved by laser absorption experiments now permit the identification of refined mechanisms which affect local features of the spectral shape at a level below one per thousand. Their proper modeling is a challenge which requires improved theoretical approaches (and the relevant input data), in terms of the quality of predictions for the most important processes involved but also of the number of mechanisms which must be taken into account. The urgent need for a better understanding of the different phenomena driving the Earth atmospheric system, which has led to large international efforts in the development of remote sensing experiments, is another reason for the stimulation of spectroscopic research. Today's Fourier transform satellite-borne instruments, for instance, provide very large amounts of spectral information at ever increasing quality in terms of spectral coverage, resolution, and signal-to-noise ratio. The treatment of these data and the accuracy requirements for the development of atmospheric physics models result in the need to constantly improve the quality and scope of the spectroscopic knowledge used for the simulation (and inversion) of measured spectra. This not only calls for increased accuracy of the spectroscopic parameters, but also for studies of new spectral regions and/or molecular species whose inclusion in the remote sensing process brings new or complementary information of the atmospheric state. The vastness of the problem is illustrated by Fig. I.1, where an atmospheric transmission spectrum is displayed. This plot shows the richness of information brought by modern sounding instruments and the subsequent considerable amount of spectroscopic knowledge required for the modeling of the signatures of the numerous species (including H_2O, CO_2, O_3, CH_4, HNO_3, *etc*) contributing to the measured spectrum.

The first (basic) parameters governing the interactions between molecules and electromagnetic fields are the internal energy levels and the matrix elements of the relevant tensor (electric dipole, quadrupole, polarizability, *etc*) responsible for the coupling between radiation and matter. These quantities translate into the frequencies and integrated intensities of the optical transitions. They manifest themselves, in the spectral domain, through the positions and the areas of the absorption/emission lines, whereas they are involved in the relative evolution and amplitude of the signal in time domain

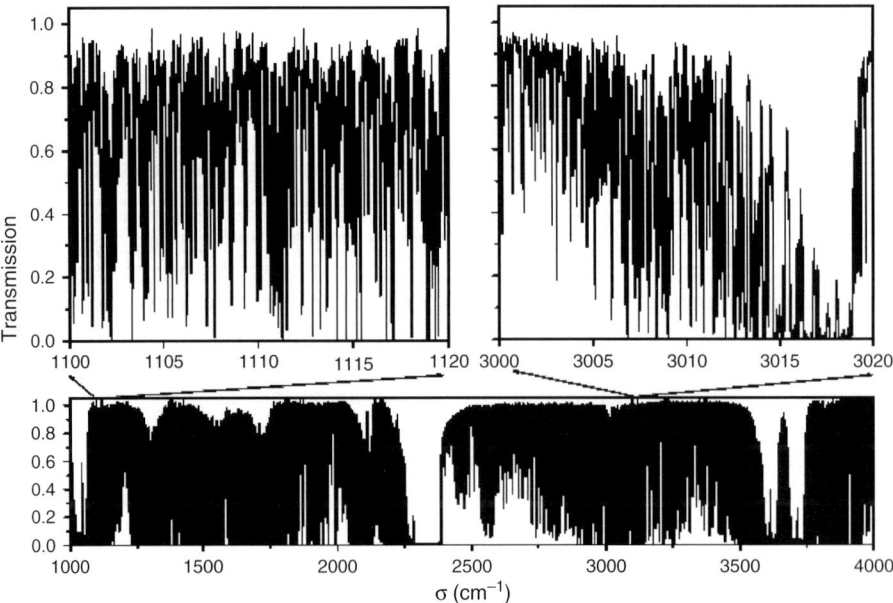

Fig. I.1: Transmission of the Earth atmosphere, *vs* wavenumber, as measured by the satellite based Fourier transform instrument of the Atmospheric Chemistry Experiment[1] when looking to a tangent height (Fig. VII.2c) of about 22 km. Courtesy of P. Bernath.

experiments. These spectroscopic data (energies and optical transition moments) are *intrinsic* characteristics of molecular species and are thus tied to the *isolated* molecule, regardless of its environment. They bring information on the geometry and charge distribution but also on the internal forces within the ensemble of particles composing the considered molecule. They are the *first* quantities to be known for any modeling of gas-radiation interactions. For these reasons, they have been, and still are, the subject of many researches leading to the constant improvement of widely used molecular spectroscopic databases.[2–5] These progresses result from experimental efforts which provide, through laser and spectrometer laboratory techniques,[6–12] the spectral signatures of the various molecular transitions (rotational, vibrational, electronic, *etc*) induced by the interaction with the electromagnetic field. From the theoretical point of view, there are basically two different approaches for the modeling of the energies and transition moments (see Refs. 13–18 and those therein). The first is *effective*, and relies on the availability of experimentally determined data on the positions and integrated intensities of the transitions. It consists in the adjustment, to measurements, of some pertinent parameters describing the Hamiltonian and the tensor that couples radiation and matter. This approach generally leads to results with an accuracy close to that of experiments (typically 10^{-5} cm^{-1} for the line positions) and it can then be used for the prediction of some not yet measured values. Nevertheless, it is hardly applicable to situations (such as highly excited states), where the number of interacting internal levels is very large.

For instance, the treatment of the polyad of methane near $13000\,\text{cm}^{-1}$, which involves almost two thousand vibrational levels and sub-levels, is clearly beyond the possibilities of effective Hamiltonian treatments. The second approach, purely theoretical and based on *ab initio calculations*, can, in principle, treat these highly excited states. However, such predictions need huge computer efforts and their accuracy (typically one cm^{-1} for the energies and positions) is still significantly poorer than the possibilities of measurements. They do not fulfill yet the accuracy needed for many practical applications. The spectral positions (or frequencies) and the optical transition moments (or intensities), which are characteristics of isolated molecules, are *not* the subject of the present book and will be considered as *known* in the following. Concerning this topic, the reader is invited to refer to the above mentioned references and to those therein for further information.

In "true" gas media, the optically active molecule (the "*radiator*") cannot be considered as alone, so that knowledge of the previously introduced intrinsic spectroscopic parameters is not sufficient for the modeling of the spectra or time-dependent signals. Except for very low pressures where considering only the Doppler effect may be sufficient, one must take into account the fact that the radiator is diluted in a "*bath*" of molecules or atoms (the "*perturbers*") with which it can interact. The effects of these *intermolecular interactions* increase with the gas density through the increase of the number of collisions per unit time between the optically active molecule and the others which compose the mixture. They lead, among other effects, to modifications of the shape of absorption/emission/scattering/refraction spectra and associated time-dependent signals. These changes are interesting for various reasons. The first one is that they are challenges for theoretical models[19–21] and for our understanding of collision processes. The second one is that they contain information on intermolecular interactions and can thus be used for the test, and improvement, of potential energy surfaces.[22–26] Finally, they must be properly modeled for accurate determinations of gas media properties from optical sounding techniques (Refs. 27,28 and chapter VII).

It is precisely this influence of intermolecular collisions (*ie* of the pressure) on the signatures, in the spectral or time domain, tied to the interaction of (neutral) gases with radiation that is the central subject of the present book. As a start and basis, in *chapter II*, some *general expressions* for the description of the effects of intermolecular interactions on the spectral shape are given. They contain the various effects associated with collisions but are essentially *formal* and often difficult to use in practical calculations. They are nevertheless of interest since they enable the introduction and discussion of some simplifications making calculations tractable in specific cases. These *approximations* naturally lead to the division of the general spectral-shape problem into particular (and more restricted) situations that are the subjects of the following parts of this book.

In order to introduce the various manifestations of intermolecular collisions discussed in chapters III to VI, let us consider the case of the transmission/emission of infrared radiation in the atmosphere of the Earth. The simple situation of *isolated transitions* is illustrated in Fig. I.2. In this case, the transmissions at elevated (tangent) altitudes, which involve low pressures, mostly depend on the position, integrated intensity and Doppler broadening of the lines, and on the instrument response [appearing here through a broadening and oscillations due to the instrumental function tied to the Fourier transform spectrometer used]. As the tangent height of the observation decreases, lower atmospheric

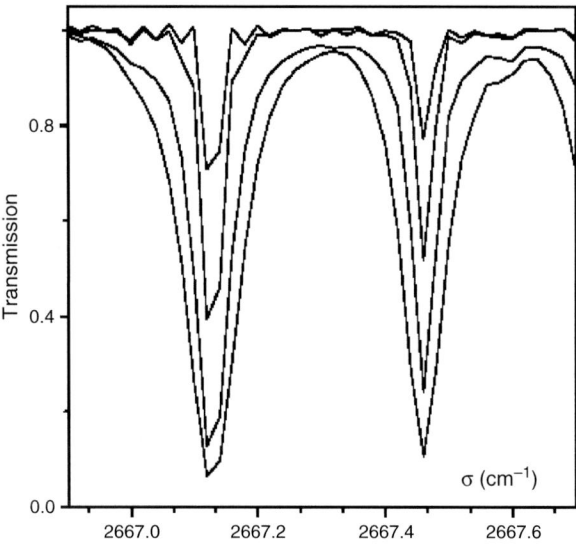

Fig. I.2: Transmissions (scaled by multiplicative factors in order to be unity in the least absorbing region) of the Earth atmosphere as measured by the satellite based Fourier transform instrument of the Atmospheric Chemistry Experiment[1] when looking down to tangent heights (cf Fig. VII.2c) (from top to bottom curve) of about 22, 16, 10, and 5 km. Courtesy of P. Bernath.

layers at higher pressures are sounded and the absorption features broaden due to the increasing contribution of collisional broadening. This progressively induces a switch from the Doppler to the Lorentz line profile. This is the first and simplest spectral manifestation of the effects of interactions between the optically active molecule (here CH_4) and the others (mostly N_2 and O_2). Together with the basic Doppler and Lorentz shapes and their combination leading to the Voigt profile, more refined collisional effects on the *shape of isolated transitions*, associated with velocity changes and velocity averaging, are the subject of *chapter III*.

Besides sparse and isolated transitions, spectra also contain clusters of *closely spaced* lines which differ only by the rotational states involved (thus belonging to the same vibrational band for instance). It is the case in Q branches or manifolds in the P or R branches. Whereas the lines may be isolated at low pressures, their contributions to the spectrum can overlap significantly at higher pressures. This situation is illustrated by the emission in the region of a Q branch of atmospheric CO_2 displayed in Fig. I.3. In this case, the Q lines are quite separated when high altitude (low pressure) layers are sounded. The measured emission then mostly depends on the positions, integrated intensities, and Doppler and Lorentz broadenings of the transitions and on the instrument function. As the tangent height decreases, higher pressures are sounded, and the Q lines overlap more and more while they increasingly broaden due to collisions. In this situation, the contributions of the various transitions are *not additive* anymore and the spectrum cannot be predicted by simply summing up their individual profiles. Indeed, intermolecular collisions induce transfers of populations among the internal levels

Introduction

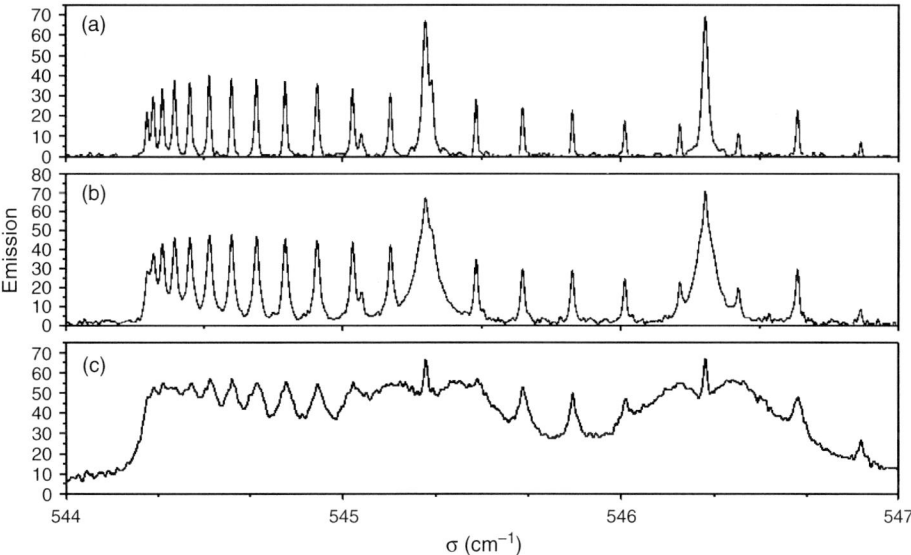

Fig. I.3: Emission of the Earth atmosphere as measured by the Smithsonian Astrophysical Observatory balloon-borne Fourier transform instrument[29] when looking down to tangent heights (cf Fig. VII.2c) of about 36 (a), 17 (b), and 12 (c) km. The region around the Q branch of the CO_2 $\nu_1+\nu_2-\nu_1$ band (centered at 544.3 cm^{-1}) is shown. The two intense lines are H_2O transitions. Courtesy of K. Jucks.

defining the optical transitions, thus resulting in intensity exchanges between the various spectral components. This process, called *collisional line mixing*, is the subject of *chapter IV* which is limited, due to the assumed *impact approximation* (cf § II.3.4), to spectral regions not too far away from the transitions of predominant contributions.

In optically thick situations (*ie* long paths with significant total pressures and intense bands of species present in large amounts) the absorption by the *far wings* of the lines can be important. It is the case, in the Earth atmosphere, of some H_2O and CO_2 bands, as illustrated by the transmission of the CO_2 ν_3 region in Fig. I.4. In this example, the absorption is essentially localized *within* the band when high altitude low pressure layers are sounded. The contribution of the line wings are very small and absorption beyond the band head (above 2390 cm^{-1} where the highest frequency transition of the ν_3 band lies) is negligible while the low frequency side is dominated by contributions of local lines. As the tangent altitude decreases, the absorption increases rapidly, largely due to the low and high frequency wings of the very intense CO_2 lines of the ν_3 band. In this situation, *none* of the approaches discussed in chapters III and IV is suitable. Since large distances from the line centers correspond to short times, specific models accounting for the finite duration of collisions are needed in *the far wings*, as discussed in *chapter V*.

Besides absorption by far line wings, large optical paths can also make the contribution of *collision-induced absorption* (CIA) significant. This process is due to the fact that intermolecular interactions give rise to a (transient) collisionally induced dipole (*collision-induced light scattering*, CILS, being associated with the induced polarizability). One of

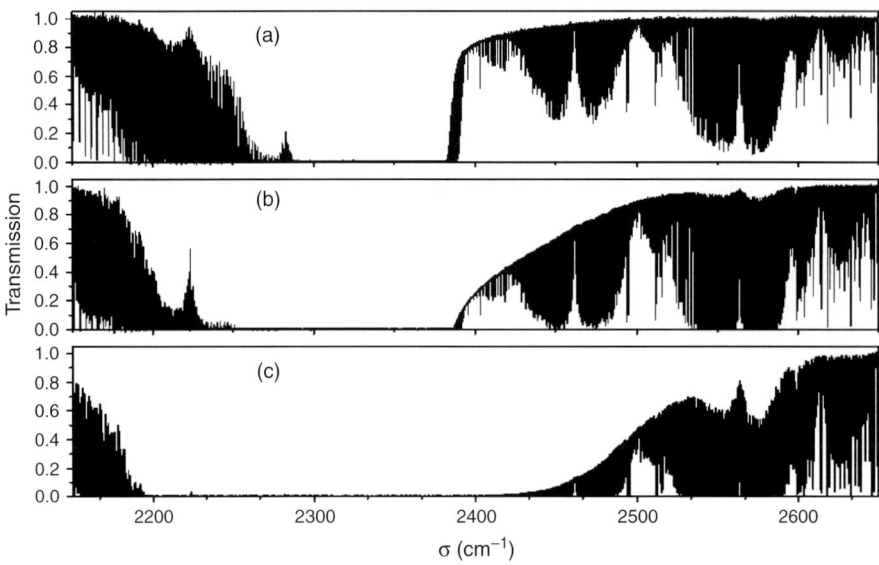

Fig. I.4: Transmissions (scaled by multiplicative factors in order to be unity in the least absorbing region) of the Earth atmosphere as measured by the satellite based Fourier transform instrument of the Atmospheric Chemistry Experiment[1] when looking down to tangent heights (cf Fig. VII.2c) of about 22 (a), 16 (b), and 10 (c) km. The region around the intense CO_2 v_3 band is shown. Courtesy of P. Bernath.

its obvious manifestations is the existence, for homonuclear diatomic molecules such as N_2, O_2, and H_2, of infrared absorption in the vibrational bands (forbidden in the absence of a permanent dipole moment). Contrary to the integrated intensities of "allowed" bands[1] which are proportional, for a pure gas, to the number density n, those of CIA bands are proportional to n^2, translating the (binary) collisional nature of the absorption process. Although quite weak, CIA cannot be neglected for optical paths involving large integrated amounts of radiators and significant total densities. This is illustrated in Fig. I.5, in the case of oxygen in the atmosphere of the Earth. These plots show narrow absorption features corresponding to allowed lines of O_2 and H_2O, carried by a broad continuum due to collision-induced absorption by O_2. The latter, which is hardly discernable in the case of small zenith angles (ZA, cf Fig. VII.2b) for which the optical path is moderate, increases rapidly with the ZA (by a factor of more than 3 from case "B" to case "D"). Specific models, different from those (chapters III–V) suitable for the description of the spectral shape of allowed transitions, are needed for the modeling of *collision-induced absorption and light scattering*. They are the subject of *chapter VI*.

In summary, after a *general* presentation of the spectral-shape problem giving the *basic equations* (chapter II), the pressure induced changes of the spectral shape will be considered going from very narrow to broad spectral regions, *ie: isolated transitions*

---

[1] tied to radiators that possess (when isolated) a non-zero tensor coupling them to the radiation field so that absorption is observed, even in the absence of significant collisional processes

$\omega_j = 2\pi\nu_j$, polarization vector $\vec{e}_j$, wave vector $\vec{k}_j$). The Hamiltonian of this mode is denoted H_j, while H_M is the Hamiltonian of the remaining part S_M of S. H_M thus contains not only contributions from the (colliding) particles and from their interactions with the field, but also from the radiation modes other than the jth. However, this last contribution, *ie* the "thermal bath of photons", is disregarded here since both *natural and radiative broadenings are neglected*. This is valid, in most practical situations, since the natural broadening is generally much smaller than those due to collisions and to the Doppler effect (*cf* § III.2.1). The radiative broadening can also be disregarded[9] since only weak external electromagnetic fields are considered here. The Hamiltonian H_{j-M} of the interaction between S_j and S_M is assumed to be switched-on at time $t = 0$ and switched-off at $t = \tau$. If $<F_j(0)>$ and $<E_j(\tau)>$ are respectively the averages over the whole system S of the energy of S_j at $t = 0$ and $t = \tau$, the absorption coefficient is defined by:[49]

$$\alpha(\omega_j) = \frac{1}{V}\frac{1}{P(\omega_j)}\lim_{\tau\to\infty}\left[<E_j(0)> - <E_j(\tau)>\right], \tag{II.1}$$

where $P(\omega_j)$ is the incoming intensity of the jth mode (its energy per unit surface perpendicular to the propagation direction at the entrance of the gas medium). Introducing the density operator $\rho(t)$ representing the statistical state of the entire system $S = S_M + S_j$ at time t, the average energy variation of the jth mode can be written as:

$$<E_j(\tau)> - <E_j(0)> = \text{Tr}_S\{H_j[\rho(\tau) - \rho(0)]\}, \tag{II.2}$$

where $\text{Tr}_S\{\ldots\}$ means the trace over all the states of the total system S. The problem of calculating the absorption coefficient is thus "reduced" to the determination of the evolution of the density operator from its initial distribution $\rho(0)$. The latter, which must be conveniently chosen according to the specific system of interest, generally coincides (thanks to LTE) with the equilibrium distribution for all variables describing the particles. The evolution of $\rho(t)$ is determined by the Liouville-von Neumann equation:[11,50,51]

$$i\hbar\frac{\partial}{\partial t}\rho(t) = [H, \rho(t)], \quad \text{where} \quad H = H_j + H_M + H_{j-M}, \tag{II.3}$$

$[A,B] = AB - BA$ being the commutator of A and B. We now consider the spectral distribution of radiation absorbed by molecular species "a", these *radiators* being in a "bath" of *perturbers* "p" of other species. We assume (*high dilutions*) that the number of radiators is a very small fraction of the total number of particles, so that interactions between molecules of species a can be neglected. The gas mixture is then considered as being divided into a large number of cells, each of them containing *a single* radiator and *many* perturbers (see the discussion in App. II.B1). For simplicity, we assume that these perturbers do not have any natural optical transition frequencies near those of the radiator. We thus consider them as *radiatively inactive* and hence ignore "resonant exchange" of radiation (see Refs. 47,52 and the discussion in appendix IV.C). *Note that*, provided that the (very) small effects of resonance exchange are neglected, the following equations do apply to pure gases and, more generally, to any mixture. Within these assumptions, the interaction H_{j-M} is given, in the dipolar approximation,[51] by:[(3)]

$$H_{j-M} = -\vec{d}\cdot\vec{E}_j(\vec{r}), \tag{II.4}$$

where $\vec{d}$ is the dipole moment of the radiator whose center of mass is located at $\vec{r}$ in the laboratory frame. The operator $\vec{E}_j(\vec{r})$, associated with the transverse electric field, is written in terms of the familiar creation (a_j^+) and annihilation (a_j) operators (within the Coulomb gauge) as:[53]

$$\vec{E}_j(\vec{r}) = iE_j^{(0)}\vec{e}_j\omega_j\left[a_j e^{i\vec{k}_j\cdot\vec{r}} - a_j^+ e^{-i\vec{k}_j\cdot\vec{r}}\right]. \tag{II.5}$$

Note that $\vec{d}$ is, a priori, the sum of the *intrinsic* dipole moment (permanent or induced by vibration for instance) of the (isolated) radiator and of the *dipole induced* in this molecule by its surrounding medium. For simplicity, we presently assume that the consequences of such induction effects are negligible with respect to the spectral influence of the intrinsic dipole. Situations where collision-induced absorption processes are not neglected are introduced at the end of this chapter and discussed in more detail in chapter VI. Writing $\rho(0) = \rho_j(0)\rho_M(0)$, and using a second order iteration in H_{j-M} for the resolution of Eq. (II.3) (linear response theory,[11,54] see App. II.A), the following expression is obtained for the absorption coefficient:

$$\alpha(\omega_j) = n_a \frac{4\pi^2\omega_j}{3\hbar c}\left(1 - e^{-\hbar\omega_j/k_BT}\right)F(\omega_j). \tag{II.6}$$

In this equation, n_a ($n_a = N_a/V$) is the number density (per unit volume) of radiators, c is the speed of light, and k_B is Boltzmann's constant. T is the absolute temperature, which drives all population distributions and is equal to the kinetic one since we assume the *Local Thermodynamic Equilibrium* (LTE). The case of non-LTE situations is a complex general problem which is beyond the scope of this book. It is nevertheless discussed in appendix II.B2, from the limited point of view of the use of collisional spectral shapes for the analysis of atmospheric spectra. Omitting the index j of the radiation mode, the *spectral density* $F(\omega)$ in Eq. (II.6) is the Fourier transform:

$$F(\omega) = \frac{1}{2\pi}\int_{-\infty}^{+\infty} e^{i\omega t}\Phi(t)dt, \tag{II.7}$$

of the *dipole autocorrelation function* $\Phi(t)$, given by[(4)]:

$$\Phi(t) = \text{Tr}_{S_M}\left\{\rho_M(0)e^{iH_Mt/\hbar}e^{-i\vec{k}\cdot\vec{r}}\vec{d}^+ e^{-iH_Mt/\hbar}\vec{d}\,e^{+i\vec{k}\cdot\vec{r}}\right\}, \tag{II.8}$$

[(3)] In the absence of a permanent dipole moment, higher order contributions to H_{j-M} must be taken into account. Among them one finds, for instance, that coming from the product of the quadrupole moment with the gradient of the electromagnetic field,[53] leading to the so-called "quadupolar absorption" (cf App. II.A).
[(4)] With respect to the case of emission,[55] in absorption, the passage from the molecular to the observer frame implies the change $\vec{r} \rightarrow -\vec{r}$ in the correlation function.[9]

General equations

where $\vec{d}^+$ is the adjoint of $\vec{d}$. The trace $\text{Tr}_{S_M}\{\ldots\}$ and the density operator $\rho_M(0)$ are now associated with all the states of the total system S except those of the radiation field. Introducing the following time-dependent quantities:

$$\vec{d}(t) = e^{iH_M t/\hbar}\vec{d}e^{-iH_M t/\hbar} \quad \text{and} \quad \vec{r}(t) = e^{iH_M t/\hbar}\vec{r}e^{-iH_M t/\hbar}, \tag{II.9}$$

Eq. (II.8) takes the more compact form:

$$\Phi(t) = \text{Tr}_{S_M}\left\{\rho_M(0)e^{-i\vec{k}\cdot\vec{r}(t)}\vec{d}^+(t)\vec{d}(0)e^{+i\vec{k}\cdot\vec{r}(0)}\right\}. \tag{II.10}$$

This equation is the quantum version of the more familiar classical expression:[56]
$\Phi_{cl}(t) = <\exp\{-i\vec{k}\cdot[\vec{r}(t)-\vec{r}(0)]\}\vec{d}(t)\cdot\vec{d}(0)>$, where $<\ldots>$ is a classical average over the initial conditions and all possible evolutions between times 0 and t. Let us now introduce the eigenstates (denoted by |I>, |F>) and associated eigenvalues (E_I, E_F) of H_M. Using the closure relation, Eq. (II.8) transforms to:

$$\Phi(t) = \sum_{I,F} \langle I|\rho_M(0)|I\rangle \langle I|e^{-i\vec{k}\cdot\vec{r}}\vec{d}^+|F\rangle e^{iE_I t/\hbar}\langle F|\vec{d}e^{+i\vec{k}\cdot\vec{r}}|I\rangle e^{-iE_F t/\hbar}. \tag{II.11}$$

The introduction of this expression into Eq. (II.7) then leads to:

$$F(\omega) = \sum_{I,F} \delta(\omega - \omega_{FI})\langle I|\rho_M(0)|I\rangle |\langle F|\vec{d}e^{+i\vec{k}\cdot\vec{r}}|I\rangle|^2, \tag{II.12}$$

where $\omega_{FI} \equiv (E_F - E_I)/\hbar$ and $\delta(\ldots)$ is the Dirac delta function. Since S_M is a macroscopic system, the eigenvalues of its Hamiltonian H_M are very close to each other and the spectrum is the juxtaposition of many closely spaced sharp singularities. However, finding the eigenstates and eigenvalues of H_M is totally hopeless due to the macroscopic nature of S_M and to the concomitant very large number of its degrees of freedom. Equation (II.12) is hence of no practical interest and a further decomposition of S_M must be made.

II.2.2 THE HAMILTONIAN OF THE MOLECULAR SYSTEM

In order to go further, thanks to the LTE assumption, we can introduce the canonical density operator:

$$\rho_M(0) = e^{-H_M/k_B T}/Z_M, \quad \text{with} \quad Z_M = \text{Tr}_{S_M}\left\{e^{-H_M/k_B T}\right\}. \tag{II.13}$$

Its introduction into Eq. (II.8) yields:

$$\Phi(t) = \frac{1}{Z_M}\text{Tr}_{S_M}\left\{e^{iH_M(t+i\hbar/k_B T)/\hbar}e^{-i\vec{k}\cdot\vec{r}}\vec{d}^+e^{-iH_M t/\hbar}\vec{d}e^{+i\vec{k}\cdot\vec{r}}\right\}. \tag{II.14}$$

Remember that H_M is the Hamiltonian of a system composed of *one* absorbing molecule and of N_p collision partners with which the radiator interacts. These perturbers will be considered as statistically *independent*, thus neglecting the influence of interactions between the perturbers on the absorption/emission processes. This assumption is clearly valid within the binary collision approximation, introduced later on, that underlies all the models presented in this book. Within this frame, H_M can be decomposed as:[55]

$$H_M = V_{a-p} + H_a + K_a + \sum_{i=1}^{N_p} [H_p(i) + K_p(i)], \qquad (II.15)$$

where:

→ H_a is the Hamiltonian for an unperturbed radiator which is at rest with respect to the observer. It describes the discrete spectrum associated with the internal states (rotational, vibrational, *etc*) of the optically active molecule;

→ K_a is the kinetic energy which describes the translational motion of the radiator relative to the observer;

→ Similarly, $H_p(i)$ describes the internal states of the ith perturber and $K_p(i)$ gives its kinetic energy relative to the observer;

→ Finally, V_{a-p} is the interaction potential between the radiator and *all* the perturbers. Since the latter are considered as independent, V_{a-p} is given by the sum of binary interactions V_{a-i} between the radiator and each of the perturbers, *ie*:

$$V_{a-p} = \sum_{i=1}^{N_p} V_{a-i}. \qquad (II.16)$$

Following Ref. 57, V_{a-p} is then divided into two parts: the first one (V_0) commutes with the internal coordinates of the molecules while the second one (V'_{a-p}) does not. In the rigid rotor limit, *ie* when one disregards the other (vibrational, electronic, *etc*) internal degrees of freedom, this distinction coincides with the separation of the interaction into isotropic and anisotropic forces. When one introduces the vibrational degrees of freedom, for instance, the part of the isotropic potential which depends on the vibrational coordinates must be grouped with the anisotropic interactions in V'_{a-p}. We thus decompose H_M according to:

$$H_M = H_0 + V'_{a-p}, \quad \text{with} \quad H_0 = V_0 + H_a + K_a + \sum_{i=1}^{N_p} [H_p(i) + K_p(i)]. \qquad (II.17)$$

The Hamiltonian H_0 contains the unperturbed internal and translational energies of the molecules together with the part of the radiator-perturbers interactions which commutes with the internal coordinates of the molecules. Hence, V'_{a-p} contains the remaining "*non commuting*" perturbing forces. Accordingly to Ref. 57, we introduce the time *evolution operator* $U(t, 0)$ [and its adjoint $U^+(t, 0)$] of the system between times 0 to t, defined in the interaction representation by:

$$e^{-iH_M t/\hbar} = e^{-iH_0 t/\hbar} U(t, 0) \quad \text{and} \quad e^{iH_M t/\hbar} = U^+(t, 0) e^{iH_0 t/\hbar}, \qquad (II.18)$$

where

$$U(t,0) = \exp_t[-i\hbar^{-1} \int_0^t \widetilde{V}_{a-p}(t')dt']. \quad (II.19)$$

In this expression, $\exp_t[\ldots]$ is a *time-ordered* exponential and:

$$\widetilde{V}_{a-p}(t) \equiv e^{iH_0t/\hbar} V'_{a-p} e^{-iH_0t/\hbar}. \quad (II.20)$$

Note that Eq. (II.18) can be extended to complex times z through:[57]

$$e^{-iH_Mz/\hbar} = e^{-iH_0z/\hbar} U(z,0), \quad (II.21)$$

with the adjoint relation:

$$e^{iH_Mz^*/\hbar} = U^+(z,0)e^{iH_0z^*/\hbar}, \quad (II.22)$$

where z^* is the complex conjugate of z. With these notations, the correlation function in Eq. (II.14) becomes:

$$\Phi(t) = \widetilde{v}\, \mathrm{Tr}_{S_M}\left\{\rho(H_0)e^{iH_0t/\hbar}U(-t-i\hbar/k_BT,0)e^{-i\vec{k}\cdot\vec{r}}\vec{d}^+ U^+(-t,0)e^{-iH_0t/\hbar}\vec{d}\,e^{+i\vec{k}\cdot\vec{r}}\right\}, \quad (II.23)$$

where

$$\rho(H_0) = e^{-H_0/k_BT}/Z_0, \quad \text{with} \quad Z_0 = \mathrm{Tr}_{S_M}[e^{-H_0/k_BT}], \quad (II.24)$$

and

$$\widetilde{v} \equiv Z_0/Z_M = \mathrm{Tr}_{S_M}[e^{-H_0/k_BT}]/\mathrm{Tr}_{S_M}[e^{-H_M/k_BT}]. \quad (II.25)$$

It is now convenient to separate the internal states of the radiator (Hamiltonian H_a) out of the expression of H_0, ie:

$$H_0 = H_a + \widetilde{H}_0. \quad (II.26)$$

According to Eq. (II.17), $\widetilde{H}_0$ describes the internal coordinates of all the perturbers, the translational coordinates of all the particles (including the radiator) and the commuting forces between the radiator and the perturbers. $\rho(H_0)$[Eq. (II.24)] then factors as:

$$\rho(H_0) = \rho(H_a)\rho(\widetilde{H}_0) \equiv \rho_a \rho(\widetilde{H}_0). \quad (II.27)$$

Introducing the eigenstates ($|i>$, $|f>$) and corresponding eigenvalues, (E_i, E_f) of the Hamiltonian H_a associated with the *internal states* of the radiator *only*, the correlation function in Eq. (II.23) can be rewritten as:

$$\Phi(t) = \sum_{i,f} e^{-i\omega_{fi}t}\langle i|\vec{D}(t)|f\rangle\langle f|\vec{d}|i\rangle\langle i|\rho_a|i\rangle, \quad (II.28)$$

with $\omega_{fi} \equiv (E_f - E_i)/\hbar$, and:

$$\vec{D}(t) = \tilde{v}\text{Tr}_{\widetilde{H_0}}\left\{\rho(\widetilde{H}_0)e^{i\widetilde{H}_0 t/\hbar}U(-t - i\hbar/k_B T, 0)e^{-i\vec{k}\cdot\vec{r}}\vec{d}^+U^+(-t, 0)e^{-i\widetilde{H}_0 t/\hbar}e^{+i\vec{k}\cdot\vec{r}}\right\}, \quad \text{(II.29)}$$

where $\text{Tr}_{\widetilde{H_0}}\{\ldots\}$ is now a trace limited to the internal and translational states of the perturbers and to the translational states of the radiator. Equations (II.28) and (II.29) provide the *starting point* for most of the formalisms presented in this book. Of course, these expressions are not yet amenable to practical calculations and simplifying approximations must still be introduced, according to the physical conditions and experimental data that one wishes to analyze. This is the subject of the next sections.

II.3 TOWARD "CONVENTIONAL" IMPACT THEORIES

When compared with Eq. (3.2) of Ref. 57, the correlation function defined by Eqs. (II.28) and (II.29) differs by the change $(+t) \rightarrow (-t)$ tied to the chosen definition [Eq. (II.7)] for the Fourier transform [*cf* Eq. (II.34)], and by the presence of the $\vec{k}\cdot\vec{r}$ exponentials. These terms are disregarded in most pressure-broadening theories, denoted here as "conventional", which thus neglect the influence of the radiator translational motion on the spectral profile. In this case, $\vec{r}$ is a scalar that does not operate on the eigenstates and the phase shifts $\exp(\pm i\vec{k}\cdot\vec{r})$ factor out as unity in Eq. (II.29). The spectral manifestations of the $\exp(\pm i\vec{k}\cdot\vec{r})$ terms are introduced in Sec. II.5 and discussed in chapter III, in the particular case of collisionally *isolated* lines [*ie* for i = i' and f = f' in Eq. (II.30)]. Extensions toward situations with collisionally coupled optical transitions are considered in Sec. VIII.2. Back to the neglect of the radiator translational motion, one can simplify Eq. (II.29) which leads to:

$$\Phi(t) = \tilde{v} \sum_{i,f,i',f'} e^{-i\omega_{fi}t}\langle i|\rho_a|i\rangle\langle f|\vec{d}|i\rangle\langle i'|\vec{d}^+|f'\rangle$$
$$\times \text{Tr}_{\widetilde{H_0}}\left\{\rho(\widetilde{H}_0)\langle i|U(-t - i\hbar/k_B T, 0)|i'\rangle\langle f'|U^+(-t, 0)|f\rangle\right\}. \quad \text{(II.30)}$$

II.3.1 GENERAL PROPERTIES OF THE CORRELATION FUNCTION

Before introducing further simplifications and derivations of the spectral-shape expression, we recall some general symmetry properties of the spectral density $F(\omega)$ and of the autocorrelation function $\Phi(t)$. Changing ω to $-\omega$ in Eq. (II.12), omitting the $\vec{k}\cdot\vec{r}$ exponentials, permuting the dummy indices I and F, and using the relations $\langle F|\rho_M|F\rangle = \langle I|\rho_M|I\rangle\exp(-\hbar\omega_{FI}/k_B T)$ and $\omega_{IF} = -\omega_{FI}$, one gets:

$$F(-\omega) = F(+\omega)e^{-\hbar\omega/k_B T}. \quad \text{(II.31)}$$

This equality is the expression of the so-called *fluctuation-dissipation theorem* (FDT), characterizing the linear response properties (here the spectral density) of a thermodynamic system at equilibrium to a small perturbation (here the electromagnetic field).[58] The importance of the application of the FDT to line-shape theory has been demonstrated by Kubo[59] and was emphasized later by Huber and Van Vleck.[60,61] From Eqs. (II.7) and (II. 31), the FDT translates, in the time domain, to:

$$\Phi(-t) = \Phi(t - i\hbar/k_B T), \tag{II.32}$$

and

$$\Phi(-t) = \Phi(t)^*, \tag{II.33}$$

which guarantees that the absorption coefficient $\alpha(\omega)$ is real. From these three relations, the following equivalent expressions can be easily derived: [(5)]

$$F(\omega) = \frac{1}{\pi}\text{Re}\left\{\int_0^\infty e^{i\omega t}\Phi(t)dt\right\} = \frac{1}{\pi}\text{Re}\left\{\int_0^\infty e^{-i\omega t}\Phi(-t)dt\right\}, \tag{II.34}$$

$$[1 - e^{-\hbar\omega/k_B T}]F(\omega) = \frac{1}{2\pi}\int_{-\infty}^{+\infty} e^{i\omega t}[\Phi(t) - \Phi(-t)]dt, \tag{II.35}$$

$$[1 - e^{-\hbar\omega/k_B T}]F(\omega) = \tanh(\hbar\omega/2k_B T)\frac{1}{2\pi}\int_{-\infty}^{+\infty} e^{i\omega t}[\Phi(t) + \Phi(-t)]dt, \tag{II.36}$$

where Re{..} denotes the real part. The demonstration that $\Phi(t)$, as given by Eq. (II.30), does satisfy the FDT is not trivial and can be found in Ref. 57.

II.3.2 THE BINARY COLLISION APPROXIMATION

A key approximation, assuming *binary collisions*, can now be introduced, thanks to the moderate perturber density (*cf* App. II.B3). It assumes[63] that strong collisions [for which an expansion to second order in $\widetilde{V}_{a-p}(t)$ of the evolution operator of Eq. (II.19) is insufficient] are well separated in time, and that the contribution of a weak collision overlapping a strong one is negligible in comparison. For simple notations, we consider the case of a linear radiator and denote by $|\beta Jm\rangle$ the eigenstates of H_a, where J is the rotational quantum number (magnitude of the angular momentum $\vec{J}$), m being the magnetic quantum number (projection of $\vec{J}$ on a spaced fixed z axis), and β stands for all other (*eg* vibrational, electronic, *etc*) quantum numbers necessary to specify the radiator internal state. Applying the

---

[(5)] Implicit in Eqs (II.34)–(II.36) and (II.7), a factor $\exp(-\varepsilon t)$ (ε being a small real positive quantity, $\varepsilon \to 0^+$) has to be added in order to ensure the convergence of the time integral.[62]

Wigner-Eckhart theorem[64] to the $\vec{d}$ matrix elements allows one to introduce the associated reduced matrix elements $\langle ...\|d\|...\rangle$, which originate from the spatial isotropy of the radiating dipole in a gas:

$$\langle \beta_f J_f m_f | \vec{d} | \beta_i J_i m_i \rangle \langle \beta'_i J'_i m'_i | \vec{d}^{+} | \beta'_f J'_f m'_f \rangle = \langle \beta'_f J'_f \| d \| \beta'_i J'_i \rangle^* \langle \beta_f J_f \| d \| \beta_i J_i \rangle$$

$$\times \sum_{q=-1,0,1} (-1)^{J_f+m_f+J'_f+m'_f} \sqrt{(2J'_i+1)(2J_i+1)} \begin{pmatrix} J'_i & 1 & J'_f \\ m'_i & q & -m'_f \end{pmatrix} \begin{pmatrix} J_i & 1 & J_f \\ m_i & q & -m_f \end{pmatrix}, \quad \text{(II.37)}$$

where (:::) is a 3J symbol.[64,65] Inserting this relation into Eq. (II.30) leads to:

$$\Phi(t) = \tilde{v} \sum_{i,f,i',f'} e^{-i\omega_{fi}t} (2J_i+1) \langle \beta_i J_i | \rho_a | \beta_i J_i \rangle \langle \beta'_f J'_f \| d \| \beta'_i J'_i \rangle^*$$

$$\times \langle \beta_f J_f \| d \| \beta_i J_i \rangle \text{Tr}_{\widetilde{H}_0} \left\{ \rho(\widetilde{H}_0) C_{f'i',fi}(t) \right\}, \quad \text{(II.38)}$$

with

$$C_{f'i',fi}(t) \equiv \sum_{\substack{q=-1,0,1 \\ m_i,m_f,m'_i,m'_f}} (-1)^{J_f+m_f+J'_f+m'_f} \sqrt{\frac{(2J'_i+1)}{(2J_i+1)}} \begin{pmatrix} J'_i & 1 & J'_f \\ m'_i & q & -m'_f \end{pmatrix} \begin{pmatrix} J_i & 1 & J_f \\ m_i & q & -m_f \end{pmatrix}$$

$$\times \langle \beta_i J_i m_i | U(-t-i\hbar/k_B T, 0) | \beta'_i J'_i m'_i \rangle \times \langle \beta'_f J'_f m'_f | U^{+}(-t,0) | \beta_f J_f m_f \rangle. \quad \text{(II.39)}$$

The binary collision approximation enables a great simplification of Eqs. (II.38) and (II.39). Indeed, having assumed that the perturbers are statistically independent, that their number N_p is large, and that strong collisions do not overlap, one can express the average in Eq. (II.38) as:[57,63]

$$\text{Tr}_{\widetilde{H}_0} \left\{ \rho(\widetilde{H}_0) C(t) \right\} = \exp_t \left\{ -n_p V \times \text{Tr}_{\widetilde{H}_0^1} \{ \rho(\widetilde{H}_0^1)[I_d - q(t)] \} \right\}, \quad \text{(II.40)}$$

where I_d is the identity operator and n_p is the number density (per unit volume) of perturbers ($n_p \equiv N_p/V$). In Eq. (II.40), the trace and density operator in the right hand side are now associated with a *single* perturber through $\widetilde{H}_0^1$, contrary to $\text{Tr}_{\widetilde{H}_0}$ and $\rho(\widetilde{H}_0)$ which are for all N_p independent perturbers. They carry out an average over the position of the radiator inside the considered volume (that will remove the V term), over the perturber internal states and over all parameters describing the *relative* motion (with respect to the center of mass) of the radiator-perturber couple. Within a classical description, the latter include the impact parameter b, the initial relative speed v_r and the time t_0 of closest approach describing the collision (in space, momentum and time).

General equations 19

The operator $[I_d - q(t)]$ is formally analog to C(t) in Eq. (II.39) and its corresponding elements are given by:

$$[I_d - q(t)]_{f'i',fi} = \delta_{\beta_i,\beta'_i}\delta_{\beta_f,\beta'_f}\delta_{J_i,J'_i}\delta_{J_f,J'_f} - \sum_{\substack{q=-1,0,1 \\ m_i,m_f,m'_i,m'_f}} (-1)^{J_f+m_f+J'_f+m'_f}\sqrt{\frac{(2J'_i+1)}{(2J_i+1)}}$$

$$\times \begin{pmatrix} J'_i & 1 & J'_f \\ m'_i & q & -m'_f \end{pmatrix} \begin{pmatrix} J_i & 1 & J_f \\ m_i & q & -m_f \end{pmatrix} \langle \beta_i J_i m_i | U_1(-t - i\hbar/k_B T, 0) | \beta'_i J'_i m'_i \rangle$$

$$\times \langle \beta'_f J'_f m'_f | U_1^+(-t, 0) | \beta_f J_f m_f \rangle. \quad \text{(II.41)}$$

Due to properties of the 3J symbols,[64,65] this is sometimes alternatively written as:

$$[I_d - q(t)]_{f'i',fi} = \sum_{\substack{q=-1,0,1 \\ m_i,m_f,m'_i,m'_f}} (-1)^{J_f+m_f+J'_f+m'_f}\sqrt{\frac{2J'_i+1}{2J_i+1}} \begin{pmatrix} J'_i & 1 & J'_f \\ m'_i & q & -m'_f \end{pmatrix} \begin{pmatrix} J_i & 1 & J_f \\ m_i & q & -m_f \end{pmatrix}$$

$$\times \{\delta_{i,i'}\delta_{f,f'} - \langle \beta_i J_i m_i | U_1(-t - i\hbar/k_B T, 0) | \beta'_i J'_i m'_i \rangle \quad \text{(II.42)}$$

$$\times \langle \beta'_f J'_f m'_f | U_1^+(-t, 0) | \beta_f J_f m_f \rangle\},$$

with $\delta_{i,i'}\delta_{f,f'} \equiv \delta_{\beta_i,\beta'_i}\delta_{\beta_f,\beta'_f}\delta_{J_i,J'_i}\delta_{J_f,J'_f}\delta_{m_i,m'_i}\delta_{m_f,m'_f}$. In these equations, U_1 refers now to a *binary* collision between *one* radiator and *one* single perturber. Introducing Eq. (II.40) into (II.38) and then into (II.34), the spectral density straightforwardly becomes:

$$F(\omega) = \tilde{v} \sum_{\substack{\beta_i,J_i \\ \beta_f,J_f}} (2J_i+1)\langle \beta_f J_f \|d\| \beta_i J_i \rangle \sum_{\substack{\beta'_i,J'_i \\ \beta'_f,J'_f}} \langle \beta'_f J'_f \|d\| \beta'_i J'_i \rangle^*$$

$$\times \frac{1}{\pi} \text{Re}\left[\int_0^{+\infty} dt e^{i(\omega-\omega_{fi})t}\exp_t\left\{-n_p V \times \text{Tr}_{\tilde{H}_0}^{-1}\{\rho(\tilde{H}_0^1)[I_d - q(t)]\}\right\}_{\beta'_f J'_f \beta'_i J'_i,\beta_f J_f \beta_i J_i}\right]. \quad \text{(II.43)}$$

Although significant simplifications have been made since Eq. (II.30), this last expression of the spectral density is still of difficult practical use and further simplifications must be made in some specific conditions. These are detailed below, naturally introducing some following chapters.

II.3.3 INITIAL STATISTICAL CORRELATIONS

If one is interested only by spectral regions "close" to the centers of the radiator optical transitions, it is then possible to neglect the influence of the initial correlations between the radiator and the perturbers contained in $\langle i|U_1(-t - i\hbar/k_B T, 0)|i'\rangle$. Let us introduce, in advance, a criterion that is part of those defining the impact approximation (*cf* Apps. II.B4 and II.C1). It assumes that $|\omega - \omega_{fi}|\tau_c \ll 1$, where τ_c is a collision duration taking the typical value (for usual temperatures and interactions) of 1 ps (10^{-12} s). Hence $|\omega - \omega_{fi}|$ is smaller than a few 10^{11} s^{-1}, value associated with times t greater than a few ps, thus much greater than $\hbar/k_B T$ (a few 10^{-14} s for "usual" temperatures) so that the effects of initial correlations can be neglected. This can be alternatively demonstrated, in the spectral domain, by the fact that $k_B T/(hc) \approx 200$ cm^{-1} is much greater than $|\omega - \omega_{fi}|/2\pi c = |\sigma - \sigma_{fi}| \approx$ a few cm^{-1}. Consequently, the matrix element $\langle i|U_1(-t-i\hbar/k_B T, 0)|i'\rangle$ in Eq. (II.42) can be replaced by $\langle i|U_1(-t,0)|i'\rangle$. Consistently, $\tilde{v}$ is well approximated by unity ($\tilde{v} = 1$) since one assumes, in Eq. (II.25), that $\rho(H_M) \approx \rho(H_0)$. Noting that:[66]

$$U(-t,0) = U^{-1}(0,-t) = U^{+}(0,-t), \qquad (II.44)$$

and changing $(0, -t)$ to $(+t, 0)$ since the correlation function is independent of the chosen initial time, Eq. (II.42) becomes:

$$[I_d - q(t)]_{f'i',fi} = \delta_{\beta_i,\beta'_i}\delta_{\beta_f,\beta'_f}\delta_{J_i,J'_i}\delta_{J_f,J'_f} - \sum_{\substack{q=-1,0,1 \\ m_i,m_f,m'_i,m'_f}} (-1)^{J_f+m_f+J'_f+m'_f} \sqrt{\frac{(2J'_i+1)}{(2J_i+1)}}$$

$$\times \begin{pmatrix} J'_i & 1 & J'_f \\ m'_i & q & -m'_f \end{pmatrix} \begin{pmatrix} J_i & 1 & J_f \\ m_i & q & -m_f \end{pmatrix} \langle \beta'_i J'_i m'_i | U_1(t,0) | \beta_i J_i m_i \rangle^*$$

$$\times \langle \beta'_f J'_f m'_f | U_1(t,0) | \beta_f J_f m_f \rangle$$

$$= \delta_{\beta_i,\beta'_i}\delta_{\beta_f,\beta'_f}\delta_{J_i,J'_i}\delta_{J_f,J'_f} - \sum_{\substack{q=-1,0,1 \\ m_i,m_f,m'_i,m'_f}} (-1)^{J_f+m_f+J'_f+m'_f} \sqrt{\frac{(2J'_i+1)}{(2J_i+1)}} \qquad (II.45)$$

$$\times \begin{pmatrix} J'_i & 1 & J'_f \\ m'_i & q & -m'_f \end{pmatrix} \begin{pmatrix} J_i & 1 & J_f \\ m_i & q & -m_f \end{pmatrix}$$

$$\langle\langle \beta'_f J'_f m'_f \beta'_i J'_i m'_i | \hat{U}_1(t,0) | \beta_f J_f m_f \beta_i J_i m_i \rangle\rangle.$$

In this equation we have introduced the Liouville (*cf* App. II.D) tetradic operator $\hat{U}_1(t,0)$ corresponding to $U_1(t,0)$. This is done for consistency with the alternative way to derive the spectral shape, based on the resolvent approach described in appendix II.E and used in many previous studies (*eg* Refs. 46,67). Note that the spectral density calculated from Eqs. (II.43) and (II.45) *does not* satisfy the FDT any more, due to the neglect of initial correlations, as demonstrated in Ref. 57.

II.3.4 THE IMPACT APPROXIMATION

We now introduce a variety of widely used approaches denoted as "impact models". These rely on a criterion, already mentioned, which limits the validity of the resulting spectral shape to regions close to the line centers of the most intense transitions. More precisely, we consider only "small" detuning $|\omega - \omega_{fi}|$ from the centers ω_{fi} of these lines, defined by $|\omega - \omega_{fi}|\tau_c \ll 1$. Together with the assumption that τ_c is much smaller than

General equations 21

the average time interval τ_0 between successive collisions, this is the impact approximation which is discussed in Apps. II.C and II.B4. In summary, it assumes that *all* times of interest (ie $|\omega-\omega_{fi}|^{-1}$, $|\omega_{f'i'}-\omega_{fi}|^{-1}$, and τ_0) are much greater than τ_c. Therefore, as suggested by Gordon[68,69] in a rate equation formulation of pressure broadening, and demonstrated later by Smith *et al*[63,70,71] and Albers and Deutch,[62] replacing $\text{Tr}_p\{\rho_p[I_d-q(t)]\}$ by the following long time behavior:

$$\exp_t\left\{-n_p V \times \text{Tr}_{\underset{H_0}{\sim 1}}\{\rho(\tilde{H}_0^1)[I_d-q(t)]\}\right\} \to e^{-(n_p W^* + iL_a)t} e^{+iL_a t}, \quad \text{for} \quad t \gg \tau_c, \quad (II.46)$$

where L_a is the Liouville operator constructed from $H_a/\hbar$ [cf Eq. (II.49)], yields the now familiar impact limit of the spectral shape. In Eq. (II.46), W^* is the complex conjugate of the so-called *impact* (frequency-independent) density-normalized *relaxation operator* W. Its matrix elements can be expressed in terms of the usual *scattering* (or diffusion) operator $S \equiv U_1(+\infty, -\infty)$:

$$\langle\langle \beta'_f J'_f \beta'_i J'_i | W | \beta_f J_f \beta_i J_i \rangle\rangle = \Big\langle \delta_{\beta_i,\beta'_i} \delta_{\beta_f,\beta'_f} \delta_{J_i,J'_i} \delta_{J_f,J'_f}$$

$$- \sum_{\substack{q=-1,0,1 \\ m_i, m_f, m'_i, m'_f}} (-1)^{J_f + m_f + J'_f + m'_f} \sqrt{\frac{(2J'_i+1)}{(2J_i+1)}}$$

$$\times \begin{pmatrix} J'_i & 1 & J'_f \\ m'_i & q & -m'_f \end{pmatrix} \begin{pmatrix} J_i & 1 & J_f \\ m_i & q & -m_f \end{pmatrix} \langle \beta'_f J'_f m'_f | S | \beta_f J_f m_f \rangle^*$$

$$\times \langle \beta'_i J'_i m'_i | S | \beta_i J_i m_i \rangle \Big\rangle_p. \quad (II.47)$$

Here $\langle\ldots\rangle_p$ means a proper average over the remaining variables that are the initial perturber internal state and collision parameters (impact parameter and relative velocity within a classical treatment). Introducing Eq. (II.46) into Eq. (II.43) and taking the Laplace transform leads to the following expression of the impact spectral density function:[6]

$$F^{\text{Impact}}(\omega) = \sum_{\substack{\beta_i, J_i \\ \beta_f, J_f}} \rho_{\beta_i, J_i} \langle \beta_f J_f \| d \| \beta_i J_i \rangle \sum_{\substack{\beta'_i, J'_i \\ \beta'_f, J'_f}} \langle \beta'_f J'_f \| d \| \beta'_i J'_i \rangle^*$$

$$\times \frac{1}{\pi} \text{Im}\left\{\langle\langle \beta'_f J'_f \beta'_i J'_i | [\omega I_d - L_a - in_p W]^{-1} | \beta_f J_f \beta_i J_i \rangle\rangle\right\}, \quad (II.48)$$

---

[6] In some studies, W is replaced by its complex conjugate, and the spectral density is no longer expressed from $\text{Im}[(\omega I_d - L_a - in_p W)^{-1}]$ as in Eq. (II.48), but from $-\text{Im}[(\omega I_d - L_a + in_p W)^{-1}]$, thus introducing two sign changes. As a result, the imaginary part of the diagonal elements of W are then the pressure shifting coefficients and not their opposites as in Eq. (II.52).

where Im{..} denotes the imaginary part. In this equation, $\rho_{\beta_i J_i} \equiv (2J_i + 1)\langle \beta_i J_i | \rho_a | \beta_i J_i \rangle$ is the population of the level $\beta_i J_i$ and we have introduced the Liouville (App. II.D) operators ωI_d and L_a, such that:

$$\langle\langle f'i'|\omega I_d|fi\rangle\rangle = \omega \times \delta_{i,i'}\delta_{f,f'},$$

$$\langle\langle f'i'|L_a|fi\rangle\rangle = \frac{\langle f|H_a|f\rangle - \langle i|H_a|i\rangle}{\hbar}\delta_{i,i'}\delta_{f,f'} = \frac{E_f - E_i}{\hbar}\delta_{i,i'}\delta_{f,f'} = \omega_{fi}\times\delta_{i,i'}\delta_{f,f'}. \quad (II.49)$$

For extensions of the present result to other types of measurements (see App. II.A), but recalling that the transition moments d *and* the relaxation operator *depend* on the order k of the tensor coupling radiation and matter, it is convenient to introduce the *complex* normalized profile $I^{Impact}(\omega)$ defined by:

$$I^{Impact}(\omega) = \frac{1}{\pi}\frac{1}{\sum_{\beta_i,J_i,\beta_f,J_f}\rho_{\beta_i J_i}\langle\beta_f J_f\|d\|\beta_i J_i\rangle^2}$$

$$\times \sum_{\substack{\beta_i,J_i,\beta_f,J_f \\ \beta'_i,J'_i,\beta'_f,J'_f}} \rho_{\beta_i J_i}\langle\beta_f J_f\|d\|\beta_i J_i\rangle\langle\beta'_f J'_f\|d\|\beta'_i J'_i\rangle^* \quad (II.50)$$

$$\times \left\{\langle\langle \beta'_f J'_f \beta'_i J'_i | [\omega I_d - L_a - in_p W]^{-1} | \beta_f J_f \beta_i J_i\rangle\rangle\right\},$$

to which the spectral density is related, for absorption, by:

$$F^{Impact}(\omega) = \left[\sum_{\beta_i,J_i,\beta_f,J_f}\rho_{\beta_i J_i}\langle\beta_f J_f\|d\|\beta_i J_i\rangle^2\right] \times \text{Im}\{I^{Impact}(\omega)\}. \quad (II.51)$$

Within this approach, *all* the effects of the collisions undergone by the radiator on the spectral shape are contained in the *impact (binary collision) relaxation matrix* W. Its diagonal elements are the usual pressure-broadening (γ_{fi}) and -shifting coefficients (δ_{fi}):

$$\gamma_{fi} \equiv \text{Re}\{\langle\langle fi|W|fi\rangle\rangle\} \text{ and } \delta_{fi} \equiv -\text{Im}\{\langle\langle fi|W|fi\rangle\rangle\}, \quad (II.52)$$

while the collision-induced transfers between the internal states of the radiator are described by the off-diagonal terms. Neglecting them straightforwardly leads to the usual (complex) Lorentzian *isolated line* normalized profile (*cf* § III.3.1), ie:

$$I^{Lorentz}_{fi}(\omega) \equiv \frac{1}{\pi}\langle\langle fi|\omega I_d - L_a - in_p W|fi\rangle\rangle^{-1} = \frac{1}{\pi}\frac{1}{\omega - \omega_{fi} - n_p\delta_{fi} - in_p\gamma_{fi}}. \quad (II.53)$$

The corresponding spectral density for a single line is then written as:

$$F^{Lorentz}_{fi}(\omega) = (\rho_i d^2_{fi}) \times \text{Im}\{I^{Lorentz}_{fi}(\omega)\} = (\rho_i d^2_{fi})\times\frac{1}{\pi}\frac{\Gamma_{fi}}{(\omega - \omega_{fi} - \Delta_{fi})^2 + (\Gamma_{fi})^2}, \quad (II.54)$$

General equations 23

with $\Gamma_{fi} = n_p\gamma_{fi}$ and $\Delta_{fi} = n_p\delta_{fi}$ and where $\text{Im}\{I^{\text{Lorentz}}(\omega)\}$ is normalized to unit area through integration over ω.

When the off-diagonal elements of W are non-zero, they give rise to the so-called *collisional line-mixing* (or -interference) effects. These are the subject of chapter IV, where their spectral manifestations and the theoretical and experimental approaches for the determination of the elements of W (*including* the diagonal terms) are described. For completeness, recall that some symmetry properties (with respect to interchanges of row and columns, *etc*) of this Zwanzig-Fano relaxation matrix have been analyzed by Ben-Reuven[45] and, more recently, by Ma *et al.*[72] Among these, let us mention the widely used so-called *detailed balance* relation (see App. II.E3), *ie*:

$$\langle\langle\beta'_f J'_f \beta'_i J'_i | W | \beta_f J_f \beta_i J_i\rangle\rangle \times \rho_{\beta_i J_i} = \langle\langle\beta_f J_f \beta_i J_i | W | \beta'_f J'_f \beta'_i J'_i\rangle\rangle \times \rho_{\beta'_i J'_i}. \tag{II.55}$$

Important remark: in some previous theoretical developments (*eg* Ref. 46) a different starting point and notations are used. For instance, $\text{Im}\{\ldots\}$ in Eqs. (II.48), (II.51), (II.54) is sometimes replaced by $-\text{Im}\{\ldots\}$. Consequently, the associated relaxation operator in these papers *corresponds to the complex conjugate* of the one introduced here and it is multiplied by $+in_p$ instead of $-in_p$. In some other studies $\text{Im}\{\ldots\}$ is replaced by $\text{Re}\{\ldots\}$ so that the complex profile I must be changed into iI^*, thus replacing $[\omega I_d - L_a - in_p W]$ by $[i(\omega I_d - L_a) + n_p W]$ in Eq. (II.50) if the real part is taken. Some studies[57,72] also start from a spectral density defined as the Fourier transform in $e^{-i\omega t}$, whereas we have made the choice of a $e^{+i\omega t}$ transform. Consequently the correlation function $\langle\vec{d}^+(t)\cdot\vec{d}(0)\rangle$ must be replaced by $\langle\vec{d}(0)\cdot\vec{d}^+(t)\rangle$. These different choices lead to final expressions of the spectral shape which may look different and involve different definitions of the associated relaxation operator but, provided that proper identifications are made, one can check that the final expressions are formally identical.

II.4 BEYOND THE IMPACT APPROXIMATION

The impact approximation limits the validity of the resulting spectral shape to regions close to the line centers of the most intense transitions through the above mentioned criterion $|\omega - \omega_{fi}|\tau_c \ll 1$. Except at very elevated pressures making the pressure-broadened widths of transitions $n_p\gamma_{fi}$ of the order or greater than $1/\tau_c$ (but in this case, the binary collision assumption is questionable and the problem becomes extremely complex), the impact approximation only breaks down in the *far wings*. In such regions, the spectral shape is *proportional* to the perturber density n_p, as well known and evidenced by many experiments. This corresponds to the fact that the matrix elements of the operator $n_p V \text{Tr}_{\widetilde{H}_0}^{\sim 1}\{\rho(\widetilde{H}_0)[I_d - q(t)]\}$ in Eq. (II.40) are small, so that the associated exponential can be linearized, *ie*:

$$\exp_t\left\{-n_p V \times \text{Tr}_{\widetilde{H}_0}^{\sim 1}\{\rho(\widetilde{H}_0)[I_d - q(t)]\}\right\}_{f'i',fi} \approx \delta_{i,i'}\delta_{f,f'} - n_p V \times \text{Tr}_{\widetilde{H}_0}^{\sim 1}\{\rho(\widetilde{H}_0)[I_d - q(t)]\}_{f'i',fi}. \tag{II.56}$$

Introducing this expression successively in Eqs. (II.40), (II.38), and (II.34) leads to:

$$F^{Wing}(\omega) = n_p \times \tilde{v} \sum_{\substack{\beta_i, J_i \\ \beta_f, J_f}} \rho_{\beta_i J_i} \langle \beta_f J_f \| d \| \beta_i J_i \rangle \sum_{\substack{\beta'_i, J'_i \\ \beta'_f, J'_f}} \langle \beta'_f J'_f \| d \| \beta'_i J'_i \rangle^* \frac{1}{\pi} \text{Re}\left[-Q(\omega)_{f'i',fi}\right], \quad \text{(II.57)}$$

with

$$Q(\omega)_{f'i',fi} = \int_0^{+\infty} dt e^{i(\omega-\omega_{fi})t} V \times \text{Tr}_{\tilde{H}_0^{-1}}\{\rho(\tilde{H}_0^1)[I_d - q(t)]\}_{f'i',fi}. \quad \text{(II.58)}$$

In order to put the spectral density under a form similar to that obtained within the impact frame [Eq. (II.50)], it is then natural to introduce a relaxation matrix $W(\omega)$ which is now *dependent* on the frequency ω. The corresponding "generalized" spectral density function is given by:

$$F(\omega) = \tilde{v} \sum_{\substack{\beta_i, J_i \\ \beta_f, J_f}} \rho_{\beta_i J_i} \langle \beta_f J_f \| d \| \beta_i J_i \rangle \sum_{\substack{\beta'_i, J'_i \\ \beta'_f, J'_{f'}}} \langle \beta''_f J'_f \| d \| \beta'_i J'_i \rangle^*$$

$$\times \frac{1}{\pi} \text{Im}\left\{ \langle\langle \beta'_f J'_f \beta'_i J'_i | [\omega I_d - L_a - i n_p W(\omega)]^{-1} | \beta_f J_f \beta_i J_i \rangle\rangle \right\}. \quad \text{(II.59)}$$

In the wings, for values of ω such that $|\omega - \omega_{fi}| \gg n_p |W(\omega)_{f'i',fi}|$, the identity:

$$[\omega I_d - L_a - i n_p W(\omega)]^{-1} \approx (\omega I_d - L_a)^{-1}[I_d + i n_p W(\omega)(\omega I_d - L_a)^{-1}], \quad \text{(II.60)}$$

can be introduced into Eq. (II.59). Then, the identification to Eq. (II.57) yields:

$$\langle\langle f'i' | W(\omega) | fi \rangle\rangle = -(\omega - \omega_{fi})(\omega - \omega_{f'i'}) Q(\omega)_{f'i',fi}. \quad \text{(II.61)}$$

Using Eq. (II.42), this leads to:

$$\{\langle\langle \beta'_f J'_f \beta'_i J'_i | W(\omega) | \beta_f J_f \beta_i J_i \rangle\rangle\} = -(\omega - \omega_{\beta_f J_f \beta_i J_i})(\omega - \omega_{\beta'_f J'_f \beta'_i J'_i})$$

$$\times \sum_{\substack{q=-1,0,1 \\ m_i, m_f, m'_i, m'_f}} (-1)^{J_f + m_f + J'_f + m'_f} \sqrt{\frac{2J'_i + 1}{2J_i + 1}}$$

$$\times \begin{pmatrix} J'_i & 1 & J'_f \\ m'_i & q & -m'_f \end{pmatrix} \begin{pmatrix} J_i & 1 & J_f \\ m_i & q & -m_f \end{pmatrix}$$

$$\times \left\langle \int_0^{+\infty} dt e^{i(\omega-\omega_{\beta_f J_f \beta_i J_i})t} \right.$$

$$\times \{\delta_{i,i'}\delta_{f,f'} - \langle \beta'_i J'_i m'_i | U_1(0, -t - i\hbar/k_B T) | \beta_i J_i m_i \rangle$$

$$\left. \times \langle \beta'_f J'_f m'_f | U_1(0, -t) | \beta_f J_f m_f \rangle^* \} \right\rangle, \quad \text{(II.62)}$$

where an average $\langle \ldots \rangle$ over all initial parameters describing the collision is made.

Various approaches for the modeling of the spectral shape in the *far wings* (of allowed lines) are presented in chapter V, where the frequency dependence of W is also discussed. Some simplifications are still needed in order to be able to carry out calculations for practical systems.

In this chapter, we have chosen to work in the "time domain", focusing on the operator U(t) which describes the evolution of the radiator under the influence of the perturbers. We have then derived expressions for U(t) which are of tractable use, through the introduction of some ad hoc assumptions valid under specific conditions (but broad and valid, with satisfactory accuracy, for a large number of applications). However, another path can be followed in which one works in the "frequency domain" from the earliest steps of the theoretical derivations. This "*resolvent*" method, recalled in appendix II.E, was initiated by Zwanzig[73] and Fano,[44] and further developed by Ben-Reuven.[46] It leads to final expressions identical to those derived above.

II.5 EFFECTS OF THE RADIATOR TRANSLATIONAL MOTION

Starting from the expression of the spectral density $F(\omega)$ [Eq. (II.7)] in terms of the dipole autocorrelation function Eq. (II.10), Smith *et al*[55] have developed a quantum *impact* theory for the *combined* effects of the radiator translational motion and pressure broadening for *coupled* lines. This motion results in the presence of the $\vec{k}.\vec{r}$ exponentials in Eq. (II.10)], leading to intricacy in the line-shape calculation. Indeed, the collision operator $S \equiv U_1(+\infty, -\infty)$ [Eq. (II.47)] being not diagonal in translational radiator states, it is therefore represented by an infinite dimensional matrix. In order to achieve the formal resolution, a differential kinetic equation approach was used. No calculation from such a quantum approach was performed for coupled lines, but, for isolated lines, ab initio methods were proposed by other authors, permitting accurate numerical calculations for some observed profiles (*cf* Sec. III.5).

In order to get a more tractable expression, Smith *et al*[74] have introduced semi-classical approximations. They enable to separate the influences of the radiator motion, of pressure broadening, and of their correlation. The resulting *semi-classical impact* expression of the spectral density $F(\omega)$ for N_L coupled transitions can be written as:[75]

$$F(\omega) = \text{Re}\left\{\sum_{\ell=1}^{N_L} \int \vec{d}_\ell^* \vec{F}_\ell(\omega, \vec{v}) d^3\vec{v}\right\}, \qquad (\text{II}.63)$$

with

$$\vec{F}_\ell(\omega, \vec{v}) = \frac{1}{\pi}\int_0^\infty e^{i\omega t}\vec{F}_\ell(t, \vec{v}) dt, \qquad (\text{II}.64)$$

where the subscript ℓ represents an optical transition through a couple of initial and final radiator internal states and d_ℓ is the corresponding transition moment [*cf* Eq. (II.37)]. Note that an additional (impact) assumption is introduced here through the neglect of the coupling between the radiator translational motion and the radiator momentum.

This means that the absorbed (or emitted) photon recoil momentum does not significantly affect the translational dynamics of the radiator. $\vec{F}_\ell(t,\vec{v})$ satisfies a Boltzmann-Liouville kinetic equation:

$$\frac{\partial \vec{F}_\ell(t,\vec{v})}{\partial t} = -i(\omega_\ell + \vec{k}\cdot\vec{v})\vec{F}_\ell(t,\vec{v}) - v_{VC_\ell}(v)\vec{F}_\ell(t,\vec{v}) + \int A_l(\vec{v},\vec{v}')\vec{F}_\ell(t,\vec{v}')d^3\vec{v}' \\ - \sum_{\ell'=1}^{N} \widetilde{W}_{\ell'\ell}(v)\vec{F}_{\ell'}(t,\vec{v}) - \sum_{\ell'=1}^{N} \int C_{\ell'\ell}(\vec{v},\vec{v}')\vec{F}_{\ell'}(t,\vec{v}')d^3\vec{v}', \quad (\text{II.65})$$

with the initial conditions:

$$\vec{F}_\ell(t=0,\vec{v}) = f_M(\vec{v})\rho_\ell \vec{d}_\ell. \quad (\text{II.66})$$

In these equations, ω_ℓ is the unperturbed angular frequency of line ℓ (ie at zero density), ρ_ℓ is the lower level populations of the transition, $f_M(\vec{v})$ is the Maxwell-Boltzmann distribution for the radiator velocity $\vec{v}$ and $v_{VC}(v=\|\vec{v}\|)$) is the speed- and transition-dependent velocity-changing (VC) rate. $A_l(\vec{v},\vec{v}')$ is the probability density (per time unit) for a $\vec{v} \leftarrow \vec{v}'$ collision, $\widetilde{W}_{\ell'\ell}(v) \equiv n_p W_{\ell'\ell}(v)$ is the speed-dependent relaxation matrix [Eq. (II.47)] (uncorrelated with velocity changes) and $C_{\ell'\ell}(\vec{v},\vec{v}')$ is the correlated relaxation matrix for $\vec{v} \leftarrow \vec{v}'$ collisions changing simultaneously the velocity and the state, or phase, of the radiator. The spectral consequence of VC collisions is the so-called Dicke narrowing (see § III.2.2). The opposite sign of the correlation contribution means that this correlation between velocity- and phase-changing collisions reduces the Dicke narrowing. The (imaginary) Laplace transform in time of Eq. (II.65) leads to the vectorial equation:[75]

$$\frac{1}{\pi}f_M(\vec{v})\rho_a\vec{d} = [v_{VC}(v) - i(\omega I_d - L_a - \vec{k}\cdot\vec{v}I_d)]\vec{F}(\omega,\vec{v}) - \int A(\vec{v},\vec{v}')\vec{F}(\omega,\vec{v}')d^3\vec{v}' \\ + \widetilde{W}(v)\vec{F}(\omega,\vec{v}) + \int C(\vec{v},\vec{v}')\vec{F}(\omega,\vec{v}')d^3\vec{v}'. \quad (\text{II.67})$$

Here $\vec{d}$ and $\vec{F}$ are N_L dimensional column vectors. The $N_L \times N_L$ matrices ρ_a, L_a, v_{VC}, and A are diagonal with elements given by: $[\rho_a]_{\ell'\ell} = \delta_{\ell,\ell'} \times \rho_\ell$, $[L_a]_{\ell'\ell} = \delta_{\ell,\ell'} \times \omega_\ell$, $[v_{VC}(v)]_{\ell'\ell} = \delta_{\ell,\ell'} \times v_{VC_\ell}(v)$, and $[A(\vec{v},\vec{v}')]_{\ell'\ell} = \delta_{\ell,\ell'} \times A_\ell(\vec{v},\vec{v}')$. $\widetilde{W}(v)$ and $C(\vec{v},\vec{v}')$ are $N_L \times N_L$ matrices with transposed off-diagonal elements related by detailed balance [Eq. (II.55)]. The collisional matrices $A(\vec{v},\vec{v}')$, $\widetilde{W}(v)$, and $C(\vec{v},\vec{v}')$ can be calculated by the scattering theory from a known or model intermolecular potential for the radiator-perturber pair. However, with some simplifying assumptions about the character of velocity-changing probabilities, phenomenological spectral-shape expressions may be obtained. For collisionally coupled lines, this possibility has been explored by Ciurylo and Pine[75] assuming that each collision thermalizes the radiator velocity, whatever the pre-collision velocity $\vec{v}'$. This "hard" collision model leads to significant simplifications in the formal expression of the spectral shape. But, even such a (hard collision) semi-classical model for Dicke-narrowed speed-dependent line mixing requires important computational tasks for practical applications. This is discussed in Sec. VIII.2, in connection with experimental data for multiplet spectra where specific signatures of these combined collisional effects were observed.

General equations 27

The semi-classical impact theory of Smith *et al*[74] for collisionally coupled lines is a generalization of the *classical* approach of Rautian and Sobelman[56] for *uncoupled* transitions. Classical descriptions of *both* external and internal degrees of freedom, joined with pertinent approximations on the character of velocity-changing collisions (like the above hard collision model), enable to obtain useful analytical expressions for the isolated line shape. The speed dependences of the line width $\Gamma(v)$ and shift $\Delta(v)$, first disregarded, were introduced later by other authors in order to get more accurate models. In the work of Ref. 56, the (normalized) classical dipole autocorrelation function of the radiator is given by [*cf* Eq. (II.10)]:

$$C(t) = \left\langle e^{-i\vec{k}\cdot[\vec{r}(t)-\vec{r}(0)]}\vec{d}(t)\vec{d}(0)\right\rangle/\langle d^2(0)\rangle \equiv \Phi_{cl}(t)/\langle d^2(0)\rangle, C(t=0) = 1, \quad \text{(II.68)}$$

where $\vec{d}(t)$ is the radiator dipole at time t. The symbol $\langle\ldots\rangle$ means a classical average which can be expressed as [setting $\vec{r}(0) = \vec{0}$]:

$$C(t) = \left\langle e^{-i\vec{k}\cdot\vec{r}(t)-i\varphi(t)}\right\rangle = \int e^{-i\vec{k}\cdot\vec{r}-i\varphi}f(\vec{r},\vec{v},\varphi,t)d^3\vec{r}d^3\vec{v}d\varphi. \quad \text{(II.69)}$$

In this expression, $\varphi(t)$ is the (complex) phase shift of the radiating dipole induced by collisions between times 0 and t, and $f(\vec{r},\vec{v},\varphi,t)$ is the distribution function for the position $\vec{r}$, the velocity $\vec{v}$, and the phase shift φ of the radiator at time t. Introducing its Fourier transform in phase shift:

$$f(\vec{r},\vec{v},t) = \int e^{-i\varphi}f(\vec{r},\vec{v},\varphi,t)d\varphi, \quad \text{(II.70)}$$

leads, from Eq. (II.69), to the correlation function suitably expressed in terms of the Fourier transform in space:

$$C(t) = \int e^{-i\vec{k}\cdot\vec{r}}f(\vec{r},\vec{v},t)d^3\vec{r}d^3\vec{v}. \quad \text{(II.71)}$$

For *uncorrelated* velocity- and phase-changing collisions, the $f(\vec{r},\vec{v},t)$ distribution function satisfies the Boltzmann kinetic equation:[56]

$$\frac{\partial f(\vec{r},\vec{v},t)}{\partial t} = -\left\{\vec{v}\cdot\vec{\nabla}_{\vec{r}} + [\Gamma(v)+i\Delta(v)] + \nu_{VC}(v)\right\}f(\vec{r},\vec{v},t)$$
$$+ \int A(\vec{v},\vec{v}')f(\vec{r},\vec{v}',t)d^3\vec{v}', \quad \text{(II.72)}$$

with the initial condition:

$$f(\vec{r},\vec{v},t=0) = f_M(\vec{v})\delta(\vec{r}). \quad \text{(II.73)}$$

$\nu_{VC}(v)$ and $A(\vec{v},\vec{v}')$ have been defined above [*cf* Eq. (II.65)], $\vec{\nabla}_{\vec{r}}$ is the gradient operator, $\delta(\vec{r})$ is the Dirac function in $\vec{r}$ variables and $f_M(\vec{v})$ is the Maxwell-Boltzmann distribution given by Eq. (III.2).

For *correlated* collisions, the Fourier transform in phase shift:

$$\widetilde{A}(\vec{v},\vec{v}') = \int e^{-i\varphi} A(\vec{v},\vec{v}',\varphi) d\varphi, \qquad (\text{II}.74)$$

has to be substituted to $A(\vec{v},\vec{v}')$ in Eq. (II.72), and the velocity-changing frequency ν_{VC} then becomes a *complex effective* frequency:

$$\nu_{\text{eff}}(v) = \int \widetilde{A}(\vec{v}',\vec{v}) d^3\vec{v}' = \int e^{-i\varphi} A(\vec{v}',\vec{v},\varphi) d\varphi d^3\vec{v}', \qquad (\text{II}.75)$$

where $A(\vec{v}',\vec{v},\varphi)$ depends on velocity and phase changes. Rautian and Sobelman[56] have further suggested to introduce a *partial correlation* by considering *both* uncorrelated and correlated collisions. This extension is presented in § III.3.6.

The explicit expression of $A(\vec{v},\vec{v}')$ [or $\widetilde{A}(\vec{v},\vec{v}')$] in Eq. (II.72), through convenient assumptions on the physical nature of collisional processes, leads to analytical line-shape models. These, which have been extensively used in various applications, are described in Sec. III.3. Let us mention that the classical kinetic equation (II.72) is consistent with the general structure of the semi-classical equation (II.65). Mention also that the normalization of the correlation function in Eq. (II.68) results [through Eq. (II.34)], for isolated lines, in a profile normalized to unity area [noted $I(\omega)$]. It is systematically used in chapter III, instead of the (non-normalized) spectral density $F(\omega)$.

II.6 COLLISION-INDUCED SPECTRA

Up to now, we have considered cases in which *only* the *permanent* tensor (*eg* dipole) coupling the radiator to the electromagnetic field contributes to the (absorption, dispersion, light scattering) spectrum. This is obviously not sufficient to explain that even non-polar gases (*eg* H_2,[76] O_2,[77] N_2,[78] etc) show measurable vibrational, rotational, and translational absorption spectra if gas densities and/or path lengths are sufficiently large. A specific mechanism must then be considered which is the fact that a transient dipole is induced by intermolecular radiator-perturber forces during collisions. In some other situations, the observed spectral shape results from a subtle interplay between a (weak) permanent tensor $\vec{d}_a$ of the radiator and that $\vec{d}^I$ induced during binary collisions. This is the case, for instance, of gaseous HD where the interference between the permanent and induced dipoles leads to a Fano line shape.[79] Another example is given by the rototranslational Raman scattering in H_2 where the collision-induced contribution and the "allowed" rotational Raman spectrum due to the anisotropy of the (permanent) polarizability are superimposed.[80] In order to model such processes, a more general analysis of the correlation function must be made. It is introduced below and further discussed in chapter VI. For completeness recall that, as discussed in § VII.5.2, remote sensing and radiative transfer in the atmospheres of planets and cold stars require accurate modeling of collision-induced spectra of various molecular species. Among these one finds H_2 (Jupiter and Saturn, for instance), N_2 (Titan and Earth), O_2 (Earth), CO_2 (Venus), as well as minor species such as CH_4, *etc*.

General equations

In the presence of induction processes, and considering (in a first step) only absorption spectra, the dipole moment operator in Eq. (II.9) must be replaced by:

$$\vec{d}(t) = \vec{d}_a(t) + \sum_{j=1}^{N_p} \vec{d}_j^I(t), \qquad (II.76)$$

where $\vec{d}_a(t)$ is the permanent (allowed) dipole of the radiator and $\vec{d}_j^I(t)$ the dipole induced by the collision between the radiator and the jth perturber. Using Eq. (II.76), the correlation function [Eq. (II.10) without the terms $\exp[\pm i\vec{k}.\vec{r}(t)]$ due to the radiator translational motion] can be written as a sum of "n-body" contributions $\Phi_n(t)$, ie:

$$\Phi(t) = \Phi_1^{aa}(t) + \Phi_2^{aI}(t) + \Phi_2^{Ia}(t) + \Phi_2^{II}(t) + etc. \qquad (II.77)$$

In this expression $\Phi_1^{aa}(t) = <\vec{d}_a^+(t).\vec{d}_a(0)>$ is the autocorrelation function of the *permanent* dipole discussed in the preceding sections, while:

$$\Phi_2^{aI}(t) = \left\langle \vec{d}_a^+(t). \sum_{j=1}^{N_p} \vec{d}_j^I(0) \right\rangle = N_p <\vec{d}_a^+(t).\vec{d}_1^I(0)>,$$

and, $\qquad (II.78)$

$$\Phi_2^{Ia}(t) = \left\langle \sum_{j=1}^{N_p} \vec{d}_j^{I+}(t).\vec{d}_a(0) \right\rangle = N_p <\vec{d}_1^{I+}(t).\vec{d}_a(0)>,$$

are both *cross-terms* between the permanent dipole $\vec{d}_a$ and the one ($\vec{d}_1^I$) induced during a *binary* collision with a *single* perturber. Finally:

$$\Phi_2^{II}(t) = \left\langle \sum_{j=1}^{N_p} \vec{d}_j^{I+}(t).\vec{d}_j^I(0) \right\rangle = N_p <\vec{d}_1^{I+}(t).\vec{d}_1^I(0)>, \qquad (II.79)$$

gives rise to the *purely induced* component responsible for the infrared absorption by pure H_2, for instance. Note that Eq. (II.77) does not consider the correlations[49] between the allowed dipoles associated with different radiators since these are expected to be small in the considered case of high dilution ($N_a \ll N_p$). It also does not include the correlation between dipoles induced in successive collisions, given by:

$$\Phi_3^{II}(t) = \left\langle \sum_{j=1}^{N_p} \vec{d}_j^{I+}(t). \sum_{j'=1, j'\neq j}^{N_p} \vec{d}_{j'}^I(0) \right\rangle = N_p^2 <\vec{d}_1^{I+}(t).\vec{d}_2^I(0)>, \qquad (II.80)$$

where $\vec{d}_1^I(t)$ and $\vec{d}_2^I(t)$ correspond to two different perturbers. This is valid since we assume low densities making the $\Phi_3^{II}(t)$ and higher order contributions negligible. The influence of $\Phi_3^{II}(t)$, leading to *intercollisional interferences*, is briefly discussed in § VI.6.2 and VI.6.3. According to Eq. (II.77), the *collision-induced absorption* (CIA) spectrum comes, through Eq. (II.79), from the dipole induced during collisions. It results in broad but resolved lines in the case of H_2 or D_2[76,81] which have large rotational constants, and leads to unresolved bands of strongly overlapping lines for heavier molecules like O_2 and N_2.[77,78] Finally note that, strictly speaking, these species may also possess an allowed spectrum. It is the case of N_2, whose quadrupolar (permanent) transitions of the $1 \leftarrow 0$ vibrational band[82] are superimposed with the fundamental CIA band near 4.2 μm. However, as demonstrated in Refs. 45,46, these two contributions do not interfere since they are associated with different operators (quadrupole and induced dipole) coupling N_2 to the electromagnetic field.

In this section we focus on the most widely used approach to compute *purely* collision-induced spectra for diatomic pairs. Other cases are discussed in chapter VI, together with more refined theoretical models. Inserting Eq. (II.79) into Eqs. (II.7) and (II.6) leads to the following equation for the absorption coefficient arising from the collision-induced dipoles:

$$\alpha(\omega) = \frac{4\pi^2 \omega}{3\hbar c}(1 - e^{-\hbar\omega/k_B T}) n_a n_p \times V g(\omega, T), \qquad \text{(II.81)}$$

where V is the volume containing the considered gas. Note that, for a pure gas and thus pairs of identical species, $n_a n_p$ must be replaced by $n_a^2/2$. Following Eq. (II.12), the spectral density $Vg(\omega, T)$ can be defined in terms of the matrix elements of the induced electric dipole moment $\vec{d} \equiv \vec{d}^I$, by the "golden rule" of quantum mechanics:[51]

$$Vg(\omega, T) = \sum_{s,s'} \rho_s \sum_{t,t'} V \times \rho_t |\langle t|\vec{d}_{ss'}|t'\rangle|^2 \delta(\omega_{ss'} + \omega_{tt'} - \omega), \qquad \text{(II.82)}$$

where $|s\rangle \equiv |v_1 J_1 m_1, v_2 J_2 m_2\rangle$ and $|t\rangle \equiv |\ell, m_\ell, E_t\rangle$ designate molecular (internal) and translational states, respectively (the prime indicating final states). ρ_s and ρ_t are Maxwell-Boltzmann factors giving the relative populations. Since linear molecules are considered here, the spherical components of the induced dipole are given by:[83]

$$d_\nu^{(k)} = \frac{(4\pi)^{3/2}}{\sqrt{2k+1}} \sum_{\lambda_1,\lambda_2,\Lambda,L} A_{\lambda_1,\lambda_2,\Lambda,L}(r_1,r_2,R) \sum_{m_1,m_2,M,m} (-1)^{\lambda_1-\lambda_2+m}\sqrt{2\Lambda+1}\begin{pmatrix} \lambda_1 & \lambda_2 & \Lambda \\ m_1 & m_2 & -m \end{pmatrix}$$

$$\times (-1)^{L-\Lambda+\nu}\sqrt{2k+1}\begin{pmatrix} L & \Lambda & k \\ M & m & -\nu \end{pmatrix} Y_{\lambda_1}^{m_1}(\Omega_1) Y_{\lambda_2}^{m_2}(\Omega_2) Y_L^M(\Omega),$$

with $\nu = 0, \pm 1, \ldots, \pm k$. $\qquad \text{(II.83)}$

The Ω, Ω_1, and Ω_2 define the orientations of the vector joining the colliding partners centers of mass, separated by R, and of the internuclear axes of molecules 1 and 2, respectively. r_1 and r_2 denote the corresponding vibrational coordinates. Multipole expansion expressions of the $A_{\lambda_1,\lambda_2,\Lambda,L}(r_1,r_2,R)$ are given in Ref. 84. The rank k of the tensor coupling radiation and matter is $k=1$ in the case of CIA, and $k=0, 2$ for collision-induced light scattering (CILS). Most CIA/CILS computations are based on Eq. (II.82) which, in fact, assumes that the Hamiltonian for a pair of molecules can be (approximately) written as the sum of two independent terms. One describes the vibrational and rotational motions of the molecules, while the other describes the translational motion of the pair. The latter is driven by the isotropic part $V_0(R)$ of the interaction potential or, in a more refined version, by a vibrational average $\langle v_1v_2|V_0(R,r_1,r_2)|v_1v_2\rangle$ of the potential, instead of $V_0(R)$. In other words, this "*isotropic approximation*", further discussed in Sec. VI.3, neglects the anisotropic part of the intermolecular potential which couples the internal and translational degrees of freedom. More refined formalisms are presented in Sec. VI.4. After some algebra,[84] the expression of the spectral density, within the isotropic approximation, is:[7]

$$V \times g(\omega, T) = \sum_{\lambda_1,\lambda_2,\Lambda,L} \sum_{\substack{v_1,J_1,v'_1,J'_1 \\ v_2,J_2,v'_2,J'_2}} \rho_{v_1J_1}\rho_{v_2J_2}(2J'_1+1)\begin{pmatrix} J_1 & \lambda_1 & J'_1 \\ 0 & 0 & 0 \end{pmatrix}^2$$

$$\times (2J'_2+1)\begin{pmatrix} J_2 & \lambda_2 & J'_2 \\ 0 & 0 & 0 \end{pmatrix}^2 V \quad \text{(II.84)}$$

$$\times G_{\lambda_1,\lambda_2,\Lambda,L}(\omega - \omega_{v'_1J'_1v_1J_1} - \omega_{v'_2J'_2v_2J_2}, T),$$

where ρ_{vJ} is the relative population of level vJ. The translational profiles are given by:

$$V \times G_{\lambda_1,\lambda_2,\Lambda,L}(\omega, T) = \lambda_0^3 \hbar \sum_{\ell,\ell'} (2\ell+1)(2l'+1)\begin{pmatrix} l & L & l' \\ 0 & 0 & 0 \end{pmatrix}^2$$

$$\times \Biggl\{ \int_0^\infty dE_t e^{-E_t/k_BT} |\langle \ell E_t v|B_{\lambda_1,\lambda_2,\Lambda,L}(R)|\ell' E_t + \hbar\omega v'\rangle|^2$$

$$+ \sum_{n,n'} e^{-E_{nl}/k_BT} |\langle \ell E_{n\ell} v|B_{\lambda_1,\lambda_2,\Lambda,L}(R)|\ell' E_{n'\ell'} v'\rangle|^2 \delta(E_{n'\ell'} - E_{nl} - \hbar\omega) \quad \text{(II.85)}$$

$$+ \sum_n e^{-E_{nl}/k_BT} |\langle \ell E_{n\ell} v|B_{\lambda_1,\lambda_2,\Lambda,L}(R)|\ell' E_{n\ell} + \hbar\omega v'\rangle|^2$$

$$+ \sum_{n'} e^{-(E_{n'\ell'}-\hbar\omega)/k_BT} |\langle l E_{n'\ell'} - \hbar\omega v|B_{\lambda_1,\lambda_2,\Lambda,L}(R)|\ell' E_{n'\ell'} v'\rangle|^2 \Biggr\}$$

where λ_0 is the thermal de Broglie wavelength $[\lambda_0 = h/(2\pi m^* k_B T)^{1/2}$, m^* being the reduced mass], $v \equiv (v_1, v_2)$ represents the vibrational quantum numbers of the molecules, n and ℓ designating the vibrational and rotational quantum numbers *of the complex* and:

$$B_{\lambda_1,\lambda_2,\Lambda,L}(R) = \langle v_1v_2|A_{\lambda_1,\lambda_2,\Lambda,L}(r_1,r_2,R)|v'_1v'_2\rangle, \quad \text{(II.86)}$$

---

[7] Equation (II.84) is written for pairs of dissimilar, and thus distinguishable, molecules. In the case of identical collision partners, an additional weight $w(\ell\ell'J_1J'_1J_2J'_2v_1v'_1v_2v'_2)$ must be introduced in order to account for the molecular exchange symmetry.[84]

where, for simplicity, we have neglected the (small) vibration-rotation coupling. The radial wavefunctions $|\ell E_t v\rangle$ needed for computations using Eq. (II.85) are the solutions of the Schrödinger equation of relative motion:

$$-\frac{\hbar^2}{2m^*}\frac{d^2\Psi}{dR^2} + \left[V_0(R) + \frac{\hbar^2 \ell(\ell+1)}{2m^*R^2} - E_t\right]\Psi = 0, \qquad (II.87)$$

where $V_0(R) = \langle v_1 v_2| V_0(r_1,r_2,R)| v_1 v_2\rangle$. As well known, the isotropic potential has a well region of negative values and a repulsive positive core at smaller R distances. Consequently, Eq. (II.87) has three types of solutions. For small values of ℓ, the centrifugal term $\hbar^2\ell(\ell+1)/(2m^*R^2)$ is almost negligible, and the effective potential is nearly the same as $V_0(R)$. For most systems, the well is deep and wide enough to support "truly bound" states of negative energies, which are subscripted with vibrational quantum numbers n and rotational ones ℓ corresponding to the end-over-end rotation of the complex. Solutions of positive and continuous energies that are non vanishing even at large R distances correspond to "free" (ie collisional) states. For large values of ℓ, the effective potential is a monotonically decreasing positive function of R. Under such conditions, only "free" states exist. At intermediate values of ℓ, discrete states of positive energy exist, which are more or less coupled to the continuum across the centrifugal barrier. In scattering theory they are called "scattering resonances" while in spectroscopy they are known as "predissociating", "quasibound", or "metastable" states. They are, in some sense, the continuation of the bound states in the continuum, and are considered as free or bound states depending on their widths due to the coupling with the opened channels. Let us consider, as an example, the rotovibrational levels of the $(N_2)_2$ dimer discussed in Sec. VI.5. The calculation of Ref. 85, within the isotropic approximation, has led to six vibrational levels and 42 rotational states for the vibrational ground state. 128 rovibrational levels were obtained with, among them, 31 predissociating states of positive energies. Note that $(N_2)_2$ energy levels can also be obtained from more sophisticated approaches which take into account the anisotropy of the potential.[86]

Equation (II.85) consists in a sum of four components. The integral (first term), which corresponds to "free-free" transitions, is, in general, the dominant contribution. The second term (the double sum), gives the "bound-bound" transitions of the van der Waals complex. Finally the last two sums correspond to "bound-free" and "free-bound" transitions. From an experimental point of view, bound-bound and bound-free/free-bound contributions may be discernable, at least at low temperature, since they sometimes lead to line structures superimposed on the broad free-free contribution.[87–89] However, in most practical situations, transitions involving bound states are completely "drowned" in the broad free-free profile, due to the experimental conditions (high pressures and temperatures, low spectral resolution).

Apart from a few exceptions, most calculations of CIA/CILS spectra have been performed within the isotropic approximation [Eqs. (II.81), (II.84), and (II.85)], since it considerably reduces the computational efforts. However, this commonly used approach is not appropriate for molecular systems with a substantial anisotropy.[90] In order to introduce the effects of the anisotropy of the potential on spectra, various

General equations 33

models have been proposed. Finally, attempts to describe the interference of permanent and induced dipoles (CIA) or polarizabilities (CILS) in the calculation of the spectra have been made. These topics are discussed in more detail in chapter VI.

II.7 CONCLUSION

This chapter, although formal, has introduced and discussed the main general equations and some approximations which enable practical computations in specific cases. From these bases, further developments are presented in the following chapters which lead to direct comparisons with measurements.

The present book is resticted to molecular radiators and disregards atomic spectra in order to keep the bibliography of reasonable size. One should nevertheless remember that the "molecular line-shapes theoretical community" has been largely fed by earlier progresses in atomic and plasma physics. This is illustrated by the biennal International Conference on Spectral Line Shapes (I.C.S.L.S.), and its proceedings, since 1973.

APPENDIX II.A: SPECTRAL AND TIME DOMAIN PROFILES IN VARIOUS SPECTROSCOPIES

II.A1 ABSORPTION, EMISSION, AND DISPERSION

For a neutral gas at thermal equilibrium submitted to an electromagnetic (e.m.) field $\vec{E}$, the resulting *linear macroscopic* electric polarization $\vec{P}$ is, from classical electrodynamics through Maxwell equations:

$$\vec{P} = \varepsilon_0(\varepsilon - 1)\vec{E} = \varepsilon_0 \chi \vec{E}, \quad \text{with} \quad \varepsilon = \varepsilon' + i\varepsilon'' \quad \text{and} \quad \chi = \chi' + i\chi''. \tag{II.A1}$$

In this equation, ε_0 is the vacuum dielectric constant, ε is the (relative) complex permittivity and χ the (dimensionless) electric susceptibility (ε and χ being scalars due to the gas isotropy). Let us consider e.m. waves passing through a gas layer in the z direction. The light intensity $I(\omega,z)$ at the angular frequency ω and depth z is given by the Beer-Lambert law:

$$I(\omega, z) = I(\omega, z = 0) \times \exp[-\alpha(\omega)z], \tag{II.A2}$$

where $\alpha(\omega)$ is the absorption coefficient. It can be related to the imaginary part ε'' of the dynamical permittivity ε, or alternatively to the susceptibility χ'':

$$\alpha(\omega) = \frac{\omega}{c}\varepsilon''(\omega) = \frac{\omega}{c}\chi''(\omega), \tag{II.A3}$$

where c is the light speed in vacuum. Let us introduce the *dipolar approximation*, which assumes that the variation of the e.m. field amplitude due to its spatial modulation over the radiator size is negligible. Thus $\varepsilon''(\omega)$ [or $\chi''(\omega)$] is the Laplace transform in time of the (absorption) *dipole autocorrelation function* $\Phi(t)$, yielding the following general relation for the electric *dipolar absorption coefficient* (see § II.2.1):

$$\alpha(\omega) = n_a \frac{4\pi^2 \omega}{3\hbar c} [1 - \exp(-\hbar\omega/k_B T)] \times \text{Re}[\mathsf{F}(\omega)], \qquad (\text{II.A4})$$

where n_a is the number density of radiators and T is the temperature. $\mathsf{F}(\omega)$ is the *complex spectral density* given by:

$$\mathsf{F}(\omega) = \frac{1}{\pi} \int_0^\infty e^{i\omega t} \Phi(t) dt, \qquad (\text{II.A5})$$

with

$$\Phi(t) = \left\langle \vec{d}^+(t).\vec{d}(0) \right\rangle. \qquad (\text{II.A6})$$

In Eq. (II.A6), for simplicity, the radiator translational motion [through the terms $\exp(\pm i\vec{k}.\vec{r})$, see Eq. (II.10)] is not explicitly introduced. <...> denotes a quantum equilibrium average, *ie* (see § II.2.1) a trace over the states of the molecular system S_M of the autocorrelation function $\vec{d}^+(t).\vec{d}(0)$ of the radiator dipole weighted by the density operator $\rho_M(0)$ at the initial time $t=0$. The negative term $-\exp(-\hbar\omega/k_B T)$ in the "net" absorption coefficient of Eq. (II.A4) comes from the contribution [*cf* Eq. (II.35)] of emission through the correlation function:

$$\Phi_e(t) = \Phi(-t) = \left\langle \vec{d}^+(-t).\vec{d}(0) \right\rangle = \left\langle \vec{d}^+(0).\vec{d}(t) \right\rangle, \qquad (\text{II.A7})$$

where the time translational invariance has been used.

The real part $\chi'(\omega)$ of the (dimensionless) electric susceptibility [Eq. (II.A1)] characterizes the *dispersion* of the e.m. wave in the gas (*ie* the dependence of the phase velocity on the angular frequency ω). Like $\alpha(\omega)$, the dispersion $\chi'(\omega)$ can be related to the complex spectral density:

$$\chi'(\omega) = \varepsilon'(\omega) - 1 = n_a \frac{4\pi^2}{3\hbar} [1 - \exp(-\hbar\omega/k_B T)] \times \text{Im}[\mathsf{F}(\omega)]. \qquad (\text{II.A8})$$

Equations (II.A4) and (II.A8) show that the complex spectral density enables calculations of absorption (through the real part) as well as of dispersion (through the imaginary part) spectra. In fact, as well known, absorption and dispersion are tied by Kramers-Kronig relations which can be written, in this particular case, under the form:

$$\chi'(\omega) = \frac{2c}{\pi} P \int_0^{+\infty} \frac{\omega' \alpha(\omega')}{\omega(\omega'^2 - \omega^2)} d\omega', \qquad (\text{II.A9})$$

General equations 35

$$\alpha(\omega) = \frac{-2\omega^2}{\pi c} P \int_0^{+\infty} \frac{\chi'(\omega')}{(\omega'^2 - \omega^2)} d\omega', \qquad (II.A10)$$

where P denotes the Cauchy principal value of the integral.

Beyond the dipolar approximation, Eq. (II.A4) must take into account the coupling between the gradient of the e.m. field and the electric quadrupolar tensor $\vec{Q}$, the associated Hamiltonian being:[91]

$$H_{e.m.-M} = -\vec{d}\cdot\vec{E}(\vec{r}) - \frac{1}{6}\vec{\vec{Q}}:\vec{\nabla}_{\vec{r}}\vec{E}(\vec{r}) + \ldots = -\sum_i d_i E_i(\vec{r}) - \frac{1}{6}\sum_{i,j} Q_{ij}\frac{\partial}{\partial r_i}E_j(\vec{r}) + \ldots,$$

(II.A11)

where

$$Q_{ij} = \int \rho(\vec{r})(3r_i r_j - r^2 \delta_{i,j}) d^3\vec{r}, \quad \text{with} \quad i,j = (x,y,z), \qquad (II.A12)$$

is the traceless quadrupole tensor, $\rho(\vec{r})$ being the distribution function of the electric charge. In Eq. (II.A11), $\vec{\nabla}_{\vec{r}}$ is the gradient operator in $\vec{r}$ variables and the symbol : means the contraction of the tensorial product. The second term in Eq. (II.A11) yields the quadrupolar absorption. The expression of associated absorption coefficient is obtained along the same procedure used for the dipole moment in § II.2.1, leading to:

$$\alpha^{(2)}(\omega) = n_a \frac{\pi^2 \omega^3}{3^4 \hbar c^3}[1 - e^{-\frac{\hbar\omega}{k_B T}}] \times \text{Re}\left\{\frac{1}{\pi}\int_0^\infty e^{i\omega t} <\vec{\vec{Q}}^+(t):\vec{\vec{Q}}(0)> dt\right\}. \qquad (II.A13)$$

Here, the exponent (2) indicates the order k of the coupling tensor involved [k = 1 being implicit in Eq. (II.A4)]. It is worth noting that the cross terms between the k = 1 and k = 2 contributions vanish (see Ref. 45 and App. II.E2). A first difference between quadrupolar and dipolar absorptions comes from the transition moments, resulting in unlike integrated intensities and selection rules. Note that the ratio of dipolar to quadrupolar intensities is generally very large, so that the quadrupolar contributions can be neglected in most cases, except for molecules such as H_2 and N_2 which show no dipolar spectrum. The second difference concerns the influence of collisions on the spectral shape since the associated collisional quantities depend on the tensor order k.

In conclusion, the absorption/emission and dispersion coefficients are related to the *same* quantity through the real and imaginary parts of the complex spectral density $F(\omega)$. The proper tensor coupling the molecules to radiation must be included in the transition moments and in the collisional parameters governing the spectral shape (through the tensor *order*).

II.A2 RAYLEIGH AND SPONTANEOUS RAMAN SCATTERINGS

If absorption essentially occurs when incident radiations have frequencies ω_{inc} close to those ω_{fi} of the radiator optical transitions, this is not the case for light scattering since ω_{inc} is generally very far away from ω_{fi}. Although familiar, the physical mechanisms underlying the various scattering processes are briefly recalled here in the simple classical picture. The (weak) oscillating e.m. field $\vec{E}$ induces a radiating dipole through the molecular polarizability $\vec{\alpha}$. The linear (L) induced dipole, given by:

$$\vec{d}^L = \vec{\alpha}.\vec{E} \Leftrightarrow d_i^L(\vec{E}) = \sum_j \alpha_{ij} E_j, (i = x, y, z), \quad \text{(II.A14)}$$

rotates, vibrates and varies in time with both $\vec{\alpha}$ and $\vec{E}$. Therefore, it emits the scattered radiation at the ω_{scat} angular frequency equal to ω_{inc} (Rayleigh scattering) but also at $\omega_{scat} = \omega_{inc} \pm \omega_{fi}$ (the Raman scattering, whose quantum theory was developed by Placzek[91b]). The differential cross-section σ for scattering into a solid angle $d\Omega$ and a frequency range $d\omega$ can be expressed in terms of a correlation function:[92]

$$\frac{\partial^2 \sigma}{\partial \Omega \partial \omega} = n_a \frac{\omega_{inc} \omega_{scat}^3}{c^4} \text{Re} \left\{ \frac{1}{\pi} \int_0^\infty e^{i\omega t} \left\langle [\vec{E}_{inc}^0 . \vec{\alpha}^+(t) . \vec{E}_{scat}^0][\vec{E}_{inc}^0 . \vec{\alpha}(0) . \vec{E}_{scat}^0] \right\rangle dt \right\}, \quad \text{(II.A15)}$$

where $\omega = \omega_{inc} - \omega_{scat}$ and $\vec{E}_{inc}^0$ and $\vec{E}_{scat}^0$ are unit vectors of the incident and scattered radiations along the directions of the electric vectors. It is convenient to split the scattered spectrum into two parts, called polarized and depolarized components. Experimentally, such a separation is obtained by choosing $\vec{E}_{inc}^0$ and $\vec{E}_{scat}^0$ either parallel (//) or perpendicular ($\perp$), leading to:[93]

$$\frac{\partial^2 \sigma_{pol}}{\partial \Omega \partial \omega} \equiv \left(\frac{\partial^2 \sigma}{\partial \Omega \partial \omega} \right)_{//} - \frac{4}{3} \left(\frac{\partial^2 \sigma}{\partial \Omega \partial \omega} \right)_\perp, \quad \text{(II.A16)}$$

and

$$\frac{\partial^2 \sigma_{depol}}{\partial \Omega \partial \omega} \equiv \frac{1}{10} \left(\frac{\partial^2 \sigma}{\partial \Omega \partial \omega} \right)_\perp. \quad \text{(II.A17)}$$

In a similar way, the polarizability can be divided into the isotropic part α_{iso} (a tensor of rank k = 0), through the rotational average of $\vec{\alpha}$, and its complementary anisotropic part $\vec{\alpha}_{aniso}$ (a tensor of rank k = 2):

$$\alpha_{iso} = \frac{1}{3} \text{Tr}_{rot}(\vec{\alpha}) \quad \text{and} \quad \vec{\alpha}_{aniso} = \vec{\alpha} - \alpha_{iso} I_d, \quad \text{(II.A18)}$$

where I_d is the identity operator. Substituting Eq. (II.A18) into Eq. (II.A15) results in the following relation between the *polarized spectrum* and the *correlation function of the isotropic polarizability*:

$$\frac{\partial^2 \sigma_{pol}}{\partial \Omega \partial \omega} = n_a \frac{\omega_{inc} \omega_{scat}^3}{c^4} \text{Re} \left\{ \frac{1}{\pi} \int_0^\infty e^{i\omega t} < \alpha_{iso}^+(t) \alpha_{iso}(0) > dt \right\}. \quad \text{(II.A19)}$$

Table II.A1: General characteristics of various linear spectroscopies

Spectroscopy	Radiation-radiator coupling	Tensor order	Spectral density
Absorption/Emission	Electric dipole or quadrupole	$k=1$ or 2	Re(F)
Dispersion	Electric dipole	$k=1$	Im(F)
Polarized Rayleigh Scattering	Isotropic polarizability	$k=0$	Re(F)
Depolarized Rayleigh Scattering	Anisotropic polarizability (with *no* rovib. transition)	$k=2$	Re(F)
Isotropic Raman Q branch	Isotropic polarizability (with $\Delta J = 0$ and vib. transition)	$k=0$	Re(F)
Rotational Raman lines	Anisotropic polarizability ($\Delta J = \pm 2$ and *no* vib. transition)	$k=2$	Re(F)
Anisotropic Raman Q branch and O/S branches	Anisotropic polarizability (with $\Delta J = 0, -2, +2$ and *with* a vib. transition)	$k=2$	Re(F)

A similar relation yields between the *depolarized spectrum* and the *anisotropic polarizability*:[92]

$$\frac{\partial^2 \sigma_{depol}}{\partial \Omega \partial \omega} = n_a \frac{\omega_{inc} \omega_{scat}^3}{c^4} \text{Re} \left\{ \frac{1}{\pi} \int_0^\infty e^{i\omega t} < \vec{\alpha}_{aniso}^+(t) . \vec{\alpha}_{aniso}(0) > dt \right\}. \quad \text{(II.A20)}$$

As explained above in the classical picture, these spectra contain several types of components, as summarized in Table II.A1.

If there is *no induced vibrational transition* in the scattering, the polarized spectrum corresponds to the Rayleigh scattering[8], and the depolarized spectrum includes two different components. Associated with the $\Delta J = 0$ (J being the rotational quantum number) selection rule tied to the anisotropic tensor $\vec{\alpha}_{aniso}$ is the depolarized Rayleigh (DPR) scattering.[95] DPR spectra, being proportional to the squared matrix transition elements of $\vec{\alpha}_{aniso}$, much smaller than α_{iso}^2, have a weak intensity and were observed only rather recently.[96,97] The $\Delta J = \pm 2$ selection rule leads to the rotational Raman scattering which appears on both sides of the Rayleigh spectrum. If the rovibrational transitions are excited in the scattering, the polarized spectrum corresponds to the *isotropic* Raman Q branch ($\Delta J = 0$ transitions), while the depolarized spectrum contains the anisotropic Raman Q branch ($\Delta J = 0$) and the O and S branches tied to $\Delta J = -2$ and $\Delta J = +2$, respectively. These three branches can be observed either at frequency $\omega_{inc} - \omega_{fi}$ (the

---

[8] In fact, the Rayleigh scattering has an additional fine structure, called the Brillouin component, arising from the Doppler effect through relativistic corrections to the scattered frequency which require to include correlation in space as well as in time.[94,95]

Stokes band) or at $\omega_{inc} + \omega_{fi}$ (the anti-Stokes band). Since depolarized spectra are generated by the anisotropic polarizability, a tensor of rank k = 2 as for the electric quadrupole in quadrupolar absorption, the comments made in II.A.1 also apply.

Alike extensions in terms of the correlation function for the molecular motion, such as magnetic resonance absorption, which generally occurs at radio or microwave frequencies, can be done,[69] but they are out of the scope of this book.

Rayleigh and spontaneous Raman effects are *linear* scattering processes since the intensity of the scattered light I_{scat} is *proportional* to the intensity of the incident light [see Eqs. (II.A14) and (II.A15)]. The next two paragraphs are devoted to nonlinear scattering processes.

II.A3   NONLINEAR RAMAN SPECTROSCOPIES

Raman spectroscopy was revolutionized by the availability of intense laser sources, leading to efficient nonlinear (NL) techniques such as Coherent Anti-Stokes Raman Scattering (CARS) and Stimulated Raman Scattering (SRS) that we now examine. Only basic formal expressions useful for gas phase spectral-shape studies are presented. For more details, the reader can refer to Refs. 9-11,34,98, for instance.

For large incident laser intensities, the linear approximation for the induced dipole [Eq. (II.A14)] is no longer valid, and NL contributions must be taken into account:

$$d_i^{NL}(\vec{E}) = \sum_{j,k} \beta_{ijk} E_j E_k + \sum_{j,k,l} \gamma_{ijkl} E_j E_k E_l + \ldots . \quad \text{(II.A21)}$$

Here, β_{ijk} and γ_{ijkl} are the components of the *hyper-polarizability* and the *second-hyper-polarizability* tensors, respectively. The first contribution in the right hand side of Eq. (II.A21) yields the hyper-Raman effect, and the hyper-Rayleigh if no rovibrational transition takes place. These effects are of little use in gas phase molecular spectroscopy due to the higher efficiency of other NL techniques (in particular SRS and CARS), and also because they cancel for molecules with a center of inversion. Thus the NL Raman spectroscopic techniques essentially come from the second term in the right hand side of Eq. (II.A21). The corresponding *third order macroscopic electric polarization*, which is the quantum average over the (microscopic) dipoles induced by the second-hyper-polarizability is given by:

$$\begin{aligned} P_i^{(3)}(\omega_4) &= \frac{1}{2\pi} \int_{-\infty}^{+\infty} e^{i\omega_4 t} P_i^{(3)}(t) dt \\ &= \frac{3\varepsilon_0}{2q!} \sum_{j,k,l} \chi_{ijkl}^{(3)}(-\omega_4; \omega_1, -\omega_2, \omega_3) E_j(\omega_1) E_k(\omega_2) E_l(\omega_3), \end{aligned} \quad \text{(II.A22)}$$

where ω_n (n = 1, . . . ,4) are the frequencies of the *four* waves mixed in the third order NL scattering processes, and q is the number of degenerate frequencies. Energy and momentum conservation requires that:

$$\omega_4 = \omega_1 - \omega_2 + \omega_3, \quad \text{and} \quad \vec{k}_4 = \vec{k}_1 - \vec{k}_2 + \vec{k}_3. \quad \text{(II.A23)}$$

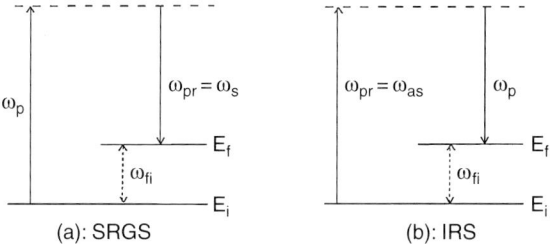

Fig. II.A1: Configuration of the lasers frequencies in SRS.

Stimulated Raman Scattering (SRS) uses *one* intense (pump) laser of intensity I_p and frequency ω_p to excite a significant fraction of radiators from the initial level of energy E_i to the final level of energy E_f. The incident pump wave induces a strong Stokes (s) e.m. wave at $\omega_s = \omega_p - \omega_{fi}$, whose intensity is larger than the linear Raman scattering by several orders of magnitude. It induces also an anti-Stokes (as) wave at frequency $\omega_{as} = \omega_p + \omega_{fi}$. Under these conditions, the radiators interact with *two* e.m. waves, that are the pump and the Stokes (or anti-Stokes) waves. This parametric interaction results in an energy exchange between these two waves, probed by a second (weakly intense) tunable laser at ω_{pr} frequency. Depending on the frequency configuration of the pump and probe lasers (*cf* Fig. II.A1), one obtains either a gain or an attenuation of the probe wave. In the stimulated Raman *gain* spectroscopy (SRGS), the pump wave is applied following the configuration of Fig. II.A1a, leading to a gain at the probed scattered Stokes frequency ω_s. In the (stimulated) *inverse* Raman spectroscopy (IRS, Fig. II.A1b), an attenuation is measured at the probed scattered anti-Stokes frequency ω_{as}, the strong pump laser being then tuned on the ω_{fi} rovibrational transition frequency.

Choosing both $\vec{E}_p$ and $\vec{E}_s$ polarized along the same x-axis orthogonal to the z-axis of propagation leads, for the Stokes polarization, to:

$$P_x^{(3)}(\omega_s) = \tilde{\chi}_{xxxx}^{(3)}(-\omega_s; \omega_s, -\omega_p, \omega_p) E_x(\omega_s) |E_x(\omega_p)|^2 e^{ik_s z}, \quad \text{with} \quad \tilde{\chi}_{xxxx}^{(3)} = \frac{3\varepsilon_0}{2} \chi_{xxxx}^{(3)}. \tag{II.A24}$$

This equation shows that a polarization wave of wave vector $\vec{k}_s$ oscillating at frequency ω_s travels through the medium with an amplitude proportional to $I_p I_s^{1/2}$, which therefore amplifies the Stokes wave (a similar result stands for the anti-Stokes wave). The remaining key term is the third order susceptibility $\chi^{(3)}(-\omega_s; \omega_s, -\omega_p, \omega_p)$, whose explicit calculation is needed to relate SRS spectral profiles to the gas properties. Assuming that the incident e.m. fields are far from electronic resonances, the resonant part of the third order susceptibility may be written as:[99]

$$_r\tilde{\chi}_{xxxx}^{(3)}(-\omega_4; \omega_1, -\omega_2, \omega_3) = \frac{n_a}{\hbar} \frac{\rho_i(0) - \rho_f(0)}{\omega_{fi} - (\omega_1 - \omega_2) - i\Gamma_{fi}} \langle f|\alpha_{xx}|i\rangle^2, \tag{II.A25}$$

where n_a is the number density of radiators, $\rho_i(0)$ and $\rho_f(0)$ are the equilibrium relative populations of levels i and f, Γ_{fi} is the collisional broadening for the isolated (collisionally uncoupled) $f \leftarrow i$ transition, and α_{xx} is the xx component of the radiator polarizability tensor [cf Eq. (II.A14) and Table II.A1]. Equation (II.A25) only accounts for the impact broadening, the shift being implicit.

Starting from the wave equation for the generation and propagation of e.m. fields with slowly varying amplitudes in dielectric media (from Maxwell's equations), and introducing the third order polarization [Eq. (II.A22)], yields the spatial evolution of the SRS (Stokes) signal intensity:

$$I(\omega_s, z) = I(\omega_s, z = 0) \times \exp\left\{ -\frac{2\omega_s z}{\varepsilon_0^2 c^2 n(\omega_s) n(\omega_p)} {_r\tilde\chi}''^{(3)}_{xxxx}(-\omega_s; \omega_s, -\omega_p, \omega_p) I(\omega_p, z) \right\}.$$

(II.A26)

Here, $n(\omega_s)$ and $n(\omega_p)$ are refractive indexes at frequencies ω_s and ω_p [$n(\omega_{s,p}) = k_{s,p} c/\omega_{s,p}$], and ${_r\tilde\chi}''^{(3)}$ is the *imaginary part* of ${_r\tilde\chi}^{(3)}$ [see Eqs. (II.A1) and (II.A25)]. The *real part* [${_r\tilde\chi}'^{(3)} + {_{nr}\tilde\chi}^{(3)}$] which includes the non-resonant contribution ${_{nr}\tilde\chi}^{(3)}$, dephases the Stokes wave but *does not change* its intensity. In SGRS, since $\rho_i(0) < \rho_f(0)$ (cf Fig. II.A1), $\tilde\chi''^{(3)}_{xxxx}$ is negative, resulting in a gain G given by:

$$G = |I(\omega_s, z = L) - I(\omega_s, z = 0)|/I(\omega_s, z = 0)$$
$$\approx -\frac{2\omega_s L}{\varepsilon_0^2 c^2 n(\omega_s) n(\omega_p)} {_r\tilde\chi}''^{(3)}_{xxxx}(-\omega_s; \omega_s, -\omega_p, \omega_p) I(\omega_p, L),$$

(II.A27)

where the interaction length L is assumed sufficiently small (as is usually the case) to justify a first-order expansion of the exponential in Eq. (II.A26). A similar expression holds for the attenuation A in IRS [$\rho_i(0) > \rho_f(0)$] by only changing ω_s in ω_{as}. From the last three equations, it results that SRS profiles are very similar to those obtained from spontaneous Raman scattering [ie in the simple case of an isolated line in the impact approximation, a Lorentzian line shape centered at frequency $\omega_1 - \omega_2$, as in Eq. (II.53)]. Apart from the integrated intensity, another difference comes from the wave vectors for which, like for ω, one has $\vec{k} = \vec{k}_s - \vec{k}_p$ [with the concomitant Doppler shift $(\vec{k}_s - \vec{k}_p) \cdot \vec{r}$].

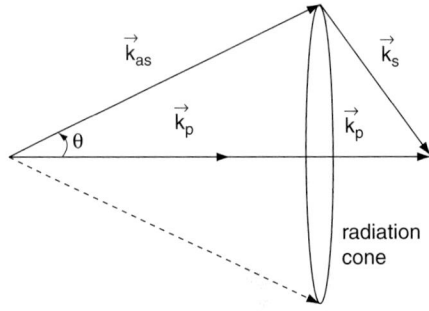

Fig. II.A2: Phase-matching and radiation cone.

General equations 41

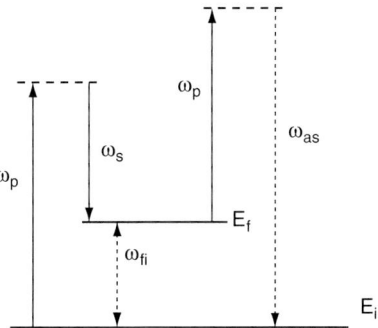

Fig. II.A3: Configuration of lasers frequencies in (conventional) CARS.

Macroscopic Stokes and anti-Stokes waves can build up only if the following phase-matching condition (*cf* Fig. II.A2) is verified:

$$2\vec{k}_p = \vec{k}_{as} - \vec{k}_s. \qquad (II.A28)$$

This condition shows that these waves are emitted within a cone whose axis is parallel to the pump beam propagation direction (Fig. II.A2). This SRS scattering geometry leads to a Doppler width different from that in spontaneous scattering.

In the *coherent anti-Stokes Raman scattering*, two strong incident lasers at frequencies ω_p and ω_s are used, such that their difference is tuned on the rovibrational transitions frequencies (*ie* $\omega_p - \omega_s = \omega_{fi}$, see Fig. II.A3).

Thus, the Stokes wave at frequency ω_s is *already* present and does not have to be generated in the nonlinear medium as in SRS. A strong anti-Stokes radiation is obtained at $\omega_{as} = 2\omega_p - \omega_s$, whose intensity can be derived along the same procedure as for the SRS signal, leading to:

$$I(\omega_{as}, L) = \frac{\omega_{as}^4}{\varepsilon_0^4 c^4 n(\omega_s) n(\omega_{as}) n(\omega_p)^2} \left| \tilde{\chi}^{(3)}_{xxxx}(-\omega_{as}; \omega_p, -\omega_s, \omega_p) \right|^2 I_p^2 I_s L^2 \text{sinc}^2(L \times \Delta k/2).$$

(II.A29)

Here, $\tilde{\chi}^{(3)}_{xxxx}(-\omega_{as}; \omega_p, -\omega_s, \omega_p)$ is the CARS susceptibility [Eq. (II.A25)] component for aligned polarizations of the incident laser beams along the x-axis and Δk is the momentum mismatch. Note that, in gas phase, $\Delta k = 2k_p - k_s - k_{as}$, since the dispersion can then be neglected for $\omega_{fi} \ll \omega_s$ and $\omega_{fi} \ll \omega_p$ at optical frequencies, and the anti-Stokes wave is generated with the same direction as the incoming beams (*cf* Fig. II.A2 with $\theta \approx 0$). The main features of CARS signals [Eq. (II.A29)] are the proportionality to the product of the lasers intensities $I_p^2 I_s$, and to the square of the radiator number density n_a (through the modulus of $\tilde{\chi}^{(3)}$). CARS shapes are more intricate than SRS ones, since:

$$\left| \tilde{\chi}^{(3)} \right|^2 = \left| {}_{nr}\tilde{\chi}^{(3)} + {}_r\tilde{\chi}'^{(3)} + i\, {}_r\tilde{\chi}''^{(3)} \right|^2 = {}_{nr}\tilde{\chi}^{(3)^2} + 2\, {}_{nr}\tilde{\chi}^{(3)} \cdot {}_r\tilde{\chi}'^{(3)} + {}_r\tilde{\chi}'^{(3)^2} + {}_r\tilde{\chi}''^{(3)^2}. \qquad (II.A30)$$

Indeed, in Eq. (II.A30), the first (constant) nonresonant term has to be adjusted to experimental data, the second nonresonant + resonant term induces a distortion of the profile through the dispersive character of $_r\widetilde{\chi}'^{(3)}$, and the third term gives rise to interferences between dispersive contributions of adjacent lines (if not sufficiently well separated). However, the nonresonant susceptibility differs in its symmetry properties from the resonant part. This can be used to cancel its contribution [*ie* the first and second terms in Eq. (II.A30] through careful arrangement of the planes of polarization of the incident fields.[100] The treatment of the leading CARS profile is then simplified since the results of the various spectral-shape models presented in this book can be directly used. Nevertheless, the presence of the real part of $_r\widetilde{\chi}^{(3)}$ requires additional cautions with respect to the spontaneous Raman or SRS profiles.

II.A4 TIME-RESOLVED RAMAN SPECTROSCOPIES

In time-resolved Raman spectroscopy, the pertinent time scales are given by the collision duration τ_c and the time of free flight τ_0, which are about 10^{-12} s and 10^{-10} s (under standard conditions), respectively (*cf* App. II.B3). The soundings thus require sub-picosecond lasers, to study at the pertinent time scale in the pump-and-probe Raman techniques considered thereafter. These NL techniques are complementary to SRS and CARS in the frequency domain, with specific advantages (and disadvantages). As an example, the *nonresonant* contribution appears at *zero time delay*, and therefore does not influence the transients.[101] Spectral-shape models cannot be directly applied to analyze the time-resolved Raman signals, but only partially transposed according to the considered technique as shown below. The use of these alternative Raman techniques for optical diagnostics are essentially prospective, and thus still limited to few physical situations, very far from the extensive use of conventional CARS. Only basic expressions are given here for time-resolved Raman signals, as illustrations.

Raman-induced polarization spectroscopy (RIPS) is a pump-and-probe technique[102,103] that uses one fast (non-resonant) pump laser pulse, linearly polarized, in order to generate a *coherent superposition of rotational states* (considered here to belong to the same vibrational state). After a time delay τ, a second short laser pulse (also linearly polarized, generally at 45° of the pump polarization) probes the quantum beats of the rotational wave packet through a recoupling scheme symmetric of the pump coupling (Fig. II.A4).

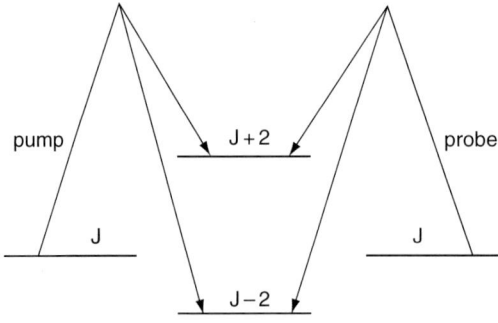

Fig. II.A4: Configuration of lasers frequencies in RIPS.

RIPS is a third order nonlinear process in which the radiator alignment is induced by the anisotropic polarizability tensor through the third order susceptibility [see Eq. (II.A25)], the quantum beats resulting in *transient rotational periodic alignments*. These beats are detected through the concomitant polarization rotation of the probe pulse using crossed polarizers. The RIPS signal can be represented[104,105] by the following expression (here restricted to a gas of linear radiators in the impact approximation):

$$I_{RIPS}(\tau) \propto n_a^2 \Delta_p^2 I_{p0} \int_{-\infty}^{+\infty} I_{probe}^2(t) \left\{ \sum_J T_J \exp\left[-\Gamma_J(t+\tau) - \Delta_p^2(\omega_J^2 - \Gamma_J^2)/4\right] \right.$$

$$\left. \times \sin\left[\omega_J(t+\tau) - \frac{\omega_J \Gamma_J \Delta_p^2}{2}\right] \right\}^2 dt, \quad \text{with,} \quad \text{(II.A31)}$$

$\omega_J = (E_{J+2} - E_J)/\hbar$ and $T_J = g_J(\rho_J - \rho_{J+2})\gamma^2(J+1)(J+2)/(2J+3)$,

g_J being the nuclear spin degeneracy and $\gamma = (\alpha_{//} - \alpha_\perp)$ the anisotropic polarizability of the radiator. In Eq. (II.A31), the detection is *homodyne* (such that the local electric field can be neglected), and $I_{probe}(t)$ is the time-dependent intensity of the probe. The shape of the pump intensity is assumed Gaussian, with a maximum I_{p0} and a temporal half width $\Delta_p(2\ln 2)^{1/2}$ shorter than the delay τ. The only line-shape parameter in this RIPS signal expression is the collisional broadening Γ_J of the states involved in the induced wave packet. In such quantum beat spectroscopy experiments resulting from a coherent superposition of *stationary* rotational eigenstates, the Doppler effect can be neglected.

In *time-resolved CARS* experiments (Fig. II.A5), two femtosecond pulses (pump and Stokes) excite the rovibrational transitions. After a delay τ, a *third* short laser probe pulse at frequency ω_{pr} generates an anti-Stokes signal through the interaction with $\chi^{(3)}(-\omega_{as};\omega_p,-\omega_s,\omega_{pr})$.[106,107] For isotropic Raman Q branch resonances, assuming Lorentzian line shapes in the frequency domain, the CARS signal in the collisional broadening regime at a given time delay is given by:[107,108]

$$I_{CARS}(\tau) = \left| \sum_J f_J \exp\{[i\Delta E_J/\hbar - \Gamma_J]\tau\} \right|^2, \quad \text{(II.A32)}$$

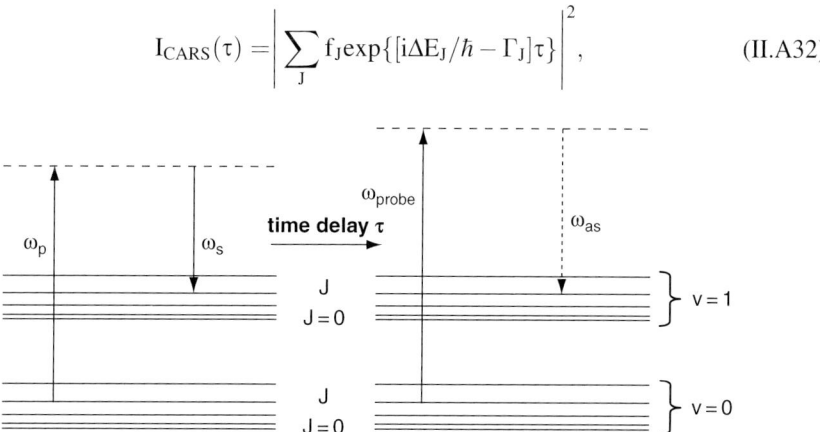

Fig. II.A5: Configuration of lasers frequencies in time-resolved CARS.

where $\Delta E_J = E(v=1,J) - E(v=0,J)$, Γ_J is the broadening of the Q(J) line, and the oscillation amplitudes f_J are proportional to the rotational populations and to the convolution of the pump and Stokes pulses in the frequency domain.[108] The sum over J is carried out over all significantly populated levels. The CARS signal in Eq. (II.A32) shows *oscillations* coming from the interferences between the Q branch transitions inside their coherent superposition, with a *decay* due to collisions. The influence of collisions appears through the collisional broadening of the Q(J) transitions only. Note that radiator speed inhomogeneous effects have been introduced by Tran et al[109] for isolated lines (*cf* § III.4.4), and that line mixing has also been investigated recently[110] (see Fig. IV.22).

This appendix has been restricted to a few spectroscopies in the frequency and time domains in direct connection with the applications considered in chapter VII. There is, of course, a broader variety of techniques suitable for studies of collisional processes for which the reader can refer to Refs. 9–11,111.

APPENDIX II.B: SOME CRITERIA FOR THE APPROXIMATIONS

II.B1 THE LARGE NUMBER OF PERTURBERS

The assumption here is that the considered medium can be represented by cells, each containing one radiator and *many* perturbers so that the spectral shape is only governed by their interactions. Hence, from an experimental point of view, the (linear) dimension L of the (cubic) cell containing the gas must be examined.[6] It must be much larger than the mean free path ℓ_0 in order to allow a sufficient number of radiator-perturber collisions before any interaction with the walls. A criterion, based on $L \gg \ell_0$, leads to the typical value $L \gg 10^{-5}/\rho_p$ cm, where ρ_p is the perturber density in amagat units.[(9)] The collisional effects on the core region of the (infrared) spectral shape being negligible below typically 10^{-3} Am at room temperature, this implies that the more drastic constraint in usual physical situations is $L \gg$ a few 10^{-2} cm, which is not a real constraint in practice. Readers interested by the reversed situation where the cell length is of the order of, or shorter than the mean free path can refer to Refs. 112,113 and those cited therein. In these papers it is shown that, starting from the usual Doppler profile observed at low density for $L \gg \ell_0$, the line width progressively reduces as L/ℓ_0 decreases and becomes smaller than unity. Collisions with the walls thus induce a narrowing of the observed line shape. This sub-Doppler broadening in a thin cell (somehow similar to the Dicke effect discussed in § III.3.2) results from the enhancement of the contributions of radiators (here atoms) with slow normal (with respect to the wall) velocities.[112,113]

[(9)] The gas density in amagat unit (Am), frequently denoted ρ, for total pressure P and temperature T, is given by the ratio of the molar volumes V: $\rho(P,T) = V(P_0, T_0)/V(P,T)$, with $P_0 = 1$ atm, $T_0 = 273.15$ K. For an ideal gas, it is given by $\rho(P,T) = (P/T)/(P_0/T_0)$. The number n_0 of molecules per unit volume at *one* amagat is then Loschmidt's number, ie $n_0 = 2.68719\ 10^{19}$ molec/cm^3, so that the number density $n(\rho)$ is given by $n(\rho) = n_0 \rho$.

II.B2 THE LOCAL THERMODYNAMIC EQUILIBRIUM

The study of departures from the Local Thermodynamic Equilibrium (LTE) is a vast problem which, in its generality, is out of the scope of this book. Various processes can lead to non-LTE, among which one finds optical pumping and chemical reactions. An entire field of research, on "coherent control",[114] aims at preparing molecular systems in ad hoc non-LTE situations (essentially through properly shaped laser pulses), in order to favor a desired evolution (eg a chosen chemical reaction). In these processes, intermolecular collisions tend to bring the system back to equilibrium so that departures from LTE are significant at sufficiently low densities (or pressures), depending on the strength of the mechanism (eg the radiation field) responsible for the non-equilibrium. Considering the broadness of the subject and the limited scope of this book, we only consider here non-LTE from the point of view of atmospheric applications (Chapt. VII) of collisional spectral shapes. A brief overview is given here but, for more details, the reader can refer to the review of Ref. 115.

From a general point of view, the time evolution of the average number $N_{s,j}$ of molecules of species s in state j is given by the following formal equation:

$$\frac{d}{dt}\vec{N}(t) = K^{Relax}\vec{N}(t) + \vec{S}[\vec{N}(t)], \tag{II.B1}$$

where $\vec{N}$ is a (column) vector whose coordinates are the various $N_{s,j}$'s, $\vec{S}$ is a vector describing the influence of the processes leading to the departures from equilibrium, and K^{Relax} is a matrix accounting for the population exchanges among the various internal states of species s. Its elements are directly related to the state-to-state rate $K_{s,j \leftarrow s,j'}$ for the collisional transfer from state s,j' to state s,j, by:

$$(K^{Relax})_{s,j,s',j'} = \delta_{s,s'}K_{s,j \leftarrow s,j'} \quad \text{for} \quad j' \neq j, \quad \text{and} \quad K_{s,j \leftarrow s,j} = -\sum_{j' \neq j} K_{s,j' \leftarrow s,j}. \tag{II.B2}$$

When $\vec{S} = \vec{0}$, the stationary equilibrium distribution is obtained. Indeed, since the state-to-state rates verify the detailed balance relation:

$$K_{s,j \leftarrow s,j'} \rho^0_{s,j'}(T) = K_{s,j' \leftarrow s,j} \rho^0_{s,j}(T), \tag{II.B3}$$

where $\rho^0_{s,j}(T)$ is the equilibrium relative population of state s,j, one straightforwardly obtains $N_{s,j}(t) = \rho^0_{s,j}(T) \times N_s$ where N_s is the total number of molecules of species s. In planetary atmospheres, a number of processes can give rise to non zero values of $\vec{S}$ that are mostly initiated by the presence of an external electromagnetic field (eg solar radiation). The first is obviously optical pumping, ie the direct excitation of molecules due to absorption of the incoming radiation. In this case, $\vec{S}$ takes the simple form $\vec{S}[\vec{N}(t)] = E.\vec{N}(t)$ where the elements of the matrix E (diagonal with respect to s) can be expressed in terms of the Einstein coefficients and of the spectral and energy characteristics of the incoming radiation. A second source of non-equilibrium, consequence of the first one, is collision-induced population exchanges by collisions where some excitation initially brought by the field is involved. The simplest is that when a species is excited by absorbing a photon before some of this energy is transferred, by collision, to another species (eventually to another state of the same species). Finally, other mechanisms are possible, such as photolysis, photo-induced chemical reactions, recombination, *etc*.[115]

A simple way to look at non-LTE in atmospheres is to compare the kinetic temperature T_{kin} with those driving the relative populations of the internal states associated with the various degrees of freedom (T_{rot}, T_{vib}, *etc*). If one considers rotational relaxation (*ie* rotation-rotation and rotation-translation transfers), the associated rates K are of the order of the pressure-broadened line width. For usual temperatures and the total pressure P (in atm), they are typically $10^{10} \times P$ s^{-1} and hence very efficient except at extremely low pressure. As a result, deviations from the equilibrium rotational distribution only occur at very high altitudes. Examples are given by NO and CO in the Earth atmosphere, where T_{rot} deviates from T_{kin} above about 100 km.[116,117] At such altitudes, the pressure is lower, for Earth, than a few Pa and the spectral shape is practically governed by the Doppler effect. When pressures are sufficient to make collisional effects on the spectral shape significant, the altitudes are relatively low and rotational levels are in equilibrium. If one now considers vibrational populations, the situation is similar, but with quantitative differences. In fact, vibration-translation relaxation processes, which tend to make T_{vib} and T_{kin} equal, are much less efficient than rotation-translation exchanges. The associated rate constant depends on the molecule, the vibrational state, and the collision partner. For collisions with N_2, it is of the order of several $10^4 \times P$ s^{-1} for the $v_2=1$ mode of CO_2, whereas it is several orders of magnitude smaller for the relaxation from the first vibrational state of O_2 (*eg* Ref. 118). The vibrational population distributions deviate from equilibrium above altitudes from a few tens of km to a hundred km, depending on the considered atmosphere and molecule. For CO_2, departures appear at atmospheric altitudes above about 40 km for Mars,[119] 60 km for Earth,[118] and 80 km for Venus.[120] Similar limits (above a few 10 km, *ie* below 10^{-2} atm in the Earth atmosphere) for the validity of the vibrational LTE assumption are obtained for various molecular species of atmospheric interest.[115] Hence, there are situations where non-LTE and collisional effects *simultaneously* affect the absorption or emission spectra. Nevertheless, these problems are not significantly entangled since neglecting the influence of the vibrational non-LTE on the spectral shape is generally a good approximation. Indeed, recall that the only internal state populations involved in the collisional quantities (broadening, shifting, relaxation operator, *etc*) are those of the perturber. These appear through an average:

$$\langle Q \rangle = \sum_p \rho_p \langle p|Q|p \rangle, \qquad (\text{II.B4})$$

where Q is the considered collisional quantity, p designates the perturber internal state and ρ_p the associated relative population. Introducing vibrational (v) and rotational quantum numbers (r) for the perturber, two quite reasonable approximations can be introduced which assume that:

$$\rho_p = \rho_{r,v} \approx \rho_v \times \rho_r = \rho_v \times \rho_r^0(T_{kin}) = \rho_v \times e^{-E_r/k_B T_{kin}} / \sum_r e^{-E_r/k_B T_{kin}}, \qquad (\text{II.B5})$$

and $\quad \langle v,r|Q|v,r \rangle \approx \langle v=0,r|Q|v=0,r \rangle,$

General equations 47

where we have taken into account the fact that rotational levels are at LTE through the equilibrium distribution $\rho_r^0(T_{kin})$ of the (purely) rotational levels. In other words, one assumes that the vibrational state does affect the interactions and the rotational energy structure enough to induce significant changes in the value of the overall average. Since this should be true for the significantly populated levels only (*ie* the lowest vibrational states for usual perturbers and temperatures), it is generally a very good approximation. Within these assumptions, the collisional quantity is given by:

$$\langle Q \rangle \approx \left[\sum_v \rho_v\right] \times \left[\sum_r \rho_r^0(T_{kin})\langle v=0,r|Q|v=0,r\rangle\right] = \sum_r \rho_r^0(T_{kin})\langle v=0,r|Q|v=0,r\rangle,$$

(II.B6)

showing that it is not affected by the non-LTE of vibrational levels. The only spectral manifestation of non-equilibrium thus comes *from the radiator* through the populations of the levels associated with the optical transitions. Non-LTE is then easily taken into account by introducing proper radiator vibrational populations (or temperatures) and using the Einstein coefficients.[115]

II.B3 THE BINARY COLLISIONS

A simple criterion for the validity of the binary collision assumption can be derived by comparing the average time of free flight τ_0 (mean time between successive collisions) with the collision duration τ_c. The value of τ_0 can be estimated from the simple case of *rigid spheres* within the gas kinetic theory. Then $\tau_0 \approx \ell_0/<v_r>$, where $\ell_0 = (\sqrt{m/m^*}\pi\sigma^2 n_p)^{-1}$ is the mean free path and $<v_r> = \sqrt{8k_B T/\pi m^*}$ is the average relative speed, σ being the diameter of the rigid spheres, m the radiator mass, m^* the reduced mass, and n_p the perturber density (assuming high dilution). For temperatures not too different from the ambient and "usual" radiators and perturbers (CO, N_2, O_2, etc), τ_0 is close to $10^{-10}/\rho_p$ s, where the density ρ_p is expressed in amagat units. Whereas the magnitude of τ_0 is quite well defined through the mean free path and relative speed, the estimation of τ_c is more difficult since it is tightly related to the collision "efficiency" for the spectral shape of interest (the broadening of isolated lines near their centers, the far wings, *etc*). This efficiency depends on the strength and range of the intermolecular potential, as well as on the dynamical efficiency of the internal energy exchanges. In order to estimate τ_c, one can introduce a characteristic length ℓ_c for the considered collisions, leading to $\tau_c \approx \ell_c/<v_r>$. Two limiting cases can then be considered for illustration. In the first one, a long range intermolecular potential and quasi-resonant rotational transfers are involved. This situation corresponds to the impact (close to line center) self broadening of low rotational quantum number lines of CO or CO_2 at room temperature, for instance. By using the same values for the parameters σ, m^*, and T as for τ_0, this leads to $\tau_c \approx 10^{-12}$ s, which, contrary to τ_0, is independent of the density. Thus $\tau_0 \approx \tau_c$ for a density of the order of one hundred amagat. The opposite case corresponds to a short range intermolecular potential and very non-resonant rotational energy transfers. It is exemplified by the effects of collisions with He on the broadening of high rotational quantum number lines of HCl and HF, or on the absorption in the very far line wings. In these cases, only interactions at very short distances are efficient and the relevant characteristic length ℓ_c is a small fraction of the molecular diameter σ. The collision time duration τ_c is significantly shortened, this being further amplified when high relative speeds are considered, as for small reduced masses and elevated temperatures. It may become of the order of 10^{-13} s or less so that τ_0 only reaches τ_c for densities of about a thousand amagat. This shows that the limit of validity of the impact approximation is *closely related* to the collision *efficiency* and requires specific studies in each situation to get a reliable criterion. Furthermore, $\tau_0 \gg \tau_c$ also defines the validity of the impact approximation (App. II.B4) so that, if the binary collision assumption breaks down, so do impact models. This sensitivity to the collision efficiency, the entanglement with the impact approximation and the fact that $\tau_0 \gg \tau_c$ is a strictly mathematical criterion[10] make the problem very complex. In practice, for a given accuracy, the density limit of a binary collision model can only be determined from experimental results pointing deviations from the expected behavior with density. A first example can be found in Ref. 121 and Fig. III.7 where Ar-broadened HD Raman lines are studied. The widths of these isolated transitions increase with the argon density, with departures from the expected linear variations becoming significant above about 200 Am at room temperature. Another illustration is given by absorption in the Ar-broadened wing of the CO_2 v_3 band, which deviates from the expected proportionality to the product $n_{CO2} \times n_{Ar}$.[122] Finally, Ref. 123 shows that the high density evolution of the width of the isotropic Raman Q branch of N_2 cannot be interpreted using simple binary collision impact models. Note that the validity of such approaches can be extended by taking "*excluded volume*" effects[35,124] into account. This introduces a correction to the number density of perturbers n_p (in the case of high dilution) which must be replaced, for small corrections, by:

$$\tilde{n}_p = n_p \left[1 + \left(\frac{2}{3} \pi \sigma_p^3 \frac{\sigma_p + 4\sigma_a}{4\sigma_p + 4\sigma_a} \right) n_p \right], \tag{II.B7}$$

where σ_a and σ_p are the collisional diameters of the radiator and perturber. This reduces, for a pure gas, to:

$$\tilde{n} = n \left[1 + \left(\frac{5}{12} \pi \sigma^3 \right) n \right], \tag{II.B8}$$

which is the low correction limit of the more general expression:

$$\tilde{n} = n \frac{1 - \pi n \sigma^3 / 12}{(1 - \pi n \sigma^3 / 6)^3}. \tag{II.B9}$$

---

[10] As often in physics, a given approximation whose strict validity is defined by the mathematically derived inequality $a \ll b$, generally stands with reasonable accuracy in a much broader range, often up to a of the order of b. This may here express the fact that a more realistic binary collisions criterion than $\tau_0 \gg \tau_c$ considers that only "strong" collisions do not overlap in time.[63]

As demonstrated in Refs. 122,123, with the use of these corrected densities, the observed data show the expected (linear) dependences on $\tilde{n}_p$. Hence, accounting for the excluded volume significantly extends the applicability of binary collision models.

II.B4 THE (FULL) IMPACT ASSUMPTION

As shown in appendix II.C1, the widely used impact approximation relies on various criteria. The first considers that collisions are largely completed. It is valid, as the binary collision assumption, for $\tau_0 \gg \tau_c$. The second criterion, associated with the neglect of the influence of the finite duration of collisions, is $|\omega - \omega_{fi}|^{-1} \gg \tau_c$, corresponding to wavenumbers such that $|\sigma - \sigma_{fi}| \ll (2\pi c \tau_c)^{-1}$. As seen above, this last quantity strongly depends, through τ_c, on the nature of the collisions and on their efficiency for the considered process. Its value lies between a few cm^{-1} for quasi-resonant collisions involving strong long range interactions, and several tens of cm^{-1} when only very close, and short, collisions are efficient. As for the binary collision assumption, general and quantitative criteria are difficult to derive and, *in fine*, only experiments can provide the true validity range. Remember that the impact treatment of line mixing also requires $|\omega_{f'i'} - \omega_{fi}|\tau_c \ll 1$ for collisionally coupled fi and f'i' transitions, this last criterion being somehow included in the above mentioned $|\omega - \omega_{fi}|\tau_c \ll 1$ one.

There are essentially two situations where impact approaches cannot be used. The first one is obviously the far wings of lines and bands for which $|\omega - \omega_{fi}|\tau_c$ is (significantly) greater than unity. In this case, finite duration of collision effects can lead to large spectral modifications, through the influence of initial correlations and of the frequency dependence of the relaxation matrix (*cf* § II.3.3 and Sec. II.4). This is illustrated by Figs. IV.11, V.8 and VIII.3 and discussed in chapter V. The second situation is that of spectrally isolated lines at elevated density. The influence of the finite duration of collisions then mainly manifests itself through two changes with respect to the impact Lorentzian profile.[125-128] As analyzed in appendix II.C1 and Ref. 129, the diagonal terms of the operator in Eq. (II.45), must then be written as:

$$Q_{fi,fi}(t) = V \times \mathrm{Tr}_{\tilde{H}_0^{-1}}\left\{\rho(\tilde{H}_0^1)[I_d - q(t)]\right\}_{fi,fi} \approx [a_{fi} + \gamma_{fi}t] + i[b_{fi} + \delta_{fi}t], \quad (\text{II.B10})$$

where all x_{fi} quantities on the right hand side are real, γ_{fi} and δ_{fi} being the pressure-broadening and -shifting coefficients. Introducing this equation into the spectral density [Eq. (II.43)], and taking the Laplace transform at frequency $(\omega - \omega_{fi})$ straightforwardly leads to the following complex isolated line shape:

$$I_{fi}(\omega) = \frac{1}{\pi} \frac{\exp\left[-(n_p a_{fi} - in_p b_{fi})\right]}{\omega - \omega_{fi} - n_p \delta_{fi} - in_p \gamma_{fi}}, \quad (\text{II.B11})$$

which differs from the corresponding purely Lorentzian impact profile of Eq. (II.53). If one considers absorption processes (and thus the imaginary part of l_{fi}), assuming moderate values of $n_p b_{fi}$, the following expression is obtained:

$$F_{fi}(\omega) \approx (\rho_i d_{fi}^2) e^{-n_p a_{fi}} \times \frac{1}{\pi} \frac{n_p \gamma_{fi} + (\omega - \omega_{fi} - n_p \delta_{fi}) n_p b_{fi}}{(\omega - \omega_{fi} - n_p \delta_{fi})^2 + (n_p \gamma_{fi})^2}, \qquad \text{(II.B12)}$$

which has to be compared with Eq. (II.54). This shows that non-impact effects on the core regions of isolated transitions lead to both an *asymmetry* of the line profile (*eg* Fig. III.8), through b_{fi}, and a *decrease* of the integrated intensity [area below $F_{fi}(\omega)$], through a_{fi}. Experimental evidence and analysis of these effects can be found in Refs. 130 and 131, respectively and those cited therein (see also Sec. III.1). Note that paths toward the accounting for both line mixing and collision-time asymmetry are given in Ref. 132.

APPENDIX II.C: THE IMPACT RELAXATION MATRIX

We study here the relaxation matrix within the *impact* limit and introduce the conditions required in order to derive it. This can be done either by looking at the time dependence of a matrix element of $[I_d - q(t)]$ in Eqs. (II.43) and (II.45), or by analyzing the frequency dependence of the scattering transition operator, as given by Eq. (55) of Ref. 44.

II.C1 ANALYSIS THROUGH THE TIME DEPENDENCE

In order to simplify notations, we consider a linear radiator and atomic perturbers and use a classical path treatment of the relative translational motion. We neglect initial correlations since, as will be shown, only angular frequencies much closer to line centers than $k_B T/\hbar$ are considered within the impact approximation. Let us first consider the case of collisionally isolated lines and thus study only the diagonal elements of the $[I_d - q(t)]$ operator. Starting from Eqs. (II.43) and (II.45), the relevant matrix element is then given by:

$$Q_{fi,fi}(t) \equiv V \times \mathrm{Tr}_{\widetilde{H}_0^1}\{\rho(\widetilde{H}_0^1)[I_d - q(t)]_{fi,fi}$$

$$= <v_r> \int_0^\infty 2\pi b\, db \times \int_0^\infty x e^{-x} dx \times \int_{-\infty}^{+\infty}$$

$$\left[1 - \sum_{\substack{m_i,m_i' \\ m_f,m_f',q}} (-1)^{m_f+m_{f'}} \begin{pmatrix} J_i & 1 & J_f \\ m_i' & q & -m_f' \end{pmatrix} \begin{pmatrix} J_i & 1 & J_f \\ m_i & q & -m_f \end{pmatrix} \right.$$

$$\left. \times \langle v_f J_f m_f' | U(t,0,t_0,b,x) | v_f J_f m_f \rangle \langle v_i J_i m_i' | U(t,0,t_0,b,x) | v_i J_i m_i \rangle^* \right] dt_0,$$

$$\text{(II.C1)}$$

General equations

where the trace over the position of the radiator in the volume V has been carried out, and the remaining averages are expressed in terms of three classical path parameters (t_0, b, x) defined below. In Eq. (II.C1), $U(t, 0, t_0, b, x)$ is the operator describing the evolution from time 0 to time t for a binary collision centered at time $t = t_0$ and with initial impact parameter b and reduced relative kinetic energy $x = m^* v_r^2/(2k_B T)$. It is then possible to transform Eq. (II.C1) by centering any collision at time $t = 0$, thanks to the following property of the evolution operator:[133]

$$U(t, 0, t_0, b, x) = e^{-\frac{i}{\hbar}H_a t_0} U(t - t_0, -t_0, 0, b, x) e^{\frac{i}{\hbar}H_a t_0}$$
$$\equiv e^{-\frac{i}{\hbar}H_a t_0} U(t - t_0, -t_0, b, x) e^{\frac{i}{\hbar}H_a t_0}, \quad \text{(II.C2)}$$

where $U(t-t_0,-t_0,b,x)$ now corresponds to the evolution from time $-t_0$ to $t-t_0$ for a collision *centered at* $t = 0$. Equation (II.C1) then becomes:

$$Q_{fi,fi}(t) = <v_r> \int_0^\infty 2\pi b db \times \int_0^\infty xe^{-x}dx \int_{-\infty}^{+\infty} f(t, t_0, b, x)dt_0, \quad \text{(II.C3)}$$

with

$$f(t, t_0, b, x) \equiv 1 - \sum_{\substack{m_i, m'_i \\ m_f, m'_f, q}} (-1)^{m_f + m_{f'}} \begin{pmatrix} J_i & 1 & J_f \\ m'_i & q & -m'_f \end{pmatrix} \begin{pmatrix} J_i & 1 & J_f \\ m_i & q & -m_f \end{pmatrix}$$

$$\times \langle v_f J_f m'_f | U(t - t_0, -t_0, b, x) | v_f J_f m_f \rangle \langle v_i J_i m'_i | U(t - t_0, -t_0, b, x) | v_i J_i m_i \rangle^* \}. \quad \text{(II.C4)}$$

The typical behavior of the $f(t, t_0, b, x)$ function *vs* t_0 is shown in Fig. II.C1 for absorption by HCl diluted in Ar. These results were calculated, using the intermolecular potential of

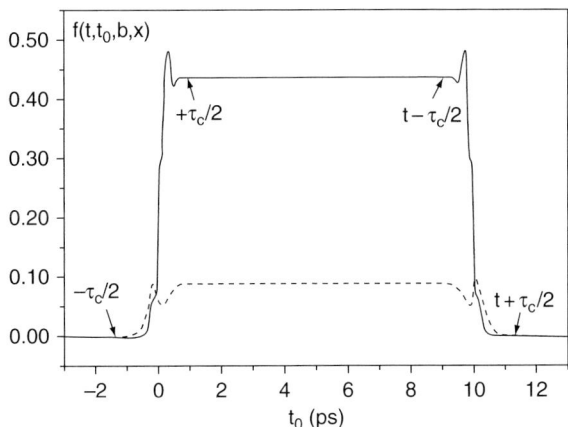

Fig. II.C1: Calculated $f(t, t_0, b, x)$ function for the R(3) line of the $v = 1 \leftarrow v = 0$ band of HCl in Ar and the values t = 10 ps, b = 4 Å, and a relative kinetic energy = 262 cm^{-1}. (——) and (- - -) give the real and imaginary parts.

Ref. 134, with the approach described in Refs. 131,135. The value t = 10 ps chosen is significantly longer than the collision duration τ_c [here less than 2 ps, as defined by $V(t) \approx 0$ unless $-\tau_c/2 \leq t \leq +\tau_c/2$ where V(t) is the intermolecular potential].

In Fig. II.C1, the finite duration of collision appears through the transients in the time intervals $[-\tau_c/2, +\tau_c/2]$ and $[t-\tau_c/2, t+\tau_c/2]$. The integral over t_0 in Eq. (II.C3) can be split into five domains, namely: $]-\infty, -\tau_c/2]$, $[-\tau_c/2, +\tau_c/2]$, $[+\tau_c/2, t-\tau_c/2]$, $[t-\tau_c/2, t+\tau_c/2]$, and $[t+\tau_c/2, +\infty[$. After some algebra, using the fact that, for any t>0 $U(t+\tau_c/2, \tau_c/2, b, x) = U(-\tau_c/2, -t-\tau_c/2, b, x) = I_d$, this integral can be performed, leading, after integration over b and averaging over the relative kinetic energy, to:

$$Q_{fi,fi}(t) = A_{fi} \times t + B_{fi} + g_{fi}(t), \qquad (II.C5)$$

where A_{fi}, B_{fi}, and g_{fi} are complex quantities and $g_{fi}(t)$ takes non zero values only for $t < \tau_c$. Note that A_{fi} and B_{fi} are directly related to the parameters of Eqs. (II.B10)–(II.B12) by $A_{fi} = \gamma_{fi} + i\delta_{fi}$ and $B_{fi} = a_{fi} + ib_{fi}$. The behavior of the diagonal elements of the Q(t) matrix is illustrated by the results in Fig. II.C2.

The parameters of the linear part of Q(t) in Eq. (II.C5) can be expressed[129,131] in terms of specific evolution operators. The slope A_{fi} is directly given by

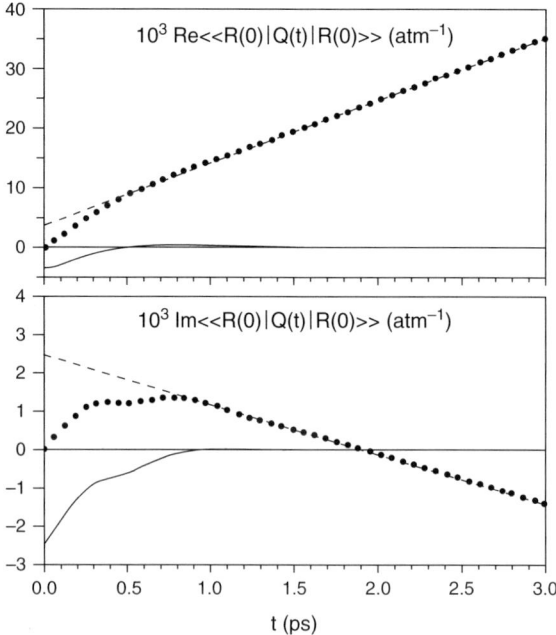

Fig. II.C2: Calculated real and imaginary diagonal elements of Q(t) [*cf* Eq. (II.C3)] for the R(0) line of the 1−0 band of HCl broadened by Ar at room temperature. (•), (- - -), and (___) give respectively the total $Q_{fi}(t)$, the linear part $A_{fi}t + B_{fi}$, and the short time function $g_{fi}(t)$, see Eq. (II.C5). After Ref. 135.

f($+\infty, -\infty, b, x$), and is thus determined from the evolution for a *completed* collision centered at time 0, *ie* from the scattering operator S. The time zero value B_{fi} can be obtained[129,131] from f($+\infty, -\infty, b, x$) and the integrals over t_0 of f($+\infty, t_0, b, x$) and f($t_0, -\infty, b, x$).

Introducing Eq. (II.C5) into Eq. (II.43), the profile of an isolated line is given by:

$$\frac{1}{\pi} \text{Re} \left[\int_0^{+\infty} e^{i(\omega - \omega_{fi})t} e^{-n_p Q_{fi}(t)} dt \right] \equiv \text{Re}[I_{fi}(\omega)^*], \qquad \text{(II.C6)}$$

with

$$I_{fi}(\omega)^* = \frac{-i}{\pi} \int_0^{+\infty} e^{i(\omega - \omega_{fi})t} e^{-n_p [A_{fi}t + B_{fi} + g_{fi}(t)]} dt. \qquad \text{(II.C7)}$$

The full impact approximation neglects the influence of the terms B_{fi} and $g_{fi}(t)$ resulting from the finite collision duration, straightforwardly leading to the purely Lorentzian (complex) profile of Eq. (II.54) [with $\Gamma_{fi} + i\Delta_{fi} = n_p A_{fi}$]. In order to derive criteria for the validity of this approximation, Eq. (II.C7) can be transformed, leading to:

$$I_{fi}(\omega)^* = e^{-n_p B_{fi}} \times \frac{-i}{\pi} \left\{ \int_0^{+\infty} e^{i(\omega - \omega_{fi})t} e^{-n_p A_{fi}t} dt \right\} \otimes \left\{ \int_0^{+\infty} e^{i(\omega - \omega_{fi})t} e^{-n_p g_{fi}(t)} dt \right\}, \qquad \text{(II.C8)}$$

i.e

$$I_{fi}(\omega)^* = e^{-n_p B_{fi}} \times I_{fi}^{\text{Lorentz}}(\omega - \omega_{fi})^* \otimes \left\{ \int_0^{+\infty} e^{i(\omega - \omega_{fi})t} e^{-n_p g_{fi}(t)} dt \right\}. \qquad \text{(II.C9)}$$

Here we have introduced the complex Lorentzian profile of Eq. (II.53) resulting from the first Laplace transform in Eq. (II.C8), $\otimes$ denoting a convolution product.

For the *full impact* (Lorentzian) profile to be valid [*ie* $I_{fi}(\omega) = I_{fi}^{\text{Lorentz}}(\omega - \omega_{fi})$], two conditions must be fulfilled. The first one is that $\exp[-n_p B_{fi}]$ must be close to unity. The second is that the Laplace transform in Eq. (II.C9) should be well approximated by a Dirac delta function. Starting from expressions given in Refs. 129,131, it is easy to show that B_{fi} is of the order of $A_{fi}\tau_c$ [*ie* $(\gamma_{fi} + i\delta_{fi})\tau_c$] as can be checked from Fig. II.C2. Now recall that $\Gamma_{fi} = n_p \gamma_{fi}$ is of the order of the inverse of the time interval $\tau_0(n_p)$ between successive collisions (rigorously equal for rigid spheres). From these remarks, the *first* criterion ($n_p B_{fi} \ll 1$) for the impact approximation leads to:

$$\tau_c \ll \tau_0(n_p). \qquad \text{(II.C10)}$$

In order to derive the second condition, note that $g_{fi}(t)$ is equal to $-B_{fi}$ for $t=0$ and drops to zero for $t > \tau_c$. According to the previous approximation, $n_p B_{fi}$ is small and thus so is $n_p g_{fi}(t)$, allowing to write:

$$\int_0^{+\infty} e^{i(\omega - \omega_{fi})t} e^{-n_p g_{fi}(t)} dt \approx \int_0^{+\infty} e^{i(\omega - \omega_{fi})t} dt - \int_0^{+\infty} e^{i(\omega - \omega_{fi})t} n_p g_{fi}(t) dt. \qquad \text{(II.C11)}$$

The first term on the right hand side is the $\delta(\omega - \omega_{fi})$ delta function, so that the Lorentzian profile is obtained from Eq. (II.C9) if the second term can be neglected. Since the magnitude of $n_p g_{fi}(t)$ is small and this function being non-zero only for $t < \tau_c$, it is obvious from the properties of Fourier/Laplace transform, that the second term in Eq. (II.C11) can always be neglected if:

$$|\omega - \omega_{fi}| \times \tau_c \ll 1. \quad \text{(II.C12)}$$

This inequality is the second criterion for the validity of the full impact approximation.

An analysis for collisionally coupled lines was made in Refs. 70,71. It leads to the more general long time behavior of Eq. (II.46), and thus to the impact expression of Eq. (II.50), provided that the following additional condition is verified:

$$|\omega_{f'i'} - \omega_{fi}| \times \tau_c \ll 1. \quad \text{(II.C13)}$$

indicating that (significantly) coupled transitions should not be centered too far apart.

These features show that the spectral shape can be described by an impact theory (*ie* in terms of scattering S matrix elements), when it originates from a number of *nearby* ($|\omega - \omega_{f'i'}| \tau_c \ll 1$) *closely spaced* ($|\omega_{fi} - \omega_{f'i'}| \tau_c \ll 1$) transitions and when collisions can be considered as *completed* [$\tau_c/\tau_0(n_p) \ll 1$]. A quantitative discussion of these conditions can be found in App. II.B4. The preceding criteria are restrictive, in particular when the separation $|\omega_{fi}-\omega_{f'i'}|$ between transitions is large. The interested reader can find in Ref. 136 a comparison between the usual impact model, conditioned by Eqs. (II.C10), (II.C12), (II.C13), and an alternative approach constrained by Eqs. (II.C10) and (II.C12) only. This last model is limited to a second-order expansion of the evolution operator in terms of the intermolecular potential and leads to significant differences with the purely impact limit for molecules with large rotational constants.

II.C2 ANALYSIS THROUGH THE FREQUENCY DEPENDENCE

Let us now consider Eq. (55) of Ref. 44 in order to establish the impact limit in the "language" of formal scattering theories (see also App. II.E). Considering only completed collisions (*ie* $\tau_c \ll \tau_0$) corresponds to keeping only totally on-shell elements of the transition matrix T, *ie* those for which $E_{I'} = E_I = E$ for $<I|T(E)|I'>$, where $|I>$ and $|I'>$ are eigenstates of the free molecule Hamiltonian with energies E_I and $E_{I'}$. Let us consider, as an example, the first of the four terms in the collisional operator $<m_c(\omega)>_p$ in Eq. (II.E8). When "going on-shell" it is given by:

$$\langle F'|T(E_I + \hbar\omega)|F\rangle\langle I'|I\rangle = \left\langle f'\vec{k}'\left|T(E_i + \frac{k^2}{2m^*} + \hbar\omega)\right|f\vec{k}\right\rangle \delta(E_i + \frac{k^2}{2m^*} - E_{i'} - \frac{k'^2}{2m^*}), \quad \text{(II.C14)}$$

where, for simplicity, we still consider an atomic perturber. In this equation, f and i denote internal states of the radiator while $\vec{k}$ and $\vec{k}'$ stand for the radiator-perturber relative translational states. The on-shell condition requires that:

$$E_{f'} + \frac{k'^2}{2m^*} = E_f + \frac{k^2}{2m^*} \approx E_i + \frac{k^2}{2m^*} + \hbar\omega, \tag{II.C15}$$

leading to $\omega = (E_f - E_i)/\hbar = \omega_{fi}$, which, combining Eq. (II.C15) with the delta function in Eq. (II.C14), yields the generalized requirement:

$$\omega \approx \omega_{fi} \approx \omega_{f'i'}. \tag{II.C16}$$

Applying the on-shell condition and the delta function restriction to the other terms of Eq. (II.E8) lead to the same criterion. Equation (II.C16) is nothing else but Eqs. (II.C12) and (II.C13) taken together. In fact, by means of the properties of the Fourier/Laplace transform, it has been possible to obtain, in App. II.C1, more refined criteria defining the validity of the impact approximation, *ie* by comparing the various angular frequencies with the inverse of the collision time duration.

APPENDIX II.D: THE LIOUVILLE SPACE

The Liouville (line) space **L** (see also Refs. 11,67) is defined as the product $H \otimes H^+$ where **H** is the Hilbert space and H^+ is its associated dual space.

Vectors in L. If the ket vectors $|\alpha\rangle$ (and associated bra's $\langle\alpha|$) form a basis of **H**, the $|\alpha\rangle\langle\beta|$'s, denoted $|\alpha\beta\rangle\rangle$, (and $\langle\langle\alpha\beta| \equiv |\beta\rangle\langle\alpha|$) which are operators in **H**, form a basis in **L**. The product of two vectors of **L** is defined by:

$$\langle\langle A'|A\rangle\rangle = \text{Tr}[A'^\dagger A], \tag{II.D1}$$

where Tr[...] designates the trace. One thus has:

$$\langle\langle\alpha'\beta'|\alpha\beta\rangle\rangle = \text{Tr}\big[|\beta'\rangle\langle\alpha'|\alpha\rangle\langle\beta|\big] = \sum_\gamma \langle\gamma|\beta'\rangle\langle\alpha'|\alpha\rangle\langle\beta|\gamma\rangle. \tag{II.D2}$$

Hence, if the $|\alpha\rangle$'s form an orthonormalized basis of **H** (*ie* $\langle\alpha'|\alpha\rangle = \delta_{\alpha,\alpha'}$), the $|\alpha\beta\rangle\rangle$'s are also orthonormalized in **L**, since Eq. (II.D2) straightforwardly leads to:

$$\langle\langle\alpha'\beta'|\alpha\beta\rangle\rangle = \delta_{\alpha,\alpha'}\delta_{\beta,\beta'}. \tag{II.D3}$$

Furthermore, one also has the following "closure" relation:

$$\sum_{\alpha,\beta} |\alpha\beta\rangle\rangle\langle\langle\alpha\beta| = 1. \tag{II.D4}$$

If A designates an operator in H, and hence a vector in L, one has:

$$|A\rangle\rangle = \sum_{\alpha,\beta} \langle\langle\alpha\beta|A\rangle\rangle \times |\alpha\beta\rangle\rangle = \sum_{\alpha,\beta} \text{Tr}[|\beta\rangle\langle\alpha|A] \times |\alpha\beta\rangle\rangle = \sum_{\alpha,\beta,\gamma} \langle\gamma|\beta\rangle\langle\alpha|A|\gamma\rangle \times |\alpha\beta\rangle\rangle$$
$$= \sum_{\alpha,\beta} \langle\alpha|A|\beta\rangle \times |\alpha\beta\rangle\rangle, \qquad (\text{II.D5})$$

yielding:

$$\langle\langle\alpha\beta|A\rangle\rangle = \langle\alpha|A|\beta\rangle. \qquad (\text{II.D6})$$

Operators in L. If $\hat{O}$ is a (super) operator in L, constructed from the operator O of H, the action of $\hat{O}$ on any vector $|A\rangle\rangle$ of L is given by (A being an operator in H):

$$\hat{O}|A\rangle\rangle = [O, A], \qquad (\text{II.D7})$$

where [,] is the commutator. One can then define the matrix elements in the Liouville space. Indeed, Eq. (II.D5) leads to:

$$\hat{O}|\alpha\beta\rangle\rangle = [O, |\alpha\rangle\langle\beta|] = O|\alpha\rangle\langle\beta| - |\alpha\rangle\langle\beta|O, \qquad (\text{II.D8})$$

leading, from Eq. (II.D6), to:

$$\langle\langle\alpha'\beta'|\hat{O}|\alpha\beta\rangle\rangle = \langle\alpha'|O|\alpha\rangle\langle\beta|\beta'\rangle - \langle\alpha|\alpha'\rangle\langle\beta|O|\beta'\rangle. \qquad (\text{II.D9})$$

Hence, if O is a Hermitian operator in H and the $|\alpha\rangle$'s are its orthonormalized eigenvectors with associated eigenvalues o_α, one has:

$$\langle\langle\alpha'\beta'|\hat{O}|\alpha\beta\rangle\rangle = (o_\alpha - o_\beta) \times \delta_{\alpha,\alpha'}\delta_{\beta,\beta'}. \qquad (\text{II.D10})$$

This equation is often used in the present book in the particular case of the operator $\hat{O} = L_a$ of L constructed from the operator $O = H_a/\hbar$, where H_a is the Hamiltonian describing the internal states of the radiator. If $|i\rangle$ and $|f\rangle$ designate eigenvectors of H_a associated with the (energy) eigenvalues E_i and E_f, then the matrix element:

$$\langle\langle fi|L_a|fi\rangle\rangle = (E_f - E_i)/\hbar \equiv \omega_{fi}, \qquad (\text{II.D11})$$

directly gives the frequency ω_{fi} [or, in wavenumbers, $\sigma_{fi} = \omega_{fi}/(2\pi c)$] of the line associated with the $f \leftarrow i$ optical transition.

Evolution operators in L. In the Hilbert space, the Heisenberg representation of the time evolution of an operator O is given by:

$$O(t) = U^+(t)OU(t), \qquad (\text{II.D12})$$

where U(t) is the evolution operator (U^+ being its adjoint), given by:

$$U(t) = e^{-iHt/\hbar}, \tag{II.D13}$$

H being the total Hamiltonian for the considered system. Let us introduce the Liouville operator L constructed from $H/\hbar$. Note that, due to Eq. (II.D7), the use of L in the equation giving the evolution of the density operator $\rho(t)$, ie

$$i\hbar \frac{\partial}{\partial t}\rho(t) = [H, \rho], \tag{II.D14}$$

leads to the following compact form:

$$i\frac{\partial}{\partial t}\rho(t) = L\rho, \tag{II.D15}$$

which naturally calls for the introduction of the Liouville operator $\hat{U}(t)$, defined by:

$$\hat{U}(t) = e^{-iLt}. \tag{II.D16}$$

From Eq. (II.D12) and the definition of L, it is then straightforward to show that:

$$O(t) = \hat{U}(t)^+ O. \tag{II.D17}$$

One thus has, successively using Eqs. (II.D17), (II.D4) and (II.D6):

$$\langle\langle \beta'\alpha' | O(t) \rangle\rangle = \langle\langle \beta'\alpha' | \hat{U}^+(t) O \rangle\rangle = \sum_{\alpha,\beta} \langle\langle \beta'\alpha' | \hat{U}^+(t) | \beta\alpha \rangle\rangle \langle\langle \beta\alpha | O \rangle\rangle$$
$$= \sum_{\alpha,\beta} \langle\langle \beta'\alpha' | \hat{U}^+(t) | \beta\alpha \rangle\rangle \langle \beta | O | \alpha \rangle. \tag{II.D18}$$

From Eqs. (II.D6), (II.D12), and the closure relation in the Hilbert space, one can write:

$$\langle\langle \beta'\alpha' | O(t) \rangle\rangle = \langle \beta' | O(t) | \alpha' \rangle = \langle \beta' | U^+(t) O U(t) | \alpha' \rangle = \sum_{\alpha,\beta} \langle \beta' | U^+(t) | \beta \rangle \langle \beta | O | \alpha \rangle \langle \alpha | U(t) | \alpha' \rangle. \tag{II.D19}$$

Identifying Eqs. (II.D18) and (II.D19) leads to:

$$\langle\langle \beta'\alpha' | \hat{U}^+(t) | \beta\alpha \rangle\rangle = \langle \beta' | U^+(t) | \beta \rangle \langle \alpha | U(t) | \alpha' \rangle = \langle \beta | U(t) | \beta' \rangle^* \langle \alpha | U(t) | \alpha' \rangle. \tag{II.D20}$$

This shows that $\langle\langle fi | \hat{U}^+(t) | f'i' \rangle\rangle = \langle f' | U(t) | f \rangle^* \langle i' | U(t) | i \rangle$, and, finally:

$$\langle\langle f'i' | \hat{U}(t) | fi \rangle\rangle = \langle f' | U(t) | f \rangle \langle i' | U(t) | i \rangle^*, \tag{II.D21}$$

in agreement with Eq. (II.45).

APPENDIX II.E: THE RESOLVENT APPROACH

II.E1 SPECTRAL-SHAPE EXPRESSION

The starting point of the Fano and Ben Reuven formalism[44–46,137] is Eqs. (II.7) and (II.8), which define the spectral density $F(\omega)$ as the Fourier transform of the autocorrelation function $\Phi(t)$ of the radiator dipole moment $\vec{d}$ (when considering, of course, electric dipole absorption spectra):

$$F(\omega) = \frac{1}{\pi} \text{Re}\left\{ \int_0^\infty dt e^{i\omega t} \text{Tr}_{S_M}\left[\rho_M(0) e^{+iH_M t/\hbar} \vec{d}^+ e^{-iH_M t/\hbar} \vec{d}\right] dt \right\}. \qquad \text{(II.E1)}$$

Here, H_M and $\rho_M(0)$ are the Hamiltonian and the initial density operator of the system S_M composed of one radiator in a bath of N_p perturbers. Note that, implicit in Eq. (II.E1), is the assumption that the translational motion of the radiating molecule can be neglected. For a correct treatment of Doppler broadening and motional narrowing, this approximation must not be made (see chapter III). The interested reader will find a possible solution in Refs. 138,139. Introducing the Liouville space operator L_M corresponding to the Hamiltonian $H_M/\hbar$ (*cf* App. II.D) gives:

$$F(\omega) = \frac{1}{\pi} \text{Re}\left\{ \int_0^\infty e^{i\omega t} \text{Tr}_{S_M}\left[\vec{d}^+ e^{-iL_M t} \vec{d} \rho_M(0)\right] dt \right\}. \qquad \text{(II.E2)}$$

Then the Laplace transform can be carried out formally, leading to:

$$F(\omega) = -\frac{1}{\pi} \text{Im}\left\{ \text{Tr}_{S_M}\left[\vec{d}^+ \frac{1}{\omega I_d - L_M} \vec{d} \rho_M(0)\right] \right\}. \qquad \text{(II.E3)}$$

In order to go further, Fano assumed[44] that initial correlations can be neglected, so that $\rho_M(0)$ can be divided into two contributions, *ie*:

$$\rho_M(0) = \rho_a(0)\rho_p(0), \qquad \text{(II.E4)}$$

where $\rho_a(0)$ and $\rho_p(0)$ are density operators for the internal degrees of freedom of *one* radiator and that for the N_p perturbers. For gases, this equation is a good approximation provided that one does not consider the far wings of the lines, *ie* for $\hbar|\omega - \omega_{fi}| \ll k_B T$, as discussed in § II.3.3. Recall that an extension of the formalism beyond this limit has been proposed by Ma *et al*[72,140] and Royer,[141,142] leading to a more general, but much more complex, theoretical formulation. Nevertheless, for simplicity, we assume here that Eq. (II.E4) is valid, and thus follow on with the approach of Fano.[44] Thanks to a frequency-dependent collisional operator $<M_c(\omega)>$, first introduced by Zwanzig,[73] which, being averaged over the bath, operates only in the space of the internal degrees of the radiator, the spectral density in Eq. (II.E3) becomes:

General equations

$$F(\omega) = -\frac{1}{\pi} \text{Im} \left\{ \text{Tr}_a \left[\vec{d}^+ \frac{1}{\omega I_d - L_a - <M_c(\omega)>_p} \vec{d}\rho_a(0) \right] \right\}, \qquad \text{(II.E5)}$$

where $\text{Tr}_a[\ldots \rho_a(0)]$ represents an average over the variables of the radiator only, while $<\ldots>_p$ is an average over all bath variables. This equation "reduces" the spectral-shape problem to the calculation of $<M_c(\omega)>_p$. Fano carried out an expansion of this operator in powers of the perturber density n_p ($=N_p/V$), that is limited, within the binary collision approximation, to the lowest order in n_p. This leads to:[44]

$$<M_c(\omega)>_p \approx n_p <m_c(\omega)>_p = n_p \text{Tr}_p \left[m_c(\omega)\rho_p(0) \right]$$
$$\equiv n_p \sum_{p,p'} \langle\langle p'p' | m_c(\omega) | pp \rangle\rangle \langle p | \rho_p(0) | p \rangle \equiv -in_p W(\omega)^*, \qquad \text{(II.E6)}$$

where $\text{Tr}_p[\ldots]$ now denotes a trace over all possible states (p) of a *single* perturber. In Eq. (II.E6), we have also introduced the complex conjugate of the widely used relaxation operator $W(\omega)$, with which the spectral density takes the general form:

$$F(\omega) = +\frac{1}{\pi} \text{Im} \left\{ \text{Tr}_a \left[d^+ \frac{1}{\omega I_d - L_a - in_p W(\omega)} d\rho_a(0) \right] \right\}, \qquad \text{(II.E7)}$$

where d is the dipole operator or, *more generally*, that describing the interaction of the radiator with the e.m. field (dipole, quadrupole, or polarizability) according to the type of spectroscopy considered (see App. II.A). Let us mention that Eq. (II.E7) is formally identical to Eq. (II.59) derived within the frame of the so-called unified time-dependent approach. Furthermore, the equivalence of these two formalisms, within the limit of binary collisions with independent perturbers, was proved by Smith et al.[63] Otherwise, as discussed in Refs. 141,142, things are much more complicated.

Fano has derived an expression for the operator $<m_c(\omega)>_p$ in terms of the *transition operator* T of the scattering theory, a decisive step for computational procedures. It is given by Eq. (55) of Ref 44 which, due to the large number of terms composing it, is not entirely reproduced here. A matrix element of $<m_c(\omega)>_p$ is given by:

$$\langle\langle F'I' | <m_c(\omega)>_p | FI \rangle\rangle \approx \langle F' | T(E_{I'} + \hbar\omega) | F \rangle \langle I' | I \rangle - \langle F' | F \rangle \langle I' | T(E_F - \hbar\omega) | I \rangle^*$$
$$+ \pi i \delta(E_{F'} - E_I - \hbar\omega) \langle F' | T(E_{F'}) | F \rangle \langle I' | T(E_I) | I \rangle^*$$
$$+ \pi i \delta(E_{I'} - E_F + \hbar\omega) \langle F' | T(E_{I'}) | F \rangle \langle I' | T(E_{I'}) | I \rangle^* \qquad \text{(II.E8)}$$
$$+ \ etc\text{(many terms)},$$

where $|I\rangle$ and $|F\rangle$ correspond to quantum states in the joint perturber-radiator space. For instance, in the case of (rigid) linear molecule-atom collisions, one has: $|I\rangle = |k\lambda m_\lambda, Jm\rangle$, where $k\lambda m_\lambda$ denotes a relative translational state while Jm defines the internal rotational state of the radiator.

Back to the impact limit. The four terms in Eq. (II.E8) are those identified by Fano as giving the most important contributions to the spectral shape. The others, given and analyzed in Refs. 44,143, cancel out in the impact limit and only represent the effects of uncompleted collisions. This limit (App. II.C2) implies, within the "resolvent" formalism, that the transition matrices T(E) have to be evaluated *on-the-energy-shell*, ie at an energy $E \approx E_I \approx E_F \approx E_{I'} \approx E_{F'}$ according to Eq. (II.C15) (also see Ref. 62). Since one now only considers on-the-energy-shell T matrix elements, one may introduce the standard relation in terms of the scattering S matrix elements:

$$\langle I'|S(E)|I\rangle = \delta_{I,I'} - 2\pi i \langle I'|T(E)|I\rangle, \quad \text{where} \quad E = E_I = E_{I'}, \tag{II.E9}$$

and the equivalent relation for $\langle F'|S(E)|F\rangle$. Starting from Eq. (II.E8) and after some algebra, the matrix elements of the impact collisional operator $\langle m_c(\omega = \omega_{fi})\rangle_p$, alternatively denoted $\langle m_c(0)\rangle_p$, can be expressed in terms of elements of the scattering matrix.[46,144] This leads to expressions identical to those given in Secs. II.3.4 and IV.3.5.

II.E2 ROTATIONAL INVARIANCE

Ben-Reuven[45,46,137] has analyzed in detail the symmetry properties of the relaxation operator, based on the isotropy of the environment of the radiator. This means that, even when anisotropies exist, such as that induced by an external electromagnetic field, the collision dynamics is not affected since Zeeman and Stark energies are usually much smaller than thermal energies. Let us briefly recall the main steps of this work. The basis vectors for the irreducible representations in the Hilbert space are $|\beta J m\rangle$, where J designates the radiator angular momentum $\vec{J}$, m its projection on a space-fixed axis, and β a shorthand for all other quantum numbers. In the corresponding Liouville space, invariant sets are given by:

$$|\beta_f J_f \beta_i J_i; kq\rangle\rangle = \sum_{m_i,m_f} (-1)^{J_f - m_f} \sqrt{2k+1} \begin{pmatrix} J_f & J_i & k \\ -m_f & m_i & q \end{pmatrix} |\beta_f J_f m_f \beta_i J_i m_i\rangle\rangle. \tag{II.E10}$$

Invariance under space rotations, which applies to systems in an isotropic environment, may be extended to the parity, ie to the inversion of the coordinates. Equation (II.E10) can thus be extended in order to include parity by just adding the labels Π_i, Π_f, and Π, where Π_i and Π_f are the parities of levels i and f, respectively, and $\Pi = \Pi_i \times \Pi_f$ ($\Pi = \pm 1$). In order to simplify notations, we will omit these labels. A photon state with angular frequency ω can be identified by the quantum numbers π, k, q.[53] A photon in this state interacts with the radiator through the multipole operator $X_q^{\pi k}$, where k labels the 2^k-pole with $\pi = (-1)^k$ for electric multipoles and $\pi = (-1)^{k+1}$ for magnetic multipoles, and where q stands for the $(2k+1)$ irreducible component of X. For dipolar absorption $k=1$, while, for Raman scattering, the spectra due to the isotropic and anisotropic parts of the polarizability correspond to $k=0$ and $k=2$, respectively (cf App. II.A).

Let us now come back to the collisional operator $\langle m_c(\omega)\rangle_p$, whose invariance under rotations and coordinates inversion has been demonstrated in Ref. 45 for isotropic gases. This means that, in the basis defined by Eq. (II.E10), $\langle m_c(\omega)\rangle_p$ is reduced to separate submatrices, each repeated $(2k+1)$ times within a given invariant subspace k (π). In other words, it is independent of q. Consequently, optical transitions associated with

General equations 61

different multipoles [*eg* a line due to the electric quadrupole (k = 2) and another due to the electric dipole (k = 1)] *do not collisionally interfere*. Hence, for a *given* type of spectroscopy (k and π fixed), Eq. (II.E7) becomes:

$$F(\omega) = \frac{1}{\pi} \text{Im} \sum_{\beta_f, J_f, \beta_i, J_i} \sum_{\beta'_f, J'_f, \beta'_i, J'_i} \rho_{\beta_i J_i} \langle \beta_f J_f \| d_k \| \beta_i J_i \rangle \langle \beta'_i J'_i \| d_k \| \beta'_f J'_f \rangle^*$$
$$\times \langle\langle \beta'_f J'_f \beta'_i J'_i | \frac{1}{\omega I_d - L_a - i n_p W(\omega)} | \beta_f J_f \beta_i J_i \rangle\rangle, \quad \text{(II.E11)}$$

where $\rho_{\beta_i J_i}$ is the relative population of level $\beta_i J_i$ and $\langle \beta'_i J'_i \| d_k \| \beta'_f J'_f \rangle$ is the reduced matrix element [Eq. (II.37)] of the tensor d_k (not necessarily the dipole) coupling radiation and matter. Note that the relaxation matrix $W(\omega)$ (and $<m_c(\omega)>_p$) also depends on k. Finally the rotationally invariant form of the relaxation operator is given by:

$$\langle\langle \beta'_f J'_f \beta'_i J'_i | W(\omega) | \beta_f J_f \beta_i J_i \rangle\rangle \equiv \langle\langle \beta'_f J'_f \beta'_i J'_i; kq | W(\omega) | \beta_f J_f \beta_i J_i; kq \rangle\rangle/(2k+1)$$
$$= \sum_{m_f, m_i, m'_f, m'_i} (-1)^{J'_f - m'_f + J_f - m_f}$$
$$\times \begin{pmatrix} J'_f & J'_i & k \\ m'_f & -m'_i & -q \end{pmatrix} \begin{pmatrix} J_f & J_i & k \\ m_f & -m_i & -q \end{pmatrix} \quad \text{(II.E12)}$$
$$\times \langle\langle \beta'_f J'_f m'_f \beta'_i J'_i m'_i | W(\omega) | \beta_f J_f m_f \beta_i J_i m_i \rangle\rangle.$$

Equation (II.E11) can be obtained in a time-dependent formalism, as shown in Ref. 63 or 145. Since the final result is identical to Eq. (II.62), it is not given here.

II.E3 DETAILED BALANCE

The detailed balance relation was derived in Ref. 46, within the impact approximation, from the two following properties. Firstly, the symmetry of the T transition matrix for (usually) real intermolecular potentials. secondly, a relation between relative populations for the radiator-perturber pair:

$$\rho_{i'} \times \rho_{p'} = \rho_i \times \rho_p, \quad \text{(II.E13)}$$

which results from the on-the-energy-shell impact approximation (see App. II.C2). Here, i and p stand for pre-collisional states of the radiator and perturber, respectively, while i' and p' are for the post-collisional ones. From these properties straightforwardly results the detailed balance relation for the impact relaxation matrix:[46]

$$\langle\langle f'i' | W(\omega_{fi}) | fi \rangle\rangle \times \rho_i = \langle\langle fi | W(\omega_{fi}) | f'i' \rangle\rangle \times \rho_{i'}. \quad \text{(II.E14)}$$

The impact limit implies that the kinetic energy of the radiator and the total energy of the perturber are the same before and after the radiative transition. This is inconsistent with *strict* energy conservation for an inelastic collision, as required, at the microscopic level (before thermal averaging), for the detailed balance (except for isolated lines and exactly resonant overlapping transitions). Using time-reversal and general symmetry considerations, Monchick[146] has shown that the non-impact relaxation matrix $W(\omega)$ obeys detailed balance exactly, *ie*:

$$\langle\langle f'i'|W(\omega)|fi\rangle\rangle \times \rho_i = \langle\langle fi|W(\omega)|f'i'\rangle\rangle \times \rho_{i'}. \qquad (\text{II.E15})$$

This relation, being no longer restricted to on-the-energy-shell as Eq. (II.E13), accounts for uncompleted collisions. In order to analyze to what extent line-mixing cross sections (off-diagonal elements of W) satisfy the detailed balance, let us first consider the semi-classical impact approach of Neilsen and Gordon[147] (§ IV.3.4b). The examination of the corresponding expression [Eq. (IV.126)] shows that it only obeys the approximate relation of Eq. (IV.107). Indeed, in the limit of high kinetic energies required by the semi-classical approximation (*ie* $|E_i - E_{i'}| \ll k_B T$), the relative populations in a collisionally induced i' ← i transition are such that $\rho_{i'}/\rho_i \approx (2J_{i'} + 1)/(2J_i + 1)$. If one now considers the fully quantum impact approach of Shafer and Gordon[144] (§ IV.3.5b), it is not possible to get a direct conclusion concerning the detailed balance. Nevertheless, the results of calculations for CO-He[148] are interesting. They show that, whereas the real parts of the line-mixing cross sections accurately verify Eq. (II.E14), this is not the case for their imaginary parts for which a sign change appears, *ie*:

$$\langle\langle f'i'|W(\omega_{fi})|fi\rangle\rangle \times \rho_i = \langle\langle fi|W(\omega_{fi})|f'i'\rangle\rangle^* \times \rho_{i'}. \qquad (\text{II.E16})$$

This discrepancy was attributed by Monchick[146] to the limitations, mentioned just above, inherent in the impact limit. He further suggested from several firm arguments that *out-of-shell* calculations would lead to *imaginary* parts of line-coupling cross sections which vanish *more rapidly* with kinetic energy than the real parts, in contrast with the quantum impact results.[148] Besides the generally much smaller amplitude of the imaginary contributions to the relaxation matrix with respect to the real ones, this expected general behavior could be why, till now, no clear evidence was found, from experiment, for the need to take the imaginary off-diagonal elements of W into account. Nevertheless, it is clear that off-shell scattering calculations are of more immediate importance than was expected for predicting accurate imaginary parts.

III. ISOLATED LINES

III.1 INTRODUCTION

Stricto sensu, the concept of *isolated* line implies that the contribution of all the other lines is negligible in the spectral range covered by its core and wings, within the experimental noise or desired computation accuracy. A less restrictive and more useful definition may be that these contributions are additive (*ie* with no line mixing, *cf* Chapt. IV) with a sufficiently weak amount to allow an accurate determination of the line shape of interest. As shown in this chapter, although a line profile does not generally reduce to the knowledge of a few parameters, it is convenient to a priori introduce some basic characteristics. Let ω_{fi} be the angular frequency (in rad/s) for the considered $f \leftarrow i$ optical transition at zero pressure (*ie* in the absence of any collisional effects). The corresponding frequency ν_{fi} (in Hz) and wavenumber σ_{fi} (in cm^{-1}) are defined by $\omega_{fi} = 2\pi\nu_{fi} = 2\pi c\sigma_{fi}$. If ν_{Max} is the line frequency at intensity maximum, the line shift frequency is $\nu_{Max}-\nu_{fi}$. The full width at half maximum (FWHM) gives the spectral range of the line core. If the line is symmetric, the half width (HWHM) is more usually used. If not, it is convenient to introduce an asymmetry parameter A:

$$A = [\nu_{HF} + \nu_{LF} - 2\nu_{Max}]/[(\nu_{HF} - \nu_{LF})/2], \qquad (III.1)$$

where ν_{HF} and ν_{LF} are respectively the high and low frequencies at half maximum. An example of asymmetric line whose peak shifts toward lower frequencies more than the center of gravity is shown in Fig. III.1, together with its characteristic difference with a symmetric profile having the same center of gravity.

Let us consider a gas in thermal equilibrium at temperature T and pressure P. At low pressure, such that collisional effects on the line shape are negligible, only subsist the inhomogeneous features due to the thermal distribution of the radiator velocity $\vec{v}$. The corresponding (Doppler) frequency distribution is the spectral representation of the (Gaussian) Maxwell-Boltzmann isotropic probability given, for the range $[\vec{v}, \vec{v} + d^3\vec{v}]$, by:

$$f_M(\vec{v})d^3\vec{v} = (m/2\pi k_B T)^{3/2} \exp[-(v/\tilde{v})^2]d^3\vec{v}, \qquad (III.2)$$

where m is the radiator mass, $v = \|\vec{v}\|$ the radiator speed, and $\tilde{v} = (2k_B T/m)^{1/2}$ the most probable speed. The (symmetric) spectral representation is then not shifted with respect to ω_{fi} and its width is an increasing function of T and a decreasing one of m.

Let us consider now the collisional effect due to the radiator velocity changes which modify the thermal distribution. A direct examination of such velocity changes can be done by coherent optical methods, without the complication of Doppler broadening.[149-151] From two-pulse photon echo experiments, the longitudinal (*ie* along the light beam) velocity change per collision was measured, for instance, to be 85 cm/s for the R(4) line

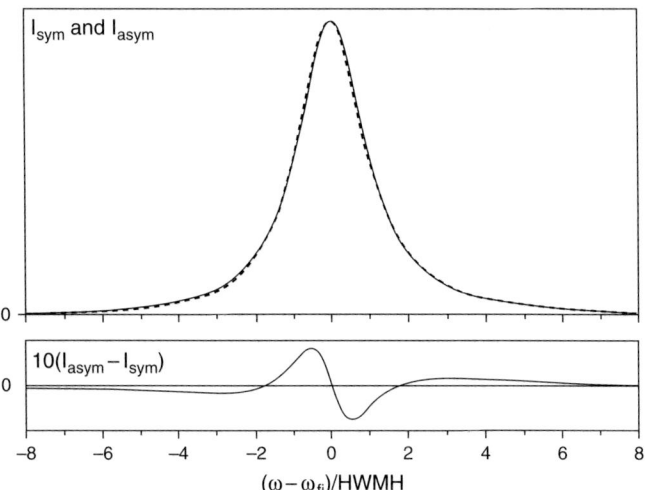

Fig. III.1: Characteristic signature for an asymmetric line profile with A > 0 (A = 0.04). The asymmetric (- - -) and symmetric (—) profiles having the same center of gravity are displayed in the top panel and the magnified difference is shown in the lower part.

of the fundamental band of pure CH_3F at room temperature.[150] If $P(v'_z \leftarrow v_z)$ is the probability of velocity changing for the z-component, the detailed balance principle prescribes that $P(v'_z \leftarrow v_z) f_M(v_z) = P(v_z \leftarrow v'_z) f_M(v'_z)$. $f_M(v_z)$ being a decreasing function of $|v_z|$ [Eq. (III.2)], it results that $P(|v_z| > |v'_z| \leftarrow |v_z|)$ is greater than $P(|v_z| \leftarrow |v'_z|)$. The velocity-changing collisions thus favor transfers from large velocities toward small ones. In other words, the (continuous) velocity distribution and, concomitantly, the spectral Doppler distribution narrow. This is the well known *Dicke narrowing*.[152] Similarly to collisionally induced transfers inside the various velocity groups, intensity transfers (called *line mixing*, cf Chapt. IV) arise between the various lines inside their discrete distribution within a rovibrational band through population migrations. Thus the Dicke narrowing may be considered as an example of line mixing inside the velocity continuum.[153]

Besides the above mentioned Dicke narrowing, the direct collisional effect mainly comes from energy exchanges between radiators and perturbers which include vibrational, rotational, and relative translational motions. These exchanges shorten the life time Δt of the initial (i) and final (f) optical levels. Following the Heisenberg principle,[51] the collisional width for the f ← i line is $\approx (\Delta t_i)^{-1} + (\Delta t_f)^{-1}$. Energy transfers occurring during collisions, the life times Δt_i and Δt_f may be approximated by the free flight duration τ_0 of the radiator. The collisional half width at half maximum (HWHM) is thus $\Gamma \approx 1/\tau_0$. In this simple picture, each collision is considered to be efficient in the energy transfer process. This is effectively pertinent for most molecular gases, due to the efficiency of the rotation to rotation (and relative translation) transfers. This results from the fact that the involved energy defect is generally much smaller than the thermal

Isolated lines 65

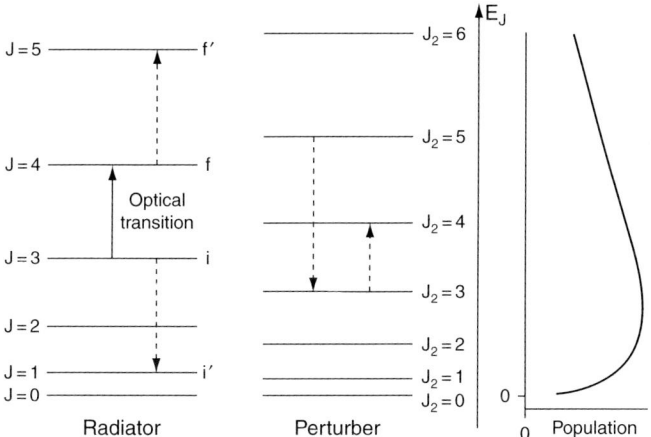

Fig. III.2: Schematic view of collisionally induced rotation-rotation transfers (- - - >) between optical levels of the radiator and the populated levels of the perturber.

energy k_BT for many collisionally induced transitions (Fig. III.2). Notice that, in contrast, vibrational energy exchanges play a marginal role in the broadening process, except in specific situations. Their large energy amount combined with the less intense intermolecular interactions make such exchanges weakly efficient. The time of free flight τ_0 can be estimated from the simple case of rigid spheres within the gas kinetic theory (*cf* App. II.B3). Then $\tau_0 \approx \ell_0/<v_r>$, where ℓ_0 is the mean free path and $<v_r>$ the average relative speed. At standard temperature and pressure conditions and for "usual" radiating molecules (CO, N_2, O_2, CO_2, *etc*) with various perturbers, τ_0 is close to 10^{-10} s. The corresponding magnitude for the broadening coefficient is $\gamma = \Gamma/(2\pi cP)$ $\approx 50 \; 10^{-3}$ cm^{-1}/atm. This typical value for the collisional half width is, at 1 atm, one order of magnitude larger than the Doppler width for infrared and Raman lines. This means that the observation of the "pure" Doppler profile requires subatmospheric pressures (for most molecules, $P < 10^{-3}$ atm). The colliding rigid spheres model also shows that the collisional width is proportional to the number density n_p of perturbers or, at a given temperature, to the pressure P for an ideal gas.

For atomic perturbers, the absence of rotation-rotation energy exchanges may lead to an important decrease of the collision efficiency in the broadening process. This is particularly the case for high rotational energy levels of light radiators, as seen in Fig. III.3 for a HCl–Ar mixture by comparison with pure HCl. Indeed, whereas the difference in the magnitude of the broadening coefficient is essentially due to the strength of the interaction (a strong long-range dipole-dipole for self-broadening instead of weaker and shorter range dispersive and repulsive forces for Ar-broadening), the differences in the evolution with J are largely due to rotation-rotation transfers. In particular, as analyzed below in the case of N_2, these processes are responsible for the bump around $J = 2$ in the pure HCl widths, close to the most populated level at room temperature (*cf* Fig. III.3). Furthermore, they partly explain the slower decrease of the

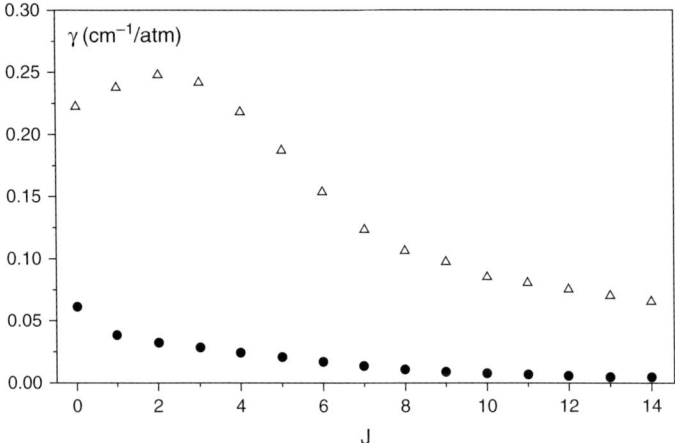

Fig. III.3: Measured broadening coefficients (HWHM), vs rotational quantum number J, due to (Δ) HCl–HCl[154] and (•) HCl–Ar[131] collisions for the R(J) lines of the $1 \leftarrow 0$ vibrational band at room temperature.

broadening with increasing J (factors of 3.4 for HCl–HCl and of more than 16 for HCl–Ar when going from $J = 0$ to $J = 14$). The efficient quasi-resonant rotation-rotation transfers (Fig. III.2) are temperature dependent, through the thermal distribution of the rotational populations for the perturber.

An increase of temperature spreads out the population over a larger number of levels with a smoother variation from one rotational level to another. The concomitant smoother variation of the broadening with the rotational quantum number J is illustrated in Fig. III.4 by Q(J) lines of pure N_2. The "peaking" distribution of rotational levels around $J = 7$ at room temperature results in a bump of γ in this rotational quantum number region. This bump is the signature of maximum efficiency of collisionally induced rotational energy transfers in a nitrogen molecular pair.

Furthermore, since the number of perturbers for *a given pressure* decreases with increasing temperature (in 1/T for a perfect gas), a significant decrease of the amplitude of γ is generally observed as illustrated in Fig. III.4 (the situation being reversed if density is used). In fact, broadening coefficients (in cm^{-1}/atm) for lines involving moderate rotational energy levels generally follow the approximate T^{-n} law[157] with the value $n \approx 0.7$ for the typical case of pure N_2.[155] On the contrary, due to a "resonance overtaking mechanism"[158] with increasing temperature (*ie* the increase of the relative kinetic energy with respect to the rotational energy jumps), negative values of n have been observed for high rotational quantum number lines of molecules with large rotational constants.[159,160]

In addition to these general features of collisional widths according to the nature of the perturbers, the pressure, and the temperature, their combined effects with the Dicke narrowing has to be analyzed at intermediate pressure, between the Doppler and collisional regimes. Competition then occurs between narrowing and broadening

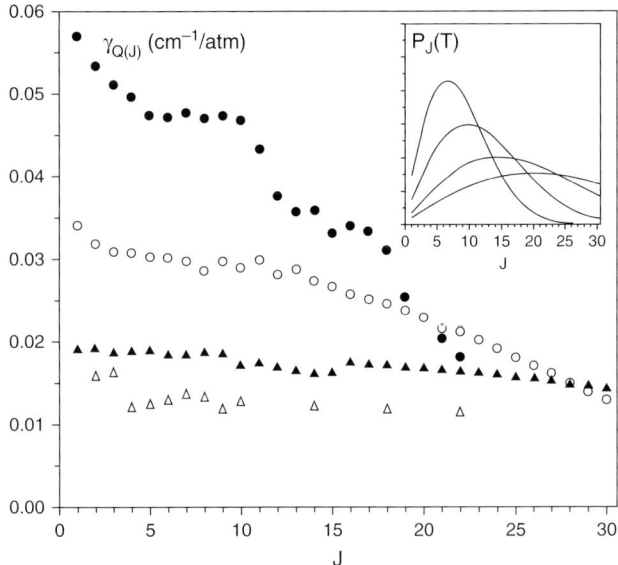

Fig. III.4: Measured line broadening coefficients (HWHM) *vs* the rotational quantum number J for isotropic Raman Q(J) lines of pure N_2 at: (•) 295 K, (○) 600 K, (▲) 1310 K, (△) 2400 K. The insert gives the associated smoothed relative population distributions. After Refs. 155,156.

mechanisms, resulting in a more or less marked minimum of Γ vs pressure.[161,162] This Dicke minimum is much deeper for light radiators due to the velocity-changing collision efficiency enhancing the competition. Figure III.5 exhibits such a Dicke minimum for various isotropic Raman Q(J) lines of pure D_2. The transition region from the Doppler broadened lines to the Dicke narrowed line occurs when the mean free path ℓ_0 becomes of the same order of magnitude as the wavelength λ associated with the photon momentum transfer k in the considered absorption or scattering process [*ie* $||\vec{k}||\ell_0 = (2\pi\ell_0)/\lambda \approx 1$, leading to a density of about 0.4 Am in Fig. III.5]. The Dicke diffusional contribution (inversely proportional to the density, see § III.2.2) is nearly equal to the collisional one (proportional to the density) in the observed HWHM at ≈ 1 Am. Their equality characterizes the Dicke minimum. Its relative contribution (*ie* Dicke/collisional) falls below 10^{-2} above 10 Am for pure D_2.[163] This example shows that, for light radiators, an accurate description of isolated line profiles must take into account *both* Dicke narrowing and collisional broadening in a *wide* density range. To a less extent, this remains true for many other molecules if one considers the performance nowadays attained by the best spectrometers (a resolution of 10^{-5} cm^{-1} and a signal-to-noise ratio of several thousand).[164,165] The insert in Fig. III.5 also points out the necessity to introduce refinements in the treatment of collisional mechanisms to improve the description of the line shape in the transition region, as discussed below.

From the same arguments used for Γ, the collisional shift Δ is also proportional to the perturber density. The Doppler profile being not shifted, no additional contributions take

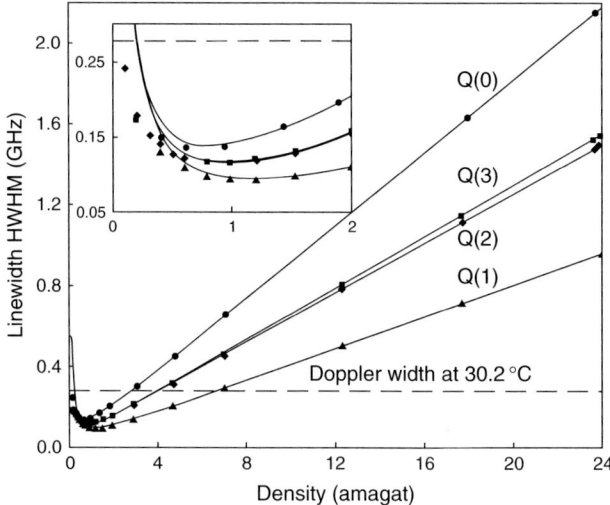

Fig. III.5: Line widths as a function of density for the Q(0) to Q(3) lines of D_2 in forward scattering at room temperature. Symbols give the experimental values, (—) is the modeling with the Dicke law HWHM = $a/\rho + b\rho$, cf Eq. (III.19). Reproduced with permission from Ref. 163.

place, giving a simpler behavior than for the HWHM. However, the determination of Δ from experimental profiles requires to be careful when the lines are asymmetric. A good approximation is to substitute the frequency ν_G of the gravity center of the line to the usual frequency ν_{Max} of the maximum.[166]

To capture, in a simple manner, the main other general features of the collisional line shift, let us consider the first-order perturbation contribution. The line frequency is given by $\tilde{\omega}_{fi} = \omega_{fi} + \hbar^{-1}[\langle f|<V>|f\rangle - \langle i|<V>|i\rangle]$, where $<V>$ is the properly averaged intermolecular potential energy for the radiator-perturber colliding pair. The shift thus appears as the *difference* of the collisional perturbation in the upper (f) and lower (i) optical levels (similar types of contributions also appear at higher orders). Consequently, Δ is only important if the amplitude of the perturbation strongly differs in the i and f levels. The line broadening being essentially the sum of the contributions in these two optical levels, the collisional shift is generally much smaller than the collisional line width. This is the case for most molecules, except hydrogen-like radiators, mainly due to the drastic reduction of the energy relaxation efficiency and the concomitant large decrease of Γ (see Table III.1).

The typical dependence of δ vs the line rotational quantum number is shown in Fig. III.6 for CO_2–He and CO_2–Ar (infinitely diluted) mixtures. Results similar to those for CO_2–Ar are also obtained for pure CO_2 and CO_2–N_2 mixtures.[168] It is worthy noting that the line shift coefficients are typically *smaller* than the corresponding line width coefficients by about *one order of magnitude* for most molecules. Besides the amplitude, let us consider the sign of Δ. At room temperature and a fortiori at lower temperatures, collisions mostly occur in the attractive range of molecular interaction so that $<V>$ is generally negative. The amplitude of the perturbation being larger in the

Table III.1: Broadening γ and shifting δ coefficients (in 10^{-3} cm^{-1}/atm) for the R(23) line of CO at 299 K[167] and the Q(1) line of H$_2$ at 295 K[166] in various highly diluted mixtures and pure H$_2$

pair	δ	γ
CO–N$_2$	−2.73	52.8
CO–He	+0.11	49.6
CO–Ar	−3.37	40.2
CO–Xe	−7.78	39.6
H$_2$–H$_2$	−3.32	0.87
H$_2$–Ne	+5.14	3.25
H$_2$–Ar	−11.8	4.42
H$_2$–Xe	−37.6	6.90

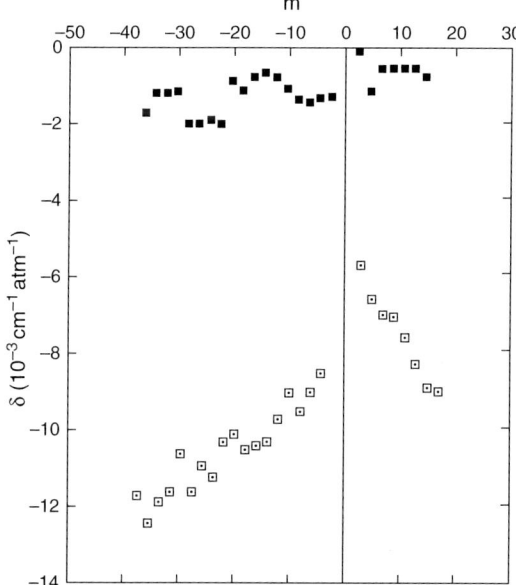

Fig. III.6: Measured room temperature pressure shift coefficients for CO$_2$ lines of the 3ν$_3$ band induced by collisions with He (■) and Ar (□). The values are plotted vs m with m = −J and J + 1 for P(J) and R(J) lines, respectively. Reproduced with permission from Ref. 168.

upper than in the lower level, the collisional line shift is negative (see Table III.1 and Fig. III.6), with the exception of CO–He and H$_2$–Ne for which the attractive contribution in the interaction potential is weaker than for the other molecular pairs (ie atomic perturbers with a weak polarizability). At higher temperature, the fraction of repulsive collisions is more important and the sign of Δ may change. This is the case for H_2–H_2 above 450 K or for H_2–Ar above 970 K.[166] These remarks concerning the amplitude and sign of Δ are based on too simple arguments to be considered otherwise as general trends. A rigorous detailed analysis for each situation has to be developed from pertinent theoretical calculations using accurate intermolecular potentials (cf § IV.3.4 and IV.3.5).

In addition to the inelastic energy transfers between colliding molecules (cf Fig. III.2), the diagonal matrix elements $\langle i|\!<\!V\!>\!|i\rangle$ and $\langle f|\!<\!V\!>\!|f\rangle$ result in other types of *elastic* relaxation mechanisms. Firstly they imply possible *molecular reorientation*. Secondly, they also include the so-called *vibrational dephasing* (cf App. IV.A) through the vibrational anharmonicity (both the mechanical anharmonicity of the vibrational states and the electrical anharmonicity coming from the even contributions of the intermolecular potential in the vibrational coordinates). Vibrational dephasing *only* arises for optical transitions where there is a *change* in the vibrational state. These two additional relaxation processes contribute, with the energy relaxation, to the line broadening and line shift.

In the above preliminary analysis of the Dicke narrowing, the velocity and the rovibrational relaxation mechanisms have been considered as independent. However, these processes may be statistically correlated (when they occur in the *same* collisional event). This *statistical correlation* modifies the spectral line shape (see § III.3.6), introducing, in particular, an asymmetry. The rovibrational states are then no longer at thermal equilibrium during the velocity relaxation and the symmetry is lost in the spectral representation of the $\vec{v}$-distribution.

The dependence of the rovibrational relaxation on the radiator speed has also to be taken into account as already mentioned (*eg* Ref. 169). Indeed, the relative translational kinetic energy participates with the rovibrational energy transfers, implying that both the radiator and perturber speeds are involved in the thermal average (see Sec. III.4). Consequently, *all the collisional profiles must be considered as inhomogeneously distributed according to radiator speed groups*, in the range [v, v + dv], each group being weighted by the (asymmetric) thermal equilibrium probability $f_M(v)dv = 4\pi v^2 f_M(\vec{v})dv$ [cf Eq. (III.2)]. The speed-changing collisions induce *exchanges* between speed groups inside this inhomogeneous distribution of line profiles. From the detailed balance principle, these exchanges favor the transfer of large speeds toward small ones, leading to an (asymmetric) inhomogeneous line shape. Since these spectral modifications result, as well from the speed dependence of the line broadening as from that of the line shift, subtle effects may occur (more especially for light radiators) according to the physical conditions (nature and concentration of the species, temperature, pressure). Hence, for light radiators, the large speed-dependent shift (cf Table III.1) leads to important inhomogeneous broadenings due to the concomitant frequency spreading (however more or less reduced by speed exchanges following their efficiency). The speed inhomogeneous features occur both in the collisional regime (*ie* at high pressure) and in the Dicke pressure range. So, even more subtle effects arise from the combination of, on one hand, the velocity-changing collisions on the Doppler distribution, and, on the other hand, the speed-changing ones on the inhomogeneous line profiles (through the speed dependence of Γ and Δ). All these aspects are treated in Sec. III.4 within the frame of various line-shape models. They are also present in ab initio approaches of the line profile (Sec. III.5). All these models are developed within the limits of the impact approximation. This means that the collisions are considered as binary with a time duration τ_c much smaller than all the other time scales of interest in the line shape. The resulting criteria being detailed in App. II.B4, only some specific relevant features for isolated transitions are now presented.

The impact approximation assumes the binary character of the collisions. To estimate the order of magnitude of ternary collisional events between a radiator and two perturbers, the translational motions of these perturbers are considered as uncorrelated. It follows that the probability of such ternary events is of the order of $(\tau_c/\tau_0)^2$ [or $(\ell_c/\ell_0)^2$]. This probability is proportional to the *square* of the perturber number density, instead of being linear for the binary collisions. This non-linear contribution is expected (*cf* App. II.B3) to be significant at densities of about and above one hundred Amagat for "efficient" collisions (*ie* in the same density range as for the finite collision-duration corrections in the core region). For short range intermolecular potentials and inefficient rotational energy transfers, the corresponding densities should be above several hundred Amagat. The direct observation of such *non-impact* effects (ternary collisions and collision-duration) requires, as well very large frequency separations between adjacent lines as a small line broadening coefficient, in order to avoid strong overlapping. Only light molecules can satisfy these two requirements (*cf* Table III.1). The expected *non-linear* behavior for the collisional broadening is illustrated in Fig. III.7 for the pure rotational $S_0(0)$ line in a HD–Ar mixture *vs* density. It is also observed for other $S_0(J)$ lines above 200 Amagat at room temperature (and above 300 Amagat at 175 K).[121]

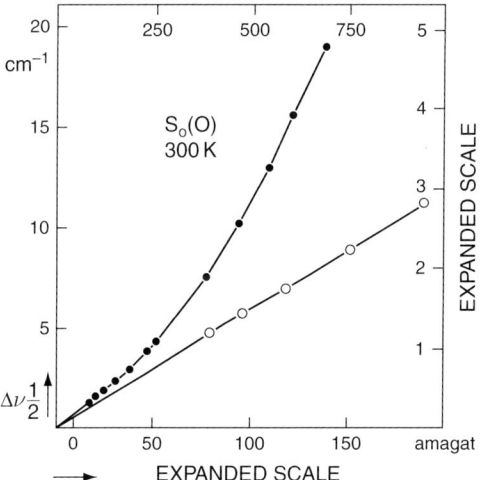

Fig. III.7: Experimental half width (HWHM) of the $S_0(0)$ Raman line for a HD–Ar mixture *vs* density (between 80 and 685 Am, bottom and upper scales) at room temperature. NB: the HD rotational $S_0(J)$ lines are separated by $4B \approx 180 \text{ cm}^{-1}$, where B is the rotational constant. Reproduced with permission from Ref. 121.

For most molecules, in the pressure range consistent with the isolated lines approximation, non-impact effects are negligible in the core and near wing regions. Nevertheless, as shown in App. II.B4, the finite duration of collisions introduces a line asymmetry. Such a spectral signature was unambiguously identified in the case of HF-Ar mixtures,[130] and later corroborated by using high resolution Fourier transform spectra.[170] The large vibrational anharmonicity of the HF molecule makes the collision duration *asymmetry* parameter b_{fi} in Eq. (II.B12) sufficiently large (several $10^{-3}\,\text{Am}^{-1}$) to permit the observation of this effect since the rotational constant is also large. Figure III.8 shows such an asymmetry for the P(3) line of HF–Ar in the $2 \leftarrow 0$ overtone band at 66 Am. Its sign is *negative* [see Eq. (III.1) and Fig. III.1], meaning that the peak shifts less than the center of gravity, leading to a sub-Lorentzian behavior in the high frequency wing (since $\Delta < 0$). A similar behavior was observed for other lines in the P and R branches. Exploring smaller pressures ($10 \leq P \leq 50$ atm) in order to establish a link with subatmospheric measurements,[171] Boissoles et al[170] concluded that collision duration asymmetry of HF-Ar lines is very weak at $P \leq 1$ atm, as confirmed by Pine and Ciurylo[172] (*cf* § III.4.3).

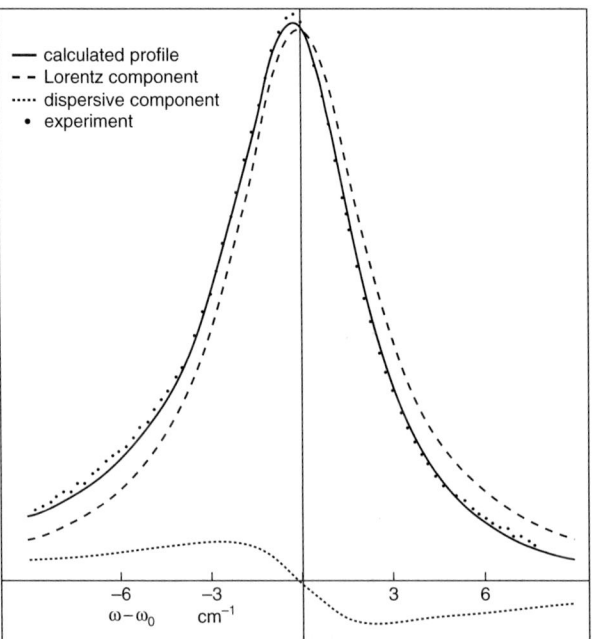

Fig. III.8: Profiles of the P(3) line of the $2 \leftarrow 0$ band of HF broadened by Ar at 66 Am. • are measured values while the lines indicate calculated results using Eq. (II.B12). The total profile (—) is composed of a Lorentzian (- - -) and of a dispersive (....) component. Reproduced with permission from Ref. 130.

Isolated lines 73

To summarize the above comments on the limitations inherent in the impact model, particular attention must be paid to the case of high density systems (*ie* typically of about and above one hundred Amagat). The finite duration of collisions and the non-binary collisional effects may then become significant and must be taken into account for an accurate description of the spectral profiles. But it is worthy noting that the lines are, in general, no longer isolated in this density range. The modeling becomes very complex since non-impact effects and line-mixing processes must be simultaneously taken into account.

Even if, in general, molecular spectra are not the additive superposition of individual lines, the knowledge of the isolated line profiles is the basic tool in order to build such spectra. A detailed analysis of the various models for isolated line shapes is given in the following sections, taking care to elaborate a synthetic and useful view of the numerous approaches throughout one century. Apart from the thermal (Doppler) distribution of the optical transition frequency, spectral modifications are induced by various collisional mechanisms (Sec. III.2). Among them, it is necessary to discuss the velocity change of the optically active molecule and the relaxation of the rovibrational states involved in the considered optical transition. The correlation between these two types of relaxation has to be also considered (Sec. III.3). Furthermore, the collisionally induced relaxation for the rovibrational states depends upon the velocity modulus (*ie* the speed) of the radiator. The influence of this speed dependence on the spectral line shape, whose (real) importance was better identified more recently, is carefully examined in Sec. III.4. The models taking into account all the physical mechanisms accurately describe any isolated line profile. However, such models are phenomenological. To overcome this limitation, alternative ab initio approaches based on the numerical resolution of a quantum kinetic equation using accurate intermolecular potentials were proposed, opening promising perspectives presented in Sec. III.5. Apart from the unperturbed resonance frequency of the considered transition, the two main parameters driving the line shape are the collisional broadening γ and shifting δ coefficients. Hence, for a long time, a particular attention has been paid to the calculation of these parameters. Recall that γ and δ are only specific terms [diagonal, *cf* Eqs. (II.52) and (IV.67)] of the impact relaxation matrix W describing the line-mixing processes analyzed in chapter IV. For this reason, a review of the methods used for their calculation is given in the following chapter (Sec. IV.3).

III.2 DOPPLER BROADENING AND DICKE NARROWING

Starting from Eqs. (II.7) and (II.30) for the spectral line shape, if any effect of the thermal motion of the radiator is disregarded, the resulting (normalized) profile is the Dirac distribution $\delta(\omega - \omega_{fi})$. This unphysical spectral distribution is due to the neglect of the role of the thermal bath played by the electromagnetic (e.m.) field (*cf* § II.2.1). For weak power sources, the average number of photons per mode being much lower than unity, the probability of induced radiative transitions between the initial $|i\rangle$ and final $|f\rangle$ states of the optical transition is negligible with respect to that of spontaneous emission.[51] This means that the life times of molecular states are predominantly governed by spontaneous photon emission leading to the "natural" width for the $f \leftarrow i$ spectral line.

This "natural" width is about 10^{-8} cm^{-1} in the infrared region, *ie* three orders of magnitude lower than the best resolutions available.[164,165] It will be disregarded in the following with respect to all the thermal motion effects analyzed now in detail.

III.2.1 THE DOPPLER BROADENING

Disregarding collisional effects between the radiator and the surrounding perturbers, let us consider the radiator free streaming. Its position $\vec{r}(t)$ with velocity $\vec{v}$ at time t [$\vec{r}(0) = \vec{0}$] is given by $\vec{r}(t) = \vec{v}t$, leading to the Doppler shift $\omega_{fi} v_z/c$ for the absorbed or scattered photon frequency (where c is the light speed and v_z the radiator velocity component along the e.m. wave propagation vector $\vec{k}$ between the radiation source and the molecule of interest: $v_z = \vec{v}\vec{k}^\circ$ ($\vec{k}^\circ = \vec{k}/\|\vec{k}\|$). The Doppler shift being v_z-dependent, the Maxwell-Boltzmann equilibrium distribution $f_M(v_z)$ [*cf* Eq. (III.2)] results in an inhomogeneous broadening of the line which is symmetric around the resonance frequency ω_{fi} since $\pm v_z$ are equally probable. The (normalized) contribution to the classical dipole autocorrelation function tied to the (free) radiator translational motion (*ie* to its external degrees of freedom) is, from Eq. (II.10):

$$C_{ext}(t) = \left\langle e^{-i\vec{k}[\vec{r}(t)-\vec{r}(0)]} \right\rangle = \left\langle e^{-i(\vec{k}\cdot\vec{v})t} \right\rangle = e^{-(\Delta\omega_D t/2)^2}, \qquad \text{(III.3)}$$

where the $\vec{v}$-thermal equilibrium average ($<\ldots>$) has been performed using the Maxwell-Boltzmann distribution $f_M(\vec{v})$ [Eq. (III.2)] and $\Delta\omega_D = k\tilde{v} = \omega_{fi}\tilde{v}/c$ ($\tilde{v} = \sqrt{2k_B T/m}$ being the most probable speed at temperature T and m the mass of the radiator). The expression of the Doppler profile is obtained from the Laplace transform [Eq. (II.34)] of the motional phase term for a freely streaming radiator [Eq. (III.3)]:

$$I_D(\tilde{\omega}) = \frac{1}{\pi}\text{Re}\left[\int_0^\infty e^{i\tilde{\omega}t} C_{ext}(t) dt\right] = \frac{1}{\sqrt{\pi}\Delta\omega_D} e^{-(\tilde{\omega}/\Delta\omega_D)^2}. \qquad \text{(III.4)}$$

Here, $\tilde{\omega} = \omega - \omega_{fi}$ is the detuning angular frequency, ω_{fi} being the zero pressure frequency for the $f \leftarrow i$ line. The Doppler profile (DP) is thus Gaussian and characterized by the parameter $\Delta\omega_D$ (in rad/s), which is the 1/e Doppler half width, different from the more commonly used half width at half maximum denoted γ_D, expressed (in cm^{-1}) as:

$$\gamma_D = \sqrt{\ln(2)}\frac{\Delta\omega_D}{2\pi c} = \omega_{fi}\sqrt{\frac{\ln(2)k_B T}{2\pi^2 mc^2}} = 3.58 \times 10^{-7}\sigma_{fi}\sqrt{T/M}. \qquad \text{(III.5)}$$

In Eq. (III.5), M is the molar mass in g/mol, and the wavenumber σ_{fi} for the molecular transition $f \leftarrow i$ is in cm^{-1}. Typical values of γ_D are of a few 10^{-3} cm^{-1} in the infrared, *ie* two orders of magnitude larger than the presently best available resolution. This means that the IR and Raman spectroscopies are highly accurate tools for the study of such a Doppler profile and of all the collisional effects on an isolated line. Of course, in order to observe the "purely" Gaussian Doppler profile, these collisional effects must be totally negligible. For most molecules considered in this book, at room temperature, the corresponding pressure is often lower than 10^{-3} atm except for light radiators (H$_2$, D$_2$, HD, *etc*)

Isolated lines

for which it is one to two orders of magnitude higher. Note that the Doppler width, determined from very high accuracy absorption measurements, can be used for a spectroscopic determination[173] of the Boltzmann constant k_B to which it is related through Eq. (III.5).

III.2.2 THE DICKE NARROWING

The collisional effect analyzed here, sometimes called motional (or confinement) narrowing, was first studied by Dicke.[152] To illustrate in a simple picture the physical mechanism underlying this Dicke narrowing of the Doppler distribution, the collisions will be first introduced only through their velocity-changing (noted VC) effect (disregarding all the other ones, in particular the collisionally induced transitions between the rovibrational states). Let us introduce the $A(\vec{v}, \vec{v}')$ probability density per unit time of velocity-changing $\vec{v} \leftarrow \vec{v}'$ collisions ($\vec{v}'$ being the precollisional radiator velocity). These VC collisions induce $\omega_{fi} v_z/c \leftarrow \omega_{fi} v'_z/c$ changes of the Doppler shift (*cf* § III.2.1) and concomitantly modify the line profile. The $A(\vec{v}, \vec{v}')$ probability density must satisfy the detailed balance principle in order to maintain the velocity equilibrium distribution, *ie*:

$$A(\vec{v}, \vec{v}') f_M(\vec{v}') = A(\vec{v}', \vec{v}) f_M(\vec{v}). \tag{III.6}$$

The VC rate is defined by:

$$\nu_{VC}(v) = \int A(\vec{v}', \vec{v}) d^3\vec{v}'. \tag{III.7}$$

Taking into account the velocity isotropy of the gas, the v_z component along the light beam must obey the following equation:

$$A(v'_z, v_z) = A(v_z, v'_z) e^{-\frac{m}{2k_B T}\left(v'^2_z - v^2_z\right)}. \tag{III.8}$$

It results that collisions inducing a decrease of $|v_z|$ are more probable than the reverse case or, equivalently, that these $|v_z|$ changes are in favor of $|v_z| > |v'_z| \leftarrow |v_z|$ (*ie* in favor of the decrease of the absolute value for the Doppler shift $\omega_{fi}|v_z|/c$). Thus, in the corresponding Doppler frequency distribution, the frequency transfers induced by collisions are more probable from the wing to the center than the reverse. This motional or Dicke *narrowing* mechanism[152] is illustrated in Fig. III.9, exhibiting the typical symmetric w-shaped signature of the collisionally induced VC mechanism from the difference between the Dicke and the Doppler profiles.

An analytical expression for the Dicke narrowed profile was established by Wittke and Dicke[174] within the *diffusional approximation* for the thermal motion of the radiator. If the square root of the mean square displacement z(t) along the light beam between times 0 and t is significantly smaller than the radiator wavelength $\lambda = 2\pi/k$, the

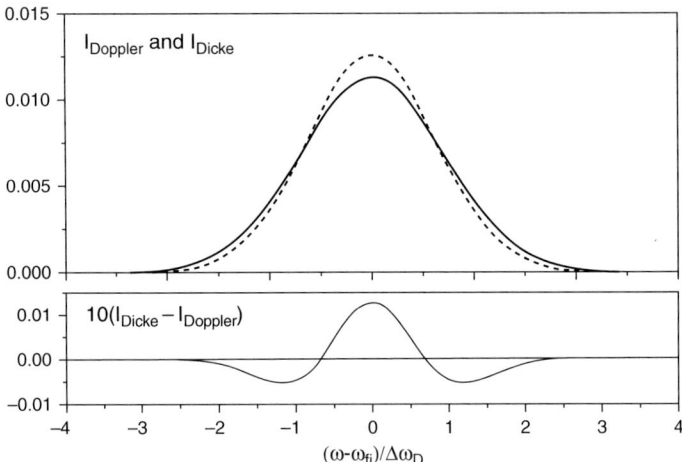

Fig. III.9: Typical line shapes for the (- - -) Dicke and (—) Doppler profiles, and their magnified difference.

translational phase term $C_{ext}(t)$ tied to the external degrees of freedom [Eq. (III.3)] can be written as:[174,175]

$$C_{ext}(t) = \left\langle e^{-ikz(t)} \right\rangle = \exp[-k^2 <z(t)^2>/2]. \tag{III.9}$$

In the diffusional limit, $<z(t)^2> = 2Dt$ (where D is the diffusion coefficient), and the time dependence of $C_{ext}(t)$ is an exponential damping given by:

$$C_{ext}(t) = e^{-k^2 Dt} \equiv C_{ext}^{Diff}(t), \tag{III.10}$$

with

$$D \equiv D_0/P = \int_0^\infty \langle v_z(0)v_z(\tau)\rangle d\tau = \frac{1}{3}\int_0^\infty \langle \vec{v}(0)\vec{v}(\tau)\rangle d\tau. \tag{III.11}$$

The D_0 value (ie the D value per unit pressure P) for the diffusion coefficient has been introduced in Eq. (III.11) since it is pressure-independent[176] within the impact collision frame considered here. The Laplace transform [see Eq. (III.4)] of Eq. (III.10) leads to the diffusional line shape:[174]

$$I_{Diff}(\tilde{\omega}) = \frac{1}{\pi} \frac{k^2 D_0/P}{\tilde{\omega}^2 + (k^2 D_0/P)^2}. \tag{III.12}$$

This *diffusional* profile has a Lorentzian shape. It is only pertinent for pressures consistent with the $<z^2(t)>^{1/2} << \lambda$ approximation underlying the diffusional motion. This means that the radiator must undergo many collisions before traveling a wavelength to justify this approximation (the alternative expression of "confinement", instead of "Dicke" narrowing, is thus sometimes used). The half width at half maximum of $I_{Diff}(\widetilde{\omega})$, $k^2 D_0/P$, decreases as the pressure increases, according to the motional narrowing mechanism explained above from the collisionally induced velocity transfers $v_z \leftarrow v'_z$, since the collisional frequency increases with pressure.

At this step, the other manifestations of the collisions (in particular the relaxation of the optical rovibrational states) must be taken into account. Indeed, they are *a priori* expected to have a magnitude comparable to the translational contribution in the pressure range for which Eq. (III.12) applies. In the next sections, the basic models for isolated line profiles are reviewed and the conditions of their validity are specified.

III.3 BASIC MODELS FOR SPECTRAL LINE SHAPES

III.3.1 THE LORENTZ PROFILE

In this paragraph, the collisional effects are introduced without explicitly describing the rovibrational states in order to be consistent with the preceding sections. Thus, these effects are considered through phase changes of the radiating dipoles.[177,178] This allows us to simply develop the basic line-shape models from the Doppler to the collisional regime. From Eq. (II.10), first neglecting the thermal motion considered just above, the (normalized) correlation function tied to the (rovibrational) internal degrees of freedom and due to the interaction between the radiator and the perturbers may be written as:

$$C_{int}(t) = e^{i\langle \varphi(t) \rangle}, \qquad (III.13)$$

where $\langle \varphi(t) \rangle$ is the averaged collisional phase shift of the radiator dipole for the optical transition (omitting the subscript "fi"), $\langle \ldots \rangle$ being a classical average defined in Sec. II.5. Within the impact approximation (*cf* App. II.B4), $\langle \varphi(t) \rangle$ is a complex quantity whose imaginary part determines the decay rate Γ of the oscillating radiator dipole and the real component is its frequency shift Δ, so that:[56]

$$C_{int}(t) = e^{-(\Gamma + i\Delta)t}, \qquad (III.14)$$

with $\quad \Gamma = n_p \int_0^\infty dv_r v_r f_M(v_r) \int_0^\infty 2\pi b db [1 - \cos \varphi(b, v_r)] \equiv n_p \gamma, \qquad (III.15)$

and $\quad \Delta = n_p \int_0^\infty dv_r v_r f_M(v_r) \int_0^\infty 2\pi b db \sin \varphi(b, v_r) \equiv n_p \delta. \qquad (III.16)$

In Eqs. (III.15) and (III.16), γ and δ do not depend on density, n_p is the number density of perturbers, and $\varphi(b, v_r)$ is the phase shift in a collision with the impact parameter b and the initial relative speed v_r (*cf* § IV.3.4), and $f_M(v_r)$ is the Maxwell-Boltzmann distribution for the radiator-perturber relative speed v_r. The corresponding quantum impact expressions for γ and δ are given in § IV.3.5. Note that the phase-changing collisions are named by the more useful *dephasing* collisions denomination (noted D) in the following. Such D collisions must be understood in a *generic* sense – ie including *all* the (elastic and inelastic) rovibrational state changes accounting not only for vibrational dephasing but also for molecular reorientation and inelastic rovibrational transitions. The Laplace transform of Eq. (III.14) at frequency $\widetilde{\omega}$ leads to the Lorentz profile LP:[177]

$$I_L(\widetilde{\omega}) = \frac{1}{\pi} \frac{\gamma P}{(\widetilde{\omega} - \delta P)^2 + (\gamma P)^2}, \qquad (III.17)$$

where $\gamma P \equiv \Gamma$ is the collisional HWHM and $\delta P \equiv \Delta$ is the collisional shift at pressure P (alternatively $n_p \gamma$ and $n_p \delta$ if densities are used). Recall that the *impact* approximation limits the applicability of this line-shape model to the *core* and *near wings* for a gas at low or mid pressure (typically lower than 100 atm at room temperature, cf Apps. II.B3 and II.B4). Let us mention that, in the microwave spectral region, an additional "antiresonant" term must be added in Eq. (III.17), leading to the Van Vleck-Weisskopf profile.[179] This "*antiresonant*" contribution corresponds to the tail of a Lorentzian profile, with the same collisional parameter Γ, but centered at the negative frequency $-(\omega_{fi} + \Delta)$ ($\omega_{fi} > 0$).

III.3.2   THE DICKE PROFILE

Assuming that no statistical correlation exists between velocity-changing VC and dephasing D collisions results in the following expression for the correlation function in the diffusional limit:

$$C(t) = C_{ext}^{Diff}(t) C_{int}(t), \qquad (III.18)$$

where $C_{ext}^{Diff}(t)$ and $C_{int}(t)$ are given by Eqs. (III.10) and (III.14), respectively. The Laplace transform at frequency $\widetilde{\omega}$ leads to the Dicke profile (DP):[174]

$$I_D(\widetilde{\omega}) = \frac{1}{\pi} \frac{\gamma P + k^2 D_0/P}{(\widetilde{\omega} - \delta P)^2 + (\gamma P + k^2 D_0/P)^2}. \qquad (III.19)$$

This (Lorentzian) Dicke profile conveniently includes as well the collisional broadening and shifting as the motional narrowing due to VC collisions. The pressure dependence of its HWHM may be analyzed within the range of pressure consistent with the diffusional approximation for the thermal motion [ie at pressures significantly above the range of validity of the (Gaussian) Doppler profile of Eq. (III.4)]. In this intermediate and high pressure regime, the HWHM first decreases with growing P down to a minimum and then increases when the collisional broadening γP becomes predominant with respect to the thermal contribution $k^2 D_0/P$ describing the Dicke narrowing. Such a behavior was

Isolated lines

observed in many molecular systems (*cf* Fig. III.5) from the beginning of the sixties' (*eg* Refs. 167,175). To accurately describe these various experimental studies in the whole pressure range, other mechanisms must also be taken into account. Among them, a more refined treatment for the velocity-changing mechanism – not limited to the above diffusional limit of Eq. (III.10) – will be first considered (§ III.3.4 and III.3.5), before introducing the statistical correlation between VC and D collisions (§ III.3.6).

III.3.3 THE VOIGT PROFILE

Let us introduce the useful and widely used Voigt profile.[169] If the VC effects (*ie* the Dicke narrowing mechanism) are disregarded, the spectral line shape is the convolution of the Lorentz profile [Eq. (III.17)] with the Doppler one [Eq. (III.4)]. Indeed, the line shape becomes the Laplace transform of the product of the two time-dependent dipole autocorrelation functions [*cf* Eq. (III.18)]: $C_{int}(t)$ [Eq. (III.14)] and $C_{ext}(t)$ [no longer limited to its first terms as in Eq. (III.10), but including *all* orders, *cf* Eq. (III.3)]. For simplicity and convenience, *dimensionless* parameters (Table III.2) are introduced following Herbert[180] and Varghese and Hanson:[181]

$$\tilde{x} = \frac{\tilde{\omega}}{\Delta\omega_D}, \quad x = \frac{\tilde{\omega} - \Delta}{\Delta\omega_D} \equiv \tilde{x} - s \quad \text{with} \quad s = \frac{\Delta}{\Delta\omega_D}, \quad \text{and} \quad y = \frac{\Gamma}{\Delta\omega_D}, \quad (III.20)$$

where $\Delta\omega_D$ is the 1/e Doppler half width given by Eq. (III.5). Furthermore, the resulting standardized profiles is *normalized here to area unity* (*ie* after integration over the x variable). The Voigt profile (VP) is then expressed as:

$$I_V(x,y) = \int_{-\infty}^{+\infty} I_D(\tilde{x}') I_L(x - \tilde{x}', y) d\tilde{x}' = \frac{1}{\sqrt{\pi}} \text{Re}[w(x,y)], \quad (III.21)$$

where w(x, y) is the complex probability function given by:[182]

$$w(x,y) = \frac{i}{\pi} \int_{-\infty}^{+\infty} \frac{e^{-t^2}}{x - t + iy} dt. \quad (III.22)$$

The expressions for Doppler [Eq. (III.4)] and Lorentz [Eq. (III.17)] profiles (respectively DP and LP), in the standardized notations of Eq. (III.20), are:

$$I_D(\tilde{x}) = \frac{1}{\sqrt{\pi}} \exp(-\tilde{x}^2), \quad I_L(x,y) = \frac{1}{\pi} \frac{y}{x^2 + y^2}. \quad (III.23)$$

Despite the fact that marked discrepancies with respect to I_V are observed for most of molecular spectra, the VP model remains a reference (*cf* Chapt. VII) and a firm basis for further more sophisticated approaches.

III.3.4 THE GALATRY PROFILE

A more general approach was developed by Galatry[183] in order to extend the Dicke model [Eq. (III.19)] toward lower pressures (*ie* down to the Doppler regime). Assuming that radiators undergo only very small velocity changes per collision, and using the Brownian movement model,[176] leads to:[183]

$$C_{ext}^S(t) = \exp\left\{-\frac{1}{2}\left(\frac{\Delta\omega_D}{\nu_{VC}^S}\right)^2\left[\nu_{VC}^S t - 1 + e^{-\nu_{VC}^S t}\right]\right\}, \qquad (III.24)$$

where ν_{VC}^S is the (speed-independent) VC collision rate in this *soft* (labeled S) collision model. The correlation function [Eq. (III.24)] for the external degrees of freedom generalizes the diffusional limit [Eq. (III.10)] to lower collisional broadening. The rate ν_{VC}^S, sometimes called the dynamical friction coefficient,[181] is connected with the diffusion coefficient D for the active species in a buffer gas (and thus denoted ν_{Diff}) by the relation:

$$\nu_{VC}^S = \frac{k_B T}{mD} \equiv \nu_{Diff}. \qquad (III.25)$$

As discussed below (§ III.4.1 and III.5.2), this relation should be used with caution. Indeed, D, through Eq. (III.25) appears in a context different from that of the conventional mass diffusion coefficient in gas kinetic theory (*eg* Refs. 181,184). Thus it does not have rigorously the same physical significance. This is the reason why some authors have introduced the terminology of "optical" diffusion coefficient D_{opt},[185] to be distinguished from the mass diffusion coefficient. Note that $C_{ext}^S(t)$ [Eq. (III.24)] consistently tends toward the diffusional limit $C_{ext}^{Diff}(t)$ [Eq. (III.10)] for large ν_{VC}^S. Furthermore, VC and D collisions are assumed to be statistically uncorrelated – and denoted (VC + D) to signify that VC and D arise only in *distinct* collisions -. Thus, the Laplace transform of the product of $C_{int}(t)$ [Eq. (III.14)] and $C_{ext}^S(t)$ [Eq. (III.24)] leads to the Galatry profile (GP), expressed in dimensionless variables x, y, z [Eq. (III.20)] by:

$$I_G(x,y,z) = \frac{1}{\pi}\text{Re}\left\{\int_0^\infty \exp\left[-(ix+y)\tilde{t} + \frac{1}{2z^2}\left(1 - z\tilde{t} - e^{-z\tilde{t}}\right)\right]d\tilde{t}\right\}$$
$$\equiv \text{Re}\{J_G(x,y,z)\}, \qquad (III.26)$$

with $z = \nu_{VC}^S/\Delta\omega_D$ and $\tilde{t} = \Delta\omega_D t$.

$I_G(x, y, z)$ can also be expressed in terms of the confluent hypergeometric function.[56,181] The GP has the proper behavior for limiting values of s, y, z, *ie* I_D for $s = y = z = 0$ (in the absence of both collisional shift, broadening and Dicke narrowing) and I_V for $z = 0$ (in the absence of the Dicke narrowing only).

In order to estimate the applicability conditions of the GP, the colliding molecules can be considered as rigid spheres. The mean persistence velocity ratio $r_{\bar{v}}$, defined as the ratio of the mean value of the post-collisional radiator velocity over the pre-collisional one, can then be expressed in terms of the masses of the colliding molecules:[186]

$$r_{\vec{v}} = \frac{1}{2}\widetilde{m} + \frac{1}{2}\widetilde{m}^2 \widetilde{m}_p^{-1/2} \ln\left[(\widetilde{m}_p^{1/2} + 1)\widetilde{m}^{-1/2}\right], \tag{III.27}$$

with $\widetilde{m} = m/\widetilde{M}$, $\widetilde{m}_p = m_p/\widetilde{M}$, $\widetilde{M} = m + m_p$, m_p being the mass of the perturber. For decreasing values of the mass ratio m_p/m from infinity to zero, $r_{\vec{v}}$ varies between zero and unity. The case $r_{\vec{v}} \approx 1$ corresponds to $m_p/m \ll 1$, *ie* to the quasi persistence of the radiator velocity memory. Hence, the soft collision approximation underlying the GP model appears to be convenient for heavy radiators and light perturbers.

III.3.5 THE NELKIN-GHATAK PROFILE

In the opposite $r_{\vec{v}} \to 0$ case, the velocity memory is lost after each collision. In this so-called *hard* (labeled H) collision approximation, the VC probability density per unit time for a $\vec{v} \leftarrow \vec{v}'$ collision is given by:[187,188]

$$A(\vec{v}, \vec{v}') = v_{VC}^H f_M(\vec{v}), \tag{III.28}$$

where v_{VC}^H is the hard collision rate, which is speed-independent as a consequence of detailed balance [see Eqs. (III.6) and (III.28)]. It is equal to the *kinetic rate* v_{Kin} since *each* VC collision is thus assumed to randomize the radiator velocity. Let us mention that the hard collision approximation does not imply $m_p/m \gg 1$, as the simple model of rigid spheres seemed to indicate through Eq. (III.27), and the usual picture of light radiators and heavy perturbers is somewhat excessive for hard collisions.

For uncorrelated hard VC and D collisions, the kinetic equation for the probability density $f(\vec{r}, \vec{v}, t)$ of finding the radiator at $\vec{r}$ with velocity $\vec{v}$, at time t [*cf* Eqs. (II.72) and (III.28)] takes the form:[56,189]

$$\frac{\partial}{\partial t} f(\vec{r}, \vec{v}, t) = -\vec{v} \vec{\nabla}_{\vec{r}} f(\vec{r}, \vec{v}, t) - (\Gamma + i\Delta) f(\vec{r}, \vec{v}, t)$$
$$- v_{VC}^H \left[f(\vec{r}, \vec{v}, t) - f_M(\vec{v}) \int f(\vec{r}, \vec{v}', t) d^3\vec{v}' \right]. \tag{III.29}$$

The first term at the right handside of Eq. (III.29) describes the free motion of the radiator. The line profile is obtained from the solution of this equation with $f(\vec{r}, \vec{v}, 0) = f_M(\vec{v})\delta(\vec{r})$, through the space-time Laplace transform and integration over $\vec{v}$, *ie*:

$$I(\widetilde{\omega}) = \frac{1}{\pi} \text{Re}\left\{ \int e^{i(\widetilde{\omega} t - \vec{k}\cdot\vec{r})} f(\vec{r}, \vec{v}, t) \, d^3\vec{r} \, d^3\vec{v} \, dt \right\}. \tag{III.30}$$

The resulting profile was originally obtained by Nelkin and Ghatak[189] by disregarding the second term in the right handside of Eq. (III.29). It describes the transition of the Mössbauer line shape from Doppler broadening to simple diffusion broadening with the characteristic narrowing due to collisions. Later, Rautian and Sobelman considered,[56] in addition, the pressure broadening. The resulting expression for the here-called Nelkin-Ghatak profile (noted NGP) is given by:

$$I_{NG}(\tilde{\omega}) = \frac{1}{\pi} \text{Re} \left\{ \left\langle \left[v_{VC}^H + \Gamma - i(\tilde{\omega} - \Delta - \vec{k}.\vec{v}) \right]^{-1} \right\rangle \right/ \left\langle 1 - v_{VC}^H \left[v_{VC}^H + \Gamma - i(\tilde{\omega} - \Delta - \vec{k}.\vec{v}) \right]^{-1} \right\rangle \right\}, \quad \text{(III.31)}$$

or

$$I_{NG}(x,y,\varsigma) = \text{Re}\{J_{NG}(x,y,\varsigma)\} = \frac{1}{\sqrt{\pi}} \text{Re}\{w(x, y+\varsigma)/[1 - \sqrt{\pi}\varsigma w(x, y+\varsigma)]\}, \quad \text{(III.32)}$$

with

$$\varsigma = v_{VC}^H/\Delta\omega_D \equiv v_{Kin}/\Delta\omega_D. \quad \text{(III.33)}$$

Note that Eq. (III.31) is alternatively called Rautian-Sobelman profile (RSP). The present NGP notation privileges the primary derivation in Ref. 189 (for $\Gamma = \Delta = 0$) and permits to keep RSP for the more general line-shape model including both hard and soft collisions (§ III.4.3). Similarly to the GP, the NGP tends toward I_D for $s = y = \varsigma = 0$ and toward I_V for $\varsigma = 0$. The simplicity of the analytical expression for $I_{NG}(\tilde{\omega})$ explains the popularity of the NGP model in the data analysis. However, if the hard collision operator [Eq. (III.28)] does not strictly describe typical situation in line-shape experiments, it properly captures some basic properties of Dicke narrowed profiles.

All the parameters for the DP, LP, VP, GP and NGP models are gathered in Table III.2.

Table III.2: Characteristic parameters for the basic models of line profile

Frequency parameters	Usually used notations and units	Dimensionless parameters
$\tilde{\omega}$: distance between the current frequency ω and that ω_{fi} of the optical transition at zero pressure	$\sigma \equiv \omega/2\pi c$ (cm^{-1})	$\tilde{x} \equiv \tilde{\omega}/\Delta\omega_D \equiv (\omega - \omega_{fi})/\Delta\omega_D$ $x \equiv (\omega - \omega_{fi} - \Delta)/\Delta\omega_D \equiv \tilde{x} - s$
$\Gamma = \gamma P$: collisional width (HWHM)	γ (cm^{-1}/atm)	$y \equiv \Gamma/\Delta\omega_D$
$\Delta = \delta P$: collisional shift	δ (cm^{-1}/atm)	$s \equiv \Delta/\Delta\omega_D$
v_{VC}^S: velocity-changing frequency for soft collisions	v_{Diff} (s^{-1})	$z \equiv v_{VC}^S/\Delta\omega_D \equiv v_{Diff}/\Delta\omega_D$
v_{VC}^H: velocity-changing frequency for hard collisions	v_{Kin} (s^{-1})	$\varsigma \equiv v_{VC}^H/\Delta\omega_D \equiv v_{Kin}/\Delta\omega_D$

Isolated lines

Let us note that the kinetic frequency ν_{Kin} introduced in Eq. (III.33) and Table III.2 has the *same* physical meaning as ν_{Diff}. Indeed, $1/\nu_{Kin}$ and $1/\nu_{Diff}$ are times characterizing the total loss of memory of the initial radiator velocity in hard and soft collision models, respectively.[56] Thus, ν_{Kin} is also connected with the diffusion coefficient [see Eq. (III.25)]

III.3.6 CORRELATED PROFILES

In Ref. 56, Rautian and Sobelman gave a thorough kinetic approach of line shapes, not only treating consistently both soft and hard collisional limits, but also opening two important channels. Firstly, they introduced the *statistical correlation* between VC and D collisions through VCD collisions (*ie* collisions changing *both* velocity and phase, but with *no* correlation between the magnitude of these changes). Implicitly, they also took a possible partial correlation into account through VCD + VC collisions considered below. Secondly, treating soft and hard collisional limits within the same kinetic frame allows to consider the occurrence of *both* soft and hard events in the collisional processes [*ie* (H + S) models, *cf* § III.4.3].

For VCD collisions, instead of (VC + D) ones considered above, the kinetic equation [*cf* Eq. (II.72)] for the (correlated) soft case is:[56]

$$\frac{\partial}{\partial t} f(\vec{r}, \vec{v}, t) = -\vec{v} \vec{\nabla}_{\vec{r}} f(\vec{r}, \vec{v}, t) - (\Gamma + i\Delta) f(\vec{r}, \vec{v}, t) \\ + \nu_{eff}^S \left[\vec{\nabla}_{\vec{v}} (\vec{v} f(\vec{r}, \vec{v}, t)) + \frac{k_B T}{m} \Delta_{\vec{v}} f(\vec{r}, \vec{v}, t) \right], \quad (III.34)$$

where $\vec{\nabla}_{\vec{v}}$ and $\Delta_{\vec{v}}$ are gradient and Laplacian operators in $\vec{v}$ variables, respectively. In Eq. (III.34), the (complex) *effective* velocity-changing soft collision frequency ν_{eff}^S is defined by [*cf* Eq. (II.75) and Ref. 56]:

$$\nu_{eff}^S = \nu_{VC}^S [1 - (\Gamma + i\Delta)/\nu_{Kin}], \quad (III.35)$$

where ν_{Kin} is the kinetic frequency. The simple modification in the kinetic equation (III.34) for the correlated soft case with respect to the uncorrelated one is the change $\nu_{VC}^S \to \nu_{eff}^S$, or in dimensionless variables $z \to z[1 - (y + is)/\varsigma]$. The resulting *correlated* Galatry profile ($G_C P$) can be simply expressed in terms of the GP as:[181]

$$I_{G_c}(x, y, z, \varsigma, s) \equiv I_G[x, y, Z \equiv z(1 - y/\varsigma) - isz/\varsigma], \quad (III.36)$$

with

$$y, s, z \leq \varsigma. \quad (III.37)$$

For the correlated hard collision case, the kinetic equation is similar[56] to that for the uncorrelated one [Eq. (III.29)] with the only change $v_{VC}^H \to v_{eff}^H$ given by:

$$v_{eff}^H = v_{VCD}^H - (\Gamma + i\Delta) = v_{Kin} - (\Gamma + i\Delta). \qquad (III.38)$$

The corresponding change, expressed in dimensionless variables, is $\varsigma \to \varsigma - (y + is)$, leading to the following expression for the *correlated* Nelkin-Ghatak profile (NG$_C$P):

$$I_{NG_c}(\tilde{x}, y, \varsigma, s) = \frac{1}{\sqrt{\pi}} \text{Re}\{w(\tilde{x}, \varsigma) / [1 - \sqrt{\pi}(\varsigma - y - is) w(\tilde{x}, \varsigma)]\}, \qquad (III.39)$$

with
$$y, s \leq \varsigma. \qquad (III.40)$$

As seen from Eqs. (III.37) and (III.40), restrictions on the relative magnitudes of y, s, z and ς must be introduced for the correlated soft and hard cases [but not for the uncorrelated cases; *cf* Eqs. (III.26) and (III.32)]. The statistical correlation for hard collisions implies that *every* collision is of VCD type (*ie* a *full* correlation), so that the corresponding collision rate v_{VCD}^H becomes equal to the kinetic collision frequency ($v_{VCD}^H = v_{Kin}$). The above restrictions result from the fact that the kinetic collision frequency must be greater than or equal to the frequency of any individual (broadening and shifting) process (so, $y, s \leq \varsigma$). A third restriction $z \leq \varsigma$ characterizes the G$_C$P, in connection with the additional possibility of soft VC collisions with *no* dephasing (*ie* SVC collisions). For such a soft (*partially*) correlated model [(SVCD + SVC) collisions], the total frequency is equal to the kinetic rate ($v_{VCD}^S + v_{VC}^S = v_{Kin}$), so that v_{VCD}^S and $v_{VC}^S \leq v_{Kin}$ [or y, s, *and* $z \leq \varsigma$ according to Eq. (III.37)]. The partial statistical correlation may be alternatively characterized by the fraction $\eta^S = v_{VCD}^S / v_{Kin}$ of correlated (SVCD) collisions among *all* the *kinetic* ones (*cf* § III.4.3). Since the (fully) correlated NG$_C$P model (labeled by the index C) only accounts for HVCD collisions, it should be preferable to take a different notation for the (*partially correlated*) G$_C$P model (as G$_{\tilde{C}}$P). However, the notation G$_C$P is so usual that it has been yet retained here (Table III.3).

A *partially* correlated hard collision model NG$_{\tilde{C}}$P can be similarly introduced[171,190] through (HVCD + HVC) collisions. The partial correlation is characterized by the fraction η^H of HVCD collisions among all the kinetic ones:

$$v_{VCD}^H = \eta^H v_{Kin}, \quad 0 \leq \eta^H \leq 1, \qquad (III.41)$$

and Eq. (III.38) becomes [*cf* Eq. (III.35)]:

$$\tilde{v}_{eff}^H = v_{Kin} - \eta^H (\Gamma + i\Delta) \quad \text{or} \quad \tilde{\varsigma} = \varsigma - \eta^H (y + is). \qquad (III.42)$$

The corresponding change from NGP to NG$_{\tilde{C}}$P is $\varsigma \to \tilde{\varsigma}$ [Eq. (III.42)], so that $I_{NG_{\tilde{C}}}(\tilde{x}, y, \tilde{\varsigma}, s)$ has the *same* formal expression as $I_{NG_c}(\tilde{x}, y, \varsigma, s)$ [Eq. (III.39)]. The *partially* correlated hard collision model NG$_{\tilde{C}}$P depends on four collisional parameters

Table III.3: Set of dimensionless parameters for the various basic profiles. The number of *collisional* parameters characterizing each profile is given into brackets

Model of profile	Notation	Equations	Dimensionless parameters
Doppler	DP	(III.4) or (III.23)	$\tilde{x}$ (0)
Lorentz	LP	(III.17) or (III.23)	x, y (2)
Voigt	VP	(III.21)	x, y (2)
Soft collisions (Galatry)	GP	(III.26)	x, y, z (3)
Hard collisions (Nelkin-Ghatak)	NGP	(III.31) or (III.32)	x, y, ς (3)
Correlated soft collisions	$G_C P$	(III.36) and (III.37)	x, y, z, ς, s (4)
Correlated hard collisions	$NG_C P$	(III.39) and (III.40)	$\tilde{x}$, y, ς, s (3)

(y, ς, s, η^H) as the $G_{\widetilde{C}}P$ one (conventionally written $G_C P$ as mentioned just above, *cf* Table III.3). For $\eta^H = 1$, the $NG_{\widetilde{C}}P$ is identical to the fully correlated $NG_C P$, and for $\eta^H = 0$, the statistically uncorrelated limit NGP is obtained, as expected. The role of this partial correlation is discussed in § III.4.3, within the frame of a more general approach including both hard and soft collisions.

III.3.7 CHARACTERISTICS OF THE BASIC PROFILES

After the pioneer work of Dicke[152,174] on the confinement narrowing, the existence of such a mechanism was suspected to occur in various optical and IR molecular spectra.[191] In 1963, Rank and Wiggins[192] presented the results of the first experimental investigation of such a narrowing in H_2 and D_2 quadrupolar absorption bands. Their results showed conclusively that Dicke narrowing did manifest itself in molecular systems at optical frequencies. Many other experimental studies followed using different methods: IR absorption,[193,194] stimulated Raman gain,[195] high-resolution Raman scattering,[96] and electric-field-induced absorption.[196] Later, the development of new efficient devices (difference frequency laser spectrometers,[197,198] controlled tunable diode lasers[199,200] and coherent anti-Stokes Raman spectrometers[201,202]) gave rise to many experimental studies of the Dicke narrowing for more than twenty molecules.[167] Sub-Doppler line shapes mainly occurring when the broadening rate is lower than the diffusion one [*cf* Eq. (III.19) and Ref. 175], the Dicke narrowing signature was first observed for light molecules (H_2, D_2, HD, HF, HCl, H_2O, *etc*). The considerable improvements of the resolution and sensitivity of various spectrometers in the last decades have enlarged the list of concerned molecules. However, simultaneously, more accurate descriptions of the speed inhomogeneous effects have significantly reduced the contribution previously attributed to the Dicke mechanism for "usual" molecular systems (see § III.4.1 and III.5.4 and Ref. 203). Besides the observation of the "Dicke minimum" from the pressure dependence of the line width for light radiators (cf Fig. III.5), the narrowed profiles exhibit more generally a non-Lorentzian behavior including a possible asymmetry. Varghese and Hanson181 have analyzed the main spectral features of the above basic profiles $I_G(x, y, z)$ [Eq. (III.26)], $I_{NG}(x, y, \varsigma)$ [Eq. (III.32)], $I_{GC}(x, y, z, \zeta, s)$ [Eq. (III.36)], and $I_{NG_c}(\tilde{x}, y, \varsigma, s)$ [Eq. (III.39)] by comparison with the standard Voigt profile [Eq. (III.21)]. The difference GP–VP (or NGP–VP) exhibits the characteristic w-shaped signature exemplified in Fig. III.9. This difference is governed[204] by the ratio $r_S = y/z = \Gamma/v_{VC}^S$ (or $r_H = y/\varsigma = \Gamma/v_{VC}^H$ for NGP–VP). For molecules with small rotational constants and for low J transitions, the maximum of the relative difference (GP–VP)/VP is typically a few percent only. For light radiators and high J transitions, when r_S (or r_H) is much lower than unity, this difference may reach several tens of percents. The role of the statistical correlation between VC and D collisions, through G_CP and NG_CP, was also analyzed in Ref. 181. It is tightly connected with the relative amount of the collisional shift with respect to the collisional width. For $\Delta = 0$ ($s = 0$), the G_CP has the same expression as the GP with the reduced narrowing parameter $Z(s=0) = z(1-y/\varsigma)$ [cf Eq. (III.36)]. The NG_CP also has, for $s=0$, the same expression as the NGP with the reduced narrowing parameter $\varsigma - y$ [instead of ς, cf Eq. (III.39)]. For molecular transitions characterized by small collisional shifts (ie, $|s| << y$) and efficient collisional broadening ($y \to \varsigma$), the Dicke narrowing is strongly reduced by the VC/D statistical correlation. Thus, both the G_CP and NG_CP are close to the VP. For significant collisional shifts and values of the r_S (or r_H) ratio smaller than unity, the G_CP (or NG_CP) is *asymmetric*, such that the peak of the line is shifted more than the center of gravity[(1)] [ie A > 0 for s < 0, cf Eq. (III.1) and Fig. III.1]. Since γ tends to decrease with increasing rotational quantum number J at high J while the absolute value of δ tends to increase (eg Figs III.3 and III.6), more pronounced asymmetric signatures arise for high J transitions. This is particularly true for light radiators and heavy perturbers due to the enhanced J-dependence of γ and δ (see Fig. III.3). On the whole, GP and NGP exhibit small differences, typically one order of magnitude smaller than their difference with the VP, unless $z, \varsigma >> y$ (ie for very weak collisional broadening). In this situation, the condition $y \leq \varsigma$ [Eqs. (III.37) and (III.40)] for the correlated models is no longer restrictive and the NG_CP may be convenient. However, the NG_CP is less flexible than the G_CP since it depends on three collisional parameters instead of four (Table III.3). In practice, the choice between the hard and soft collision models was frequently made considering the mass ratio $\Lambda = m_p/m$ (cf § III.3.4). When Λ is close or equal to unity, this choice was generally done by simple convenience. The first attempts to model narrowed isolated line shapes by the GP and NGP appear in 1972.[161,205] Progressively, fast and accurate procedures were developed to calculate these profiles, including complex narrowing parameters to take into account the VC/D correlation.[180,181,206–208] From the 1990's, these procedures have been widely used in computer routines to extract the collisional broadening and shifting parameters from observed profiles (cf App. III.A).

[(1)] Note that the sign of this VC/D correlation asymmetry is opposite to that of the collision duration asymmetry observed in HF–Ar (see Fig. III.8).

In order to illustrate the degree of pertinency of the basic profiles in the experimental context of this period of time, the isotropic Raman Q(J) lines of D_2 (in pure D_2 and in various D_2–X mixtures) are considered now from the exhaustive study of Rosasco and coworkers.[209–211] They used a continuous-wave stimulated Raman spectrometer with a spherical mirror multipass cell. Its resolution was 10^{-4} cm^{-1} and its signal-to-noise ratio of about 300. Unavoidable beam crossings within the cell gave rise simultaneously to forward- and backward-scattering Raman signals, and the observed signal was always the sum of two components. Both forward and backward scattering geometries were considered for the pump/probe laser beams (respectively co- and counter-propagating), in order to get a careful test of line-shape models through a possible large variation of the magnitude of the momentum transfer $k = ||\vec{k}_{Pump} - \vec{k}_{Probe}||$ between forward and backward components. The density dependence of the Raman line width for the Q(1) line of D_2 diluted in He is shown in Fig. III.10. The half widths for the two components were derived by fitting the spectrum to two Lorentzian line shapes (for densities above 10 Am) or to a single Lorentzian for the forward scattering and a GP for the backward scattering at lower densities. This density dependence is reasonably well described by the Dicke law [cf Eq. (III.19), with $k = k_{forward}$ or $k_{backward}$, see App. II.A]. The fitted resulting diffusion coefficient D_0 agrees within 1% with the value calculated from the mass diffusion coefficient,[35] under the assumption of an ideal gas behavior. No significant asymmetry was observed for the Q(J) lines in pure D_2 and D_2–He mixtures.

In contrast, in D_2–Ar and D_2–CH$_4$ mixtures, *asymmetric* line shapes were observed in the backscattering geometry,[211] leading to the characteristic signature shown in Fig. III.1 through difference with the (symmetric) GP theoretical model [Eq. (III.26)]. A complex narrowing parameter was introduced[211] in the GP model in order to take into account a possible VC/D statistical correlation [cf Eq. (III.35)]. Systematic errors in the residual

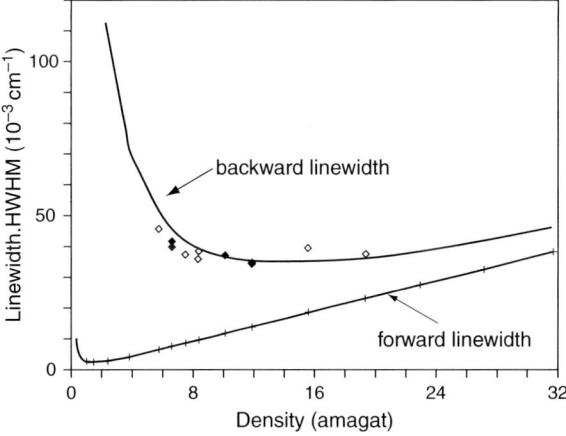

Fig. III.10: The density dependence of the Q(1) line width (HWHM) derived from fits of observed spectra for 10% D_2 in He at 297 K. The solid curves have been calculated using Eq. (III.19). Reproduced with permission from Ref. 211.

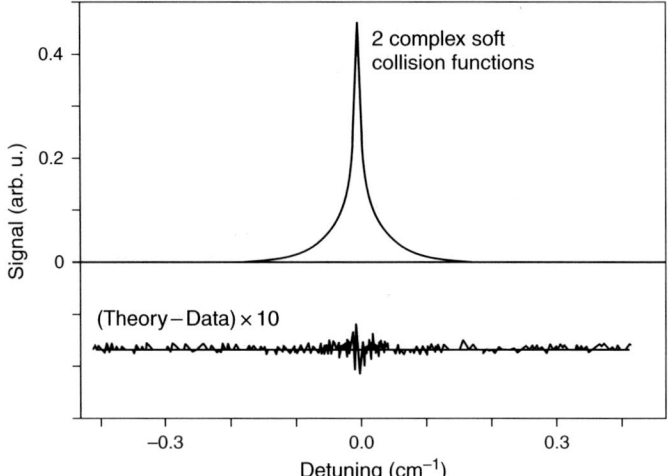

Fig. III.11: Comparison of complex soft collision line-shape functions for forward and backward components to the experimental Q(1) spectrum for a 50% D_2 + 50% Ar mixture at T = 297 K and 3.7 Am. Reproduced with permission from Ref. 211.

with respect to the GP model for each of the two components in a D_2–Ar mixture are thus completely removed (Fig. III.11).

The imaginary part of the obtained narrowing parameter (∼10%) is consistently positive, according to Eq. (III.35) (since $\Delta < 0$). A similar study for D_2–CH_4 (whose line shifts are two times larger than for D_2–Ar) gives the same agreement, corroborating the connection between the complex narrowing parameter and the collisional shift for the (correlated) G_CP. The D_0 value for D_2–CH_4 deduced from the pressure dependence of the half width (as above for D_2–He) agrees within 3% with the mass diffusion coefficient, but is 17% lower for D_2–Ar. All the D_0 values are J-independent. In spite of the good overall agreement with measurements of the Q(J) line shapes in D_2–X mixtures modeled by the soft collision model, Rosasco and coworkers[209–211] suggested that other processes could explain their observations. Indeed, at the accuracy attained in their experiments, the fitting procedure with a phenomenological model was suspected to mask other mechanisms. For instance, no attempt was made to take into account the speed dependence of the broadening and shifting coefficients, whose significant effects are considered in Sec. III.4.

A second example of asymmetric narrowed line shape is now presented, in connection with the concept of partial statistical correlation for velocity-changing and dephasing collisions. The Dicke narrowing was observed for high J transitions by Pine[212] in the fundamental absorption band of HF in neon and argon buffer gases by using a high-resolution (∼10^{-4} cm^{-1}) tunable laser difference-frequency spectrometer. Particular caution was taken for the baseline stability, crucial for this type of experiments, and for the sensitivity (through the signal-to-noise ratio ∼500). The observed P(J) line shapes were interpreted from the GP and NGP models. The general trends of the data *vs* pressure and J were realistically taken into account. A slight asymmetry was definitely identified, more important for HF–Ar than for HF–Ne due to larger collisional shifts for the former mixture. This asymmetry, negligible at atmospheric pressure, increases with decreasing pressure and exhibits a maximum at about 0.1 atm for HF–Ar. Its sign is consistent with that predicted from the statistical correlation between VC and D collisions discussed above (*ie* A > 0 for s < 0). Similar asymmetries were also observed in the fundamental vibrational band of HF and HCl in N_2 and air buffer gases.[213] A more extensive study of P(J) and R(J) lines in HF–Ar mixtures was performed later[171] in order to accurately test the basic soft and hard line-shape models (GP, NGP, G_CP, NG_CP). Typical results are given in Fig. III.12 for the P(5) line. Residuals for NGP (Fig. III.12a) and GP (Fig. III.12b) are similar (shape and amplitude) and smaller than those obtained with the VP (Fig. III.12f), but well above the noise level. The dispersion-shaped residuals indicate a positive asymmetry [Eq. (III.1)] consistent with a VC/D correlation. The correlated soft (G_CP) model leads to a decrease of the statistical errors in the residual (Fig. III.12e), but

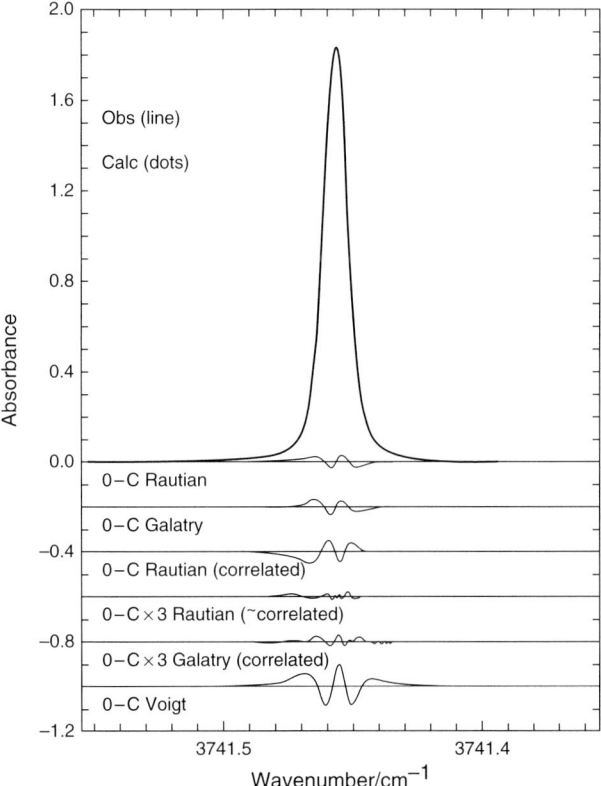

Fig. III.12: Experimental data (top curve) and meas-calc deviations with various models for the P(5) line of HF (v = 1 ← v = 0) highly diluted in Ar at P = 13.33 kPa atm and T = 296 K. Figs. III.12a to III.12f in the text correspond to the meas-calc deviations from top to bottom. Reproduced with permission from Ref. 171.

deviations still remain larger than experimental uncertainties. The NG$_C$P residuals (Fig. III.12c) exhibit an overcompensation of the asymmetry. From this overcompensation, Pine[171] introduced an empirical partially correlated hard profile by changing the complex narrowing parameter in the NGP expression [cf Eq. (III.38)], with a concomitant shift of the resonance frequency. This model is characterized by a fourth adjustable collisional parameter governing the partial correlation. Although different from the NG$_{\widetilde{C}}$P model presented in § III.3.6 [Eq. (III.42)] and developed later by Joubert et al[190] within the Rautian-Sobelman kinetic frame, the empirical model of Pine[171] captured the essential physical meaning of the partial correlation. It leads to a better fit of the line shape (Figs. III.12d). However, several anomalies remain, the main one being the opposite sign of the asymmetry for the first rotational lines P(1), R(1) and R(2). Moreover, the effective narrowing parameter v_{VC}^H determined from these fits is only about half that estimated from the mass diffusion coefficient through Eq. (III.25). It also exhibits a marked J-dependence, inconsistent with the purely kinetic character of the VC collision rate. Further refined analyses of the observed profiles[172,214,215] revealed that their modeling must take into account the speed dependence of the line-broadening and -shifting parameters introduced in the next section. This is a fortiori true for heavier radiators for which the velocity-changing collisional effects are less significant.

III.4 SPEED-DEPENDENT LINE-SHAPE MODELS

Starting from the 1990's, numerous high-resolution spectroscopic studies have been devoted to the spectral consequences of the speed dependence of the broadening and shifting parameters $\Gamma(v)$ and $\Delta(v)$ [or $\gamma(v)$ and $\delta(v)$]. Some particularly illustrative experimental results and their interpretation in terms of $\Gamma(v)$ and $\Delta(v)$ are presented below. Let us mention that the isotropy of the velocity distribution for a gas at thermal equilibrium [Eq. (III.2)] restricts the above dependence to the modulus $v = \|\vec{v}\|$ of the radiator velocity $\vec{v}$ (ie to the speed).

III.4.1 OBSERVATION OF SPEED-DEPENDENT INHOMOGENEOUS PROFILES

The speed dependence of the collisional relaxation rates was observed from the seventies, as well in the infrared using a nonlinear velocity selective spectroscopic method in NH_3,[216,217] as in the microwave region from transient decays measurements in NH_3, OCS, and H_2CO.[218] Later, speed-dependent broadening was distinguished from Dicke narrowing in the oxgen A band by using high resolution spectroscopy.[219] From the theoretical point of view, deviations from the Lorentzian line shape due to the speed dependence of the line broadening were pointed out by Luijendijk[220] from the comparison between width measurements in microwave spectra and semi-classical calculations for various molecules (OCS, CH_3Cl, H_3CCCH). Farrow et al[221] have observed much larger speed dependent effects in the isotropic Raman Q(J) lines of H_2 diluted in a heavy perturber gas in the collisional regime (ie for sufficiently high density, well above the Dicke minimum, cf § III.3.2). This regime, also called[222] hydrodynamic regime, is such

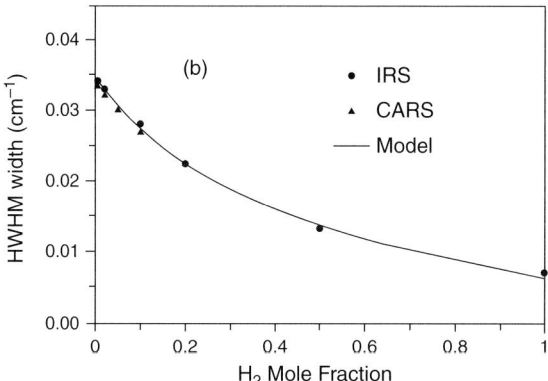

Fig. III.13: Widths (HWHM) of the Q(1) line of H_2 in Ar *vs* mole fraction at 8.3 Am and 295 K determined from fitting Lorentzian profiles to the experimental data (symbols) and to the sdNGP model (solid line) [Eq. (III.49) without $\vec{k}\cdot\vec{v}$ and substituting the speed-changing rate v_{sC}^H to v_{VC}^H], Γ being taken v-independent. Reproduced with permission from Ref. 221.

that the residual inhomogeneous Doppler broadening can be neglected with respect to the collisional broadening. Figure III.13 shows the HWHM of the Q(1) transition in H_2–Ar mixtures *vs* the H_2 mole fraction x_{H2}, obtained from inverse Raman scattering (IRS) and CARS experiments. The (homogeneous) collisional broadening Γ [deduced from its usual linear variation *vs* mole fraction $\Gamma = x_{H2}\Gamma(H_2 - H_2) + (1 - x_{H2})\Gamma(H_2 - Ar)$] is a straight line (not reported) between the same value of the HWHM for $x_{H2} = 1$ and the value $0.015\,\text{cm}^{-1}$ for $x_{H2} = 0$. This figure shows the role of the inhomogeneous broadening through the strong *non-linear* variation of the line width *vs* mole fraction. For highly diluted H_2, the inhomogeneous contribution is of the same magnitude as the homogeneous broadening and significantly increases with increasing temperature. Furthermore, the corresponding observed profiles are *asymmetric*, especially at high temperature, with the same type of residual signature as for the VC/D correlation (*cf* Fig. 1c of Ref. 221 and comment below). Using an intuitive speed-dependent hard collision model through Δ(v), Farrow *et al*[221] gave an essentially correct interpretation of all observed features. For H_2 highly diluted in heavy perturber gases, the line shift is much larger than the broadening (by a factor of several units, *cf* Table III.1) and consequently plays a predominant role in the inhomogeneous spectral features.

Exhaustive further studies of Q(J) line profiles in various H_2–X mixtures (X = He, Ne, Ar, Xe, N_2) were performed by Berger *et al*[166,223] and Sinclair *et al*,[224] always in the high density collisional regime. The experimental data at any concentration for a wide range of temperature (up to 1 200 K) were accurately modeled by using the speed-dependent (sd) correlated hard collision model, including both Γ(v) and Δ(v), derived by Robert *et al*[225] This model [noted 1DsdNGP, *cf* Eq. (III.73)] is presented within the frame of the speed-dependent Rautian-Sobelman generalized approach in § III.4.3.

In order to explain the physical mechanisms underlying the observed non-Lorentzian shape and non-linear variation of the width *vs* mole fraction, let us consider Fig. III.14. The Lorentzian Q(1) line shape is calculated[226,227] for various H_2 speed values, each

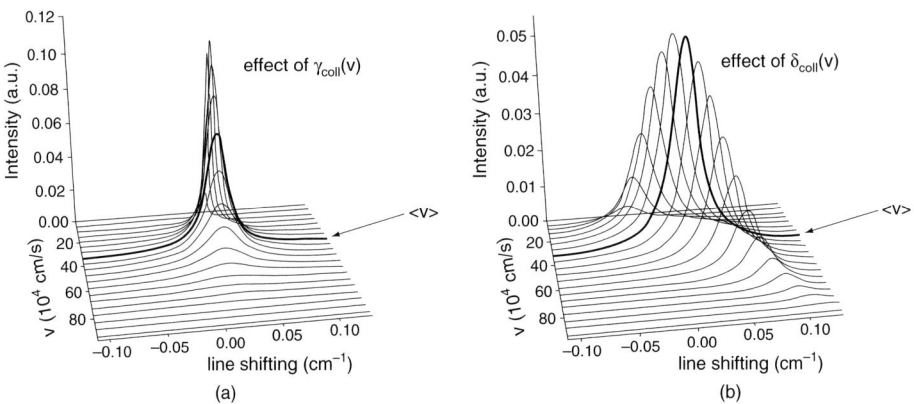

Fig. III.14: H_2 Q(1) line shapes calculated at 1200 K and 1 Am for 5% H_2 in N_2 vs H_2 speed groups: (left:a) with the (only) effect of the speed dependence of the collisional line broadening Γ(v); (right:b) with the (only) effect of the speed dependence of the collisional line shift Δ(v). The profile calculated for v = <v> is indicated by the arrow. Reproduced with permission from Ref. 227.

profile having an intensity proportional to the associated speed probability according to the Maxwell-Boltzmann distribution. If only the speed dependence of the collision broadening is first considered, most of the intense spectral components have a width smaller than Γ(<v>), where <v> is the mean speed (Fig. III.14a). After averaging over the thermal distribution, the resulting profile is symmetric and narrower than the speed-independent Lorentzian line corresponding to Γ(<v>). This narrowing effect was clearly identified in H_2-N_2 mixtures at high temperature (T = 1200 K),[224] although much less important than the inhomogeneous distribution of the shift (Fig. III.14b) leading, after speed averaging, to an asymmetric strongly broadened line. The sign and magnitude of the asymmetry are connected with the speed-derivative Δ'(v) = dΔ(v)/dv of Δ(v). This derivative is positive with a large amplitude for H_2 with heavy perturbers. This results in a wide spreading of the v-dependent Q(1) line, increasing with increasing speed and weighted by the Maxwell-Boltzmann distribution favorable to the high values of the speed (Fig. III.14b). Since Δ' > 0, the resulting spectral profile exhibits an enhancement of the high frequency wing and an attenuation of the low frequency one [ie an asymmetry characterized by A > 0, cf Eq. (III.1) and Fig. III.1]. The inhomogeneous signature of the radiator speed is thus similar to that of the VC/D correlation in a Dicke narrowed line for Δ < 0. For Δ' < 0, the increasing spread of the v-dependent line would be oriented toward decreasing frequency (opposite to the case of Fig. III.14b), leading to a reversed asymmetry (similar to the signature of VC/D correlation for Δ > 0).

More generally it must be emphasized that the *amplitude* and the *sign* of the *slope* of the curves Γ(v) and Δ(v) play a crucial role in the observed speed inhomogeneous features and consequently in the asymmetry. Furthermore, an increase of temperature increases this amplitude and spreads out the thermal distribution, resulting in an enhancement of these inhomogeneous signatures.[166,221,223,224] In fact, to quantitatively model these features, the crucial role of the *speed-changing* (sC) collisions must also be taken into account. They induce transfers between the various speed groups, leading to a more or less drastic reduction of the inhomogeneous spectral effects following the magnitude of the sC collision rate v_{sC} with respect to that of the collisional rate. As stated later,[228] for *pure* H_2, velocity- and speed-changing rates are equal so that the above mentioned reduction becomes a *total collapse* with *no* residual inhomogeneity. In contrast, for heavy perturbers in H_2–X mixtures, only a *small fraction* of VC collisions (~10%) changes the radiator speed, leading to a *weak* reduction of the inhomogeneous distribution of the line shape.

It is not surprising that these speed inhomogeneous effects were first observed for light radiators, for which the spectral signatures are the most pronounced. These effects were later evidenced in many other molecular systems, following the development of more and more efficient spectroscopic devices. Using a high-resolution difference frequency spectrometer (resolution of 10^{-4} cm^{-1} and signal-to-noise ratio of 2000), Duggan et al[185,229] carefully studied the shape of IR lines of CO buffered by He, Xe and N_2. The atmospheric importance of CO and the theoretical tractability of its IR spectrum make the CO–X mixtures prototype systems. These studies were performed at subatmospheric pressures, in order to capture the spectral effects of both the Dicke narrowing and the inhomogeneous speed distribution. As seen in § III.3.7, a common but empirical analysis method of spectral profiles was to fit the diffusion coefficient D and to compare the result (frequently called "optical" diffusion coefficient) to the calculated mass diffusion coefficient. Significant departures between these values were previously mentioned by Pine for HF[212] and by Varghese and Hanson for HCN.[181] This problem was carefully investigated for CO–N_2[185] in order to propose an explanation for the physical origin of these departures. Assuming a statistical uncorrelation of VC and D collisions, the time correlation C(t) of the radiating dipole for an isolated line is the product of the contributions tied to the external and internal degrees of freedom [Eq. (III.18)]. For light molecules like H_2 or D_2, the collisional relaxation rate being much lower than the VC one, the decay of $C_{int}(t)$ is so slow that it permits to accurately analyze the translational dynamics.[163] The case of CO–N_2 corresponds to the *more common opposite* situation since the collisional relaxation rate is about three times larger than the VC one. Thus, the time evolution of the translational correlation function $C_{ext}(t)$ plays a marginal role and experimental data essentially express the dynamics of the rovibrational states. Consequently, the type of model (hard or soft) used to describe this evolution becomes subsidiary. Using the soft (GP) model [Eq. (III.26)] for the translational contribution, Duggan et al[185] extracted the time decay $C_{int}(t)$ of the internal degrees of freedom for CO highly diluted in N_2. It was found that the line shape agrees with measurements at low pressure (below ~0.1 atm). At higher pressure, when the collisional relaxation becomes predominant, the soft model fails through a divergence of its narrowing parameter z (*cf* Fig. III.15). This drastic discrepancy was attributed to a non-exponential decay of the rovibrational time correlation $C_{int}(t)$ (or equivalently to a non-Lorentzian behavior of its Laplace transform) for CO–N_2.

The effects of two other perturbers (He, Xe) were also investigated[229] and new experiments with improved resolution (~6 10^{-5} cm^{-1}) and signal-to-noise ratio (3000) were performed[230] to clarify the origin of this non-exponential decay. A simple speed-dependent line-shape model, including the Doppler broadening and based on the

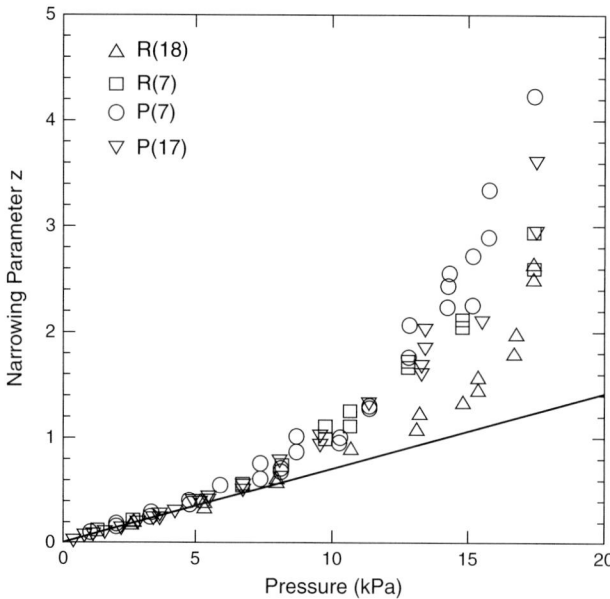

Fig. III.15: Narrowing parameter z in the soft (GP) model [Eq. (III.26)] derived from fits of CO–N_2 spectra with a floating z value. The relative uncertainty in the individual points is approximately 10%. The straight line is the narrowing parameter calculated from the mass diffusion coefficient [Eq. (III.25)]. Reproduced with permission from Ref. 185.

soft collision approximation, was used to analyze the data. On account of the weakness of the line shifts (typically one order of magnitude smaller than the broadening, cf Table III.1), only the speed dependence of the line broadening $\Gamma(v)$ was introduced through straight line collision paths. An interaction potential for CO–X molecular pairs varying as R^{-q} was chosen (R being the intermolecular distance). The q-exponent goes from zero for the speed-independent limit, up to infinity for hard spheres, with the intermediate Lennard-Jones typical values $q = 6$ and $q = 12$. These studies brought strong arguments in favor of the role of $\Gamma(v)$ in the non-exponential decay of the correlation function observed for CO–N_2. The experimental results for CO–He and CO–Xe[229] confirm this statement, through the reduced sensitivity to the speed dependence for He and its enhancement for Xe. An additional argument is that the closest correspondence between "optical" and mass values for the diffusion coefficient D was obtained for $q = 6$, as expected for a non-polar perturber through van der Waals and permanent dipole-induced dipole interactions. A high resolution study of CO IR profiles in various mixtures was performed later,[167] leading to similar conclusions. In particular, they found no evidence for an "optical" diffusion coefficient D_{opt} (cf § III.3.4) since their empirical speed-dependent hard and soft models removed the departures (with respect to the mass diffusion coefficient value) observed for heavy perturbers at increasing pressure. Using a correlated speed-dependent soft model,[231] Duggan et al showed[230] that the speed dependence of $\Gamma(v)$ may apparently reduce the

Dicke narrowing. They also pointed out the limitations of this model, which accounts for the VC/D correlation on the external degrees of freedom (*ie* on the Dicke narrowing), but not on the internal relaxation (*ie* with *no* possible reduction of the speed inhomogeneous distribution).

To summarize the results of the studies of CO in various mixtures,[167,185,229,230] it may be mentioned that, if the effect of the speed dependence of the line broadening coefficient on the Dicke narrowing was clearly established, the question of the pertinence of an "optical" diffusion coefficient D_{opt} remained open. Indeed, among the speed-dependent line-shape models used, none was free of criticism. Further improvements of these models were required to get a firm answer. Beyond the above mentioned effect of speed-changing collisions on the internal relaxation, a more realistic description of the speed dependence of Γ (and, in general, Δ) was also necessary.

Besides these spectral studies of the effects of collisional processes, alternative observations of the effects of speed inhomogeneous effects combined with the Dicke narrowing mechanism were done in the time domain. By comparison with the frequency domain spectroscopies, the *time domain experiments* enable a more direct observation of the speed-dependent signatures with an excellent sensitivity.[218,232–235] Let us mention that in the microwave spectral region of pure rotational transitions, the Doppler width is considerably reduced with respect to IR regions due to its proportionality to the transition frequency [*cf* Eq. (III.5)]. This means that the pressure range where the collisional broadening is of the same order as the Doppler broadening is reduced by the same scale (typically 10^{-3}, *cf* caption of Fig. III.16). The time-domain techniques exploit the transient decay signal of the induced sample polarization. In the so-called "*optical*

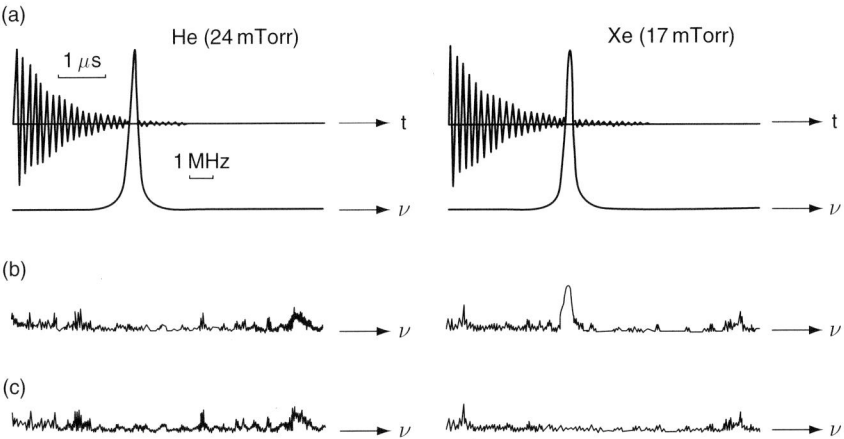

Fig. III.16: (a) Time domain signals and corresponding frequency domain power spectra (linear scale) for the $(J,K) = (2,1) \leftarrow (1,1)$ rotational transition of CH_3F at $1.05 \, 10^{-6}$ atm partial pressure and 136 K in mixtures with He [P(He) = $31.6 \, 10^{-6}$ atm] and Xe [P(Xe) = $22.4 \, 10^{-6}$ atm].; (b) residuals from fits using a time domain (speed-independent) Voigt model [Eq. (III.43)]; (c) residuals from fits using a time domain *speed-dependent* Voigt model [Eq. (III.46)]. The (b) and (c) lower spectra are magnified by 10^3. NB. Reproduced with permission from Ref. 234.

free precession" experiments, the polarization is induced by a pulsed coherent radiation field. The subsequent monitored decay signal of the sample coherence contains the signatures of the molecular relaxation processes including both external and internal degrees of freedom. When only speed-independent rotational relaxation is present, the observed signal is simply an exponential decay with a rate Γ (the HWHM in the frequency domain). When the Doppler broadening is also significant, an extra damping is detected, the Voigt time domain signal being given by:

$$S(t) = A \exp\left[-\Gamma t - \frac{(k\tilde{v}t)^2}{4}\right] \cos[(\omega - \omega_{fi} - \Delta)t + \Psi], \qquad (III.43)$$

where t is the time elapsed since the end of the pump pulse, $\tilde{v}$ is the most probable radiator speed, A and Ψ being respectively the initial amplitude and phase of the signal. If $\Gamma(v)$ is the speed-dependent rate of the signal decay, the signal is no longer exponential after speed-averaging. Such non-exponential signatures were first observed by Coy et al[218,232] for various molecular systems (OCS, H_2CO, NH_3, C_2H_4O and N_2O) and later by Mäder and coworkers[233] (N_2O, HCCF and HCN). Rohart et al[234] have further studied the two sources of inhomogeneous deviation from an exponential decay. Experimental evidence for speed-dependent relaxation of the $(J,K) = (2,1) \leftarrow (1,1)$ rotational transition of CH_3F in He and Xe mixtures is shown in Fig. III.16. An exponential decay is clearly observed for the He perturber, but not for Xe. To remove this discrepancy, a (useful) quadratic model for $\Gamma(v)$ and $\Delta(v)$ was introduced:[234,236,237]

$$\Gamma(v) + i\Delta(v) = \Gamma\left[1 + a_w\left(\frac{v^2}{\tilde{v}^2} - \frac{3}{2}\right)\right] + i\Delta\left[1 + a_s\left(\frac{v^2}{\tilde{v}^2} - \frac{3}{2}\right)\right], \qquad (III.44)$$

with

$$\Gamma = \langle\Gamma(v)\rangle \quad \text{and} \quad \Delta = \langle\Delta(v)\rangle. \qquad (III.45)$$

In Eq. (III.44), the empirical parameters a_w and a_s characterize the (quadratic) speed dependence of the width and shift respectively (both amplitude and sign). Using Eq. (III.44) for $\Gamma(v)$ [Δ was assumed v-independent, *ie* $a_s = 0$, since $\Delta \ll \Gamma$ for CH_3F], the signal becomes, after Maxwell-Boltzmann averaging over the radiator speed:[234]

$$\langle S(v,t)\rangle = A\left\{\frac{\exp\left[-\Gamma(1-\frac{3}{2}a_w)t\right]}{(1+\Gamma a_w t)^{3/2}}\right\} \\ \times \exp\left[-\frac{(k\tilde{v}t)^2}{4(1+\Gamma a_w t)}\right]\cos[(\omega - \omega_{fi} - \Delta)t + \Psi]. \qquad (III.46)$$

The first term in brackets describes the speed-dependent collisional relaxation contribution to the signal. At $t = 0$, the initial decay rate is equal to the mean value Γ of $\Gamma(v)$ [Eq. (III.45)]. At later times, the decay is slower than the exponential one in connection with the positive sign of a_w. Note that the Doppler effect also reduces with increasing time through the factor $(1 + \Gamma a_w t)$ in the second term of Eq. (III.46). To this slower decay of the $<S(v, t)>$ signal corresponds a narrowing of the spectral distribution in the frequency domain. Calculations of $<S(v, t)>$ lead to a consistent modeling of experimental data for CH_3F–Xe (Fig. III.16c).

From the knowledge of the $<S(v, t)>$ signal, it is possible to get a more direct signature of the speed dependence effects. If the initial observation time is shifted toward a later delay t_0, by ignoring the part of the signal between $t = 0$ and t_0, an "apparent" exponential decay rate $\widetilde{\Gamma}(t_0)$ can be deduced for each value of t_0 from the first derivative at time t_0. The resulting relaxation rates vs t_0 are displayed in Fig. III.17. As expected, for the He buffer gas the relaxation rate is not affected by this initial time delay shift. It remains constant vs t_0. In contrast, a marked decrease of the "apparent" relaxation rate $\widetilde{\Gamma}$ is observed for CH_3F–Xe. This decrease is the signature of the following mechanism. Since the slowest molecules relax less rapidly [cf Eq. (III.44) with $a_w > 0$], their contributions to the signal $<S(v, t)>$ at longer times are enhanced. The initial delay time shift thus induces an increasing role of their contribution for increasing time, leading concomitantly to more and more reduced "apparent" decay rates. In the above analysis,[234] no attempt was done to include the role of the velocity-changing collisions on the Doppler distribution (ie the Dicke narrowing). Their effect was only estimated from the temperature dependence of the relaxation rates (through the fitting parameter a_w) between 120 K and 300 K and was shown less significant than the observed narrowing. In a similar study of CH_3F in polar buffer gases (allowing the use of the ATC model,

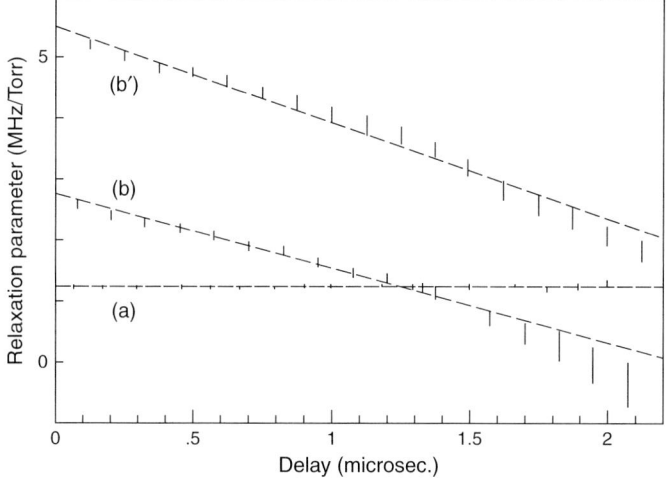

Fig. III.17: "Apparent" relaxation rate $\widetilde{\Gamma} = (2\pi)^{-1} d\Gamma(t_0)/dP(X)$ for the CH_3F transitions of Fig. III.16 (where P(X) is the X buffer gas pressure) from experimental time domain signals vs the *shifted* time delay t_0 (see text). (a) in a mixture with He at 296 K; (b) in a mixture with Xe at 296 K; (b') in a mixture with Xe at 128 K. Reproduced with permission from Ref. 234.

cf § IV.3.4d, to calculate the speed dependence of relaxation rates), the velocity-changing collisional effect was found to be much weaker than the speed inhomogeneous one.[238] Evidence was also obtained in these time domain experiments of a speed dependence for the shift.

To conclude this section, it must be emphasized that if, in the collisional (or hydrodynamic) regime, a consistent interpretation of the speed inhomogeneous features for H_2 perturbed by heavy perturbers was obtained, several problems subsisted in the Dicke narrowing pressure range. The situation is indeed more intricate since it requires a simultaneous treatment of VC collisions for the Dicke narrowing and sC collisions for the speed inhomogeneous distribution. Moreover, these inhomogeneous effects being sensitive to the slopes of $\Gamma(v)$ and $\Delta(v)$, a *refined* description of the speed dependence of these parameters is needed to get steady models for the *whole* range of pressure. The basic speed-dependent models (hard and soft) are first presented in the next section and more elaborated approaches are described in § III.4.3 and III.4.4.

III.4.2 BASIC SPEED-DEPENDENT PROFILES

Mizushima[239,240] and Edmonds[241] are the authors who fisrt introduced ad hoc modifications in order to take into account the speed-dependent widths to interpret pressure-broadened spectral line shape. The formulation of the speed-dependent Voigt profile (sdVP), from firm physical basis, was done by Berman,[242] leading to:

$$I_{sdV}(\widetilde{\omega}) = \text{Re}\{J_{sdV}(\widetilde{\omega})\} = \frac{1}{\pi}\text{Re}\left\{\int f_M(\vec{v})\left[\Gamma(v) - i(\widetilde{\omega} - \Delta(v) - \vec{k}.\vec{v})\right]^{-1} d^3\vec{v}\right\}. \quad \text{(III.47)}$$

This expression is the Maxwell-Boltzmann average of the (velocity-dependent) Doppler pressure-broadened and -shifted (speed-dependent) profiles, describing the absorption in the direction $\vec{k}^0$ from an ensemble of radiators with velocity $\vec{v}$ and speed $v = \|\vec{v}\|$. If the Doppler $\vec{k}.\vec{v}$ term is dropped out, Eq. (III.47) becomes a weighted sum of Lorentzian profiles (WSLP). Performing angular integrations in Eq. (III.47) and introducing dimensionless variables [*cf* Eq. (III.20)], the sdVP is given by:[242]

$$I_{sdV}(x,y) = \frac{1}{2\sqrt{\pi}}\left\{1 - \frac{4}{\pi}\int_0^{+\infty} \tan^{-1}\left[\frac{x(u)^2 + y(u)^2 - u^2}{2uy(u)}\right] e^{-u^2} u\, du\right\}, \text{ with}$$
$$u = v/\widetilde{v} \quad \text{and} \quad \tan^{-1} \in [-\pi/2, +\pi/2]. \quad \text{(III.48)}$$

The sdVP model was thoroughly discussed by Ward et al[243]. It reduces to the original VP[169] when the speed dependences of Γ and Δ are ignored.

A speed-dependent hard collision model was introduced by Farrow et al[221] for the collision regime (*ie* when the Doppler term $\vec{k}.\vec{v}$ can be neglected in the leading profile).

The speed-dependent Nelkin-Ghatak profile (sdNGP) was derived from the kinetic equation,[244,245] leading to:

$$I_{sdNG}(\widetilde{\omega}) = \text{Re}\{J_{sdNG}(\widetilde{\omega})\}$$

$$= \frac{1}{\pi}\text{Re}\left\{\frac{\left\langle\left[v_{VC}^H + \Gamma(v) - i(\widetilde{\omega} - \Delta(v) - \vec{k}.\vec{v})\right]^{-1}\right\rangle}{1 - v_{VC}^H\left\langle\left[v_{VC}^H + \Gamma(v) - i(\widetilde{\omega} - \Delta(v) - \vec{k}.\vec{v})\right]^{-1}\right\rangle}\right\}. \quad \text{(III.49)}$$

The statistically correlated speed-dependent Nelkin-Ghatak profile (sdNG$_C$P) can be obtained from Eq. (III.73) for $\varepsilon = 1$, by including the $\vec{k}.\vec{v}$ term:[225]

$$I_{sdNG_C}(\widetilde{\omega}) = \text{Re}\{J_{sdNG_C}(\widetilde{\omega})\}$$

$$= \frac{1}{\pi}\text{Re}\left\{\frac{\left\langle\left[v_{VCD}^H - i(\widetilde{\omega} - \vec{k}.\vec{v})\right]^{-1}\right\rangle}{1 - \left\langle\frac{v_{VCD}^H - \Gamma(v) - i\Delta(v)}{v_{VCD}^H - i(\widetilde{\omega} - \vec{k}.\vec{v})}\right\rangle}\right\}, \quad \text{with} \quad v_{VCD}^H = v_{Kin}. \quad \text{(III.50)}$$

Angular integrations over the radiator velocity orientation can be analytically performed in Eqs. (III.49) and (III.50), leading[245] to expressions for the sdNGP and sdNG$_C$P in terms of an integral over the speed, as in Eq. (III.48).

In contrast with the sdNGP for hard collisions, it is not possible to get a general analytical expression for the sdGP. Indeed, the solution of the Boltzmann kinetic equation is generally not tractable for speed-dependent relaxation rates, due to the structure of the (soft) Fokker-Planck collision operator. Neglecting the derivatives of $\Gamma(v)$ and $\Delta(v)$ in this operator[246] results in an *approximate* analytical expression for the speed-dependent Galatry profile (denoted asdGP[215]). After angular integrations over the velocity distribution, the asdGP is expressed as:[230,231]

$$I_{asdG}(x, y, z) = \frac{1}{\pi}\text{Re}\left\{\int_0^{+\infty} d\widetilde{t}\, \exp\left[-\frac{1}{4z^2}(2z\widetilde{t} - 3 + 4e^{-z\widetilde{t}} - e^{-2z\widetilde{t}})\right]\right.$$

$$\left. \times \frac{4}{\sqrt{\pi}}\int_0^{+\infty}\text{sinc}\left[\frac{(1-e^{-z\widetilde{t}})u}{z}\right]\exp\{-[ix(u) + y(u)]\widetilde{t}\}e^{-u^2}u^2 du\right\} \quad \text{(III.51)}$$

$$\equiv \text{Re}\langle J_{asdG}[x(u), y(u), z]\rangle,$$

where $\text{sinc}(X) = \sin(X)/X$ and $\langle ...\rangle$ is the Maxwell-Boltzmann average over the reduced speed $u = v/\widetilde{v}$. $J_{asdG}[x(u), y(u), z]$ is the (uncorrelated) *complex* approximate speed-dependent Galatry profile which can be straightforwardly obtained from Eq. (III.51) by identification [*ie* omitting the symbol Re, the integral over u, and the weighting factor $4u^2\exp(-u^2)/\sqrt{\pi}$]. The asdGP *does not take into account transfers induced by VC collisions between speed groups* and, concomitantly, the resulting modifications in the speed inhomogeneous spectral features (cf § III.4.1). In the limit of very large VC collision rates, the asdGP tends to the WSLP [cf Eq. (III.48) disregarding the Doppler shift term]. The asdGP, being the resulting profile of soft velocity-changing collisions with *no* speed change, may be named "*hypersoft*" model[247] (*ie* for collisions weakly changing the radiator velocity orientation but *not* its modulus).

An *exact* analytical solution for the Fokker-Planck equation was obtained[248] for the speed-dependent collisional width and shift model of Eq. (III.44). Numerical tests from this specific exact sdGP indicate that the asdGP may be accurately (within about 1%) used to fit the spectral line shape for molecular systems for which the soft collision approximation is relevant. In the opposite case, very large differences can be obtained, as for H_2 perturbed by Ar.[249]

The statistical correlation between velocity-changing and dephasing collisions is taken into account by the substitution of VCD collisions to (VC+D) ones. This leads to the substitution, in Eq. (III.51), of the complex speed-dependent effective v_{eff}^S frequency [Eq. (III.35)] to the (speed-independent) v_{VC}^S, *ie* in dimensionless variables:

$$z \to Z(u) = z[1 - y(u)/\varsigma] - is(u)z/\varsigma, \quad \text{with} \quad u = v/\tilde{v}. \quad \text{(III.52)}$$

This means that, after this substitution and speed averaging, one gets $I_{asdG_C}[x(u), y(u), z, \varsigma, s(u)]$ from Eq. (III.51). The asdG$_C$P depends on four collisional parameters (y, z, ς, s), instead of three (y, z, s) for the asdGP [Eqs. (III.51) and (III.20)], as for the (speed-independent) G$_C$P and GP, respectively (*cf* Table III.3).

The above speed-dependent line-shape models (*ie* sdNGP, sdNG$_C$P, asdGP and asdG$_C$P) require a realistic description of the speed dependence of the line broadening and shift. These relaxation rates can be related to the collision cross-sections $\sigma_\Gamma(v_r)$ and $\sigma_\Delta(v_r)$, which depend on the relative speed $v_r = \|\vec{v}_r\|$ between the radiator and the perturber:[243]

$$\Gamma(v, T) + i\Delta(v, T) = n_p \int v_r f_M(\vec{v}_r + \vec{v}, T)[\sigma_\Gamma(v_r) + i\sigma_\Delta(v_r)] d^3\vec{v}_r, \quad \text{(III.53)}$$

where $\vec{v}_r = \vec{v}_p - \vec{v}$, $\vec{v}_p$ being the perturber velocity. After integration over the orientation of $\vec{v}_r$, Eq. (III.53) is expressed[220,250] as an integral over the relative speed v_r of the cross-sections weighted by the conditional probability $f(v_r|v, T)$:

$$\Gamma(v, T) + i\Delta(v, T) = n_p \int_0^\infty v_r f(v_r|v, T)[\sigma_\Gamma(v_r) + i\sigma_\Delta(v_r)] dv_r, \quad \text{(III.54)}$$

where:[250]

$$f(v_r|v, T) = \frac{4}{\pi} \frac{v_r}{v\langle v_p \rangle} \text{sh}\left[\frac{8}{\pi} \frac{vv_r}{\langle v_p \rangle^2}\right] \exp\left[-\frac{4}{\pi} \frac{v^2 + v_r^2}{\langle v_p \rangle^2}\right], \quad \text{(III.55)}$$

$\langle v_p \rangle$ being the mean perturber speed. Equation (III.54) shows that $\Gamma(v,T)$ and $\Delta(v,T)$ *can be deduced* from the relative speed dependence of the corresponding cross-sections $\sigma_\Gamma(v_r)$ and $\sigma_\Delta(v_r)$. Figure III.18 illustrates the behavior of the conditional probability

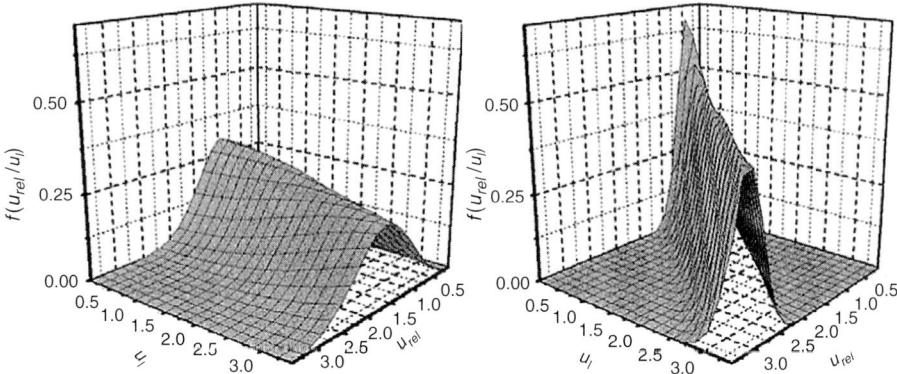

Fig. III.18: Conditional probability of the relative speed at given radiator speed $f(u_{rel}|u_l)$ at $T = 295$ K, with $u_{rel} = v_r/<v_r>$ and $u_l = v/<v>$, for two values of the mass ratio parameter $\Lambda = m_p/m$. $\Lambda = 0.2$ (left: a) and $\Lambda = 5$ (right: b).[245]

[Eq. (III.55)] for two characteristic values of the mass ratio parameter $\Lambda = m_p/m$. For perturbers much lighter than radiators (Fig. III.18a), the conditional probability is almost independent of the radiator speed and close to the Maxwell-Boltzmann distribution for the relative speed. For perturbers much heavier than radiators (Fig. III.18b), the distribution "peaks" along the diagonal $v = v_r$. Thus, collisions of a radiator with a much heavier perturber induce slight changes of its speed v (cf § III.3.4), and these changes arise about $v \approx v_r$ since v_p is then very weak.

The simplest (but not realistic) model for the line broadening and shift considers the colliding molecules as rigid spheres, resulting in the absence of speed dependence for the associated cross-sections. A more realistic description was introduced by Berman[242] and Ward et al[243] through straight-line trajectories and a spherically symmetric interaction potential varying as the inverse q^{th} power of the radiator-perturber separation R of their centers of mass. The intermolecular potential is then:

$$V(R) = \frac{\text{Constant}}{R^q}, \quad q \geq 3. \tag{III.56}$$

The speed dependence of Γ and Δ can be then expressed analytically as:[243]

$$\Gamma(v,T) + i\Delta(v,T) = (\Lambda+1)^{-\frac{q-3}{2q-2}} M\left(-\frac{q-3}{2q-2}, \frac{3}{2}, -\Lambda\frac{v^2}{\tilde{v}^2}\right)[\Gamma(T) + i\Delta(T)]. \tag{III.57}$$

In Eq. (III.57), $\Gamma(T) = \langle\Gamma(v,T)\rangle$ and $\Delta(T) = \langle\Delta(v,T)\rangle$ are the mean values of Γ and Δ, $\tilde{v}$ is the most probable speed and $M(...)$ the confluent hypergeometric function.[182] Notice that, in this simple model, the shift-to-width ratio is independent of the radiator speed and of the temperature:[243]

$$\frac{|\Delta(v,T)|}{\Gamma(v,T)} = \tan\left(\frac{\pi}{q-1}\right). \tag{III.58}$$

More generally, the collisional cross sections $\sigma_\Gamma(v_r)$ and $\sigma_\Delta(v_r)$ can be calculated from more refined intermolecular potentials by using semi-classical or quantum approaches (cf § IV.3.4 and IV.3.5). The resulting numerical data can then be expanded in a series of α_i^{th} and β_j^{th} powers of the relative speed v_r:

$$\sigma_\Gamma(v_r) = \sum_{i=1}^{N} a_i v_r^{\alpha_i} \quad \text{and} \quad \sigma_\Delta(v_r) = \sum_{j=1}^{N'} b_j v_r^{\beta_j}. \tag{III.59}$$

Reporting Eq. (III.59) in Eq. (III.54) leads, after integration over v_r, (see Ref. 238 and discussion in § VIII.4.1) to:[250]

$$\Gamma(v,T) + i\Delta(v,T) = n_p \left\{ \sum_{i=1}^{N} a_i \left(\frac{4}{\pi}\right)^{-\alpha_i/2} \hat{\Gamma}\left(\frac{\alpha_i+4}{2}\right)(\Lambda+1)^{-\frac{\alpha_i+1}{2}} \langle v_r \rangle^{\alpha_i+1} \right.$$
$$\times M\left(-\frac{\alpha_i+1}{2}, \frac{3}{2}, -\Lambda\frac{v^2}{\tilde{v}^2}\right) + \sum_{j=1}^{N'} b_j \left(\frac{4}{\pi}\right)^{-\beta_j/2} \tag{III.60}$$
$$\left. \times \hat{\Gamma}\left(\frac{\beta_j+4}{2}\right)(\Lambda+1)^{-\frac{\beta_j+1}{2}} \langle v_r \rangle^{\beta_j+1} M\left(-\frac{\beta_j+1}{2}, \frac{3}{2}, -\Lambda\frac{v^2}{\tilde{v}^2}\right) \right\},$$

with α_i and $\beta_j > -4$. In Eq. (III.60), $\hat{\Gamma}(\ldots)$ is the Euler Gamma function[182] and $\langle v_r \rangle = (8k_B T/\pi m^*)^{1/2}$, $m^* = m_p m/(m+m_p)$ being the reduced mass. After performing the Maxwell-Boltzmann average over the speed v in Eq. (III.60) [or, equivalently, in Eq. (III.54) through Eq. (III.59)], the temperature dependence of the mean broadening and shifting rates is given by:

$$\Gamma(T) + i\Delta(T) \equiv \langle \Gamma(v,T) \rangle + i\langle \Delta(v,T) \rangle = n_p \left[\sum_{i=1}^{N} A_i T^{\frac{\alpha_i+1}{2}} + i \sum_{j=1}^{N'} B_j T^{\frac{\beta_j+1}{2}} \right],$$

$$\text{with} \quad A_i = a_i \left(\frac{4}{\pi}\right)^{1/2} \hat{\Gamma}\left(\frac{\alpha_i+4}{2}\right)(2k_B/m^*)^{\frac{\alpha_i+1}{2}} \quad \text{and} \tag{III.61}$$

$$B_j = b_j \left(\frac{4}{\pi}\right)^{1/2} \hat{\Gamma}\left(\frac{\beta_j+4}{2}\right)(2k_B/m^*)^{\frac{\beta_j+1}{2}}.$$

Let us consider again the simple intermolecular potential model of [Eqs. (III.56) in order to get a first application of Eqs. (III.59) to (III.61). The identification of Eq. (III.57) with the general expression (III.60) allows one to deduce the following relation between the radial q power [Eq. (III.56)] and the α_1 and β_1 parameters relative to velocity powers [Eq. (III.59)]:

$$\alpha_1 = \beta_1 = -\frac{2}{q-1}, \quad \alpha_{i \neq 1} = \beta_{j \neq 1} = 0. \tag{III.62}$$

Consequently, the temperature dependence of $\Gamma(T)$ and $\Delta(T)$ in Eq. (III.57) is, per number density unit:

$$\Gamma(T)/n_p \quad \text{and} \quad \Delta(T)/n_p \propto T^{\frac{\alpha_1+1}{2}} = T^{\frac{q-3}{2q-2}}, \quad \text{or, per pressure unit,} \quad \text{(III.63)}$$

$$\gamma(T) = \Gamma(T)/P \quad \text{and} \quad \delta(T) = \Delta(T)/P \propto T^{\frac{\alpha_1-1}{2}} = T^{\frac{q-3}{2q-2}-1} \equiv T^{-n}, \quad \text{with} \quad n = \frac{q+1}{2q-2}.$$
(III.64)

For rigid spheres ($q \to \infty$), Eq. (III.64) leads to the expected $n = 0.5$ result for the broadening and shifting coefficients γ and δ (cm^{-1}/atm) [cf Eq. (III.54) with σ_Γ and σ_Δ independent of v_r]. Values for n can be deduced for other q-values, eg q = 5, corresponding to a quadrupole-quadrupole interaction, leads to n = 0.75 when γ and δ are in pressure unit. The corresponding expression for the speed dependence of Γ and Δ immediately results from Eq. (III.57). Because of its simplicity, this speed-dependent model [Eqs. (III.56), (III.57) and (III.63) or (III.64)] has been largely used. Let us mention that, as long as accurate ab initio calculated data for $\sigma_\Gamma(v_r)$ and $\sigma_\Delta(v_r)$ are not available [permitting to exploit the *general equations* (III.59) to (III.61)], additional phenomenological parameters are required in simplified models [one for Eqs. (III.56) and (III.57) or two for Eq. (III.44) for instance]. Considering the number of characteristic collisional parameters already used in the models of profiles (Table III.3), these additional ones in fits of experimental profiles may obscure the meaning of the resulting line shape (eg Ref. 214). Indeed, the deficiencies of (too) simple models for the speed dependence of Γ and Δ may introduce severe ambiguities in the real adequacy of the model of profile.[251] Hence, if the goal is not the usefulness of these models for specific applications, but their physical adequacy, great caution must be taken in order to get pertinent conclusions.

Since the speed-dependent profiles take into account two sources of inhomogeneous broadening (Doppler and speed distribution of Γ and Δ), attention must be paid to the frequency migration induced by collisions within these two inhomogeneous distributions. A protocol is to isolate, in a first step, the second source of inhomogeneity by considering pressures well above the Dicke minimum, so that no significant Doppler contribution subsists in the line profile. This is precisely the case of the isotropic Raman Q(J) lines of H_2 in presence of heavy perturbers at high density. Indeed, let us recall that such molecular systems are particularly interesting due to the very large inhomogeneous speed effects. The first analysis[221] was based on the sdNGP [Eq. (III.49)], disregarding the Doppler shift. When this equation is used together with the previous definition of v_{VC}^H [ie the radiator VC rate in the hard collision approximation, cf Eq. (III.29)], the sdNGP is inadequate since the inhomogeneous features are thus deleted by VC collisions, as explained for pure H_2 in § III.4.1. In fact, Farrow et al[221] pertinently substituted the appropriate speed changing sC rate v_{sC}^H to v_{VC}^H. The sdNGP thus leads to a realistic description of all the observed speed inhomogeneous features. v_{sC}^H was found[221] smaller than v_{VC}^H by about one order of magnitude for H_2–Ar, implying that the H_2 speed is conserved after many VC collisions with Ar. This was further justified from molecular dynamics simulations.[228] The substitution of v_{sC}^H to v_{VC}^H in the sdNGP thus considerably reduces the collisionally induced frequency migration tied to the speed groups exchanges, resulting in a strongly broadened asymmetric inhomogeneous spectral line shape consistent with experimental data (*cf* Fig. III.13).

Before examining the joined effects of Doppler and speed distributions of Γ and Δ on the line shape in the Dicke pressure range (through a comparison of the above basic speed-dependent models with experiments), more general profiles (including the previous ones as special cases) are first presented.

III.4.3 THE RAUTIAN-SOBELMAN MODEL

General analytical profiles were first introduced by Ciurylo[246] and by Lance and Robert.[252,253] The aim was to overcome the limitations inherent in the use of either the hard *or* soft collision approximations. This is required when the speed dependence of Γ and Δ is taken into account. Indeed, the differences between the hard and soft models of profile are then much greater, especially when a speed-derivatives approximation for soft collisions is introduced to save the analytical solution (see below).[246] The basic concept is to consider sequences of collisional events including *both* hard and soft collisions. The kinetic rate is now defined by the sum of the hard and soft rates:

$$v_{Kin} = v_{VC}^H + v_{VC}^S, \quad \text{with} \quad v_{VC}^H = \varepsilon v_{Kin}, 0 \leq \varepsilon \leq 1, \tag{III.65}$$

where ε is the fraction of hard collisions. It characterizes the "hardness" of the VC collisions and (1-ε), the concomitant "softness". When Γ and Δ are considered as *speed-independent*, the Rautian-Sobelman kinetic formalism[56] leads to the following equation for the probability in the hard/soft statistically *uncorrelated* case [*ie* for (HVC + SVC + D) collisions:

$$\frac{\partial}{\partial t} f(\vec{r}, \vec{v}, t) = -\vec{v}\vec{\nabla}_r f(\vec{r}, \vec{v}, t) + v_{VC}^S \left\{ \vec{\nabla}_{\vec{v}}[\vec{v} f(\vec{r}, \vec{v}, t)] + \frac{v^2}{2} \Delta_{\vec{v}} f(\vec{r}, \vec{v}, t) \right\} \\ - v_{VC}^H \left\{ f(\vec{r}, \vec{v}, t) - f_M(\vec{v}) \int f(\vec{r}, \vec{v}', t) d^3\vec{v}' \right\}. \tag{III.66}$$

For $\varepsilon = 0$ (*ie* $v_{VC}^H = 0$), Eq. (III.66) reduces to the (soft) GP model. The $\varepsilon = 1$ case (*ie* $v_{VC}^S = 0$) leads to the (hard) NGP model. The solution of the uncorrelated hard/soft model [called Rautian-Sobelman (RS) profile in the following] can be written in terms of the GP.[56,246] In dimensionless variables, the RSP is given by:

$$I_{RS}(x, y, \varsigma, \varepsilon) = \text{Re} \left\{ \frac{J_G[x, Y(y, \varsigma, \varepsilon), z(\varsigma, \varepsilon)]}{1 - \pi \varepsilon \varsigma \times J_G[x, Y(y, \varsigma, \varepsilon), z(\varsigma, \varepsilon)]} \right\}, \tag{III.67}$$

where $J_G(x,y,z)$ is the *complex* Galatry profile [Eq. (III.26)] and:

$$Y(y, \varsigma, \varepsilon) = y + \varepsilon \varsigma \quad \text{and} \quad z(\varsigma, \varepsilon) = (1 - \varepsilon)\varsigma. \tag{III.68}$$

Isolated lines 105

In Eqs. (III.67) and (III.68), $z(\varsigma, \varepsilon)$ is the narrowing parameter for the fraction $(1-\varepsilon)$ of soft collisions. In these equations, the parameter ς is always defined as the reduced kinetic rate $\varsigma = v_{Kin}/\Delta\omega_D$ [cf Eq. (III.33)], but it is *no* longer equal to $v_{VC}^H/\Delta\omega_D$, as in Sec. III.3, since *only* a fraction ε of collisions is of hard type.

The statistically *correlated* RS$_C$P profile corresponds to sequences of combined HVCD and SVCD collisional events, the dephasing arising either in hard collisions (HVCD) or in soft collisions (SVCD). The substitution procedure of a complex effective collision rate v_{eff} to v_{VC}[56] can be done [cf Eqs. (III.35) and (III.38)]. Let us note that in the RS$_C$P model, v_{Kin} is equal to the total collision frequency $v_{VCD}^H + v_{VCD}^S$ [Eq. (III.65)]. Furthermore, besides the restrictions on y and $s \leq \varsigma$ [Eq. (III.37)], the third $z \leq \varsigma$ (or $v_{VCD}^S \leq v_{Kin}$) condition is then automatically satisfied. The (complex) effective hard and soft collision rates are defined by:

$$\widetilde{v}_{VC}^H = v_{VCD}^H - (\Gamma_{VCD}^H + i\Delta_{VCD}^H) = \varepsilon[v_{Kin} - (\Gamma + i\Delta)],$$
$$\widetilde{v}_{VC}^S = v_{VCD}^S - (\Gamma_{VCD}^S + i\Delta_{VCD}^S) = (1 - \varepsilon)[v_{Kin} - (\Gamma + i\Delta)],$$
(III.69)

or, in dimensionless variables,

$$\varepsilon\widetilde{\varsigma} = \varepsilon[\varsigma - y - is] \quad \text{and} \quad \widetilde{z} = (1 - \varepsilon)(\varsigma - y - is) = (1 - \varepsilon)\widetilde{\varsigma}. \quad \text{(III.70)}$$

Notice that, in Eq. (III.69), the partial broadening $\Gamma_{VCD}^{S\,or\,H}$ *and* shift $\Delta_{VCD}^{S\,or\,H}$ are assumed to be proportional to the VC hard or soft rates with the same ratios ε and $(1-\varepsilon)$. Since the only change introduced by the VC/D statistical correlation is the substitution of $\widetilde{\varsigma}$ to ς, the leading RS$_C$P may be formally expressed in terms of RSP as:

$$I_{RS_C}(x, y, \varsigma, s, \varepsilon) = I_{RS}(x, y, \widetilde{\varsigma}, \varepsilon), \quad \text{(III.71)}$$

where I_{RS} is defined through Eqs. (III.67) and (III.68). Equation (III.71) for $\varepsilon = 1$ is identical to the NG$_C$P. For $\varepsilon = 0$, the G$_C$P is not recovered, because the (HVCD + SVCD) collisional model is *fully* correlated and does not take into account possible additional independent VC collisions, as for the (partially correlated) G$_C$P (cf § III.3.6). Such a partial correlation is examined below.

The role of the *speed dependence* through $\Gamma(v)$ and $\Delta(v)$ must then be considered. The case of the RSP is not explained further, since the physical assumptions underlying this line-shape model are too crude. When both hard and soft collisions proceed in collisional sequences, the hypothesis of statistically independent dephasing collisions D through (HVC + SVC + D) collisions is not very realistic. Indeed, *all* velocity memory processes between the two H and S limits are phenomenologically included through the RS model, requiring to take the VC/D correlation into account. An approximate (labeled a) expression for the *speed-dependent* (fully) *correlated* Rautian-Sobelman profile (asdRS$_C$P) for (HVCD + SVCD) collisions can be obtained by *neglecting* the velocity derivatives of $\Gamma(v)$ and $\Delta(v)$ in the (soft) Fokker-Planck collision operator [through the second contribution in the right hand side of Eq. (III.66)]. The asdRS$_C$P thus straightforwardly follows[246,253] from Eqs. (III.66) to (III.71):

$$I_{asdRS_C}(x,y,\varsigma,s,\epsilon) = \text{Re}\left\{\frac{\langle J_{asdG}[x(u),\widetilde{Y}(u),\widetilde{z}(u)]\rangle}{1 - \pi\epsilon\langle\widetilde{\varsigma}(u) \times J_{asdG}[x(u),\widetilde{Y}(u),\widetilde{z}(u)]\rangle}\right\}, \quad (III.72)$$

with $\widetilde{Y}(u) = y(u) + \epsilon\widetilde{\varsigma}(u)$, $\widetilde{z}(u) = (1-\epsilon)\widetilde{\varsigma}(u)$, and $\widetilde{\varsigma}(u) = \varsigma - y(u) - is(u)$.

In Eq. (III.72), $J_{asdG}[x(u),\widetilde{Y}(u),\widetilde{Z}(u)]$ is the *complex* approximate speed-dependent Galatry profile [Eq. (III.51)] with the substitution of $\widetilde{Y}$ to y, and $\widetilde{z}$ to z. It should be mentioned that, as for the VC collisions in the asdGP, the soft VCD collisions do not induce frequency migration in the speed inhomogeneous distribution but *only* the hard VCD collisions do.

In order to justify, from experiments, the need to introduce a pertinent *partial*, instead of "fully", VC/D correlation, let us again consider H_2 highly diluted in heavy perturbers in the collisional (or hydrodynamic) regime (*ie* at pressures such that the residual Doppler broadening becomes negligible with respect to the collisional one). It was shown in § III.4.1 that the sdNGP [Eq. (III.49)], with v_{sC}^H instead of v_{VC}^H, is essentially correct to interpret the speed inhomogeneous spectral signatures in this regime (*ie* the non-linear dependence of the observed HWHM *vs* concentration and the line asymmetry). In fact, a careful reexamination of the results has evidenced a weak residual inhomogeneous contribution ($\approx 10\%$) to the broadening coefficient of pure H_2, when modeling the isotropic Raman Q(J) lines by the sdNGP model.[166] This discrepancy is totally removed by accounting for the partially correlated hard (sCD + D) model,[225] instead of the statistically uncorrelated hard (sC + D) model through the sdNGP.[221] Let us recall that only the speed-changing (sC) collisions, and not the VC ones, are pertinent in the collisional regime since the velocity-orientation plays no significant role in the spectral line shape. So, the 3D kinetic equation [Eq. (III.66)] reduces to 1D (only speed-dependent). Since the (hyper) soft collisions in the asdRS$_C$P model do not induce speed groups exchanges,[252] the SVCD contribution tends to the D one. Moreover, the HVCD contribution tends to the hard sCD one. Hence, the hard (sCD + D) model is the limit of the (HVCD + SVCD) one [*ie* of the (3D) asdRS$_C$P, *cf* Eq. (III.72)] in the collisional regime. This (sCD + D) model (thus denoted 1DasdRSP) may be written as:[225]

$$I_{1DasdRS}(\widetilde{\omega}) = \frac{1}{\pi}\text{Re}\left\{\frac{\langle\{\epsilon v_{Kin} + (1-\epsilon)\Gamma(v) - i[\widetilde{\omega} - (1-\epsilon)\Delta(v)]\}^{-1}\rangle}{1 - \langle\epsilon[v_{Kin} - \Gamma(v) - i\Delta(v)]\{\epsilon v_{Kin} + (1-\epsilon)\Gamma(v) - i[\widetilde{\omega} - (1-\epsilon)\Delta(v)]\}^{-1}\rangle}\right\}.$$

(III.73)

Velocity-orientation changes are *implicitly* involved in sCD and D collisions, so that:

$$v_{Kin} = v_{sCD}^H + v_D, v_{sCD}^H = \epsilon v_{Kin}, v_D = (1-\epsilon)v_{Kin}, 0 \leq \epsilon \leq 1. \quad (III.74)$$

For $\epsilon = 1$, Eq. (III.73) is identical to the sdNG$_C$P [Eq. (III.50), neglecting the Doppler contribution $\vec{k}\cdot\vec{v}$]. In this fully correlated limit, each collision is of hard sCD type and the profile is *purely* homogeneous. This is precisely the case for pure H_2. $\varepsilon = 0$ corresponds to the (sd) D collisions. For $0 < \varepsilon < 1$, the 1DasdRSP model is partially correlated through sCD and D collisions, but non labeled to avoid ambiguity with 3D notations for the velocity. The case $\varepsilon = 0$ is the WSLP, since the speed inhomogeneous distribution is not reduced by collisions. The line shape of H_2 highly diluted in Ar or N_2 nearly corresponds to the second limit ($\varepsilon \approx 0.1$).[166,224] The consistency of the 1DsdRSP in the collisional regime was checked by Chaussard and others from experimental studies of H_2 Q(J) line shapes in various binary and ternary mixtures for wide ranges of pressure and temperature.[20,254–256] The "hardness" parameter ε in Eq. (III.73) is thus defined in terms of ε_{H2}, ε_X, ε_Y and of the kinetic rates (per unit of pressure) $\tilde{v}_{Kin}^{H_2}, \tilde{v}_{Kin}^{X}, \tilde{v}_{Kin}^{Y}$ for each H_2–H_2, H_2–X, and H_2–Y colliding pairs, by:

$$\varepsilon v_{Kin} = \left(x_{H_2} \varepsilon_{H_2} \tilde{v}_{Kin}^{H_2} + x_X \varepsilon_X \tilde{v}_{Kin}^{X} + x_Y \varepsilon_Y \tilde{v}_{Kin}^{Y} \right) \times P, \qquad (III.75)$$

where x_{H_2}, x_X and x_Y are the mole fractions of the various species and P is the total pressure. The fitted values of ε for various H_2–X mixtures (X = H_2, He, Ne, N_2, Ar) are presented in Table III.4. These "experimental" values ε_{fitted} are compared with molecular dynamics simulation (MDS) results[228] using realistic intermolecular potentials. These ε_{MDS} values were deduced from the simulation of the velocity- and speed-autocorrelation functions at various concentrations and temperatures, through the ratio of the correlation times ($\varepsilon_{MDS} = \tau_{\tilde{v}}^{MDS} / \tau_v^{MDS}$). This "hardness" parameter may also be estimated from the rigid sphere or billiard-ball (BB) model. Indeed ε_{BB} is then simply related (*cf* § III.5.4) to the relative mass parameter $\Lambda = m_p/m$ by:[249]

$$\varepsilon_{BB} = 2/(1 + \Lambda). \qquad (III.76)$$

The ε_{BB} values are very close to the MDS data (Table III.4). A slight difference can be denoted for H_2–N_2 due to the role of the anisotropic quadrupole-quadrupole interaction,

Table III.4: Comparison between values of the "hardness" parameter (ε_{fitted}) fitted from experimental data obtained at high density (*ie* in the collisional regime) using the 1DsdRSP model[20,254–256] with those predicted from molecular dynamics simulations (ε_{MDS})[228] and from the billiard ball model (ε_{BB}) [Eq. (III.76)]

	H_2–H_2	H_2–He	H_2–Ne	H_2–N_2	H_2–Ar
ε_{fitted}	1.0	0.68	0.14	0.14	0.10
ε_{MDS}	1.0	0.66	0.18	0.15	0.10
ε_{BB}	1.0	0.67	0.18	0.13	0.095

disregarded in the BB model. Table III.4 underlines the consistency of the 1DasdRSP for the (high density) collisional regime.

To estimate further the degree of this consistency, a particular attention must be paid to the speed-dependent model used for $\Gamma(v)$ and $\Delta(v)$ (cf § III.4.2). Let us recall that, if the shift of the H_2 Q(J) lines is measured vs concentration in mixtures with heavy perturbers, a non-linearity appears at high dilution, especially at high temperature. This non-linearity, due to the line asymmetry, can be removed by considering the frequency of its center of gravity instead of its maximum.[166] This enabled to get experimental values for the homogeneous line shifting coefficient $\delta(cm^{-1}/Am)$ at infinite dilution vs temperature, accurately represented by:[166,224,254]

$$\delta(T) = A_\delta \sqrt{T} + c_\delta. \qquad (III.77)$$

The corresponding expression for the shift cross-section vs the relative speed v_r immediately follows from Eqs. (III.59) and (III.61), ie:

$$\sigma_\Delta(v_r) = a_\delta + c_\delta/v_r. \qquad (III.78)$$

This equation leads, through Eq. (III.60), to an explicit expression for the speed dependence of δ:

$$\delta(u',T) = A_\delta \sqrt{\frac{m^*}{m_p}} \sqrt{T} \left[\frac{e^{-u'^2}}{2} + \left(\frac{\sqrt{\pi}}{4u'} + \frac{u'\sqrt{\pi}}{2} \right) \Phi(u') \right] + c_\delta, \; u' = \sqrt{\Lambda} v/\tilde{v} \equiv v/\tilde{v}_p,$$

$$(III.79)$$

where $A_\delta = a_\delta \sqrt{8k_B/\pi m^*}$, $\tilde{v}_p$ is the most probable perturber speed, and Φ is the error function.[182] Equation (III.79) can be used to get, from measurements, a speed-dependent law for the shift $\Delta(v,T) = n_p \delta(v,T)$ (which plays a very predominant role with respect to $\Gamma(v,T) = n_p \gamma(v,T)$ in the speed inhomogeneous features for H_2-heavy perturbers). A similar procedure for the broadening coefficient is much more delicate. To get $\gamma(T) = \langle \gamma(v,T) \rangle$ from experiments, the inhomogeneous contribution must be removed. This requires, in general, a careful protocol to avoid significant bias through the model used for the profile analysis. As seen below for HF–Ar,[214] an attractive alternative is to use accurate ab initio data for $\sigma_\Gamma(v_r)$ and $\sigma_\Delta(v_r)$ if available. For H_2 highly diluted in Ar or N_2, the inhomogeneous spectral signature of $\gamma(v)$ was found significant only above T = 1000 K.[224] Accurate values of the homogeneous broadening coefficient were obtained at high pressure from measurements between 300 and 1000 K. The observed linear dependence of the line broadening coefficient vs T, $\gamma(T) = A_\gamma T + c_\gamma$, thus leads, from Eqs. (III.59) to (III.61), to $\sigma_\Gamma(v_r) = a_\gamma v_r + c_\gamma/v_r$ (with $a_\gamma = (m^*/3k_B)A_\gamma$) and to an explicit expression for $\gamma(v,T)$.[166,223] This rigorous determination for $\Gamma(v,T)$ and $\Delta(v,T)$ speed-dependent laws completes the critical analysis of the pertinence of the 1DasdRSP for the collisional regime (ie for pressure above several atmospheres at room temperature and above ten atmospheres at 1000 K for H_2–X mixtures). At lower pressure, when both Doppler and speed inhomogeneities are present, not only the speed-changing collisions but also those changing the velocity-orientation have to be taken into account through VC collisions.

Isolated lines 109

In order to test the (3D) asdRS$_C$P, stimulated Raman scattering measurements were performed[20,254] in a wide pressure range (from subatmospheric up to several tens of atmospheres) for the isotropic Raman Q(1) line shape of H$_2$ in Ar and N$_2$ mixtures. An accurate modeling of all the spectral profiles in a wide temperature range was obtained with a single set of fitting parameters, *ie* $\varepsilon(T)$, $\nu_{Kin}(T)$, $\Gamma(T)$, and $\Delta(T)$. A careful analysis of these results, through a comparison with the set of fitted parameters obtained by using only high pressure profiles (*ie* in the collisional regime) revealed[257] significant residual discrepancies. These discrepancies were attributed to insufficiencies of the (fully correlated) asdRS$_C$P in the description of the collisionally induced frequency migration inside the Doppler and speed inhomogeneous distributions at intermediate pressure (*ie* in the "Dicke narrowing regime").

Further tests of the asdRS$_C$P [Eq. (III.72)] were made[172] from the above analyzed accurate infrared spectra[171] for HF-Ar (*cf* § III.3.7). These tests evidenced the necessity to take into account a partial correlation through (HVCD + SVCD + HVC + SVC) instead of (HVCD + SVCD) collisions. In this *partially* correlated speed-dependent RS model (noted asdRS$_{\widetilde{C}}$P),[215] the partial correlation is defined from the fractions η^H and η^S of correlated collisions among the kinetic ones (*cf* § III.3.6) as:

$$\nu_{Kin} = \nu_{VCD} + \nu_{VC} = (\nu_{VCD}^H + \nu_{VCD}^S) + (\nu_{VC}^H + \nu_{VC}^S), \qquad (III.80)$$

with [*cf* Eq. (III.41)]:

$$\nu_{VCD}^H = \eta^H \nu_{Kin}, \quad \nu_{VCD}^S = \eta^S \nu_{Kin}, 0 \le \eta^H \quad \text{and} \quad \eta^S \le 1. \qquad (III.81)$$

The (complex) speed-dependent effective frequencies for hard and soft collisions are:

$$\widetilde{\nu}_{VCD}^H(v) = \varepsilon\{\nu_{Kin} - \eta^H[\Gamma(v) + i\Delta(v)]\}, \quad \widetilde{\nu}_{VCD}^S(v) = (1-\varepsilon)\{\nu_{Kin} - \eta^S[\Gamma(v) + i\Delta(v)]\},$$
$$\text{with } \varepsilon = (\nu_{VCD}^H + \nu_{VC}^H)/\nu_{Kin}. \qquad (III.82)$$

Assuming a *unique* correlation parameter[215] $\eta = \eta^H = \eta^S$ leads, in dimensionless variables, to:

$$\varepsilon\widetilde{\varsigma}(u) = \varepsilon\{\varsigma - \eta[y(u) + is(u)]\}, \quad \widetilde{z}(u) = (1-\varepsilon)\{\varsigma - \eta[y(u) + is(u)]\},$$
$$0 \le \eta \text{ and } \varepsilon \le 1. \qquad (III.83)$$

The expression of the *partially correlated* Rautian-Sobelman profile asdRS$_{\widetilde{C}}$P in terms of the complex approximate speed-dependent Galatry profile [Eq. (III.51)] is, from Eqs. (III.70) to (III.72):

$$I_{asdRS_{\widetilde{C}}}(x, y, \varsigma, s, \varepsilon, \eta) = \text{Re}\left\{\frac{\langle J_{asdG}[x(u), \widetilde{Y}(u), \widetilde{z}(u)]\rangle}{1 - \pi\varepsilon\langle\widetilde{\varsigma}(u) \times J_{asdG}[x(u), \widetilde{Y}(u), \widetilde{z}(u)]\rangle}\right\}, \qquad (III.84)$$

with

$$\widetilde{Y}(u) = y(u) + \varepsilon \widetilde{\varsigma}(u), \ \widetilde{z}(u) = (1-\varepsilon)\widetilde{\varsigma}(u), \ \widetilde{\varsigma}(u) = \varsigma - \eta[y(u) + is(u)], \quad \text{(III.85)}$$
$$u = v/\widetilde{v}, 0 \leq \eta \text{ and } \varepsilon \leq 1.$$

The approximate partially correlated speed-dependent Rautian-Sobelman profile asdRS$_{\widetilde{C}}$P depends, besides the zero pressure line frequency, on five characteristic collisional dimensionless parameters tied [*cf* Eqs. (III.20) and (III.33)] to the kinetic frequency (ς), the collisional broadening (y), the collisional shift (s), the "hardness" parameter (ε) and the correlation parameter (η). This asdRS$_{\widetilde{C}}$P contains the previous models in particular limits, for $\eta = 0$ or 1 and/or $\varepsilon = 0$ or 1 [*ie* the uncorrelated (or correlated) speed-dependent hard and/or soft models, Table III.5]. A further improvement will be introduced for completeness, although of negligible consequences for most molecular systems at atmospheric and, a fortiori, subatmospheric pressure.

Since the asymmetry is important in the inhomogeneous signatures of broadened lines, other sources of asymmetry must be taken into account. Apart from those arising from the mixing of overlapping lines (§ IV.2.2a) and collision-induced transitions (§ VI.6.2), a dispersive asymmetry may arise from the finite collision duration (Fig. III.8 and App. II.B4). The resulting so-called "*collision-duration asymmetry*" leads to the following modification in the correlation function tied to the internal degrees of freedom of the radiator [Eqs. (III.13), (III.14), and (II.B10)]:

$$C_{int}(t) = e^{-[(a'+ib')+(\Gamma+i\Delta)t]}. \quad \text{(III.86)}$$

The above expression is obtained as an asymptotic behavior for $t \gg \tau_c$, with a *finite* value of τ_c. As Γ and Δ, a' = n_pa and b' = n_pb, which are also speed-dependent, can be explicitly related to the intermolecular potential (*cf* App. II.C1 and Ref. 128). From Eq. (III.86), the collision-duration effects result in multiplying the numerator of Eq. (III.84) by exp{$-$[a'(u) + ib'(u)]} inside the brackets.[215] In practice, the weakness of the spectral modifications due to the collision-duration permits to neglect the speed dependence of a' and b' and to use a first-order expansion in b'. Moreover, the exp($-$a') term can be absorbed in the overall integrated intensity of the line, since it does not affect the shape. The correction due to the finite duration of collisions is then an asymmetric dispersive

Table III.5: Various line-shape models (see acronyms list) as special cases of the (3D)asdRS$_{\widetilde{C}}$P model [Eq. (III.84)]

asdRSP	asdRS$_C$P	sdNGP
asdRS$_{\widetilde{C}}$P with $\eta = 0$	asdRS$_{\widetilde{C}}$P with $\eta = 1$	asdRS$_{\widetilde{C}}$P with $\varepsilon = 1, \eta = 0$
sdNG$_C$P	asdGP	asdG$_C$P
asdRS$_{\widetilde{C}}$P with $\varepsilon = 1, \eta = 1$	asdRS$_{\widetilde{C}}$P with $\varepsilon = 0, \eta = 0$	asdRS$_{\widetilde{C}}$P with $\varepsilon = 0, \eta = 1$

factor (1-ib'), through a generally justified first-order expansion (see App. II.B4 and Ref. 172). The resulting asymmetrized (labeled A) profile becomes:

$$I_{AasdRS_{\widetilde{C}}}(x, y, \varsigma, s, \varepsilon, \eta, b') = \text{Re}[(1 - ib')J_{asdRS_{\widetilde{C}}}(x, y, \varsigma, s, \varepsilon, \eta)], \qquad (III.87)$$

where $J_{asdRS_{\widetilde{C}}}$ is the complex asdRS$_{\widetilde{C}}$P given by Eq. (III.84) without "Re".

In § III.3.7, it was shown that the modeling of the P(J) and R(J) rovibrational lines of HF diluted in Ar by a partially correlated speed-independent hard collision model (NG$_{\widetilde{C}}$P) is not satisfactory.[171] Pine and Ciurylo[172] reinvestigated these line shapes with the asdRS$_{\widetilde{C}}$P. Since, for HF-Ar, collision-duration effects were previously detected[130,170] at high pressure (above 10 atm, see Fig. III.8), the AasdRS$_{\widetilde{C}}$P [Eq. (III.87) expressed through Eqs. (III.84) and (III.85)] was used. A careful multispectrum fitting procedure (§ IV.4.1) was applied in order to force the parameters to be consistent throughout and, thus, to obtain a greater sensitivity. Furthermore, possible bias tied to the use of a simple model for the speed dependence of the collisional line broadening and shift was avoided. Indeed, $\Gamma(v)$ and $\Delta(v)$ were deduced[214] from accurate quantum calculations of $\sigma_\Gamma(v)$ and $\sigma_\Delta(v)$ with a very reliable intermolecular potential[23] [cf Eqs. (III.59) to (III.61)]. The multifits were performed with six floated quantities v_{Kin}, Γ, Δ, ε, η, and b', corresponding respectively to the reduced parameters ς, y, s, ε, η, and b' in Eq. (III.87), the remaining parameter ω_{fi} (through x) being the (known) frequency of the line at zero pressure. Due to the weakness of the collision-duration effect at subatmospheric pressures, dispersive corrections were barely determinable. Disregarding them [ie setting b' = 0 in Eq. (III.87)] leads only to slight increases of the fits residuals without significant modifications of the five other parameters. Their relative variations are less than 0.01% for v_{Kin}, a few 0.01% for Γ, a few 0.1% for Δ, about one percent for ε, and a few percent for η. Figure III.19 displays the multifit results for the R(6) line from the AasdRS$_{\widetilde{C}}$P and various limiting models. Collision-duration effect is barely discernible (Figs. III.19a and III.19b) and speed dependence is not prominent (Figs. III.19a and III.19c). Fixing $\varepsilon = 1$ ("purely" hard, Fig. III.19d) and $\varepsilon = 0$ ("purely" soft, Fig. III.19e) shows the predominance of hard collisions. Figure III.19g demonstrates the role of the correlation through "purely" hard collisions ($\varepsilon = 1$), the residual error strongly increasing for $\eta = 0$ with respect to the fit with floated η (Fig. III.19d). Comparisons between Figs. III.19d, f and g confirm that correlation is much more important than the speed dependence. The main anomaly concerning the sign of the asymmetry for the P(1), R(1) and R(2) lines (cf § III.3.7) is well removed, with no negative values neither for the correlation parameter η nor for the hardness one ε. The higher J values (J ≥ 3) have preponderantly hard collisions ($\varepsilon > 0.5$) and the shapes of the broadest lines R(0) and P(1) almost only result from soft collisions.

From the study of HF rovibrational lines broadened by Ar,[172] several remarks can be made in connection with the asdRS$_{\widetilde{C}}$P model. At subatmospheric pressure, the Dicke narrowing has a large influence on the line widths and shapes. The rate and "hardness" of VC collisions affect the asymmetries. The role of the speed dependences of line broadening and shifting is also obvious, especially for low J transitions. The modeling with the asdRS$_{\widetilde{C}}$P is highly accurate, and the only remaining discrepancy concerns the kinetic rate. Its fitted values remain 30% to 50% lower than the calculated one from the mass diffusion coefficient[258] in Eq. (III.25). A marked J-dependence subsists for the first rovibrational

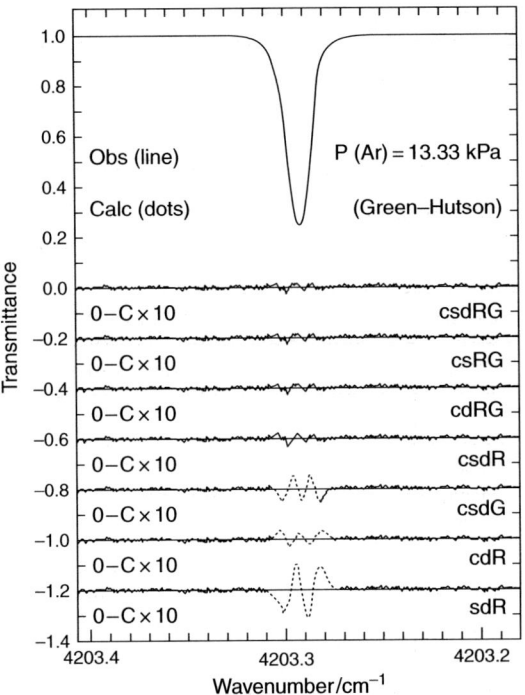

Fig. III.19: Experimental profile and 10x(meas-calc) deviations for the R(6) line of the fundamental band of HF broadened by Ar at 13.33 kPa and room temperature. The correspondences between the models indicated on the plot and the notations used in this chapter are the following (from top to bottom, labeled Figs. III.19a to III.19g in the text): csdRG→AasdRS$_{\tilde{C}}$P[0.73, 0.204] csRG→asdRS$_{\tilde{C}}$P[0.72, 0.194] cdRG→AaRS$_{\tilde{C}}$P[0.55, 0.242] csdR→AasdRS$_{\tilde{C}}$P[1(fixed), 0.193] csdG→AasdRS$_{\tilde{C}}$P[0(fixed), 0.236] cdR→AaRS$_{\tilde{C}}$P[1(fixed), 0.217] sdR→AasdRS$_{\tilde{C}}$P[1(fixed), 0(fixed)]. The numbers in brackets are values of [ε, η]. Reproduced with permission from Ref. 172.

lines which may be related to their larger broadening cross-sections and smaller energy gaps. This suggests that v_{Kin} is a transition-dependent quantity and that only its thermal average should be associated with the macroscopic diffusion constant.[172]

Among the approximations underlying the asdRS$_{\tilde{C}}$P, a particular attention has to be paid to the assumed speed independence of the hard and soft velocity changing rates. This assumption [cf Eq. (III.7)] can be related to the detailed balance principle [Eq. (III.6)]. In the uncorrelated case, ie for $\eta=0$ in Eq. (III.85), Eqs. (III.6) and (III.28) for hard collisions imply that v_{VC}^H must be *speed independent* [$v_{VC}^H(v) = v_{VC}^H(v'), \forall(v, v')$]. For soft collisions, in the approximation of the model asdRS$_{\tilde{C}}$P through the asdGP in Eq. (III.84) (ie with no speed exchange between speed groups, cf supra), the kernel $A(\vec{v}, \vec{v}')$ can be written in the "hypersoft" limit as:

$$A(\vec{v}, \vec{v}') = A(\vec{v}^o, \vec{v}'^o) v_{VC}^S(v') \delta(v - v'), \int A(\vec{v}^o, \vec{v}'^o) d^3\vec{v}'^o = 1, \vec{v}^o = \vec{v}/\|\vec{v}\|. \quad \text{(III.88)}$$

The detailed balance thus implies no mathematical restriction on $v_{VC}^S(v)$ [through Eqs. (III.6) and (III.88)]. But, since no speed change occurs during hypersoft collisions, one has $v_{VC}^S(v) = v_{VC}^S$, justifying, once more, the above assumption. When the correlation is considered through VCD collisions, only the statistical correlation is taken into account (VC and D being simultaneous collisional events) but the VC and D changes are assumed to be unrelated in magnitude.[56] Consequently, the detailed balance principle independently applies to VC and D collisions and the above conclusions concerning v_{VC}^H and v_{VC}^S remain valid. An attempt was made[214] to test an eventual speed dependence of the v_{VC} rate in HF-Ar mixtures from the asdRS$_{\widetilde{C}}$P (with $\varepsilon = 1$), showing no evidence for a v dependence. The possible role of $v_{VC}(v)$ was also theoretically explored,[259] by using the rigid sphere model[260] for the collision kernel $A(\vec{v}, \vec{v}')$. A further analysis was achieved[261] from the so-called kangaroo generalization[262,263] of the hard collision model.

The case where the energy and moment exchanges between internal (rovibrational) and external (translational) degrees of freedom are taken into account was thoroughly discussed by Berman[264–266] and Nienhuis.[267] Equation (III.7) is no longer applicable since the kernel becomes state-dependent and $A_{fi}(\vec{v}, \vec{v}')$ must be substituted to $A(\vec{v}, \vec{v}')$. If the intermolecular potentials V_i and V_f strongly differ in the initial and final quantum states, the assignment of trajectories becomes problematic and the classical picture fails. This quantum state-selecting effect results in destructive interference between $|i\rangle$ and $|f\rangle$ scatterings, leading to a decrease of the Dicke narrowing. Semi-classical approaches tend to underestimate the line widths. This effect was shown to be small for soft collisions.[265] For hard collisions, this is only true for close V_i and V_f potentials. Under this condition, the $|i\rangle$ and $|f\rangle$ trajectories nearly coincide and the classical picture is saved. This specific validity condition was checked for H_2-X mixtures.[228]

Another approximation of the asdRS$_{\widetilde{C}}$P model is the absence of a priori explicit discrimination between velocity-orientation $\vec{v}^o$ and velocity-modulus v changes induced by VC collisions. Such a discrimination, introduced by Nienhuis,[267] may become important when both Doppler and speed inhomogeneous effects are significant and when v_{VC} and v_{sC} strongly differ. The asdRS$_{\widetilde{C}}$P model phenomenologically accounts for distinct velocity- and speed-changing mechanisms through hard and soft collisions. Indeed, hard collisions induce velocity-orientation *and* speed changes, but soft collisions only induce velocity-orientation changes and *no* speed ones [J_{asdG} in Eq. (III.84) being the (complex) *approximate* speed-dependent Galatry profile]. Thus, the "hardness" parameter ε indirectly governs the relative contribution of these two mechanisms. Generally, the description of VC and D collisions (including the statistically correlated VCD ones) through a linear combination of hard and soft collisions is likely the most important weakness of the asdRS$_{\widetilde{C}}$P.

To conclude, in spite of its limitations mainly due to its phenomenological nature, the asdRS$_{\widetilde{C}}$P model presents several advantages. It is *analytical* and contains, as limiting cases, most of all the intensively used hard and soft models [in particular the (soft) asdGP and the (hard) sdNGP, *cf* Table III.5]. It accounts for the *speed dependence* of the line-broadening and -shifting, and offers the flexibility of a *partial correlation* between velocity-changing and dephasing collisions.[268,269] Finally, efficient numerical procedures are now available and can be used in computer routines (see App. III.A). The asdRS$_{\widetilde{C}}$P [Eqs. (III.84) and (III.85)] is already (in simplified versions) a useful tool for optical diagnostics in gas media (cf Sec VII.3).

In the next section, an alternative model is presented. It enables a clearer discrimination between velocity-orientation and speed in the exchange mechanisms induced by collisions within the Doppler and speed inhomogeneous distributions of the line shape.

III.4.4 THE KEILSON-STORER MEMORY MODEL

Avoiding the linear combination of hard and soft collision rates [Eq. (III.66)] inherent in the asdRS$_{\widetilde{C}}$P model, Robert and Bonamy[270] proposed to describe the kernel $A(\vec{v}, \vec{v}')$ [Eqs. (III.6) and (III.7)] in terms of the velocity memory function introduced by Keilson and Storer[187] (KS), ie:

$$A(\vec{v}, \vec{v}') = \nu_{VC} f_{\gamma_{\vec{v}}}(\vec{v}, \vec{v}') \equiv \nu_{VC}(1 - \gamma_{\vec{v}}^2)^{-3/2} f_M\left(\frac{\vec{v} - \gamma_{\vec{v}}\vec{v}'}{\sqrt{1 - \gamma_{\vec{v}}^2}}\right), \quad -1 \leq \gamma_{\vec{v}} \leq 1, \quad (III.89)$$

where the velocity-changing collision rate ν_{VC} is assumed to be speed-independent and $f_M(...)$ means the Maxwell-Boltzmann distribution [Eq. (III.2)]. Equation (III.89) extends the definition of Eq. (III.28) (restricted to hard collisions) to *all types* of collisions from hard to soft. Indeed, $f_{\gamma_{\vec{v}}}(\vec{v}, \vec{v}')$ is the KS probability density per unit time that the radiator, with velocity $\vec{v}'$, undergoes a transition into a volume $d^3\vec{v}$ about $\vec{v}$ (the parameter $\gamma_{\vec{v}}$ characterizing the velocity memory). From this KS probability density, the average value of $\vec{v}$ after collision is equal to $\gamma_{\vec{v}}\vec{v}'$, so that the velocity transferred in average *per collision* is:

$$\Delta\vec{v}' = \vec{v}' - \vec{v} = (1 - \gamma_{\vec{v}})\vec{v}'. \quad (III.90)$$

Three particular situations may be examined, in connection with the models analyzed in § III.4.2,3. For $\gamma_{\vec{v}} = 0$, the $\vec{v}$-thermal equilibrium is established after *each* collision. These hard collisions transfer, in average per collision, the total amount of the incoming velocity and the velocity memory is then *totally* lost. For $\gamma_{\vec{v}} \rightarrow 1$, the KS velocity distribution peaks sharply about *almost* zero $\vec{v}$-change and the velocity transferred, in average per collision, is very weak. The velocity memory is lost after a *large* number of such soft collisions. For $\gamma_{\vec{v}} = -1$, the sign of $\vec{v}$ is changed but *not* the speed $v = \|\vec{v}\|$ and the velocity transferred, in average per collision, is *twice* the incoming velocity. This velocity memory process is *anticorrelated* ($\vec{v}^o = -\vec{v}'^o$, with $\|\vec{v}^o\| = \|\vec{v}'^o\| = 1$ and $v = v'$). Such an anticorrelated process is expected to play an important role for heavy radiators highly diluted in a bath of (light) molecules at very high density when the neighbors form a "cage" around the radiator like in liquids.[271] The radiator translational motion then tends toward an oscillation inside such a "cage".

Before studying the Dicke narrowing regime where both velocity-orientation and speed-exchanges have significant effects on the line shapes, Robert and Bonamy[270] first considered the collisional (or hydrodynamic) regime, where only the speed exchange

Isolated lines

mechanism arises. The leading kinetic equation becomes monodimensional (1D), since only the speed memory is relevant. The corresponding KS memory function is obtained by averaging over the $\vec{v}$-orientation, leading to:[271]

$$f_{\gamma_v}(v,v') = \frac{2}{\sqrt{\pi(1-\gamma_v^2)}\,\gamma_v \widetilde{v}v'}\,\exp\left\{-\left[\frac{v^2+\gamma_v^2 v'^2}{\widetilde{v}^2(1-\gamma_v^2)}\right]\right\}\,\mathrm{sh}\left(\frac{2\gamma_v v v'}{\widetilde{v}^2(1-\gamma_v^2)}\right),\; 0\le \gamma_v \le 1.$$
(III.91)

The mathematical procedure used to numerically solve the 1D kinetic equation will be presented below, as a particular case of the more general 3D kinetic equation (see also Ref. 272). The resulting speed-dependent Keilson-Storer profile (sdKSP) allows one to analyze the spectral modifications from hard to soft speed memory processes, with all intermediate physical situations. It must be noted that *no* approximation is thus made on the derivatives of $\Gamma(v)$ and $\Delta(v)$ in the resolution of the kinetic equation, in contrast with the asdRS$_{\widetilde{C}}$P due to (hyper) soft collisions (*cf* § III.4.3). Consequently the sdKSP model introduces *no* limitation in the frequency migration induced by collisions according to the nature of the collision (or, equivalently, the type of velocity memory process through $\gamma_{\vec{v}}$). A typical result for the uncorrelated (1D) sdKSP model [*ie* for (sC + D) collisions in the hydrodynamic limit] is shown in Fig. III.20 for H$_2$–Ar.

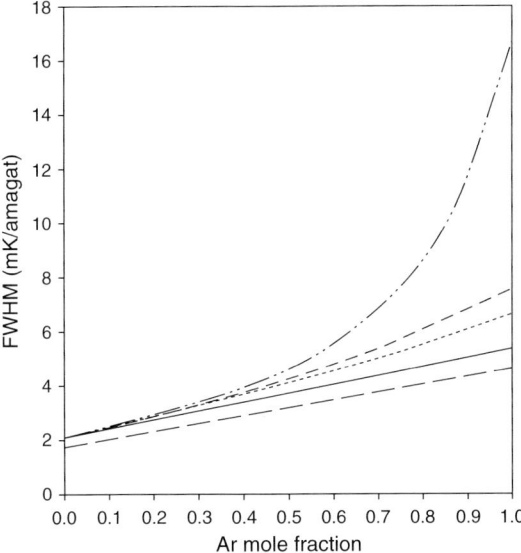

Fig. III.20: Broadening coefficient (FWHM) calculated with the 1DsdKSP model for the Q(1) line of H$_2$ in argon *vs* the Ar mole fraction at T = 295 K. The KS speed memory parameter γ_v^{Ar} for the collisional pairs has the successive values (from top to bottom curves): 1.00, 0.95, 0.80, and 0.00. The lowest curve is the homogeneous contribution to the collisional line broadening calculated from experiments[166] by using the usual linear variation *vs* mole fraction. Reproduced with permission from Ref. 270.

The speed dependences used for $\Gamma(v)$ and $\Delta(v)$ are taken from experimental data,[166] and H$_2$–H$_2$ collisions are described by the (uncorrelated) hard collision model (*ie* $\gamma_v^{H_2} = 0$). As expected, a strong inhomogeneous broadening arises at large argon mole fraction through H$_2$–Ar collisions having a *soft* speed memory, the *hypersoft* limit being $\gamma_v = 1$, *cf* Eq. (III.88). A weak discrepancy with respect to the experimental homogeneous broadening, independent of the mole fraction, appears in Fig. III.20 (by comparison of the $\gamma_v = 0$ curve with the homogeneous one below). It is due to an inhomogeneous residual broadening for pure H$_2$. It was checked later[273] that the sC/D correlation (through the sCD collisional model instead of sC+D) totally removes this discrepancy. This is fully consistent with the previous analysis using the 1DsdRSP [Eq. (III.73)] which requires $\varepsilon = 1$ (only sCD collisions) to agree with pure H$_2$ data.

A direct connection between the 1DasdKSP and the 1DsdRSP models was established, by considering the *linearized* Keilson-Storer (LKS) speed memory:

$$f_{\varepsilon_L}(v, v') = \varepsilon_L f_{\gamma_v=0}(v, v') + (1 - \varepsilon_L) f_{\gamma_v=1}(v, v') = \varepsilon_L f_M(v) + (1 - \varepsilon_L) \delta(v - v'), \quad \text{(III.92)}$$

where $0 \leq \varepsilon_L \leq 1$ is the fraction of hard collisions among (hard+hypersoft) sC ones. The 1DasdLKSP model is analytical and its expression is formally identical[270] to the 1DasdRSP.[253] But this formal identification is *only* possible thanks to the concomitant neglect of speed-derivatives in the soft collision kernel for the 1DasdRSP model (*ie* the substitution of hypersoft to soft VC collisions, *cf* § III.4.3).

The generalization to 3D of the 1D sdKSP model requires the substitution of the velocity memory function [Eq. (III.89)] to the speed memory function [Eq. (III.91)].[272,273] Let us consider, first, the case where VC and D collisions are statistically independent [*ie* the (VC+D) model]. The 3D impact kinetic equation for the (normalized) radiating dipole $d(\vec{v}, t)$ with velocity $\vec{v} = v\vec{v}^o$ at time t is:[56,273]

$$\frac{\partial}{\partial t} d(\vec{v}, t) = -\nu_{VC}[d(\vec{v}, t) - \int f_{\gamma_{\vec{v}}}(\vec{v}, \vec{v}') d(\vec{v}', t) d^3\vec{v}] - [i\vec{k}\cdot\vec{v} + \Gamma(v) + i\Delta(v)] d(\vec{v}, t), \quad \text{(III.93)}$$

where the KS velocity memory function $f_{\gamma_{\vec{v}}}(\vec{v}, \vec{v}')$ is defined by Eq. (III.89). The spectral line shape is related to $d(\vec{v}, t)$ through:

$$I_{sdKS}(\widetilde{\omega}) = \frac{1}{\pi} \text{Re}\left[\int d(\vec{v}, t) e^{i\widetilde{\omega}t} d^3\vec{v} dt \right], \quad d(\vec{v}, t=0) = f_M(\vec{v}). \quad \text{(III.94)}$$

In order to solve Eq. (III.93), it is convenient to expand the KS velocity memory function [Eq. (III.89)] in the basis formed by the product of the orthonormalized (bar labeled) associated Laguerre polynomials $\bar{L}_n^{\ell+1/2}(x)$ (also called Sonine polynomials[274]) and the Legendre polynomials $P_\ell(\vec{v}^o \cdot \vec{v}'^o)$:[182]

$$f_{\gamma_{\vec{v}}}(\vec{v}, \vec{v}') = \frac{1}{2\pi} \bar{f}_M(x) \times \sum_{n=0}^{\infty} \sum_{\ell=0}^{\infty} \frac{2\ell+1}{2} \gamma_{\vec{v}}^{2n+\ell} \bar{L}_n^{\ell+1/2}(x) \bar{L}_n^{\ell+1/2}(x') P_\ell(\vec{v}^o \cdot \vec{v}'^o), \quad \text{(III.95)}$$

Isolated lines

where $x = mv^2/2k_BT$, $x' = mv'^2/2k_BT$, and $\bar{f}_M(x) = 2\sqrt{x/\pi}e^{-x}$. The product of the spherical harmonics $Y_{\ell m}(\vec{v}^o)$ by the associated Laguerre polynomials being eigenfunctions of the 3D integral operator[271] in the kinetic equation (III.93), the radiating dipole may be developed over these normalized eigenfunctions as:

$$d(\vec{v},t) = d(x, \vec{v}^o, t) = \bar{f}_M(x) \sum_{n,\ell,m} a_{n\ell m}(t) Y_{\ell m}(\vec{v}^o) \bar{L}_n^{\ell+1/2}(x), \qquad (III.96)$$

with $a_{n\ell m}(t=0) = \delta_{n0}\delta_{\ell 0}\delta_{m0}$.

From Eqs. (III.96) and (III.94), it results that:

$$I_{sdKS}(\widetilde{\omega}) = \frac{1}{\pi} \text{Re}\left[\int a_{000}(t) e^{i\widetilde{\omega}t} dt\right], \qquad (III.97)$$

where the coefficients $a_{n\ell m}(t)$ in Eq. (III.96) are solutions of the equation:[273]

$$\frac{\partial}{\partial t} a_{n\ell}(t) + \sum_{\ell'=0,\ell\pm 1} \sum_{n'=0}^{\infty} M_{nn'}^{(\ell,\ell')} a_{n'\ell'}(t) = 0, \qquad (III.98)$$

and the matrix elements $M_{nn'}^{(\ell,\ell')}$ are given by:

$$\begin{aligned}
M_{nn'}^{(\ell,\ell-1)} &= i\Delta\omega_D \frac{\ell}{\sqrt{(2\ell-1)(2\ell+1)}} I_{nn'}^{\ell-1}, \\
M_{nn'}^{(\ell,\ell)} &= \left[\nu_{VC}(1-\gamma_{\tilde{v}}^{2n+\ell})\delta_{nn'}\right] + \left[\Gamma_{nn'}^{\ell} + i\Delta_{nn'}^{\ell}\right], \\
M_{nn'}^{(\ell,\ell+1)} &= i\Delta\omega_D \frac{\ell+1}{\sqrt{(2\ell+1)(2\ell+3)}} I_{nn'}^{\ell}.
\end{aligned} \qquad (III.99)$$

The matrix elements of Γ, Δ, and I in Eq. (III.99) are defined by:

$$\begin{aligned}
\Gamma_{nn'}^{\ell} + i\Delta_{nn'}^{\ell} &= \int \bar{f}_M(x) \bar{L}_n^{\ell+1/2}(x)[\Gamma(x) + i\Delta(x)]\bar{L}_{n'}^{\ell+1/2}(x)dx, \\
I_{nn'}^{\ell} &= \int \bar{f}_M(x) \bar{L}_n^{\ell+1/2}(x) x^{1/2} \bar{L}_{n'}^{\ell+3/2}(x)dx.
\end{aligned} \qquad (III.100)$$

The m index has been omitted in Eqs. (III.98) to (III.100) since only the m' = 0 component of the $a_{n'\ell'm'}(t)$ coefficients is implied in Eq. (III.97) and the M matrix is diagonal in m. The Laplace transform [Eq. (III.97)] of Eq. (III.98) leads to the matricial equation:

$$(-i\widetilde{\omega}I_d + M)a(\widetilde{\omega}) = 1_{00}, \qquad (III.101)$$

where the initial condition $a(t=0) = 1_{00}$ has been used [Eq. (III.96)]. If S means the matrix diagonalizing M, the resulting profile is given by:[273]

$$I_{sdKS}(\widetilde{\omega}) = \frac{1}{\pi}\text{Re}[a_{00}(\widetilde{\omega})] = \frac{1}{\pi}\sum_k \sum_q \frac{S_{00,kq}(S^{-1})_{kq,00}}{i\widetilde{\omega} + D_{kq}}, \qquad (III.102)$$

the diagonal D matrix being given by $D = S^{-1}MS$. At very low pressure, such that the collisional effects can be neglected, Eq. (III.101) leads to the Gaussian Doppler profile [Eq. (III.4)]. At high pressure, in the *hydrodynamic limit*, when the $\vec{k}.\vec{v}$ term in Eq. (III.93) can be omitted, the 3D kinetic equation reduces to a 1D equation after averaging over the velocity-orientation. The kernel is then expressed in terms of the KS speed memory function [Eq. (III.91)] and *only the $\ell = 0$ component subsists* in its expansion on the associated Laguerre polynomials [*cf* Eq. (III.95)]. Thus, the modifications in Eq. (III.102) for the 1DsdKSP are to suppress the second index q, and to substitute γ_v to $\gamma_{\tilde{v}}$ in the remaining $\ell = \ell' = 0$ contribution in Eq. (III.99).

Due to the Markovian character of the kinetic equation, the generalization of Eq. (III.102) to a mixture of two species (a) and (b), with respective mole fractions x_a and $x_b = (1-x_a)$, is simple. By taking the linear dependence of $\Gamma(v)$, $\Delta(v)$, and v_{VC} with respect to the mole fraction into account, the velocity memory function for the gas mixture [with the same type of notation as in Eq. (III.75)] is:

$$f_{\gamma_{\tilde{v}}}(\vec{v},\vec{v}') = v_{VC}^{-1}\left[x_a v_{VC}^{(a)} f_{\gamma_{\tilde{v}}^{(a)}}(\vec{v},\vec{v}') + (1-x_a) v_{VC}^{(b)} f_{\gamma_{\tilde{v}}^{(b)}}(\vec{v},\vec{v}')\right], \qquad (III.103)$$

with the concomitant modification in the expression of the sdKSP model through Eq. (III.93).

The VC/D *statistical correlation*, through VCD collisions, can be easily taken into account by conveniently modifying the kinetic equation.[273] The formal expression for the (fully correlated) sdKS$_C$P is identical to that of the sdKSP [Eq. (III.102)]. The only change concerns the M matrix, $\left(M_{nn'}^{(\ell,\ell)}\right)_C$ having to be substituted to $M_{nn'}^{(\ell,\ell)}$ with:

$$\left(M_{nn'}^{(\ell,\ell)}\right)_C = [v_{VC}(1-\gamma_{\tilde{v}}^{2n+\ell})\delta_{nn'}] + \gamma_{\tilde{v}}^{2n'+\ell}[\Gamma_{nn'}^\ell + i\Delta_{nn'}^\ell] \qquad (III.104)$$

It is possible to introduce, as for the asdRS$_{\tilde{C}}$P in § III.4.3, a *partial* correlation through (VCD + VC) collisions by substituting $\tilde{v}_{VC}(v)$ to v_{VC} in Eq. (III.93), with [*cf* Eq. (III.82)]:

$$\tilde{v}_{VC}(v) = v_{VC} - \eta[\Gamma(v) + i\Delta(v)], \qquad (III.105)$$

where η is the fraction of VCD collisions among the (VCD + VC) *kinetic* ones. The VC/D correlation parameter η is such that, putting $\eta = 0$ in Eq. (III.105) leaves unchanged Eq. (III.93) and, for $\eta = 1$, one gets the fully correlated kinetic equation. The formal expression for the *partially correlated* speed-dependent Keilson-Storer

Isolated lines 119

profile (sdKS$_{\widetilde{C}}$P) is the same as for the sdKSP [Eq. (III.102)], the *only* change being the substitution of $\left(M_{nn'}^{(\ell,\ell)}\right)_{\widetilde{C}}$ to $\left(M_{nn'}^{(\ell,\ell)}\right)_{C}$, with:

$$\left(M_{nn'}^{(\ell,\ell)}\right)_{\widetilde{C}} = [\nu_{VC}(1-\gamma_{\widetilde{v}}^{2n+\ell})\delta_{nn'}] + [1-\eta(1-\gamma_{\widetilde{v}}^{2n'+\ell})] \times (\Gamma_{nn'}^{\ell} + i\Delta_{nn'}^{\ell}), 0 \leq \eta \leq 1. \tag{III.106}$$

For $\eta = 0$ and $\eta = 1$, Eqs. (III.99) and (III.104) are consistently obtained. Like the asdRS$_{\widetilde{C}}$P, the sdKS$_{\widetilde{C}}$P is characterized by five collisional parameters [$\nu_{VC}, \Gamma, \Delta, \gamma_{\widetilde{v}}$ (instead of ε, more precisely (1-ε)), and η]. Its non-analytical expression [Eqs. (III.102) and (III.106)] makes its use more tedious that the asdRS$_{\widetilde{C}}$P. We should mention that Bonamy *et al*[273] have proposed an alternative efficient method avoiding the diagonalization procedure in Eq. (III.102). Due to the tridiagonal structure of the M matrix [*cf* Eq. (III.98)], it is possible to use a forward and backward substitution method to get, by *recurrence*, the $[M_{\widetilde{C}}^{-1}]^{(\ell,0)}$ block matrices for ℓ going from $\ell_{\max}$ to zero as required to numerically solve Eq. (III.101) and to get the sdKS$_{\widetilde{C}}$P. It should be noted that, as for the asdRS$_{\widetilde{C}}$P, an eventual correction accounting for the collision-duration (through a dispersive factor) could be introduced, if necessary [*cf* Eq. (III.87)].

An important remark must be made concerning the choice of the velocity-changing rate ν_{VC} in Eq. (III.93). Using Eq. (III.25) for ν_{VC} (*ie* ν_{Diff}) would lead asymptotically to $\nu_{VC}\delta(\vec{v}-\vec{v}')$ for the collision kernel [Eq. (III.89)] in the $\gamma_{\widetilde{v}} = 1$ limit. This means that no Dicke narrowing effects subsist in this limit. As pointed out in Ref. 237, this restriction can be removed by fixing the diffusion coefficient D. In the KS model[187] $D = k_B T/m\nu_{VC}(1-\gamma_{\widetilde{v}})$, so that fixing D implies to *substitute* $\nu_{Diff}/(1-\gamma_{\widetilde{v}})$ to ν_{VC}. Thus, for $\gamma_{\widetilde{v}} \to 1$, ν_{VC} approaches infinity while, simultaneously, the velocity transferred, in average per collision, goes to zero [*cf* Eq. (III.90)]. This captures the subtleties of the soft collision model for $\Lambda = m_p/m \to 0$ and the sdKSP then reduces to the sdGP. In the collisional regime, where only (1D) speed inhomogeneous effects occur, the corresponding pertinent choice becomes $\nu_{sC}/(1-\gamma_v)$, *ie* ν_{Diff} (see Ref. 228). Let us mention that, in Fig. III.20, the line shape was calculated with ν_{Diff} in 1DsdKSP, leading to large inhomogeneous features for $\gamma_v \to 1$ consistent with experimental data.[270]

The (uncorrelated) sdKSP model was studied by Shapiro *et al*[237] in order to analyze the combined effects of the speed inhomogeneous and Doppler distributions through careful comparisons with sdRSP results. Let us recall that the sdKSP has the same limits for $\gamma_{\widetilde{v}} = 0$ and $\gamma_{\widetilde{v}} \to 1$ [for $\nu_{VC} = \nu_{Diff}/(1-\gamma_{\widetilde{v}})$, *cf supra*] as the (non-approximate) sdRSP for $\varepsilon = 1$ and $\varepsilon = 0$, respectively. For intermediate values of the velocity memory parameter $\gamma_{\widetilde{v}}$, the differences between these two models of profile were calculated with the quadratic approximation [Eq. (III.44)] for the speed dependence of the line broadening which enables to get an exact analytical solution for the sdRSP (*cf* § III.4.2).[248] The shift was disregarded, since often much smaller than the broadening (except for light radiators). The numerical values of the collisional parameters [$\nu_{Diff}, \Gamma, a_w, \gamma_{\widetilde{v}}$, and ε(consistently taken equal to $1-\gamma_{\widetilde{v}}$)] were chosen in order to get as well an important influence of both sources of inhomogeneity, as significant distinct effects for hard and soft collisions. The expected role of the speed effects is well observed (Fig. III.21). The most important is to note that, for this typical set of parameters, the sdKSP and sdRSP are close to each other. Their signatures lie between the sdGP and the sdNGP, accordingly

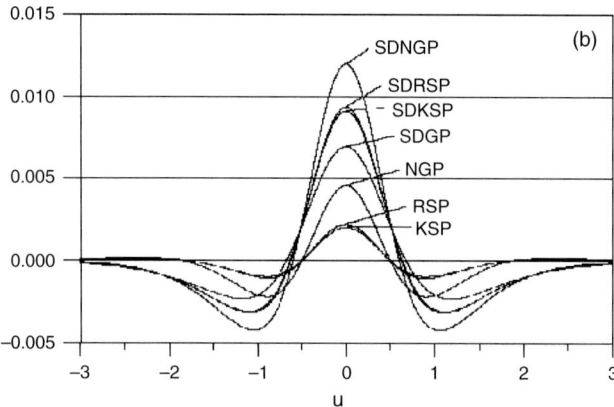

Fig. III.21: Differences between various profiles vs $u=(\omega-\omega_{fi})/\Delta\omega_D$ and the reference soft Galatry profile (GP) for the typical set of parameters $y=0.7$, $a_w=0.1$, $s=0$, $\varsigma_\varsigma=0.8$, $\varepsilon=0.5$. (SD must be read here sd). Reproduced with permission from Ref. 237.

with the choice $\gamma_{\vec{v}}=\varepsilon=0.5$. These results show that the approximation inherent in the sdRSP (ie the linear combination of hard and soft collisions) is not critical.

A particular attention was also paid[237] to the key role played, in the collisional regime, by the relative magnitude of the velocity-changing rate at high pressure with respect to the rovibrational rate. In this regime where the influence of the Doppler-shift term is negligible, only the speed-changing rate v_{sC} is thus pertinent.[221,225] If $v_{sC}\ll|\Gamma+i\Delta|$, the speed relaxes slowly and each speed group has its own Lorentzian profile, the exchanges between the speed groups being not efficient. The sdKSP and sdRSP tend toward a weighted sum of Lorentzians (WSLP). In the opposite case $v_{sC}\gg|\Gamma+i\Delta|$, the speed relaxes much more quickly than the rovibrational states and the corresponding limit is a Lorentzian profile with the average values $\Gamma=\langle\Gamma(v)\rangle$ and $\Delta=\langle\Delta(v)\rangle$ for the line broadening and shift. As seen above (§ III.3.7 and III.4.1), the Raman Q(J) lines of pure H_2, and the infrared R(J) lines of CO highly diluted in He are examples of the latter physical situation. The former one is illustrated, to some extent, by Q(J) lines of H_2 diluted in Ar or N_2. In the intermediate pressure range, between Doppler and collisional regimes, two mechanisms have to be considered simultaneously. These are the *speed memory* which governs the exchanges inside the speed inhomogeneous distribution, and the *velocity memory* through the collisionally induced transfers inside the Doppler distribution. The sdKS$_{\widetilde{C}}$P, as the sdRS$_{\widetilde{C}}$P, does not discriminate a priori between velocity and speed. This implicitly assumes that these two memory mechanisms are *not* too different. If the VC and sC rates are significantly distinct, a problem arises in the consistent description of the two sources of inhomogeneity. It has been shown in § III.4.3 that the asdRS$_{\widetilde{C}}$P phenomenologically accounts for possible distinct rates through the specific role of hard and soft collisions in frequency migration. The approximate treatment of the speed-dependent soft collisions restricts their induced frequency migration within the Doppler distribution, the speed exchanges being exclusively controlled by the hard collisions. Hence, their relative importance, through the "hardness" parameter ε also indirectly monitors the distinction between velocity and speed exchanges. In the sdKS$_{\widetilde{C}}$P model, all possible types of collisions are allowed in terms of a velocity memory mechanism, controlled by the parameter $\gamma_{\vec{v}}$ [which plays a role similar to the softness parameter $(1-\varepsilon)$ in the sdRS$_{\widetilde{C}}$P but without the linearization procedure, cf Eq. (III.65)].

When the *correlation times* for the speed memory and velocity memory are *very different*, Joubert et al[257] have introduced an approximate treatment of the KS memory function [Eq. (III.89)], allowing a *clear discrimination between velocity-orientation and speed*, through the following *alternative* choice for the velocity memory function:

$$f_{\gamma_{\vec{v}^o},\gamma_v}(\vec{v},\vec{v}') = f_{\gamma_{\vec{v}^o}}(\vec{v}^o,\vec{v}'^o) \times f_{\gamma_v}(v,v'), \tag{III.107}$$

where $\gamma_{\vec{v}^o}$ is the velocity-orientation memory parameter. In Eq. (III.107), the KS speed memory function is given by Eq. (III.91) and $f_{\gamma_{\vec{v}^o}}(\vec{v}^o,\vec{v}'^o)$ is obtained by averaging the KS velocity memory function [Eq. (III.89)] over the speed. Their expressions, on the basis of Legendre and Laguerre polynomials, are:

$$f_{\gamma_{\vec{v}^o}}(\vec{v}^o,\vec{v}'^o) = \frac{1}{2\pi}\sum_{\ell=0}^{\infty}\frac{2\ell+1}{2}\gamma_{\vec{v}^o}^{\ell} P_{\ell}(\vec{v}^o \cdot \vec{v}'^o), \tag{III.108}$$

and

$$f_{\gamma_v}(x,x') = \bar{f}_M(x) \times \sum_{n=0}^{\infty}\gamma_v^{2n}\bar{L}_n^{1/2}(x)\bar{L}_n^{1/2}(x'). \tag{III.109}$$

This *decoupling* approximation for the KS velocity memory [Eq. (III.107)] is *only valid* when the time autocorrelation functions tied to the radiator velocity and to the radiator speed exhibit very *distinct* time scales so that the speed memory quasi persists during the loss of the velocity memory. This is typically the case for a light radiator with heavy perturbers, such as $H_2 - X$ mixtures ($X = Ar, N_2, Xe$). This clearly appears in Fig. III.22 where values from molecular dynamics simulations (MDS) using realistic (isotropic plus anisotropic) intermolecular potentials[257] are reported for various perturbers as a function of the mass ratio $\Lambda = m_p/m$. Aside from the crucial role of the relative mass in the velocity- and speed-memory mechanisms, it is interesting to note the significant shortening of the speed memory (but not of the velocity memory) for H_2-N_2 collisions through the anisotropic part of the potential. This is a direct consequence of the translation-rotation energy transfers.

The velocity memory in Eq. (III.107) is controlled by *two* parameters [$\gamma_{\vec{v}^o}$ for the velocity-orientation and γ_v for the velocity-modulus (or speed)], instead of *one* ($\gamma_{\vec{v}}$) for the KS model [Eq. (III.89)]. The leading [*biparametric* (b)] model of profile, denoted (b)sdKS$_{\widetilde{C}}$P, is straightforwardly deduced from the (monoparametric) sdKS$_{\widetilde{C}}$P through Eqs. (III.107) to (III.109) instead of Eq. (III.95) *via* (III.93) and (III.94). Equation (III.102) for the profile is unchanged. The *only* modification is the substitution in the $\left(M_{nn'}^{(\ell,\ell)}\right)_{\widetilde{C}}$ matrix elements [Eq. (III.106)] of the product $\gamma_v^{2n}\gamma_{\vec{v}^o}^{\ell}$ to $\gamma_{\vec{v}}^{2n+\ell}$ (and also $\gamma_v^{2n'}\gamma_{\vec{v}^o}^{\ell}$ to $\gamma_{\vec{v}}^{2n'+\ell}$). Aside the distinction between velocity-modulus and velocity-orientation introduced in the (b)sdKS$_{\widetilde{C}}$P through Eq. (III.107) for the KS memory function when correlation times significantly differ, a suitable choice of v_{VC} has to be made. The choice $v_{VC} = v_{Diff}/(1-\gamma_{\vec{v}})$ ensures[237] that the soft

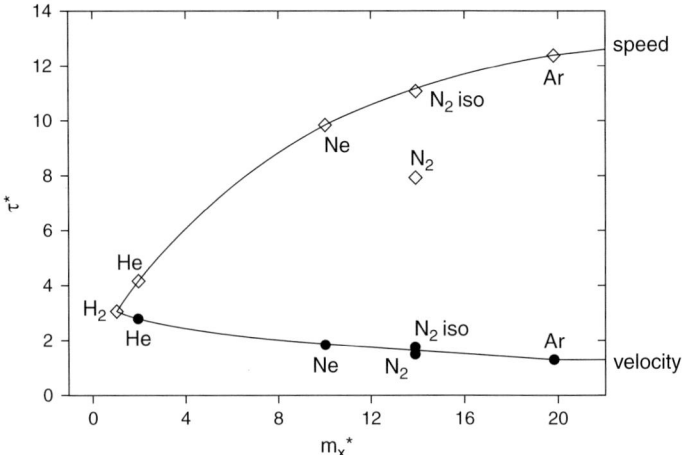

Fig. III.22: Reduced correlation times [see Eq. (III.110)] vs the relative perturber mass $\Lambda = m_p/m$ ($\equiv m_X^*$) for velocity $\tau_{\bar{v}}^* = \tau_{\bar{v}}/\tau_0$ (•) and speed $\tau_v^* = \tau_v/\tau_0$ (◇) for various infinitely diluted H_2–X mixtures (τ_0 being the average free flight time of non-attracting rigid spheres traveling at the mean relative speed $<v_r>$). "iso" for N_2 means that the potential used in the MDS is restricted to its isotropic part. Reproduced with permission from Ref. 257.

collision limit is obtained for heavy radiators with light perturbers. In the present opposite situation ($\Lambda = m_p/m \gg 1$), such a requirement is no longer pertinent. It is the correct description of the large speed inhomogeneous features, observed alone in the collisional regime,[221] which must be privileged. Thus, according to the above discussion concerning the case of the 1DsdKSP, the suitable choise for v_{VC} is then v_{Diff}.

An example of application of the (b)sdKSP (or equivalently the (b)sdKS$_{\tilde{C}}$P with $\eta = 0$, ie uncorrelated) is shown in Fig. III.23 for a H_2–N_2 mixture whose distinct velocity- and speed-memories (Fig. III.22) justify the use of the biparametric model. Note that the values for the memory parameters ($\gamma_{\bar{v}^0} = 0.41$ and $\gamma_v = 0.92$) were calculated from MDS.[228,257] Moreover, as explained in § III.4.3, the speed dependence of the collisional line broadening $\Gamma(v)$ and shift $\Delta(v)$ were directly deduced from accurate SRS experiments at high density, and v_{Diff} was calculated from the mass diffusion coefficient [cf Eq. (III.25)]. Hence, no fitting procedure was used to perform the calculation of the (b)sdKSP for H_2–N_2. Figure III.23 thus illustrates a successful test of the (b)sdKSP for the whole range of pressure (ie from the Doppler to the collisional regime).

A further important test concerns the asymmetry of the line shape. Fig. III.24 displays the observed variation of the asymmetry in a H_2–Ar mixture, for the whole range of pressure.[254,255] The calculated asymmetry A [Eq. (III.1)] from the (b)sdKSP model (without any fitting parameter) is fully consistent with experimental data, in both the (high pressure) collisional regime and the Dicke narrowing pressure region. The *two* sources of inhomogeneity are thus accurately described by this model of profile. Notice that the occurrence of a maximum for A proves the presence of *both* Doppler and speed inhomogeneous distributions in the Dicke minimum region (cf Fig. III.23). The progressive vanishing of the Doppler contribution with increasing pressure leads to an

Isolated lines

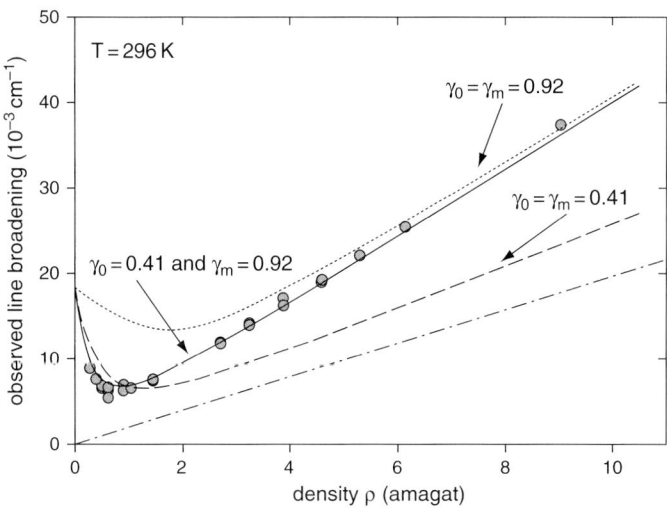

Fig. III.23: Comparison of the HWHM calculated from the (b)sdKSP model for the Q(1) line of 5% H_2 in N_2 with experimental data[20,224] (o). The two memory parameters have MDS values ($\gamma_{\tilde{v}^0} = 0.41 = \gamma_0$ and $\gamma_v = 0.92 = \gamma_m$).[257] The homogeneous collisional broadening is also reported (lowest dotted-dashed line). Reproduced with permission from Ref. 273.

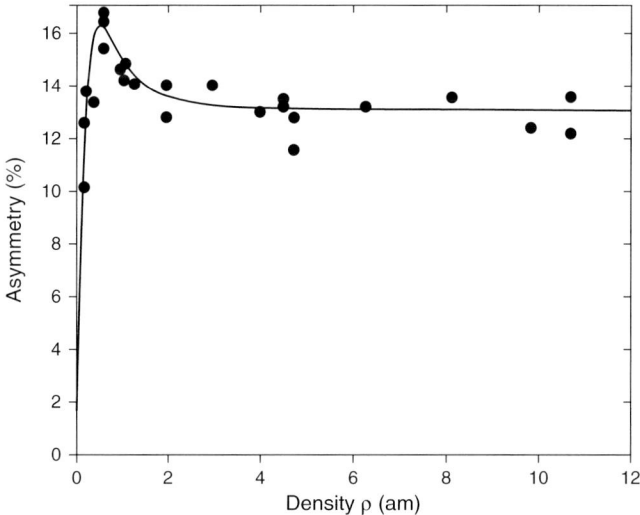

Fig. III.24: Comparison of the calculated asymmetry A [see Eq. (III.1)] from the (b)sdKSP model for the Q(1) line of 5% H_2 in Ar with experimental data[254,255] (•). The two memory parameters have MDS values ($\gamma_{\tilde{v}^0} = 0.21$ and $\gamma_v = 0.96$).[257] Reproduced with permission from Ref. 273.

asymptotic behavior in the hydrodynamic limit. In the opposite limit of very low pressure (*ie* in the Doppler regime), the expected value A = 0 tied to the (symmetric) Gaussian profile [Eq. (III.4)] is obtained.

It must be emphasized that the validity of the (b)sdKS$_{\widetilde{C}}$P model is tied to that of the decoupling approximation [Eq. (III.108)]. So, for instance, if the role of H$_2$–H$_2$ collisions becomes significant in the spectral line shape (at higher H$_2$ mole fractions), these collisions must be described by the (monoparametric) sdKS$_{\widetilde{C}}$P model since pure H$_2$ has then the same velocity and speed memory (*cf* Fig. III.22). This illustrates the precautions required when using the (b)sdKS$_{\widetilde{C}}$P model.

Let us mention that a fitting procedure can be used for practical applications (not exploited in the examples of Figs. III.23 and III.24), through the opportunity of the *partial* correlation [*cf* Eq. (III.106)]. Recall that the five collisional parameters characterizing the sdKS$_{\widetilde{C}}$P and the asdRS$_{\widetilde{C}}$P are the same in nature, except for the velocity memory parameter [$\gamma_{\widetilde{v}}$ replacing the "softness" one $(1-\varepsilon)$]. The (b)sdKS$_{\widetilde{C}}$P needs an additional parameter [$\gamma_{\widetilde{v}}$ being replaced by $\gamma_{\widetilde{v}^o}$ and γ_v, *cf* Eq. (III.107)], but it should be noticed that these memory parameters can be calculated from MDS data. Indeed, $\gamma_{\widetilde{v}^o}$ and γ_v are related to the velocity and speed correlation times $\tau_{\widetilde{v}}$ and τ_v, within the decoupling approximation [Eq. (III.107) to (III.109)], by:[257]

$$\tau_{\widetilde{v}} = \frac{\tau_0}{2}\left[\frac{1}{1-\gamma_{\widetilde{v}^o}} + \frac{1}{1-\gamma_{\widetilde{v}^o}\gamma_v^2}\right], \tau_v = \frac{\tau_0}{1-\gamma_v^2}, \quad \text{(III.110)}$$

τ_0 being the rigid sphere free flight time at $v = <v_r>$. Thus, values for γ_v and $\gamma_{\widetilde{v}^o}$ can be directly obtained from MDS predictions of $\tau_{\widetilde{v}}$ and τ_v, as results from exponential fits of the velocity and (centered) normalized speed autocorrelation functions.[228,257]

Based on the same phenomenological concepts of hard and soft collisions as the (analytical) asdRS$_{\widetilde{C}}$P, the sdKS$_{\widetilde{C}}$P is a *more rigorous* approach thanks to the introduction of the KS velocity memory in place of a *linear* combination of hard and soft collisions. Furthermore, the characteristic parameters of this memory function (including both velocity-orientation and speed considered as a whole, through $\gamma_{\widetilde{v}}$, or *separately*, through $\gamma_{\widetilde{v}^o}$ and γ_v, if their memories exhibit *different* time scales) have a direct physical meaning and can be calculated from MDS data. The price to be paid is the loss of the analytical character of the line-shape expression. For specific systems, the ability of the sdKS$_{\widetilde{C}}$P to accurately describe the spectral features in the entire range of pressure (from the Doppler up to the collisional regime) was already proved.[257,273,275] The possibility of partial VC/D correlations [Eq. (III.106)], increases its flexibility for complex physical situations occurring in many applications. Its usefulness for more intensive applications in IR and Raman molecular spectroscopies will depend on the development of efficient numerical codes (*cf* App. III.A).

Before closing this section, an illustrative example obtained from the interesting alternative of time-resolved Raman spectroscopy (see App. II.A4) is presented. In the CARS experiments with H$_2$ as optically active molecule,[109,276] two femtosecond pulses excite the Q(J) branch resonance for the v = 1 ← v = 0 vibrational transition. After a time delay τ, a probe pulse generates an anti-Stokes signal, which can be expressed, within the frame of the (3D)sdKS$_{\widetilde{C}}$P model, as:[105,273]

$$I_{\text{sdKS}_{\widetilde{C}}}^{\text{CARS}}(\tau) = \left|\sum_J f_J \exp[(i\Delta E_J/\hbar)\tau] \times \sum_k \sum_q S_{00,kq}^J D_{kq}^J (S^{J-1})_{kq,00}\right|^2. \quad \text{(III.111)}$$

Isolated lines

Here $\Delta E_J = E(v=1,J) - E(v=0,J)$, D^J_{kq} and $S^J_{00,kq}$ are the eigenvalues and eigenvector components (for the J-transition) tied to the M matrix [cf Eqs. (III.98) to (III.106)], and the oscillation amplitude factors f_J are defined in App. II.A4. The sum over J is extended to all significant J-transitions of the Q-branch. In the high density range, Tran et al[109] have tested the (uncorrelated) 1DsdKS model for H_2-N_2 mixtures at temperatures between 300 and 900 K by using the time-resolved CARS technique. Since, at the studied pressures, the Doppler contribution is negligible, the second index q in Eq. (III.111) may be suppressed (the leading index k being associated with the 1D speed). Figure III.25 displays a comparison of $I^{CARS}_{1DsdKS}(\tau)$ calculated from Eq. (III.111) with the experimental CARS signal. The observed good agreement offers the

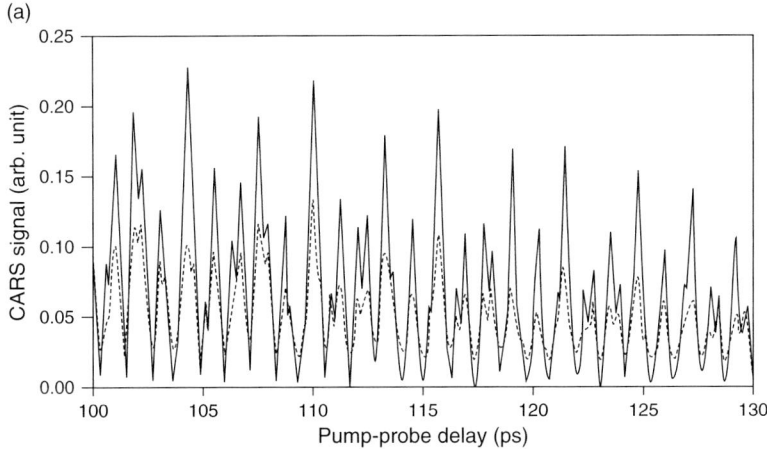

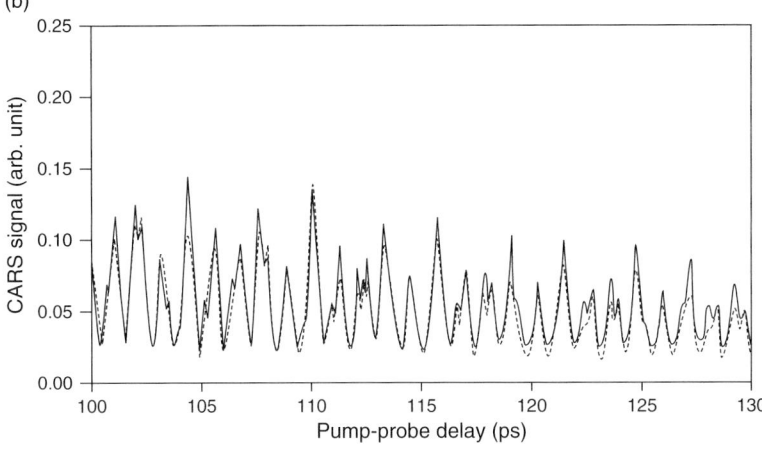

Fig. III.25: CARS signal of 20% H_2 in 80% N_2 at 900 K and 9.25 Am. The dotted line represents the experimental data. The full line represents the signal obtained with a homogeneous Lorentzian model (a) and the 1DsdKS model (b) with the memory parameter $\gamma_v = 0.93$. Reproduced with permission from Ref. 109.

opportunity of accurate temperature diagnostics from this technique, even in the presence of strong speed inhomogeneous features (*cf* § VII.3.2). The use of the more general (3D)sdKS$_{\widetilde{C}}$P model would allow to extend its applicability to lower densities down to the Doppler regime, with the additional flexibility of the partial correlation ($\widetilde{C}$) characterized by the parameter η in Eq. (III.106).

Aside from the phenomenological models asdRS$_{\widetilde{C}}$P and sdKS$_{\widetilde{C}}$P, several attempts were made to derive ab initio spectral line-shape theories permitting the calculation of the profile *directly* from intermolecular potentials. Such complete calculations should be feasible in a near future, not only limited to simple molecular systems, due to the increasing capacity of computers. The next section is devoted to these attempts.

III.5 AB INITIO APPROACHES OF THE LINE SHAPE

From a fundamental point of view, it should be noted that several different semi-empirical models can fit a measured line shape equally well, leading to possible conflicting physical interpretations. To overcome such difficulties essentially inherent in the phenomenological nature of such models, a true calculation should be free of any adjustable line-shape parameters. Up to now, only few attempts were done to *effectively* calculate *ab initio* line shapes, but this field is of interest for future progresses.

III.5.1 THE WALDMANN-SNIDER KINETIC EQUATION

The quantum mechanical theory of spectral shapes is most conveniently expressed in terms of the elegant Liouville space (see Apps. II.D and II.E). This approach is based on a density expansion of the line-shape function I(ω) for low up to moderate density regimes. An alternative point of view may be adopted by using the kinetic theory formalism (*cf* Sec. II.5). The spectrum is then related to the time-dependent density matrix through a Fourier transform in space and a (imaginary) Laplace transform in time as:[277,278]

$$I(\omega) = \frac{1}{\pi} \text{Re} \left\{ \int_0^\infty dt \int d^3\vec{r} \int d^3\vec{v} e^{i(\omega t - \vec{k}\cdot\vec{r})} \text{Tr}[T^+ \rho(\vec{r}, \vec{v}, t)] \right\}, \qquad (III.112)$$

where $\rho(\vec{r}, \vec{v}, t)$ is the *distribution* function (in $\vec{r}$ and $\vec{v}$) and *density* operator (in the internal states) at time t. The symbol Tr[...] denotes the trace over all quantum states of the radiator and perturbers (*cf* § II.2.1), and T is an orthonormalized spherical tensor defining the radiator "state multipole" (T^+ being the adjoint of T). For the state multipole of a radiative transition one has:

$$T_{fi}^{(k)q} \equiv i^k (2k+1)^{1/2} \sum_{m_f, m_i} (-1)^{J_f - m_f} \begin{pmatrix} J_f & k & J_i \\ -m_f & q & m_i \end{pmatrix} |\beta_f J_f m_f\rangle \langle \beta_i J_i m_i|, \qquad (III.113)$$

where fi is a shorthand denoting the transition $\beta_f J_f \leftarrow \beta_i J_i$ (J_i and J_f are the initial and final rotational quantum numbers and β the quantum numbers of all the other internal degrees of freedom). (:::) means the Wigner 3J symbol in the convention of Edmonds.[65] k and q are respectively the polarization quantum number (characterizing the considered spectroscopic line profile) and its magnetic projection (k = 0 for isotropic Q-lines, k = 1 for dipolar transitions, k = 2 for anisotropic Raman lines, cf App. II.A).

Within the *impact* approximation, Waldmann[279] and Snider[280] (WS) have generalized the Boltzmann equation treating quantum mechanically the collision operator. The WS equation was extended[280–282] to take into account degenerate internal states and to include the non-diagonal terms of the distribution function density matrix. The resulting line-shape expression for a spectral transition f ← i of order k is:[278]

$$I_{fi}^{(k)q}(\omega) = \frac{1}{\pi} \text{Re}\left[\int_0^\infty dt \int d^3\vec{r} \int d^3\vec{v} e^{i(\omega t - \vec{k}\cdot\vec{r})} \rho_{fi}^{(k)q}(\vec{r}, \vec{v}, t)\right], \quad \text{(III.114)}$$

with

$$\rho_{fi}^{(k)q}(\vec{r}, \vec{v}, t) = \sum_{\beta,J,m} \langle \beta Jm | T_{fi}^{(k)q} \rho(\vec{r}, \vec{v}, t) | \beta Jm \rangle. \quad \text{(III.115)}$$

Assuming that the radiator is not too far from equilibrium, one can linearize the density operator, *ie*:

$$\rho(\vec{r}, \vec{v}, t) = \rho_{th}(\vec{r}, \vec{v}, t)[1 + \Phi(\vec{r}, \vec{v}, t)]. \quad \text{(III.116)}$$

The subscript th means the thermal equilibrium value. The deviation Φ of the density operator from its equilibrium is solution of the *linearized* WS kinetic equation for an optically active molecule infinitely diluted in a thermal bath of perturbers:[278]

$$\left[\frac{\partial}{\partial t} + \vec{v}\cdot\vec{\nabla} + L_a + R\right]\Phi = 0. \quad \text{(III.117)}$$

L_a is the unperturbed Liouville operator for the internal states of the radiator ($L_a = i\hbar^{-1}[H_a,]$ where H_a is the corresponding (unperturbed) Hamiltonian and [,] means the commutator). R is the linearized WS *collision operator* in the Liouville space, involving the (binary) interaction V_{a-p} between radiator and perturber (through $L_V = i\hbar^{-1}[V_{a-p},]$). The Fourier-Laplace transform [Eq. (III.114)] of Eq. (III.117) leads to the matrix equation:

$$[(-i\omega + i\vec{k}\cdot\vec{v})I_d + L_a + R]\widetilde{\Phi} = 1_0, \quad \text{(III.118)}$$

where 1_0 is a generalized identity operator such that:

$$\widetilde{\Phi}_{fi}^{(k)q} 1_0 \equiv \widetilde{\Phi}_{fi}^{(k)q} \delta_{f,f_0} \delta_{i,i_0} \delta_{k,k_0} \delta_{q,q_0}, \quad \text{(III.119)}$$

(f_0, i_0, k_0, q_0) characterizing the initial radiative transition. $R\widetilde{\Phi}$ can be expressed[277,278] in terms of the transition matrix in the wave-vector space used to construct the relaxation matrix in the line broadening theory.[44] As done (*cf* App. II.E) for the W relaxation matrix elements in terms of S-matrix elements in the total J-representation, the rotational invariance of the WS collision operator (through V) can be similarly exploited. But an additional "*spherical*" approximation has to be introduced to save the scattering calculation methods, leading to a spectral line shape independent of the radiation propagation direction [cf Eq. (III.114)]. The "sphericalization" procedure forces the velocity portion of the collision operator to be *also* rotationally invariant. It assumes that the q-components of the state multipole are not coupled by collisions and introduces a pertinent average over these components.[184,277,278]

The main difficulty in solving Eq. (III.118) is due to the fact that the velocity dependence of the density operator is not well described by a truncated polynomial expansion at low pressure, since the $\vec{k}\cdot\vec{v}$ drift term is dominant. Therefore, ab initio calculations of $I_{fi}^{(k)}(\omega)$ use alternative methods. An approach was developed by Corey and McCourt,[184] Monchick and Hunter,[277] and Blackmore,[278] approximating the internal and translational relaxation processes in the WS operator by a two modes model, as previously proposed by Hess.[283] The Hess method was generalized to take line couplings into account through the above Liouville space representation of the linearized WS equation. This so-called Hess method (GHM) is now summarized.

III.5.2 THE GENERALIZED HESS METHOD

The GHM is an adaptation of the Bhatnagar-Gross-Krook method[284] which consists in replacing the Boltzmann equation by a finite number of relaxation terms following a suitable criterion (usually that the predicted transport properties be correct). Hess[283] retained a *two relaxation modes* model characterized by two *complex* relaxation frequencies ω_a and ω_r [defined below through Eqs. (III.123) and (III.124)], *optimized by a variational* calculation minimizing the error in the approximation. Such an approximation offers two main advantages with respect to the classical Chapman-Cowling method.[35,186] First, the GHM solution spans the whole space of solution of the original Boltzmann equation. Second, free flight and collision terms are treated on equal footing enabling solutions meaningful in the rarefied gas dynamic regime.

The GHM model can be applied to almost resonant overlapping lines.[277] The available tests[285–287] from experiments being restricted to *isolated* lines of diatomic molecules infinitely diluted in a monoatomic buffer gas, the line-shape expression is given here for this case only. This means that the "sphericalized" ω_a and ω_r frequencies are no longer tetradics [ie $(\omega_a \text{ or } \omega_r)_{f'i',fi}]^{277,288}$ but *scalars*[287] as in the original work of Hess.[283] Starting from the WS equation (III.118), the GHM equation is obtained by replacing the WS operator R by two complex frequencies:[287]

$$\left(-i\widetilde{\omega} + i\vec{k}\cdot\vec{v}\right)\widetilde{\Phi} + \omega_a <\widetilde{\Phi}> + \omega_r\left[\widetilde{\Phi} - <\widetilde{\Phi}>\right] = 1 + \Delta W(\widetilde{\Phi}), \qquad \text{(III.120)}$$

where $\widetilde{\omega} = \omega - \omega_{fi}$ is the detuning frequency, $<\widetilde{\Phi}>$ is the $\vec{v}$-averaged value of $\widetilde{\Phi}$, and $\Delta W(\widetilde{\Phi})$ is an error term which is *minimized* by requiring that:

$$\frac{1}{4\pi}\int d^2\vec{k}^o \int d^3\vec{v}(\vec{k}^o\cdot\vec{v})^y \left(\Delta W(\widetilde{\Phi})\right)_{fi}^{(k)q} = O(\vec{k}\cdot\vec{k}) \,\forall\, (i,f,k,q), y = 0, 1. \qquad \text{(III.121)}$$

Isolated lines

The resulting GHM profile [*cf* Eqs. (III.114), (III.120) and (III.121)] is given by:

$$I_{fi}^{(k)}(\tilde{\omega}) = \frac{1}{\pi} \text{Re}\left\{\omega_a - \omega_r + \left[\left\langle\left(-i\tilde{\omega} + i\vec{k}.\vec{v} + \omega_r\right)^{-1}\right\rangle\right]^{-1}\right\}^{-1} + O(\vec{k}.\vec{k}), \quad (III.122)$$

$\langle \cdots \rangle$ being the Maxwell-Boltzmann average over the radiator velocity [*cf* Eq. (III.2)]. The "sphericalization" of the ω_a and ω_r tensors was chosen to be an average over the radiation propagation direction, leading to:

$$\omega_a = n_p \omega_0^{00}(k), \quad \omega_r = n_p \tilde{m}_p \omega_1^{11}(k) + \tilde{m}\omega_a, \quad \tilde{m}_p = \frac{m_p}{\tilde{M}}, \quad \tilde{m} = \frac{m}{\tilde{M}}, \quad \tilde{M} = m_p + m, \quad (III.123)$$

with

$$\omega_\lambda^{ss'}(k) = \frac{4\pi^{1/2} \hbar^2}{m^{*3/2}(2k_B T)^{1/2}} \int_0^\infty \exp(-x^2) x^{s+s'+1} \sigma_\lambda^k(J_f J_i, x) dx, \quad (III.124)$$

where $x^2 = E_r/k_B T$, $E_r = (m^* v_r^2)/2$ being the relative kinetic energy. The collisional cross-sections $\sigma_\lambda^k(J_f J_i, x)$ are defined [Eq. (14) of Ref. 287] in terms of the S-matrix (at the $E_r + E_f$ and $E_r + E_i$ energies) through momentum recoupling algebra more complicated than for pressure broadening, except for $s = s' = \lambda = 0$. As a direct consequence, $\omega_0^{00}(k)$ [Eq. (III.124)] is *identical* to the pressure broadening coefficient given by Shafer and Gordon.[144] The line-shape expression in Eq. (III.122) is formally similar to the NGP [Eq. (III.31)], with nevertheless meaningful differences. The two relaxation frequencies ω_a and ω_r are *complex* and *explicitly* related to the S-matrix, allowing ab initio scattering calculations from potential energy surfaces. If the relaxation frequency $(\Gamma + i\Delta)$ plays a role similar to ω_a, this is not the case for the velocity-changing rate ν_{VC} with respect to ω_r. Indeed, ω_r [Eq. (III.123)] includes two contributions. The first one, which is essentially a momentum transfer cross-section, is inversely proportional to an effective diffusion coefficient (as ν_{VC}), but slightly different from the gas kinetic mass diffusion. The second contribution is proportional to the pressure broadening-shifting coefficient. Such a complex rate, intermixing translational and internal relaxations, was phenomenologically restored in the sdRS$_{\tilde{C}}$P and sdKS$_{\tilde{C}}$P models by introducing the (partial) correlation between velocity-changing and dephasing collisions (*cf* § III.4.3 and III.4.4).

Blackmore et al[285,286] performed calculations from Eqs. (III.122) to (III.124), for Q lines of the v = 1 ← 0 transition of D_2 infinitely diluted in He, at pressures where Dicke narrowing was observed.[210] An accurate ab initio potential energy surface[289] was used, leading to a consistent agreement of GHM calculations with experimental data for the position of the minimum of the Q-lines HWHM *vs* density (*cf* Fig. III.27 thereafter). Further tests were carried out for R and P lines of HF-Ar previously analyzed[171] (*cf* § III.3.7 and III.4.3). Exact close-coupled calculations of the collision cross-sections [Eq. (III.124)],[23] using an accurate interaction potential, were performed with

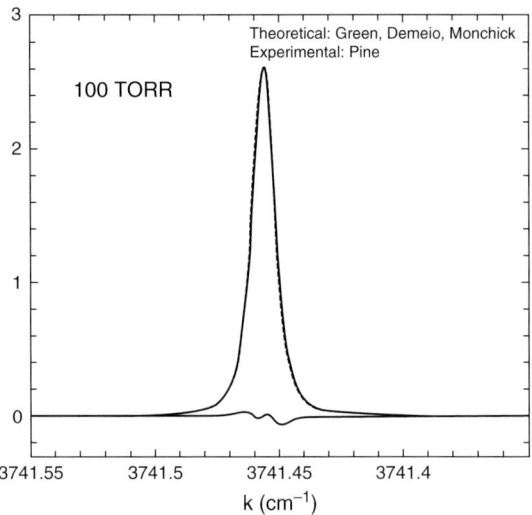

Fig. III.26: Calculated (solid line) and observed[213] (dashed line) profiles for the P(5) line of HF ($v = 1 \leftarrow 0$) highly diluted in argon at P = 0.1316 atm and T = 296 K. meas-calc deviations are also shown. Reproduced with permission from Ref. 287.

uncertainties lower than one percent. Direct comparison of GHM results with observed profiles showed a rather good consistency (Fig. III.26), even if discrepancies are significantly larger than those obtained from multifits of measurements (see Fig. III.19).

It must be emphasized that these GHM results are *completely* ab initio. Apart from remaining inaccuracies in the interaction potential, the discrepancies can be attributed to the approximations underlying the GHM model [a two relaxation modes model within the (linearized) WS kinetic frame]. But it should be recalled that it provides, according to Eq. (III.121), the *best* solution which can be written in terms of only *two* (complex) relaxation frequencies. Furthermore, this model conveniently reduces to a Lorentzian profile with the (single) pertinent collision relaxation rate at high pressure and to the suitable Doppler Gaussian profile at very low pressure. Its extension to finite concentration was done by Monchick,[290] who also generalized the WS equation beyond the impact approximation to give a consistent (formal) description of the far line wings. The Hess method could be improved by describing the collision operator with more than two relaxation frequencies. A more attractive approach consists in solving the Waldmann-Snider kinetic equation (III.118) numerically, as proposed by Blackmore[291] within the frame of the collision kernel method (CKM).

III.5.3 COLLISION KERNEL METHOD

The current method representing the collision operator as a matrix in a basis set of Sonine polynomials[292,293] is not useful to solve the line-shape problem, because of the resulting large number of collision integrals. Blackmore[291] developed a more convenient

Isolated lines 131

approach based on an *irreducible* representation of the collision operator. Its rotational invariance implies its commutation with the *total* angular momentum operator $\widetilde{J}$. So, it may be represented by uncoupled blocks, corresponding to a given $\widetilde{J}$ value, in the pertinent basis:

$$U^{\widetilde{J}M}_{J_fJ_ikL}(\vec{w}^o) = \sum_{q,s} T^{(k)q}_{J_fJ_i} Y_{Ls}(\vec{w}^o)(-1)^{k-L+M}(2\widetilde{J}+1)^{1/2} \begin{pmatrix} k & L & \widetilde{J} \\ q & s & -M \end{pmatrix}, \quad \text{(III.125)}$$

where $Y_{Ls}(\vec{w}^o)$ are spherical harmonics for the orientation $\vec{w}^o$ of the reduced radiator velocity $\vec{w}$, in the laboratory frame. The reduced velocity is given by $\vec{w} = (m/2k_BT)^{1/2}\vec{v}$ (concomitantly, $\vec{K} = (m/2k_BT)^{-1/2}\vec{k}$). Projecting the WS equation [Eq. (III.118)] on this basis gives:[285]

$$-i\widetilde{\omega}\widetilde{\Phi}^{\widetilde{J}M}_{J_fJ_ikL}(w) + iKw \sum_{\widetilde{J}'=|\widetilde{J}-1|}^{\widetilde{J}+1} A^{\widetilde{J}'}_{\widetilde{J}M} \widetilde{\Phi}^{\widetilde{J}'M}_{J_fJ_ikL}(w)$$

$$+ \sum_{J'_f,J'_i,k',L'} \int w'^2 K^{\widetilde{J}}(J_fJ_ikLw, J'_fJ'_ik'L'w')\widetilde{\Phi}^{\widetilde{J}M}_{J'_fJ'_ik'L'}(w')dw' = \delta_{i,i_0}\delta_{f,f_0}\delta_{q,q_0}\delta_{\widetilde{J},0}, \quad \text{(III.126)}$$

where the Fourier-Laplace $\widetilde{\Phi}^{\widetilde{J}M}_{J_fJ_ikL}$ components are defined by:

$$\widetilde{\Phi}^{(k)}_{J_fJ_i}(\vec{w}) = \sum_{\widetilde{J},M,L} \widetilde{\Phi}^{\widetilde{J}M}_{J_fJ_ikL}(w) U^{\widetilde{J}M}_{J_fJ_ikL}(\vec{w}^o). \quad \text{(III.127)}$$

The parameters $A^{\widetilde{J}'}_{\widetilde{J}M}$ are the expansion coefficients of the recoupling of $\vec{K}.\vec{w}$ and $U^{\widetilde{J}M}$ to form a new set of irreducible spherical tensors. The kernel $K^{\widetilde{J}}$ (independent of M by isotropy[65]) is obtained by integrating the matrix elements of the collision operator in the $U^{\widetilde{J}M}$ basis over the pre- and post-collision momenta of the radiator/perturber pair, except the scalar magnitudes p' = mv' and p = mv of the radiator. Then ten integrals were analytically reduced to a *two*-dimensional one by means of a suitable transformation of the momentum space. The resulting general expression for the kernel $K^{\widetilde{J}}$ is given by Eq. (42) of Ref. 291 for an optically active molecule infinitely diluted in a monoatomic heat bath and was extended to mixtures of linear molecules at finite concentration.[294] This expression can still be simplified by making convenient approximations. When the coupling between internal and translational modes is weak, only elastic collisions are significant. A further approximation is that the S-matrix depends separately on the orbital and rotational angular momenta (which is exact for scattering from a spherically symmetric potential). Thus, the $K^{\widetilde{J}}$ kernel [Eq. (III.126)] becomes, in this quantum elastic limit:[291]

$$K^{\widetilde{J}}(J_fJ_ikLw, J'_fJ'_ik'L'w') = K^{\widetilde{J}}(J_fJ_ikL, ww')\delta_{J_i,J'_i}\delta_{J_f,J'_f}\delta_{k,k'}\delta_{L,L'}. \quad \text{(III.128)}$$

This approximation is pertinent for *isolated* lines since their shapes are predominantly governed by elastic collisions since the coupling with other off-diagonal density matrix elements (*ie* f'i' ≠ fi, see chapter IV) may be neglected.

The main advantage of the above collision kernel method (CKM) to solve the WS kinetic equation (III.126) is that the set of collision operators (diagonal in $\tilde{J}$ and M) may be reduced to a set of *symmetric* bounded kernels which are functions of the reduced *speeds* w and w'. Thus, the CKM allows the use of *collocation numerical procedures*[295] to get accurate solutions, *even if* the drift and collision terms are of equal importance. These collocation procedures are based on a suitably chosen discrete set of points on which the density matrix is evaluated without large computational effort. The CKM was applied[285,286] to isolated Q-lines of D_2-He mixtures and compared as well to the experimental data[210] as to the GHM results obtained with the same accurate interaction potential.[289] In this application, the "sphericalization" was introduced through the approximation:

$$\widetilde{\Phi}^{(k)}_{J_f J_i} = \frac{1}{2k+1} \sum_q T^{(k)q}_{J_f J_i} \widetilde{\Phi}^{(k)q}_{J_f J_i}(\omega, \vec{K}, \vec{w}), \quad \text{with} \quad \text{(III.129)}$$

$$\widetilde{\Phi}^{(k)q}_{J_f J_i}(\omega, \vec{K}, \vec{w}) \equiv \sum_L \widetilde{\Phi}^{(k)}_{L,fi}(\omega, w)(L+1/2)^{1/2} P_L(\vec{K}^o \cdot \vec{w}^o), \quad \text{(III.130)}$$

where P_L are the Legendre polynomials.[182] Note that $\widetilde{\Phi}^{(k)}_{L,fi}$ is independent of q in this approximation. From Eqs. (III.126) to (III.130), the kinetic equation becomes:

$$-i\widetilde{\omega}\widetilde{\Phi}^{(k)}_{L,fi}(\omega,w) + iKw\Big\{(L+1)[4(L+1)^2-1]^{-1/2}\widetilde{\Phi}^{(k)}_{L+1,fi}(\omega,w)$$
$$+ L[4L^2-1]^{-1/2}\widetilde{\Phi}^{(k)}_{L-1,fi}(\omega,w)\Big\} \quad \text{(III.131)}$$
$$+ \int w'^2 K^{\tilde{J}}(J_f J_i kL, ww')\widetilde{\Phi}^{(k)}_{L,fi}(\omega,w')dw' = \delta_{i,i_0}\delta_{f,f_0}\delta_{\tilde{J},0}.$$

The kernel $K^{\tilde{J}}(J_f J_i kL, ww')$ is given in Ref. 291 [Eq. (57) with a correction factor $(2mk_B T)^{3/2}$ due to the use of the reduced speed in Eq. (III.131)].

The collocation numerical procedure used in Refs. 285,286 to solve Eq. (III.131) is to convert $(L_{max}+1)$ coupled integral equations to a $N \times (L_{max}+1)$ order matrix equation by evaluating $\widetilde{\Phi}^{(k)}_{L,fi}(\omega,w)$ and $K^J(J_f J_i kL, ww')$ at N pivot points w_i, w'_j (i,j = 1,...,N) appropriate to the weighting function $w^2 \exp(-w^2)$.[296,297] Except for the sparse adjacent blocks $L' = L \pm 1$, the matrix is concentrated in $N \times N$ blocks along the diagonal. Converged solutions were obtained for $L_{max} = 5$ and $N = 30$, leading to the inversion of 180×180 matrices, except at low density where the rate of convergence with L is less rapid requiring more tensor orders. The resulting line shape $I^{(k)}_{fi}(\omega)$ [Eqs. (III.131) and (III.114)–(III.116)] was ab initio calculated for isotropic (k = 0) Raman Q(J) lines of D_2 infinitely diluted in He, over the whole pressure range including the Dicke minimum region and the collisional regime at moderately high pressure. The CKM produced better results than the GHM as shown in Fig. III.27. In the Dicke narrowing pressure range, the agreement with experiment is excellent. At higher pressure the CKM curve (at infinite dilution) lies slightly above the measured data. Vibrationally inelastic collisions (not included in the calculation) could be the origin of this weak discrepancy.[285,286]

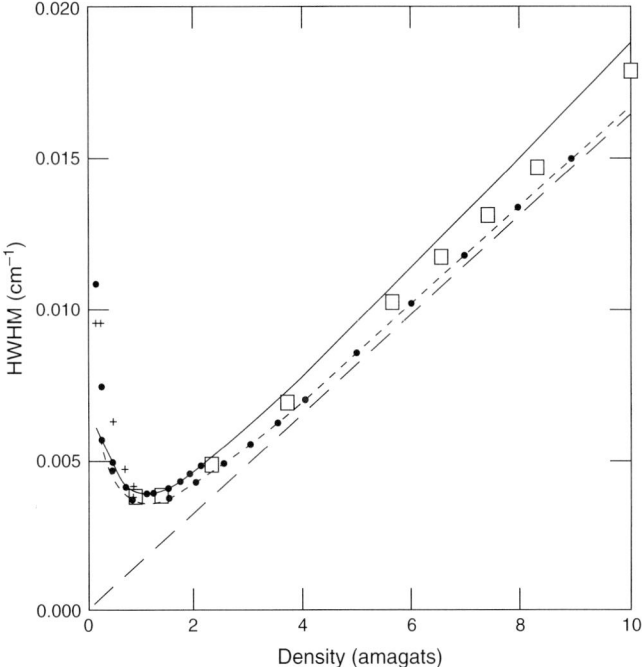

Fig. III.27: Half width (HWHM) *vs* density of the Q(2) line for 10% D_2 in 90% He at room temperature. □ are experimental values.[210] (—) CKM calculations; (•) CKM with $L_{max} = 9$ to control convergence. (- - -) GHM calculations; (– – –) and (*,+) correspond to various approximations of the GHM according to the considered density range (*ie* toward either high or low density). Reproduced with permission from Ref. 285.

Let us note that the CKM is *more* straightforward to apply than the GHM because it involves *less* tedious algebra and is *simpler* to extend to large order approximations (multiplying the number of pivot points by one or two orders of magnitude leaves the algorithms unchanged). Unfortunately, the *computer time* becomes the limitation. Therefore, though feasible in principle from Ref. 294, calculations at finite concentration for mixtures containing molecular perturbers are hardly tractable (and, to our knowledge, have not been performed till now). In order to overcome this limitation, but saving the spirit of ab initio line-shapes calculations, May and coworkers have chosen[222,298,299] to start from a simplified Waldmann-Snider kinetic equation.

III.5.4 APPROACHES FROM A SIMPLIFIED WALDMANN-SNIDER EQUATION

Ignoring the degeneracy of the internal level, the WS kinetic equation (III.118) reduces[222] to the Wang Chang-Uhlenbeck kinetic equation.[300] This equation captures all the structure of the WS one, but *neglects the polarization* features tied to the tensorial nature of the interactions of the radiator with both the probe and the perturbers. The quantum numbers k and q of the density operator matrix elements $\rho_{fi}^{(k)q}(\vec{r},\vec{v},t)$ in Eq. (III.114) can thus be omitted. Equation (III.118) remains formally unchanged, except that the initial condition is then restricted to $\delta_{ii_0}\delta_{ff_0}$ [Eq. (III.119)] and that the collision operator (noted hereafter $\bar{R}$) becomes *scalar*. This approximation restores the interest of the Sonine polynomial basis[292,293] to solve the set of complex kinetic equations since the numerous integrals tied to the spatial degeneracy disappear. As done for the Keilson-Storer model [*cf* Eq. (III.95)], let us introduce a complete set of basis functions $\varphi_{n\ell}(\vec{v})$, having the proper axial symmetry along the radial propagation direction $\vec{k}^o$:

$$\varphi_{n\ell}(\vec{v}) \equiv \varphi_{n\ell}(x,\vec{v}^o) = \bar{L}_n^{\ell+1/2}(x)P_\ell(\vec{k}^o \cdot \vec{v}^o), \qquad (III.132)$$

with $x = mv^2/2k_BT$. The matrix elements of $\bar{R}$ in this basis are given by (*cf* § III.4.4):

$$\langle n\ell|\bar{R}|n'\ell'\rangle = \int d^2\vec{v}^o \int dx \bar{f}_M(x)\bar{L}_n^{\ell+1/2}(x)P_\ell(\vec{k}^o \cdot \vec{v}^o)\bar{R}P_{\ell'}(\vec{k}^o \cdot \vec{v}^o)\bar{L}_{n'}^{\ell'+1/2}(x). \qquad (III.133)$$

The Doppler shift $i\vec{k}\cdot\vec{v}$ in Eq. (III.118) can be analytically expressed in this basis, leading to $n' = n, n+1, \ell' = \ell-1$ and $n' = n, n-1, \ell' = \ell+1$ non-vanishing matrix elements [Eq. (9) of Ref. 299]. Different approaches are then possible according to the type of description chosen for $\bar{R}$. Shapiro et al[298] considered the case of *uncorrelated* velocity-changing (VC) and dephasing (D) collisions, so that:

$$\bar{R} = \bar{R}_{VC} + \bar{R}_D, \qquad (III.134)$$

where $\bar{R}_{VC}$ is the velocity-changing collision operator and $\bar{R}_D$ is the dephasing operator, given by:

$$\bar{R}_D = \Gamma(v) + i\Delta(v). \qquad (III.135)$$

The matrix elements of the (scalar) operator $\bar{R}_{VC}$ in the $\{\varphi_{n\ell}(\vec{v})\}$ basis functions [Eq. (III.132)] were taken from a previous study[293] for *rigid sphere* (or billiard-ball, BB) gases:

$$\langle n\ell|\bar{R}_{VC}|n'\ell'\rangle = \nu_{Diff}f_D M_{n\ell,n'\ell'}^{E^*} \equiv \nu_{VC}f_D M_{n\ell,n'\ell'}^{E^*}, \qquad (III.136)$$

where ν_{Diff} is given by Eq. (III.25). The (real) coefficients $M_{n\ell,n'\ell'}^{E^*}$ were *analytically expressed* in terms of the *mass ratio* and f_D is a correction factor for the mass diffusion coefficient.[301] Its value, very close to unity (within a few percent), was easily deduced from $M_{n\ell,n'\ell'}^{E^*}$ [Eqs. (17) and (19) of Ref. 299].

This line-shape model, where the potential controlling the *translational* motion is taken as that for rigid spheres, was called the *speed-dependent billiard ball* profile (sdBBP). One of the main advantages of this model is to introduce, through the translational motion, the direct influence of the mass ratio $\Lambda = m_p/m$ on the profile. This effect was discussed by Rautian[302] in the soft, the hard and the Lorentz collision models. A first application[299] was to test the sdBBP with the main objective to analyze the influence of Λ on the sdBB profile.

Concomitantly, a reasonable (but not ab initio) modeling of the speed dependence of the line broadening was used, the shift Δ being neglected since, for the considered CO molecule, it is more than one order of magnitude smaller than Γ.[237] This speed dependence was chosen quadratic [Eq. (III.44)]. The matrix elements of $\bar{R}_D$ are then analytical (with only $n' = n, n \pm 1$, $\ell' = \ell$ elements). For all the characteristic parameters, appropriate values for the fundamental band of CO at room temperature were retained. It was preliminary checked that the (speed independent) BBP for $\Lambda \to 0$ well reduces to the (soft) GP [Eq. (III.26)], testing the excellent and rapid convergence for $n_{max} = \ell_{max} = 10$ (10^{-6} of the peak value and a few seconds on a common personal computer). At moderate pressure, when both $\bar{R}_{VC}$ and $\bar{R}_D$ play significant roles (ie for $z = v_{Diff}/\Delta\omega_D \approx 1$), departures from the $\Lambda = 0$ limit were clearly seen (Fig. III.28). These departures were strongly enhanced by the speed dependence of Γ [the relative value (sdBB-G)/G at the maximum of the line is about 2% for $\Lambda = 0$ and 4% for $\Lambda = 20$].

Numerical calculations of the sdBBP at various Λ values (for the same values of the other parameters as in Fig. III.28, except for y and z), were also performed in the hydrodynamic regime [ie for $z \gg 1$, when the Doppler term $i\vec{k}.\vec{v}$ may be dropped out from Eq. (III.118)]. Thus, as expected, the sdBBP approaches a weighted sum of Lorentzian (WSL) profiles at high Λ values. In such a physical situation, the radiator

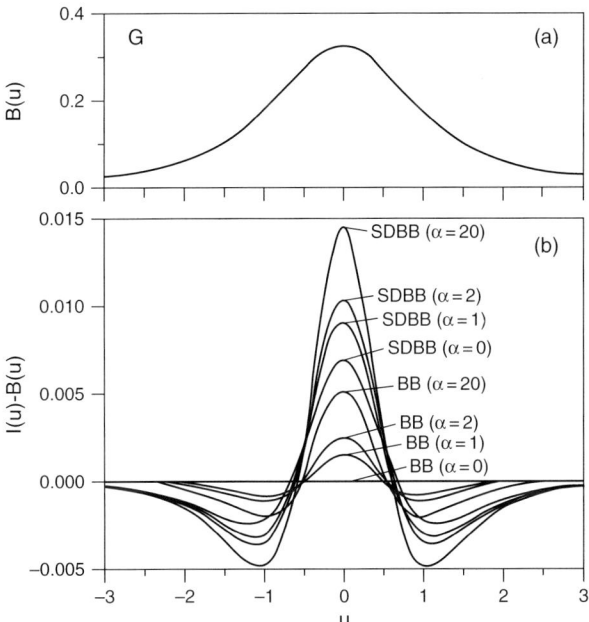

Fig. III.28: Comparison between speed independent (BB) and speed-dependent (SDBB) billiard-ball profiles for various $\alpha \equiv \Lambda = m_p/m$ values. The Galatry profile [Eq. (III.26)] is shown in the upper panel, and differences with respect to this profile are displayed in the lower panel. The parameters used in the calculations are: $v_{Diff}/\Gamma = 1.14$, $y = 0.7$, $z = 0.8$, and $a_w = 0.1$ [Eq. (III.44)]. Reproduced with permission from Ref. 299.

moves with almost constant speed and, since $v_{Diff} \approx \Gamma$, no important collisionally induced exchanges occur between the speed groups. The sdBBP was also compared with the sdNGP in the hydrodynamic regime [Eq. (III.49) without the Doppler term $\vec{k}.\vec{v}$], showing that these two profiles nearly coincide for $\Lambda \approx 1$.[299] This is in contradiction with the usual (but erroneous) picture of the hard collision (HC) model presumed to be suitable for large Λ values. In fact, in the high density collisional regime, only speed changing collisions are relevant (*cf* § III.4.1). So, since in the sdNGP, sC and VC collisions are governed by the *same* hard collisional memory process ($\vec{v}$ being considered as a whole), this requires that $m_p = m$.[228]

Ciurylo *et al*[249] introduced an "*apparent*" hardness parameter ε_{BB}, through specific matrix elements of the velocity-changing collision operator [*cf* Eq. (III.136)]:

$$\varepsilon_{BB} = |\langle 10|\bar{R}_{VC}|10\rangle|/|\langle 01|\bar{R}_{VC}|01\rangle| = M^{E^*}_{10,10}/M^{E^*}_{01,01} = 2/(1+\Lambda), \Lambda = m_p/m.$$

(III.137)

This definition is the ratio of two effective (first order) collision frequencies tied, at the denominator, to the velocity change, and, at the numerator, to the *modulus* of the velocity transferred per collision. For $0 \leq \varepsilon_{BB} \leq 1$, this numerator reduces to the more familiar (effective) speed-changing collision frequency v_{sC}, according to the excellent agreement obtained for ε_{BB} with molecular dynamics data for v_{sC}/v_{VC} (*cf* Table III.4). This allows an interpretation of the concept of "hardness" parameter in terms of the mass ratio $\Lambda = m_p/m$. The case $\varepsilon_{BB} = 1$, for $m_p = m$, is related to collisions having the *same* speed and velocity memory. The limit $\varepsilon_{BB} \to 0$, for $m_p \gg m$, holds for collisions *solely* changing the velocity-orientation, characteristic of the "Lorentz gas" profiles. For $1 < \varepsilon_{BB} \leq 2$, the numerator of Eq. (III.137) can no longer be identified to the (effective) sC collision frequency due to the necessity to take into account collisions inducing *velocity reversal* [*cf* Eq. (III.90) with $-1 \leq \gamma_{\vec{v}}, \varepsilon_{BB} = 2$ corresponding to $\gamma_{\vec{v}} = -1$].

The aim of the sdBB approach was[299] to carry out full ab initio calculations of the line shape, including a translation motion controlled by a BB potential and speed-dependent line broadening and shifting coefficients conveniently related to an accurate anisotropic intermolecular potential. Subsequent studies[165,303,304] were devoted to such an approach for isolated lines of CO highly diluted in Ar, for which accurate experimental profiles were available.[305] Quantum close-coupled calculations of the collision cross section $\sigma_\Gamma(v_r)$, as a function of the relative speed v_r, were performed[303] by using an ab initio potential energy surface.[306] The speed-dependent line broadening $\Gamma(v, T)$ was deduced from the procedure described in § III.4.2 [Eqs. (III.59)–(III.61)]. The first comparison between the sdBBP and experimental profiles concerned the P(2) line of CO at 301 K for nine pressures from 0.048 to 0.99 atm.[303] Significant discrepancies were observed at pressures lower than 0.2 atm, up to $\pm 10^{-2}$ of the peak value at 0.048 atm (such a residual was well above the noise level by about one order of magnitude). The possible sources for these discrepancies were carefully examined (the BB potential, the $\Gamma(v)$ calculation and the line mixing). All were discarded on firm arguments, unequivocally showing that the correlation between VC and D collisions must be taken into account.

This analysis was reexamined[165,304] over a much broader range of conditions by slightly modifying the collision operator to take into account the VC/D correlation. New high resolution (4.5 10^{-6} cm^{-1}), high signal-to-noise ratio (3000) experiments were performed for a 10^{-3} CO mole fraction in argon. Measurements were made for seven lines (between J = 1 and J = 16), at five temperatures between 214 and 324 K stabilized at ± 0.1 K, and for six pressures between 0.025 and 1 atm accurate within 5.10^{-4}. Over one hundred line profiles were recorded and calculated with the sdBBP model. As in Ref. 303, apart fitting parameters having no influence on the line shape (line integrated absorbance, relative line center frequency in a scan, and linear base line), two other parameters were used. The first one accounts for the weak line mixing asymmetry, with the same factor that in Eq. (III.87). A second one rescales the amplitude of the line broadening $\Gamma = <\Gamma(v)>$, calculated with an accuracy of $\pm 2\%$, in order to be consistent with the noise level ($\pm 0.1\%$). It was checked that this rescaling does not alter the intrinsic shape of the calculated line. Except this (marginal) adjustment of Γ, the sdBBP model is fully ab initio and accounts for three main physical properties: the speed dependence of Γ, the sC collisions effect on $\Gamma(v)$, and the mass effect on the translation (see Table III.6). To account for the VC/D correlation, v_{VCD} was substituted to v_{VC} in Eq. (III.136) with $v_{VCD} = v_{VC} - \eta\Gamma$, $0 \leq \eta \leq 1$ [cf Eq. (III.42)]. v_{VCD} and Γ being fitted and v_{VC} calculated from [Eq. (III.25) using the mass diffusion coefficient], the above VCD expression leads to the correlation parameter η. This parameter was found to be pressure- and temperature-independent, and varying linearly with the rotational quantum number J, from 0.31 for P(1) to 0.40 for P(16). The fitted v_{VCD} values were three to ten times smaller than v_{Diff}, leading to a *drastic reduction* (from 70% to 90% depending on the line) of the Dicke narrowing relatively to that predicted from the mass diffusion coefficient. Finally, a non-physical deviation of $\gamma = \Gamma/P$ from a constant high pressure value was observed at low pressure, which is within the uncertainty ($\pm 2\%$) of the calculated γ values but not at the noise level ($\pm 0.1\%$). This non-physical behavior was attributed to the approximate form of the collision operator $\bar{R}$ in the sdBB model [Eqs. (III.135) and (III.136)]. Since only rotationally *elastic* collisions contribute to the Dicke narrowing of an *isolated* line,[285,286,307] v_{Diff} (through $\bar{R}_{VC}$) overestimates the effective (elastic) rate of VC collisions. Moreover $\bar{R}_D$ does not represent "purely" dephasing collisions, even if it is not possible to determine its precise meaning since based rather on a mathematical than on a physical approximation (the intuitive subscript VC and D, or VCD in the correlated case, affected to $\bar{R}$, being clearly inappropriate). The exhaustive and thorough study of CO line shapes in CO-Ar mixtures[165,304] shows the limits of such an approximation for the collision operator. However, for pressure above 0.2 atm, where these limits tied to Dicke narrowing disappear, Wehr et al[165,304] deduced from their experimental data highly accurate line broadening coefficients (within the noise uncertainty $\pm 0.1\%$) by using the sdBBP model.

At the end of this section, it has to be emphasized that fully ab initio methods (GHM, CKM) based on the Waldmann-Snider kinetic equation take into account *all* the physical properties (Table III.6) involved in the Dicke narrowing mechanism. Such approaches remain computationally intensive and therefore not currently suitable for many applications. Ranging between phenomenological models (asdRS$_{\tilde{C}}$P and

Table III.6: Dicke narrowed models of isolated line profile and their physical properties. (A): radiator speed dependence of the internal relaxation ; (B): radiator speed dependence of the external relaxation ; (C): influence of speed-changing collisions on the relaxation ; (D): influence of the correlation between internal and external relaxations. (E): dependence on the perturber to radiator mass ratio through the translational motion. NB: The influence of velocity-changing collisions on the translation is taken into account in *all* these (Dicke narrowed) models

Model of profile	Collision type	Physical Properties	Equation of definition	Refs.
G	soft		(III.26)	183
NG	hard		(III.31) or (III.32)	56,189
G_C	soft	(D)	(III.36) and (III.37)	56
NG_C	hard	(D)	(III.39) and (III.40)	56
asdG	soft	(A)	(III.51)	231,245
sdNG	hard	(A)	(III.49)	244
$asdG_C$	soft	(A), (D)	(III.51) and (III.52)	231
$sdNG_C$	hard	(A), (D)	(III.50)	225,245
$asdRS_{\widetilde{C}}$	hard/soft	(A), (C), (D)	(III.84) and (III.85)	215
$sdKS_{\widetilde{C}}$	hard/soft	(A), (C), (D)	(III.102), (III.99), (III.100), (III.106)	270,272,273
GHM	general	(A)–(E)	(III.122) to (III.124)	283,277
CKM	general	(A)–(E)	(III.131) and (III.114) to (III.116)	291
sdBB	Billiard-Ball	(A), (C)–(E)	(III.118) and (III.134) to (III.136), and (III.114) to (III.116)	298,299

$sdKS_{\widetilde{C}}P$) and ab initio methods, intermediate approaches aiming at modeling the collision operator (and not the profile itself), seem a priori a good compromise. However, they lose physical properties having significant influences on the line shape, at the current noise level. Existing quantum dynamics codes such as MOLS-CAT[308] and MOLCOL[309] already calculate all the scattering amplitudes necessary to determine the *complete* collision operator.[285–287] Calculations of line shapes from GHM, CKM, or equivalent rigorous methods, should be feasible on today's computers for mixtures of molecules at finite concentration. A remaining problem for applications could be the gap of one order of magnitude between the accuracy of the line profiles calculated ab initio from potential energy surfaces, and the present experimental noise level (*cf* Fig. III.26). Recent studies[165,303,304] within the billiard ball model indicate that floating Γ successfully fills this gap *without* distorting the line shape. A similar procedure could be extended to the GHM and CKM methods to rescale the *amplitude* of the (correlated) internal and external relaxation processes, through pertinent parameters which remain to be defined.

III.6 CONCLUSION

To conclude this chapter, let us summarize the main general characteristics of the numerous approaches for isolated line shapes. Two *synthetic phenomenological* models are now available to accurately describe the inhomogeneous features tied to the Doppler distribution and to the speed dependence of the line width and shift, which are under the strong influence of velocity- and speed-changing collisions, respectively. The partially correlated approximate speed-dependent Rautian-Sobelman profile (asdRS$_{\tilde{C}}$P) presents the advantage of being analytical and gives, as limiting cases, the familiar hard and soft collision models (*cf* Table III.5). The partially correlated speed-dependent Keilson-Storer profile (sdKS$_{\tilde{C}}$P) looses this advantage but allows a more rigorous treatment of velocity- and speed-changing collisions in terms of memory mechanisms. Both approaches possess the same (restricted) physical limitations (*cf* Table III.6) and they are relatively easy to handle for many applications (computations can be carried out on a PC). For chronological reasons, the most used models were the simplified versions of the asdRS$_{\tilde{C}}$P, *ie* the uncorrelated or correlated speed-dependent Nelkin-Ghatak profile (sdNGP or sdNG$_C$P) and the uncorrelated or correlated approximate speed-dependent Galatry profile (asdGP or asdG$_C$P). Although simplified, these versions were essentially used for the modeling of laboratory experiments. For radiative heat transfer and remote sensing discussed in chapter VII, the use of the various speed-dependent models remains unusual, essentially for computational reasons discussed in App. III.A. It is then clear that the ab initio approaches, which are very promising, should be a fortiori devoted to specific situations in the mid term.

In this chapter, most of the illustrative examples have been chosen among light radiators (*ie* H_2, D_2 and HF). Indeed, they are more appropriate to clearly identify each of the mechanisms involved in the spectral line shape (velocity-changing, speed-changing and correlation effects). For heavier radiators (*eg* N_2, O_2, CO, CO_2, NO, O_3, HCN, *etc*) the Dicke narrowing contribution tied to the velocity-changing collisions is frequently weaker than the speed inhomogeneous one. This is true, not only at high pressure (*cf* § III.4.1), but also for pressures where both Doppler and collisional broadenings are significant.[203,310,311]

If isolated lines are no longer considered, but spectral bands or manifolds with many collisionally coupled lines, the above comments become obviously more acute. The next chapter precisely treats the important problem of collisional line mixing, with the goal to extend the validity of some basic (Lorentz and Voigt) models from isolated transitions to clusters of collisionally coupled lines. This is done within a frame disregarding the spectral influence of the speed dependence and velocity changes. Nevertheless, it is worthy noting that a few articles have been devoted to the inhomogeneous features within clusters of coupled lines, as discussed in Sec. VIII.2.

APPENDIX III.A COMPUTATIONAL ASPECTS

The partially correlated approximate speed-dependent Rautian-Sobelman asdRS$_{\widetilde{C}}$P model [Eq. (III.84)] is expressed in terms of the complex approximate speed-dependent Galatry profile $J_{asdG}[x(u), \widetilde{Y}(u), \widetilde{Z}(u)]$ [Eq.(III.51)]. The (dimensionless) parameters $x(u)$, $\widetilde{Y}(u)$ and $\widetilde{z}(u)$ are defined in terms of the usual line shift, line broadening, and effective kinetic rate through Eq. (III.85), u being the reduced radiator speed ($u = v/\widetilde{v}$, $\widetilde{v} = (2k_BT/m)^{1/2}$). Disregarding, in a first step, the speed dependence (sd) (*ie* u) and the (partial) correlation (*ie* $\widetilde{C}$), results in the substitution of the complex Galatry profile $J_G(x,y,z)$ to $J_{asdG}[x(u), \widetilde{Y}(u), \widetilde{z}(u)]$ in Eq. (III.84). $J_G(x,y,z)$ can be expressed as the following Laplace transform of a *correlation function* in time, [*cf* Eqs. (III.26) and (III.20)]:

$$J_G(x, y, z) = \frac{1}{\pi} \int_0^\infty e^{ix\widetilde{t}} \Phi_G(y, z, \widetilde{t}) d\widetilde{t}, \qquad (III.A1)$$

with $\quad \Phi_G(y, z, \widetilde{t}) = \exp[-y\widetilde{t} + \frac{1}{2z^2}(1 - z\widetilde{t} - e^{-z\widetilde{t}})]$, $\widetilde{t} = \Delta\omega_D t$ [*cf* Eq.(III.3)]. (III.A2)

The complex Voigt profile $J_V(x,y)$ is straightforwardly obtained in the limit $z \to 0$ from the time correlation function:

$$\Phi_V(y, \widetilde{t}) = \exp(-y\widetilde{t} - \widetilde{t}^2/4). \qquad (III.A3)$$

Recall that the GP [$=\text{Re}(J_G)$] and VP [$=\text{Re}(J_V)$] are here normalized to area unity (*cf* III.3.3), in contrast with the definition used in most of the articles referenced in this appendix where this area is taken as $\pi^{1/2}$. These two profiles cannot be evaluated in closed form and have to be computed by using efficient algorithms. The large use of the VP for radiative transfer and remote sensing applications (*cf* chapter VII) has led to the development of a wide variety of algorithms (*eg* Refs. 312–314 and those therein). Their efficiencies (*ie* both accuracy and celerity) have been tested, showing somewhat conflicting requirements because improved accuracy of an expansion of Gauss-Hermite integration is in general achieved by increasing the number of terms used. Some information on the main algorithms and on the results of these tests are now presented.

1. ALGORITHMS FOR THE VOIGT AND GALATRY PROFILES

Besides empirical fits, not considered here, the computational approaches can be essentially classified in two groups. In the first one are algorithms using series, asymptotic or continued fraction expansions, rational approximations, Gauss-Hermite integrations, or combination thereof (*eg* Refs. 207,312). The (x,y) plane for the VP, for instance, is generally divided into several regions and an appropriate method is selected for each of them. Schreier[313] has performed a careful analysis of the main algorithms within this first group, in view of applications to spectroscopic studies and to line-by-line spectra calculations. He concluded by recommending the approach proposed by Hui et al[315] and the W_4 version of Humlicek,[207] also mentioning the higher flexibility of the latter. This W_4 algorithm, which uses a "wide-mesh sampling" in the line wing and a "fine-mesh" in the core, is a good compromise between accuracy and computational efficiency. Furthermore, it can be vectorized, leading to a significant acceleration of computations.

The second type of approaches evaluate the VP in the Fourier-transformed space, where the convolution integral is just the product of the Fourier transform (FT) of the Lorentz and Gauss functions [cf Eq. (III.21)]. For spectra with many contributing lines, only two transforms are required because of the frequency shift properties of the FT.[316] The FT computational methods become especially advantageous when hardware fast FT processors are available.

The Voigt profile being a special case of the GP model, efficient codes for the GP computation can take advantage of the existing algorithms for the VP. This was done[180] by expressing the GP in terms of the incomplete Gamma function[182] and using adequate expansions in several selected regions of the (y,z) plane. As an example of this procedure, the expansion in the particular case of the "central" region ($0 \leq y \leq 1$ and $0 \leq z \leq$ several 10^{-2}) is:

$$I_G(x,y,z) = \frac{1}{\pi} \mathrm{Re}\left\{ w(q) + \sum_{n=3}^{n_{max}} \frac{c_n(z)}{i^n} \times \frac{d^n}{dq^n}[w(q)] \right\}, \qquad (\mathrm{III.A4})$$

where $w(q) \equiv w(x+iy)$ is the complex probability function [Eq. (III.22)] and $c_n(z)$ are the coefficients of the asymptotic expansion.[180] The derivatives in Eq. (III.A4) are computed from a recursion relation and $w(q)$ by using an efficient algorithm such as the W_4 one of Ref. 207. Note that the first term in Eq. (III.A4) gives the VP [cf Eq. (III.21)]. The number of terms in the expansion depends on the desired accuracy. For instance, $n_{max} = 8$ was retained[181] in order to get four significant digits accuracy with a preceding version (CPF12) of Humliceck's code.[206]

An alternative *discrete Fourier transform* (DFT) approach was developed later.[208] This faster algorithm proved useful and efficient for least squares fits of measured spectra using the GP, since these adjustments need repeated evaluations of the Fourier transform correlation function [Eqs. (III.A1)–(III.A3)] and of its derivatives with respect to all parameters. In the preceding routine,[181] the derivatives had to be estimated using finite differences making the fitting procedure much slower. In the DFT method, the GP is pre-computed from Eq. (III.A1) for an array of N equally spaced discrete points of frequency x_k, ie:

$$J_G(x_k, y, z) \approx \frac{1}{\pi} \int_0^T e^{ix_k \tilde{t}} \Phi_G(y, z, \tilde{t}) d\tilde{t} \approx \sum_{n=0}^{N-1} f_n \exp(2i\pi kn/N), \qquad (\mathrm{III.A5})$$

with $x_k = 2\pi k/T$. The coefficients f_n are integrals calculated by a second-order Simpson equation. At arbitrary frequencies $x \neq x_k$, the GP is computed by interpolations within these tabulated values. Note that these interpolations are not always required, if a fast Fourier transform (FFT) is used with a sufficient number N of points.[316] Indeed, FFT algorithms improve the efficiency since the total number of floating point operations scales as NlogN instead of N^2. Since the derivatives of the complex GP with respect to x, y, and z can be expressed as Fourier transforms, the *same* computational procedure applies. Details on the computer program (GALFIT) are given in Ref. 208. It is worthy noting that the two above mentioned types of approaches also *apply* for *complex* values of z, ie when the correlation between velocity-changing and dephasing collisions (labeled $\widetilde{C}$ in the generalized asdRS$_{\widetilde{C}}$P model) is taken into account.

The last step to be considered is the introduction of the speed dependence in the line-shape model [through x(u), $\widetilde{\widetilde{Y}}$(u), and $\widetilde{\widetilde{z}}$(u) in Eq. (III.84)].

2. COMPUTATION OF SPEED-DEPENDENT PROFILES

The inclusion of speed dependence in the line shape generally consists in incorporating a model for $\Gamma(v,T)$ and $\Delta(v,T)$ in the formula, and then to construct a computationally efficient code for the evaluation of the resulting profile. A first simple model [Eq. (III.44)] is of specific interest since, due to its *quadratic* form, it permits to get an exact analytical solution of the kinetic equation in the soft collision limit.[248] This makes easier the computation of the (speed-dependent) J_{asdG} in Eq. (III.84) for the asdRS$_{\widetilde{C}}$P evaluation, through the use of an efficient algorithm (*cf supra*). A more commonly used approach is the classical-path R^{-q} potential model,[242,243] as well as its generalization within the frame of the VP.[317–319] There are several reasons for its usefulness. The first one is that it is characterized by a single physically meaningful parameter q, which is the *potential range* [Eq. (III.56)]. The second one is that it leads to an *analytical* solution for the speed-dependence [Eq. (III.57)]. Furthermore, q is usually taken as a fitting parameter, introducing an additional flexibility of interest for applications. However, the oversimplified character of this potential model implies restrictions, exemplified below. More realistic[248] is the "quadratic" model of Eq. (III.44) for $\Gamma(v)$ and $\Delta(v)$, whose characteristic parameters a_w and a_s can be related to the potential range q and the mass ratio $\Lambda = m_p/m$ through Eq. (III.60).[311]

When more pertinent intermolecular potentials are introduced, the advantage of an analytical solution for $\Gamma(v,T)$ and $\Delta(v,T)$ is lost. However, calculating the collisional broadening and shift as a function of the relative speed v_r (or relative kinetic energy $E_r = m^* v_r^2/2$) allows one to straightforwardly deduce the concomitant radiator speed dependence through Eqs. (III.59) and (III.60). This method was first explored by Luijendijk[220,320] using a semi-empirical potential including both long and short range interactions within the frame of the ATC model (*cf* § IV.3.4d). Later on, thanks to the used of refined potentials[321] and of an improved semi-classical model[322] (*cf* § IV.3.4e), more accurate values for $\Gamma(v,T)$ and $\Delta(v,T)$ were obtained. Figure III.A1 shows a typical result obtained[323] with this method for the speed dependence of $\Gamma(v,T)/<\Gamma(v,T)>$ for C_2H_2–Xe collisions. In this figure are also reported, for comparison, results obtained from the R^{-q} potential model (with q = 6, according to the nature of the molecular pair) and a speed law directly deduced from the experimental data using a procedure described in § III.4.3. Figure III.A1 shows the consistency between the former and the latter speed laws for the line broadening, the results obtained with the R^{-q} potential model being in poor agreement with them.

Isolated lines

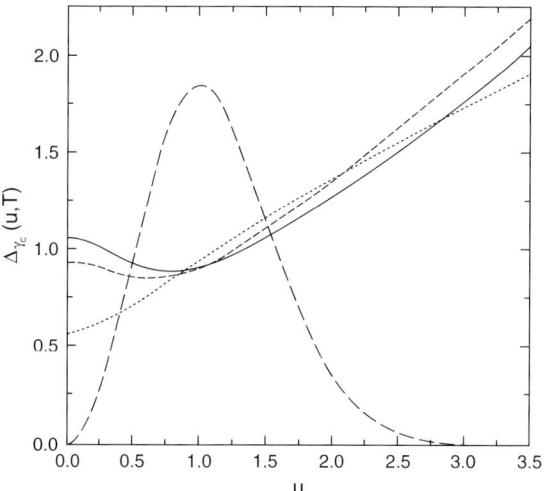

Fig. III.A1: Comparison between speed dependence laws $\Delta_\gamma(u,T) \equiv \Gamma(u,T)/<\Gamma(u,T)>$, $u = v/\tilde{v}$ obtained from RB calculations (—) with that (- - -) deduced from experimental data (see § III.4.3) for the P(14) line of C_2H_2 broadened by Xe at 293 K. The speed law (.....) corresponds to the R^{-6} potential model. The Maxwell-Boltzmann distribution is also represented. Reproduced with permission from Ref. 323.

The best accuracy is obtained when close-coupling quantum values of the broadening and shifting cross sections $\sigma_\Gamma(E_r)$ and $\sigma_\Delta(E_r)$ (*cf* § IV.3.5) calculated from an accurate potential energy surface are available. Such a situation was exploited by Pine and Ciurylo[172] in their thorough analysis of rovibrational HF−Ar line shapes (see § III.4.3). Nevertheless, the determination of the speed laws needs a careful procedure. For instance, for HF−Ar, $\sigma_\Gamma(E_r)$ and $\sigma_\Delta(E_r)$ were calculated for 15 E_r values between 30 and 1000 cm^{-1}.[23] In fact, the integration over the relative energy (or speed) through Eq. (III.54), requires a thinner grid, and the choice of the function for the fits must be made with particular caution due to instabilities.[214] The main features of the resulting speed laws are illustrated in Fig. III.A2. Aside from the poor quality of results obtained with the R^{-q} potential model, note the strong differences between the Γ and Δ speed laws.

In the "brute" force method, the computation of the asdRS$_{\tilde{C}}$P is very time consuming. One must perform m × n calculations for each point of the spectral profile (m being the number of discrete speeds v_i and n that of discrete times t_j). If N discrete frequencies x_k are chosen to sample the profile, this leads to N × m × n calculation points. A strong reduction of the computational time (by about two orders of magnitude) can be obtained by introducing *pseudo* correlation functions, in order to take advantage of the FT properties [*cf* Eq. (III.A5)]. Accounting for the specific structure of the line-shape formula,[245] one can write:

$$I_{asdRS_{\tilde{C}}}(x, y, \zeta, s, \varepsilon, \eta) = \frac{1}{\pi} \operatorname{Re} \left\{ \frac{\int_0^\infty e^{i\tilde{x}\tilde{t}} C_{pseudo}(\tilde{t}) d\tilde{t}}{1 - \pi\varepsilon \int_0^\infty e^{i\tilde{x}\tilde{t}} C'_{pseudo}(\tilde{t}) d\tilde{t}} \right\}, \qquad \text{(III.A6)}$$

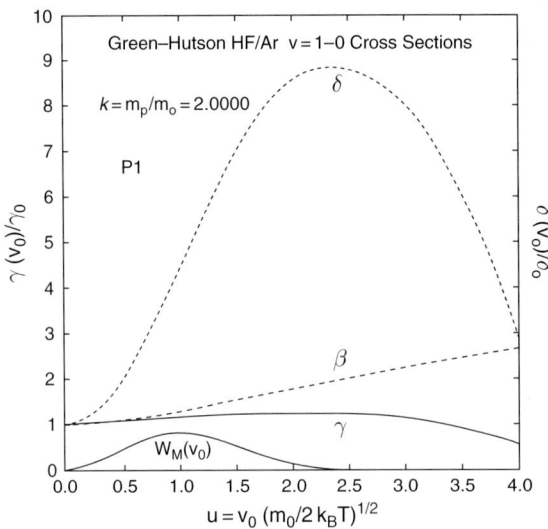

Fig. III.A2: Speed dependence laws for the P(1) line of HF broadened by Ar at 296 K, including both the broadening and shifting coefficients γ and δ, as obtained from quantum calculation.[23] The corresponding result of the classical-path R^{-6} potential model is also shown (labeled β). $W_M(v_0)$ is the Maxwell-Boltzmann distribution for the radiator speed. Reproduced with permission from Ref. 214.

where $C_{pseudo}(\tilde{t}) = <C_{pseudo}(\tilde{t},u)>$ and $C'_{pseudo}(\tilde{t}) = <\tilde{\zeta}(u)C_{pseudo}(\tilde{t},u)>$, $<\cdots>$ being the Maxwell-Boltzmann average over the reduced radiator speed ($u = v/\bar{v}$). Note that the explicit expression for $C_{pseudo}(\tilde{t})$ may be straightforwardly obtained from Eqs. (III.84) and (III.51). Thus an efficient algorithm can be used (cf supra) for the Fourier transform of each of these pseudo correlation functions, and the number of iterations can be reduced to $2[m \times n + n \times N]$. Using the simple model of Eq. (III.44) for $\Gamma(v)$ and $\Delta(v)$ makes the pseudo correlation function analytical. This has been exploited to analyze as well transient decay signals,[234,238] as frequency-modulated spectral profiles.[324]

Besides the fact that few accurate results have been published for the Γ and Δ speed laws, it must be emphasized that the complexity and computer cost of the present codes including this speed dependence have, up to now, forbidden their use in large scale spectra calculation. However, deviations from the Voigt profile are observed in atmospheric spectra (cf chapter VII) and remote sensing inversion codes will likely be extended in the near future in order to fulfill the increasing need for accuracy of applications. Despite the increase of computer power, this will require the development of very efficient algorithms. From this point of view, an interesting track was explored by Berman et al[325] who proposed to replace the considered profile by a different (but approximate) one for which fast routines are already available. This method was for the case of Lorentzian lines with speed-dependent broadening coefficients. It was found[325] that the sum of two *speed-independent* profiles accurately approximates the speed-dependent line shape, provided that ad hoc parameters are used. This methodology is not restricted to purely Lorentzian lines and was applied to more complex profiles including a speed-dependent Lorentzian part in Ref. 305.

As a conclusion, it is likely that the very high accuracy requirements of remote sensing applications will soon impose to include speed-dependence and velocity-changing line profiles in the computer codes used to treat atmospheric spectra. This will probably motivate further developments of efficient algorithms but will also strongly enlarge the scope of isolated line shape laboratory studies which will have to provide data for many transitions and species. The question of the pertinent quantities to be stored in spectroscopic databases remains opened, as discussed in Ref. 326 and Sec. VIII.4.

IV. COLLISIONAL LINE MIXING
(WITHIN CLUSTERS OF LINES)

IV.1 INTRODUCTION

This chapter, devoted to *line-mixing* effects,[327] describes situations where collisions are sufficiently numerous and efficient for some lines not to be collisionally "isolated" one from the other. This phenomenon is often encountered in practical situations such as infrared atmospheric emission and Raman scattering which cannot be predicted accurately in all regions by using the models of the preceding chapter. It is the case of narrow Q branches, for instance, which can show a significant narrowing with respect to the isolated lines behavior. The aim of this chapter is to explain the mechanism, to review the state-of-the-art concerning this problem, and to give the reader tools and bibliographical paths. In this introduction, the process giving rise to collisional line mixing is explained and its main characteristics are described. The second section is devoted to expressions of the spectral profile when lines cannot be considered as collisionally isolated. The assumptions made (mostly the *binary collision* and *impact* approximations introduced in chapter II) are recalled together with general equations. Simplified expressions are derived in asymptotic cases and computational tools and recommendations are finally given. The third section is devoted to the *relaxation matrix* W, key element of the problem since, within the approximations made, it contains *all* the influence of collisions. A review of the available theoretical approaches is given, from very simple empirical models suitable for complex molecular systems to fully quantum calculations starting from the interaction potential energy surface today limited to simple collisional pairs. The fourth part considers situations where a direct determination of some line-mixing parameters from fits of measured signals is possible. Finally, in the last section, we give a tentative review of the available literature on collisional line-mixing effects in molecular spectra. This chapter is limited to theory, laboratory experiments, and comparisons between predictions and measurements. The consequences of line mixing for optical soundings of gas media are the subject of Sec. VII.4.

Note that, contrary to what has been mostly done in the preceding chapters, we *use wavenumber units* (σ, cm^{-1}) instead of frequencies (ν, Hz or ω, rad/s). This choice is made here since wavenumbers are used in most infrared and Raman studies, and thus in the majority of the figures used and of the references cited in this chapter. The readers should simply remember than any quantity Q in cm^{-1} can be converted to frequency units by the relation: $Q(Hz) = cQ(cm^{-1})$ [or, if angular frequencies ω are used $Q(rad/s) = 2\pi c Q(cm^{-1})$] where c (cm/s) is the speed of light.

Collisional line mixing (or interference, or coupling) may significantly change the signals in the spectral and time domains when lines cannot be considered as collisionally "isolated", that is when the contributions of the various transitions overlap significantly. It can affect the core regions of closely spaced and sufficiently pressure-broadened lines, as well as the weak absorption regions (troughs between transitions, band and branch wings) where the spectrum or response signal in the time domain results from the cumulative contributions of the wings of several transitions. In order to explain the process, a simple picture can be given by considering only two optical transitions: $f \leftarrow i$ and $f' \leftarrow i'$ at frequencies ν_{fi} and $\nu_{f'i'}$, respectively (as it is the case for a doublet of closely spaced lines centered away from others, *eg* Refs. 328,329). For single photon absorption, the mechanisms involved are shown in Fig. IV.1. In the presence of radiation, a molecule in (internal) level i can be excited to level f by absorbing a photon of frequency ν_{fi}, giving rise to an absorption line at wavenumber $\sigma_{fi}(\text{cm}^{-1}) = (E_f - E_i)/hc$. It can also be transferred by collision from i to i′, then be excited to level f′ by absorbing a photon of frequency $\nu_{f'i'}$, and finally relax from f′ to f by collision. This second path from i to f *via* i′ and f′ shows that a molecule initially on level i can contribute to the absorption line at wavenumber $\sigma_{f'i'}$ through population transfers induced by the interaction of the optically active molecule (the *radiator*) with others (the *perturbers* constituting the thermal bath). The reverse path (i′ → i by collision + i → f by photon absorption + f → f′ by collision) is also possible.

These collisional processes induce mixing terms and the optical transitions thus cannot be considered as isolated one from the other since population (*ie* absorption intensity) is exchanged between them. Another way to put it is to recall that the probability of absorption is the square of the sum of matrix elements connecting upper and lower states. In the case of two lines, this gives rise to the usual isolated line contributions associated with transitions $f \leftarrow i$ and $f' \leftarrow i'$ but also to a cross interference term (leading to the expression "line-interferences") associated with the product. In some situations, this line-mixing process can be non-zero and significantly influence the spectral shape or time response signal. Some characteristics of this effect can be pointed out.

(1): For the process to exist, the $f \leftrightarrow f'$ and $i \leftrightarrow i'$ collisional transfers must be *allowed* by the nature of the molecular levels and by the radiator-perturber interaction potential. This obviously forbids line coupling between optical transitions of different molecules or of different isotopomers of a given species, for instance.

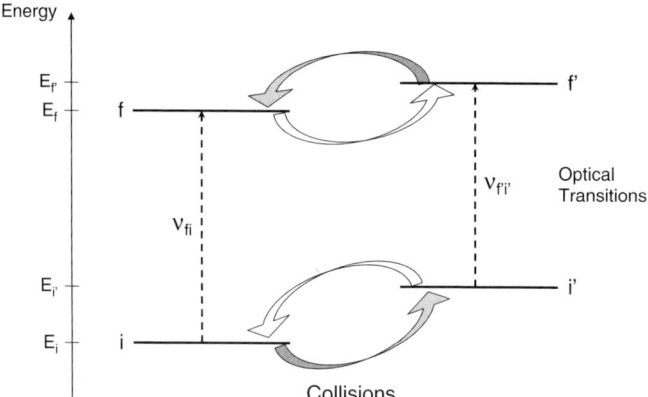

Fig. IV.1: Schematic view of the line-mixing process in the case of two optical transitions.

(2): When they are possible, the collisional population transfers have a significant influence on the spectral shape only if they are efficient. In order to derive a simple criterion (necessary but *not* sufficient), recall that the pressure-broadened width Γ gives a good idea of the magnitude of internal energy jumps that intermolecular forces can induce. Hence, since the energy balance of the collisional processes of Fig. IV.1 is $\Delta E = |E_i - E_{i'} + E_{f'} - E_f| = hc|\sigma_{f'i'} - \sigma_{fi}|$, line mixing has a significant influence on absorption near the transition centers for conditions (pressure, temperature, collision partner) making $\Gamma(\text{cm}^{-1})$ *of the order of or greater than* $|\sigma_{f'i'} - \sigma_{fi}|$. This is illustrated by the simulations made for a doublet plotted in Fig. IV.2. At low pressures, when $|\sigma_{f'i'} - \sigma_{fi}| \gg \Gamma$, the lines can be considered as isolated one from the other and are correctly modeled neglecting line mixing. As the broadening increases and the transitions overlap, the collision-induced transfers of population become increasingly important and the spectral shape cannot be modeled without taking the line-mixing process into account. Figure IV.2 shows that this mechanism leads to a reduction of absorption in the wings, to the enhancement of the trough at intermediate pressures, and to a narrowing of the profile when the two lines have merged.

The above condition in terms of the spectral vicinity of the transitions with respect to the collisional broadening is *not* sufficient. In fact, one must also consider the *efficiency* of the process. As mentioned above, even if centered very close to each other, transitions associated with different species do not interfere. Another example is given by infrared lines involving different vibrational states, but which are closely spaced due to spectral coincidence. It is the case,[2] for $^{12}\text{C}^{16}\text{O}_2$, of the P(29) transition of the $\nu_1 + 2\nu_3 \leftarrow \nu_1 + \nu_3$ band at 2276.93 cm^{-1} and the P(12) line of the $5\nu_2 + \nu_3 \leftarrow 5\nu_2$ band at 2276.94 cm^{-1}. Line mixing, which requires collisional transfers $i \leftrightarrow i'$ between the initial (and final $f \leftrightarrow f'$) levels of these transitions, negligibly affects their spectrum. Indeed, the internal

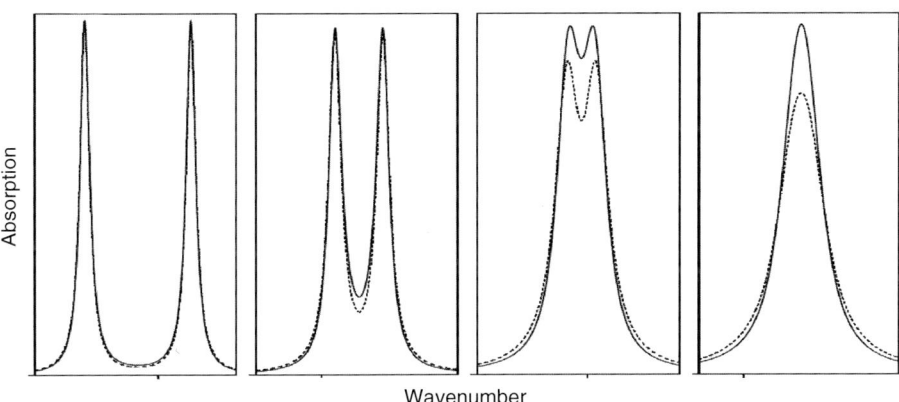

Fig. IV.2: Typical evolution of the shape of a doublet of collisionally coupled transitions with increasing total pressure. (——) and (- - -) correspond to predictions made with and without the inclusion of line mixing. The pressure increases by a factor of three when going from a plot to the next one to the right.

state changes involve jumps of $\approx 600\,\text{cm}^{-1}$ for the energy, of 17 for the rotational quantum number J, and of five quanta for the bending vibrational mode v_2. This is a much less efficient process than the collisional broadening of each of the transitions which involves purely rotational jumps (for pure CO_2, the efficient $\Delta J=\pm 2$ transfers induced by the quadrupole-quadrupole interaction). The cross terms due to collisional interferences are much smaller than the line widths and assuming isolated transitions is thus a very good approximation, even though the spectral overlapping is important. Another example where spectral vicinity is not sufficient for efficient line mixing is that of lines involving states of different nuclear spins, even though the associated $f \leftrightarrow f'$ and $i \leftrightarrow i'$ transitions may be almost resonant. It is the case[2] of the three components of the R(3) manifold of the v_3 band of CH_4 which are two F-type transitions – (4F1 15) ← (3F2 1) and (4F2 16) ← (3F1 1) – and one A-type transition – (4A1 7) ← (3A2 1) – involving levels closer than $0.1\,\text{cm}^{-1}$. Whereas the two F lines are collisionally coupled, there is no mixing,[330] despite the vicinity, with the A transition. Indeed, nuclear spin changes, through spin-rotation interactions for instance, are extremely slow[331] and can be neglected when compared with the very fast rotational changes giving rise to the broadening of the transitions. In conclusion, line mixing between closely spaced lines significantly affects the spectral shape *only* if the radiator-perturber interactions are *efficient* in inducing $f \leftrightarrow f'$ and $i \leftrightarrow i'$ changes between the upper states and the lower states of the optical transitions.

(3): Since collisions keep the equilibrium Maxwell-Boltzmann population distribution, the exchange terms are related by the detailed balance relation [Eq. (II.55)]. One can then write, with simplified notations: $(1 \leftarrow 2)\rho_2 = (2 \leftarrow 1)\rho_1$, where ρ_j is the equilibrium relative population of the initial level of line j (j = 1,2). This indicates that the more efficient of the two counter processes ($1 \leftarrow 2$ and $2 \leftarrow 1$, see Fig. IV.1) is that starting from the less populated level. From this one can infer that line mixing favors the transfer of intensity from the weak lines toward the strong ones. In a more general way, it leads to *transfers from the weak absorption regions to those of strong absorption* (toward the center of gravity). Hence, with respect to isolated lines models which neglect internal population exchanges, line mixing induces a *narrowing* of spectral structures, enhancing the intense regions and lowering the weak ones. This is spectacularly illustrated by the experimental results of Fig. IV.3, which are for the isotropic Raman Q branch of N_2. With increasing density, the Q branch not only does not broaden but it shows, besides a spectral shift, a significant reduction of its width which lowers by a factor of about two when increasing the density by the same factor.

This narrowing with increasing density, which can be particularly efficient in isotropic Raman Q branches, is not observed in infrared or anisotropic Raman Q branches. As long as they do not overlap with other branches, these structures broaden with increasing density. This is due to the non-zero couplings between Q lines and the P and R (or O and S) transitions. Indeed, the decrease of the Q-branch width with density requires *all* transfers to be efficient and it is not the case of the $Q \leftrightarrow X$ (X = P and R or X = O and S) since these transitions are centered far away [$|\sigma_Q - \sigma_{X \neq Q}| \gg \Gamma(\text{cm}^{-1})$]. This remaining broadening is exemplified in Fig. IV.4 by a N_2O infrared Q branch. Note that, although the width of the Q branch increases with density, line mixing does have an effect through the $Q \leftrightarrow Q'$ transfers, these being efficient since $|\sigma_Q - \sigma_{Q'}| \leq \Gamma$. Predictions

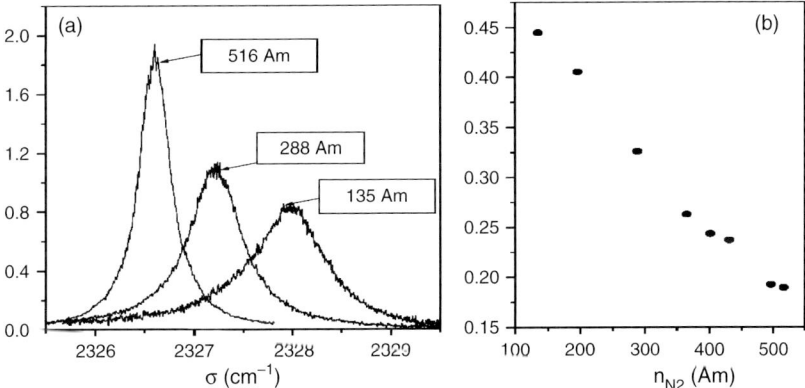

Fig. IV.3: Shape of the isotropic Raman Q branch of pure N_2 at room temperature. (a): Measured (area) normalized spectra for the densities 135 Am, 288 Am, and 516 Am. (b): Q branch half width (HWHM, in cm^{-1}) versus the N_2 density. Courtesy of B. Lavorel, after Ref. 332.

neglecting this effect overestimate the broadening at atmospheric pressure by a factor of about two (Fig. IV.4b). Let us mention that the Q branch width can be approximated by a linear dependence on pressure (or density). The slope is the *effective* broadening coefficient γ(Q), the zero pressure value being related to the weighted root mean square of the line positions.[333]

With respect to the isolated (Lorentzian) lines predictions $\gamma^{Lor}(Q)$, the reduction, due to line mixing, of the effective broadening coefficient $\gamma^{Obs}(Q)$ of IR Q branches of a given radiator *significantly* depends on the vibrational symmetry and on the perturber. In the case of CO_2 for instance, $\gamma^{Lor}(Q)/\gamma^{Obs}(Q)$ is typically 3 for collisions with Ar and N_2

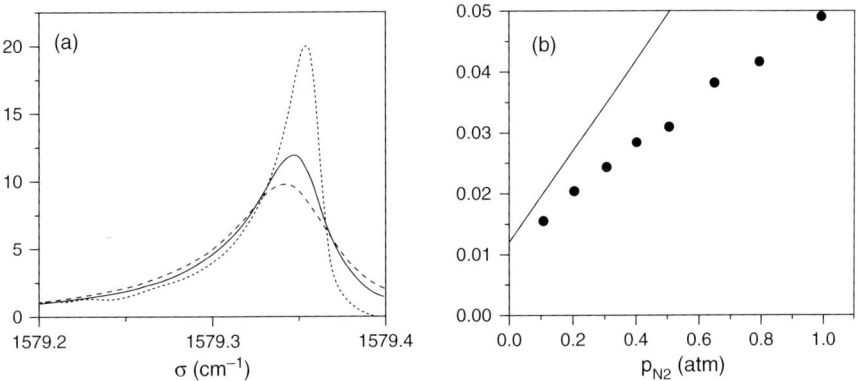

Fig. IV.4: Shape of the $2\nu_2^{0e} - \nu_2^{1f}$ infrared Q branch of N_2O broadened by N_2 at room temperature: (a): Measured (area) normalized absorption coefficients for total pressures 0.102 atm (...), 0.403 atm (——), and 0.652 atm (- - -). (b): Q branch half width (HWHM, in cm^{-1}) versus the N_2 pressure: (•) measured values, (——) calculated values neglecting line mixing. After Ref. 334.

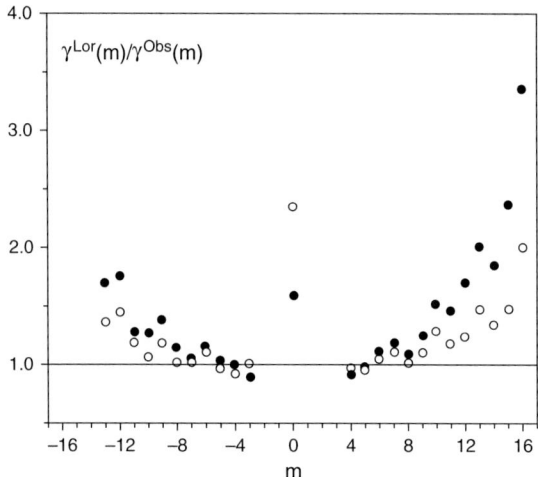

Fig. IV.5: Ratios of the effective broadening coefficient of absorption structures of the ν_3 band of methane at room temperature predicted by a purely isolated lines approach to the observed values. The results are plotted vs m, where $m = -J$ and $J+1$ for P(J) and R(J) manifolds, the results for the Q branch being shown at $m = 0$. (•) and (o) are for mixtures with Ar and He, respectively. After Ref. 336.

and 5 for mixtures with He in perpendicular bands, the corresponding numbers being about 1.5 and 2.0 for parallel transitions.[335] This relative narrowing induced by line mixing is also observed in P and R manifolds or doublets. For methane, this is illustrated by Fig. IV.5 where results for a Q branch and P(J) and R(J) manifolds are plotted. Again, the relative influence of line mixing significantly depends on the collision partner and on the considered spectral structure. Note that the increase of the effects with the rotational quantum number in P and R branches can be understood from Fig. IV.1. Indeed, as J increases, the couplings $R(J) \leftrightarrow R(J')$ between lines of a given manifold R(J) and those in other $R(J' \neq J)$ clusters involve larger and larger energy jumps $|E_J - E_{J'}|$ (about 20 cm^{-1} for $J=1 \rightarrow J'=2$ and 115 cm^{-1} for $J=10 \rightarrow J'=11$). Collisional transfers between levels of different manifolds are thus less and less probable as J increases, leading to the relative increase of line couplings (and thus narrowing) *within* the manifold where the energy jumps are very small (less than 0.3 cm^{-1} for $J=10$).

While infrared and anisotropic Raman clusters of transitions, despite line mixing, broaden with increasing pressure when spectrally isolated from others, the *entire* band can *narrow* (as isotropic Q branches) when densities such that *all* lines of different branches overlap significantly are considered. This is illustrated in Fig. IV.6 by the ν_3 infrared band of CO_2 broadened by Ar at high pressures.

(4): In the wings, line mixing generally leads to large changes of the spectral shape, regardless of pressure and of the spectral separation of the optical transitions. Indeed, far away from *all* lines, at wavenumbers σ equally distant from all of them (*ie* $|\sigma - \sigma_{fi}| \gg |\sigma_{f'i'} - \sigma_{fi}|$), all pressure broadenings are equally important. The absorption is small and,

Collisional line mixing

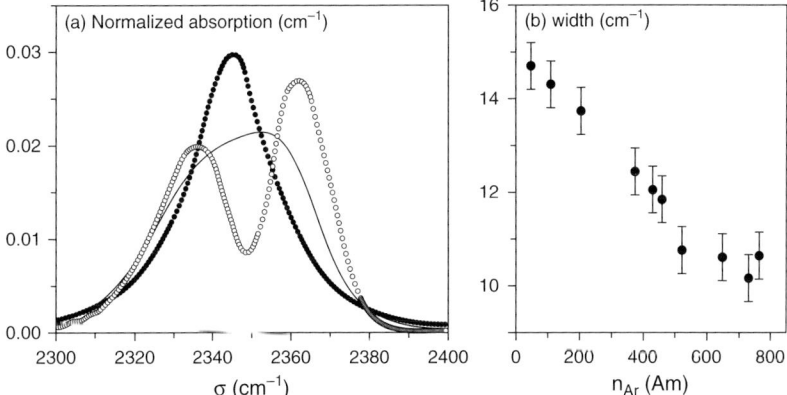

Fig. IV.6: Shape of the v_3 band of CO_2 broadened by Ar at room temperature. (a): Measured (area) normalized absorption coefficient for the densities n_{Ar} = 113 Am (o), 342 Am (——), and 766 Am (•). (b): Measured band half width Γ_B (HWHM, such that the spectral range $[\bar\sigma - \Gamma_B, \bar\sigma + \Gamma_B]$ around the center of gravity $\bar\sigma$ contains half of the total area) vs the argon density. After Ref. 337.

although weak, the efficiency of the line-mixing process becomes comparable to that of the broadening. Together with the above mentioned fact that intensity is transferred from weak to strong absorption regions, this indicates that line mixing results is a *large reduction* of the far band wings. This is illustrated by Fig. IV.7 which displays the absorption by CO_2 in 1 atm of Ar in the R branch troughs and high frequency wing of the

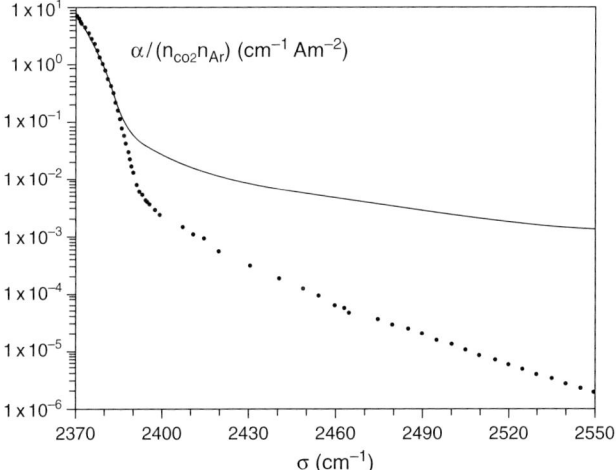

Fig. IV.7: Density (bi)normalized absorption in the R branch troughs and high frequency wing of the v_3 band of CO_2 in Ar at room temperature. (•) are measured values from Refs. 122,338,339; (——) are calculated values neglecting line mixing. After Ref. 122.

ν_3 band. Neglecting line mixing and assuming isolated lines with Lorentzian shapes lead to satisfactory results in the core region of the R branch below $\approx 2380\,\text{cm}^{-1}$ (corresponding to $J \leq 50$). This is expected since absorption is then dominated by contributions of local (and intense) transitions hence involving small line overlapping at 1 atm. When going toward higher frequencies, the relative contribution of overlapping increases, leading to more and more efficient line-mixing processes. Neglecting them induces overestimation errors reaching three orders of magnitude about $200\,\text{cm}^{-1}$ away from the band origin. Note that, at such large distances from line centers, the impact approximation breaks down (*cf* App. II.B4).

In conclusion, remember, at this step, that line mixing has the following characteristics: (i) It affects intense spectral features close to the centers of some lines provided that the radiator-perturber interactions permit efficient population exchanges between the lower (and upper) levels of the optical transitions and that the overlapping is significant. In this case, line mixing leads, with respect to isolated lines models, to a narrowing of the structures. This shows, in the spectral domain, through the decrease of the near wings, the increase of the peak and, generally, the increase of the troughs between coupled lines. The influence in the time domain is thus a slower decay with time in the wings and a stronger decay at the peaks. (ii) It always affects the wings of the lines and bands, leading to a reduction of the spectral signal (absorption, emission, scattering) which can be very significant far away from all transitions.

IV.2 THE SPECTRAL SHAPE

With the exception of few studies considering time-dependent femtosecond Raman scattering (see Ref. 110 and those therein), all studies of line-mixing effects have been made in the spectral domain. For this reason, this section focuses on the spectral shape *only*. Nevertheless, equations given in Refs. 110,340 show that the time-dependent Raman signal can be expressed, as in the spectral domain, in terms of the relaxation matrix. Following lines similar to those used below, it is quite straightforward to obtain equivalent expressions in the time domain.

IV.2.1 APPROXIMATIONS AND GENERAL EXPRESSIONS

Let us first recall some simplifying approximations introduced in chapter II, for which (quantitative) criteria are given in App. II.B. • We consider densities sufficiently low to assume *binary collisions*, *ie* an optically active molecule (radiator) interacts with a single other molecule or atom (perturber) at the time. • We work within the frame of the "*impact approximation*" and thus focus on regions not too far away from the optical transitions giving major contributions to the spectrum. • We assume *Local Thermodynamic Equilibrium*, so that all populations are driven by the kinetic temperature through Maxwell-Boltzmann statistics. • We *neglect* the influence of the changes of the radiator velocity and that, through averaging of the profile, of the speed dependence of collisional parameters. Hence, all the effects discussed in chapter III besides those leading to the purely Doppler, Lorentzian, and Voigt approaches for isolated lines, are disregarded

Collisional line mixing 155

here. Their influence on spectra affected by line mixing is the subject of Sec. VIII.2.
• Finally, in order to simplify notations, we consider the case of a mixture of *radiators* of species "a" highly diluted in a bath of optically inactive *perturbers* "p". This makes a-a collisions much more rare than a-p collisions so that only a-p interactions must be taken into account. Within this frame, all collisional quantities Q (in cm^{-1} or s^{-1}) are proportional to the density n_p of the bath molecules or atoms (*ie* $Q = n_p q$ where q is per unit density).

Within these approximations, which are valid under the conditions of many practical situations (*cf* App. II.B), the relevant quantities in the spectral domain (absorption, emission, Raman signal, *etc*) for a binary mixture of radiators "a" and perturbers "p" with densities n_a and n_p ($n_a \ll n_p$) at temperature T can all be related to the following complex normalized profile [*cf* Eq. (II.50)]:

$$I(\sigma, n_p, T) = \frac{1}{\pi} \frac{\sum_\ell \sum_{\ell'} \rho_\ell(T) d_\ell d_{\ell'} \langle\langle \ell' | [\Sigma - L_a - i n_p W(T)]^{-1} | \ell \rangle\rangle}{\sum_\ell \rho_\ell(T) \times d_\ell^2}. \quad \text{(IV.1)}$$

Note that, if the small effects of resonance exchange (App. IV.C) are neglected, this expression can be applied to pure gases and, more generally, to any mixture by simply replacing $n_p W$ by $\Sigma n_i W_{a-i}$ where the sum is made over all species i composing the gas, W_{a-i} being the relaxation matrix for species a colliding with species i. As discussed in appendix II.A, the quantity in Eq. (IV.1) enables the calculation of the profiles involved in various spectroscopic techniques.[1] One can calculate the transmission and dispersion characteristics: the refractive index [real part of $I(\sigma)$] and the absorption coefficient [imaginary part of $I(\sigma)$] resulting from the interaction of the molecular electric or magnetic dipole or quadrupole (tensors of rank $k=1$ and $k=2$, respectively) with the electromagnetic field. Similarly the signals in Raman scattering resulting from the isotropic or anisotropic polarizability (tensors of rank $k=0$ and $k=2$, respectively) and the Coherent Anti-Stokes Raman response due to the third order susceptibility are also related to $I(\sigma, n_p, T)$ by proportionality relations with its imaginary part or squared modulus. Remember that the relaxation matrix W *depends* on the order k of the radiation-matter coupling and is thus different in infrared and Raman spectroscopies, for instance.

In Eq. (IV.1), the sum extends over all lines $\ell = f \leftarrow i$ and $\ell' = f' \leftarrow i'$ of the radiator[2] (noted $\langle\langle f'i'|$ and $|fi\rangle\rangle$ in the *Liouville* space, see App. II.D), d_ℓ is the reduced matrix

---

[1] The expression of the complex profile I in Eq. (IV.1) respects the "traditional" choice of most studies devoted to line-mixing. It is different from that $J = iI^*$ made in chapter III which is consistent with the usual choice of the "isolated lines community". Consequently, the absorption and dispersion coefficients, related to the real and imaginary parts of J, respectively, are associated with the imaginary and real parts of I. Finally, note that, in some papers, $(-i n_p W)$ in Eq. (IV.1) is replaced by $(+i n_p W^*)$, where W^* is the complex conjugate of W, so that I must be changed to I^*.

[2] At *long* wavelengths, the "negative resonances" corresponding to the transitions $i \leftarrow f$ and $i' \leftarrow f'$, centered at the negative wavenumbers $-\sigma_{fi}$ and $-\sigma_{f'i'}$, must be taken into account in the sum; they can make significant contributions, particularly in the microwave domain.

element of the tensor d (*not* necessarily restricted to the electric dipole, *cf supra*) that couples radiation and matter, *ie*:

$$d_{\ell'} = \langle f' \|d\| i' \rangle, \text{ and } d_\ell = \langle f \|d\| i \rangle, \tag{IV.2}$$

and ρ_ℓ is the equilibrium relative population of the initial level of line ℓ, defined from the collisionally unperturbed radiator density operator:

$$\rho_a(T) = \exp(-H_a/k_B T)/\text{Tr}[\exp(-H_a/k_B T)], \tag{IV.3}$$

H_a being the corresponding Hamiltonian and Tr[...] meaning the trace over all its internal states. In Eq. (IV.1), Σ, L_a, and W are (super) operators in the Liouville (line) space (*cf* App. II.D). The first two, which are *diagonal* and real, are associated with the current wavenumber σ and with the positions σ_ℓ of the unperturbed lines, *ie*:

$$\langle\langle \ell' |\Sigma| \ell \rangle\rangle = \langle\langle \ell' |\sigma I_d| \ell \rangle\rangle = \delta_{\ell,\ell'} \times \sigma, \text{ and,}$$

$$\langle\langle \ell' |L_a| \ell \rangle\rangle = \frac{\langle f'|H_a|f\rangle \delta_{i,i'} - \langle i'|H_a|i\rangle \delta_{f,f'}}{hc} = \delta_{\ell,\ell'} \frac{(E_f - E_i)}{hc} = \delta_{\ell,\ell'} \times \sigma_\ell, \tag{IV.4}$$

with $\delta_{\ell,\ell'} = \delta_{i,i'} \delta_{f,f'}$. Within the approximations made, *all* the influence of radiator-perturber (a-p) collisions on the spectrum of species "a" is contained in the (complex) *relaxation matrix* W, which is *frequency independent* within the impact frame. Since the influence of the radiator velocity changes and averaging (*cf* chapter III and Sec. VIII.2) are disregarded, W(T) is the *thermally averaged* value of the radiator-perturber relative speed (v_r)-dependent matrix $W(T,v_r)$. It is density normalized (cm^{-1}/Am or cm^{-1}/atm), associated with a-p collisions, and dependent on the order k of the tensor coupling the radiators to the electromagnetic field. The number of its elements is the square of the number of lines but many terms are zero (or negligible) due to collisional selection rules (or to the weakness of some population transfers) so that, in practice, W is block-diagonal. Its off-diagonal elements describe collisional interferences between lines, and the diagonal terms are the pressure-broadening (γ_ℓ, HWHM) and -shifting (δ_ℓ) coefficients (per unit density or pressure) of lines of species a induced by collisions with species p, *ie*:

$$\langle\langle \ell |W(T)| \ell \rangle\rangle = \gamma_\ell(T) - i\delta_\ell(T). \tag{IV.5}$$

Furthermore, W verifies the detailed balance relation [Eq. (II.55)] ensuring that the total population is conserved (see App. II.E3 and Ref. 46):

$$\langle\langle \ell' |W(T)| \ell \rangle\rangle \rho_\ell(T) = \langle\langle \ell |W(T)| \ell' \rangle\rangle \rho_{\ell'}(T). \tag{IV.6}$$

When line mixing is neglected, all off-diagonal elements of W are zero and Eq. (IV.1) reduces to a sum of individual line contributions with (complex) Lorentzian shapes:

$$I^{\text{Lorentz}}(\sigma, n_b, T) = \frac{1}{\pi \sum_\ell \rho_\ell(T) \times d_\ell^2} \sum_\ell \frac{\rho_\ell(T) \times d_\ell^2}{\langle\langle \ell |\Sigma - L_a - in_p W(T)| \ell \rangle\rangle}. \tag{IV.7}$$

Close to line centers, this leads to the usual Lorentzian expression of the spectral absorption coefficient α:

$$\alpha^{\text{Lorentz}}(\sigma, n_a, n_p, T) = n_a \times \sum_\ell \frac{S_\ell(T)}{\pi} \frac{n_p \gamma_\ell(T)}{[\sigma - \sigma_\ell - n_p \delta_\ell(T)]^2 + [n_p \gamma_\ell(T)]^2}, \quad \text{(IV.8)}$$

where Eqs. (IV.4) and (IV.5) have been used and we have introduced the *integrated intensities* $S_\ell(T)$ of the lines, defined by [*cf* Eqs. (II.6) and (II.54)]:

$$S_\ell(T) = \frac{8\pi^3}{3hc} \sigma_\ell [1 - \exp(-hc\sigma_\ell/k_B T)] \rho_\ell(T) d_\ell^2. \quad \text{(IV.9)}$$

In the long wavelength region, negative resonances must be taken into account, leading, in the absence of line mixing, to isolated line contributions ($f \leftarrow i$ with $f \geq i$, *ie* $\omega_{fi} \geq 0$) with Van Vleck-Weisskopf profiles[179] (*cf* § III.3.1):

$$I^{\text{VVW}}(\sigma, n_p, T) = \frac{1}{\pi} \left\{ \frac{1}{\sigma - \sigma_\ell - n_p \delta_\ell(T) - i n_p \gamma_\ell(T)} \right. \\ \left. + \frac{1}{\sigma + \sigma_\ell + n_p \delta_\ell(T) - i n_p \gamma_\ell(T)} \right\}. \quad \text{(IV.10)}$$

Note that it is convenient to write Eq. (IV.1) in a more compact symmetric form, *ie*:

$$I(\sigma, n_p, T) = \frac{1}{\pi} \langle\langle u_0(T) | [\Sigma - L_a - i n_p{}^s W(T)]^{-1} | u_0(T) \rangle\rangle, \quad \text{(IV.11)}$$

where ${}^s W$ is the *symmetric* matrix related to W by ${}^s W \equiv \rho_a^{1/2} W \rho_a^{-1/2}$ [*cf* Eq. (IV.6)]:

$$\langle\langle \ell'|{}^s W|\ell\rangle\rangle = \frac{\sqrt{\rho_{\ell'}}}{\sqrt{\rho_\ell}} \langle\langle \ell'|W|\ell\rangle\rangle = \frac{\sqrt{\rho_\ell}}{\sqrt{\rho_{\ell'}}} \langle\langle \ell|W|\ell'\rangle\rangle = \langle\langle \ell|{}^s W|\ell'\rangle\rangle, \quad \text{(IV.12)}$$

and

$$|u_0(T)\rangle\rangle \equiv \frac{1}{\sqrt{\sum_\ell \rho_\ell(T) \times d_\ell^2}} \times \sum_\ell \sqrt{\rho_\ell(T)} d_\ell |\ell\rangle\rangle. \quad \text{(IV.13)}$$

Note a very *important* property of W, valid when species a can be considered as a *rigid rotor*. It is the case when the free radiator eigenfunctions are not perturbed by any (vibration-rotation, Coriolis, *etc*) significant coupling and when the matrix elements of the interaction potential with the collision partner only depend on the rotational states (disregarding any influence of vibration, for instance). The relaxation matrix then verifies the following *sum rule*:[338]

$$\sum_{\ell'} d_{\ell'} \langle\langle \ell'|W|\ell\rangle\rangle = 0. \quad \text{(IV.14)}$$

This equation is widely used, since it *relates* the off-diagonal elements of W to the diagonal ones (the individual line-broadening and line-shifting coefficients, quantities which are relatively easy to measure) through the equation:

$$-\frac{1}{d_\ell} \sum_{\ell' \neq \ell} d_{\ell'} \langle\langle \ell'|W(T)|\ell\rangle\rangle = \langle\langle \ell|W(T)|\ell\rangle\rangle = \gamma_\ell(T) - i\delta_\ell(T). \quad \text{(IV.15)}$$

When the rigid rotor approximation does not apply, Eq. (IV.14) breaks down and must be replaced, introducing line-dependent deviations $\Delta\gamma_\ell$ and $\Delta\delta_\ell$ to the sum rule, by:

$$\frac{1}{d_\ell} \sum_{\ell'} d_{\ell'} \langle\langle \ell'|W(T)|\ell\rangle\rangle = \Delta\gamma_\ell(T) - i\Delta\delta_\ell(T). \quad \text{(IV.16)}$$

Equation (IV.16) is often simplified assuming that the sum over each column leads to a *line-independent* quantity, ie ($\forall \ \ell$):

$$\frac{1}{d_\ell} \sum_{\ell'} d_{\ell'} \langle\langle \ell'|W(T)|\ell\rangle\rangle = \Delta\gamma(T) - i\Delta\delta(T), \quad \text{(IV.17)}$$

with

$$\Delta\gamma(T) - i\Delta\delta(T) = \sum_{\ell,\ell'} \rho_\ell(T) d_\ell d_{\ell'} \langle\langle \ell'|W(T)|\ell\rangle\rangle \bigg/ \sum_\ell \rho_\ell(T) d_\ell^2. \quad \text{(IV.18)}$$

Equation (IV.17) is valid for spectroscopically unperturbed isotropic Raman Q branches, where $\Delta\gamma$ and $\Delta\delta$ result from the *vibrational dephasing*. They can then be identified as the "vibrational" broadening and shifting coefficients γ_v and δ_v due to the dependence of the radiator-perturber isotropic potential on the vibrational state (*cf* appendix IV.A and Ref. 341, for instance). In other situations, Eq. (IV.17) is approximate but convenient since it depends on only two parameters which can be adjusted on experiments.

IV.2.2 ASYMPTOTIC EXPANSIONS

We consider here situations where the general expression of the spectral profile in Eq. (IV.1) can be simplified, leading to equations depending on a reduced number of collisional quantities. This can be done at low and high densities when the overlapping of pressure-broadened transitions is weak and strong, respectively. Simplifications are also possible in the wings, for wavenumbers that are far away from the centers of all the lines composing a manifold or an entire band.

IV.2.2a Weak overlapping

First-order approximation: At sufficiently low densities (or pressures), such that $|\sigma_\ell - \sigma_{\ell'}| \gg n_p |\langle\langle \ell'|W|\ell\rangle\rangle| \ \forall(\ell,\ell')$ with $\ell \neq \ell'$, an expansion of Eq. (IV.1) can be made.[342] It is based on the idea that the perturbation introduced by collisions is small so that the eigenvectors $|\ell\rangle\rangle + |\Delta\ell(n_p,T)\rangle\rangle$ of $L_a + in_p W$ are close to those $(|\ell\rangle\rangle)$ of L_a.

Introducing these perturbed eigenvectors into Eq. (IV.1), one can carry out a series expansion. Only keeping the terms of order zero and one, it leads to:

$$I^{1st}(\sigma, n_p, T) = \frac{1}{\pi \sum_\ell \rho_\ell(T) d_\ell^2} \sum_\ell \frac{\rho_\ell(T) d_\ell^2}{\langle\langle \ell | \Sigma - \mathbf{L}_a - i n_p W(T) | \ell \rangle\rangle} \quad (IV.19)$$

$$\times \left\{ 1 + \sum_{\ell' \neq \ell} \frac{d_{\ell'}}{d_\ell} \langle\langle \Delta\ell'(n_p, T) | \ell \rangle\rangle + \frac{d_{\ell'} \rho_{\ell'}(T)}{d_\ell \rho_\ell(T)} \langle\langle \ell' | \Delta\ell(n_p, T) \rangle\rangle \right\}.$$

Since $|\ell\rangle\rangle + |\Delta\ell(n_p, T)\rangle\rangle$ is an eigenvector of $L_a + in_p W$ and $|\ell\rangle\rangle$ is an eigenvector of L_a, one can write, to first order, for $\ell \neq \ell'$ and with simplified notations:

$$\langle\langle \ell | L_a + in_p W (|\ell'\rangle\rangle + |\Delta\ell'(n_p, T)\rangle\rangle) = \sigma_{\ell'} \langle\langle \ell | \Delta\ell' \rangle\rangle$$
$$= \sigma_\ell \langle\langle \ell | \Delta\ell' \rangle\rangle + \langle\langle \ell | in_p W | \Delta\ell' \rangle\rangle, \quad (IV.20)$$
$$(\langle\langle \Delta\ell' | + \langle\langle \ell' |) L_a + in_p W | \ell \rangle\rangle = \sigma_{\ell'} \langle\langle \Delta\ell' | \ell \rangle\rangle$$
$$= \sigma_\ell \langle\langle \Delta\ell' | \ell \rangle\rangle + \langle\langle \Delta\ell' | in_p W | \ell \rangle\rangle, \text{ i.e.:}$$

$$\langle\langle \ell | \Delta\ell' \rangle\rangle = \frac{\langle\langle \ell | in_p W | \ell' \rangle\rangle}{\sigma_\ell - \sigma_{\ell'}} \text{ and } \langle\langle \Delta\ell' | \ell \rangle\rangle = \frac{\langle\langle \ell' | in_p W | \ell \rangle\rangle}{\sigma_\ell - \sigma_{\ell'}}. \quad (IV.21)$$

Introducing these expressions into Eq. (IV.19) leads to the (complex) Rosenkranz profile:[342]

$$I^{1st}(\sigma, n_p, T) = \frac{1}{\pi \sum_\ell \rho_\ell(T) d_\ell^2} \left\{ \sum_\ell \frac{\rho_\ell(T) d_\ell^2}{\sigma - \sigma_\ell - n_p \delta_\ell(T) - i n_p \gamma_\ell(T)} [1 + i n_p Y_\ell(T)] \right\}, \quad (IV.22)$$

with

$$Y_\ell(T) = \sum_{\ell' \neq \ell} \frac{d_{\ell'}}{d_\ell} \left\{ \frac{\langle\langle \ell' | W(T) | \ell \rangle\rangle}{\sigma_\ell - \sigma_{\ell'}} + \frac{\rho_{\ell'}(T)}{\rho_\ell(T)} \frac{\langle\langle \ell | W(T) | \ell' \rangle\rangle}{\sigma_\ell - \sigma_{\ell'}} \right\}. \quad (IV.23)$$

Due to Eq. (IV.6), this reduces to:

$$Y_\ell(T) = 2 \sum_{\ell' \neq \ell} \frac{d_{\ell'}}{d_\ell} \frac{\langle\langle \ell' | W(T) | \ell \rangle\rangle}{\sigma_\ell - \sigma_{\ell'}}. \quad (IV.24)$$

The first-order line-coupling coefficient Y_ℓ is a priori complex, but it is often considered as real due to the neglecting of the off-diagonal imaginary elements of W. When applicable, the first-order approximation can be very convenient for two reasons. The first one is that it puts the spectral shape under a line-by-line form. For instance, introducing the integrated intensities given by Eq. (IV.9), Eq. (IV.8) becomes:

$$\alpha^{1st}(\sigma, n_a, n_p, T) = n_a \sum_\ell S_\ell(T) \text{Im}\left\{\frac{1 + in_p Y_\ell(T)}{\sigma - \sigma_\ell - n_p \delta_\ell(T) - in_p \gamma_\ell(T)}\right\}, \quad \text{(IV.25)}$$

which is computationally efficient and of easy interfacing with molecular databases providing the individual line parameters σ_ℓ, S_ℓ, γ_ℓ, δ_ℓ, and Y_ℓ. When Doppler effects are taken into account, the resulting expression is:

$$\alpha^{1st}(\sigma, n_a, n_p, T) = n_a \sum_\ell \frac{S_\ell(T)}{\gamma_D(\sigma_\ell, T)} \sqrt{\frac{\ln(2)}{\pi}} \text{Re}\left[(1 - in_p Y_\ell(T)) \times w(x + iy)\right], \quad \text{(IV.26)}$$

where $\gamma_D(\sigma_\ell, T)$ is the radiator Doppler HWHM [Eq. (III.5)] and the dimensionless variables x and y [see Eq. (III.20)] are recalled below:

$$x = \frac{\sigma - \sigma - n_p \delta_\ell(T)}{\gamma_D(\sigma_\ell, T)/\sqrt{\ln(2)}}, \quad y = \frac{n_p \gamma_\ell(T)}{\gamma_D(\sigma_\ell, T)/\sqrt{\ln(2)}}. \quad \text{(IV.27)}$$

$w(z = x + iy)$ is the complex probability function, defined as:[312]

$$\text{for } y > 0, \quad w(z) = \frac{i}{\pi}\int_{-\infty}^{+\infty}\frac{e^{-t^2}}{z-t}dt = e^{-z^2}\left[1 + \frac{2i}{\sqrt{\pi}}\int_0^z e^{t^2}dt\right], \quad \text{(IV.28)}$$

for which efficient computer routines are available (see App. III.A, Refs. 207,314 and those therein), including the derivatives needed for minimization procedures. The second advantage of the first-order approach is that the number of collisional quantities is significantly reduced, typically from half the square of the number of coupled lines (W matrix) to twice the number of lines (diagonal elements of W and first-order line-coupling coefficients Y). This is of interest for determinations of line-mixing parameters from fits of measured spectra, as discussed in Sec. IV.4

Although the limited number of parameters and the line-by-line form a priori make the first-order approach tempting for efficient computations and storage in databases, this is *not* really the case for two reasons. The first one is that, in the absence of reference calculations using Eq. (IV.1), the errors introduced by the first-order approximation are difficult to guess. They depend on the pressure and on the spectral structure of the considered spectral feature.[343] The second problem comes from the fact that, since W verifies Eq. (IV.6), the Y_ℓ coefficients must satisfy the following sum rule:

$$\sum_\ell \rho_\ell(T) d_\ell^2 Y_\ell(T) = 0, \quad \text{(IV.29)}$$

(and eventually some others detailed later). When Eq. (IV.29) is not accurately satisfied, calculated spectra can show large errors which quickly increase when going away in the wings (see Refs. 344,345 and § IV.2.3a). As a result, up to four parameters with up to nine significant digits had to be used in Ref. 346 in order to describe the Y's and their temperature dependences. Furthermore, slight modifications of the spectroscopic parameters (σ_ℓ, ρ_ℓ, and/or d_ℓ) may forbid the use of a former set of Y_ℓ since Eq. (IV.29) then breaks down.

Second-order approximation.

An approach similar to that leading to the first-order expression in Eq. (IV.22) has been used to extend the expansion to second order,[347] leading to the following expression:

$$I^{2^{nd}}(\sigma, n_p, T) = \frac{1}{\pi \sum_\ell \rho_\ell(T) d_\ell^2} \sum_\ell \frac{\rho_\ell(T) d_\ell^2 [1 + i n_p Y_\ell(T) + n_p^2 g_\ell(T)]}{\sigma - \sigma_\ell - n_p \delta_\ell(T) - i n_p \gamma_\ell(T) + n_p^2 \delta \nu_\ell(T)}, \quad (IV.30)$$

where the second-order line-mixing parameters are given by:[347]

$$\delta \nu_\ell(T) = \sum_{\ell' \neq \ell} \frac{\langle\langle \ell' | W | \ell \rangle\rangle \langle\langle \ell | W | \ell' \rangle\rangle}{\sigma_{\ell'} - \sigma_\ell},$$

$$g_\ell(T) = \sum_{\ell' \neq \ell} \frac{\langle\langle \ell' | W | \ell \rangle\rangle \langle\langle \ell | W | \ell' \rangle\rangle}{(\sigma_\ell - \sigma_{\ell'})^2} - \left[\sum_{\ell' \neq \ell} \frac{d_{\ell'}}{d_\ell} \frac{\langle\langle \ell' | W | \ell \rangle\rangle}{\sigma_{\ell'} - \sigma_\ell} \right]^2 \quad (IV.31)$$

$$+ 2 \sum_{\ell' \neq \ell} \frac{d_{\ell'}}{d_\ell} \frac{\langle\langle \ell' | W | \ell \rangle\rangle \langle\langle \ell | W | \ell' \rangle\rangle}{(\sigma_\ell - \sigma_{\ell'})^2} - 2 \sum_{\substack{\ell' \neq \ell \\ \ell'' \neq \ell}} \frac{d_{\ell'}}{d_\ell} \frac{\langle\langle \ell' | W | \ell'' \rangle\rangle \langle\langle \ell'' | W | \ell \rangle\rangle}{(\sigma_\ell - \sigma_{\ell'})(\sigma_\ell - \sigma_{\ell''})}.$$

This extension brings improvements in a *limited* pressure range only, when the density squared terms in Eq. (IV.30) are small but not negligible when compared to the zero- and first-order contributions. In this case, illustrated in Fig. IV.8, significantly more accurate results are obtained than when a first-order approximation is used.

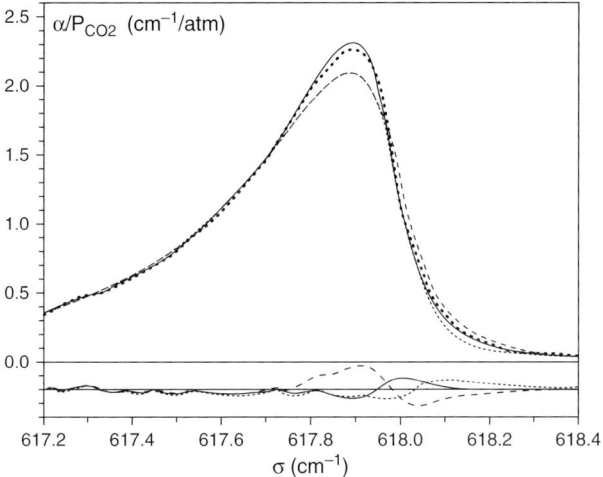

Fig. IV.8: Absorptions and in the $(10^00)_{II} \leftarrow (01^10)_I$ Q-branch of CO_2 diluted in N_2 T = 250 K and 1 atm. (•) are experimental values. Computed values and meas-calc deviations obtained using the models: (——) full W operator, (···) 2nd order approximation, (- - -) 1st order approximation. Reproduced with permission from Ref. 343.

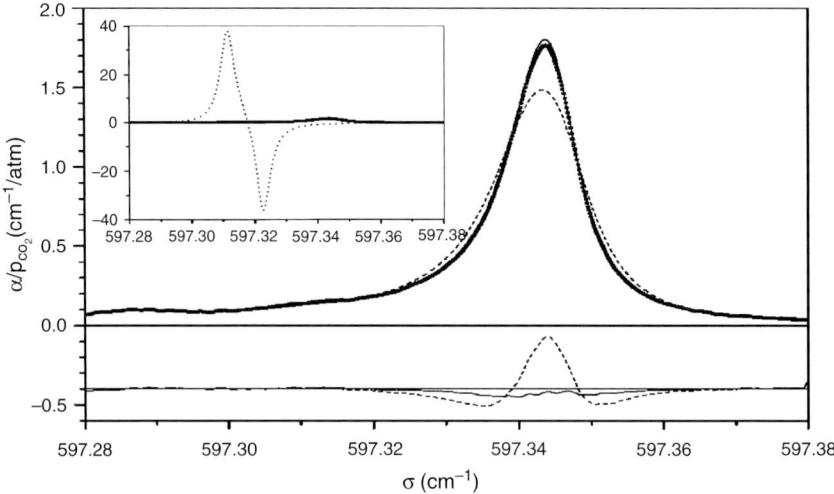

Fig. IV.9: Absorptions and meas-calc residuals in the $(11^10)_{II} \leftarrow (02^20)_I$ Q-branch of CO_2 diluted in N_2 at T = 300 K and P = 0.1 atm. The symbols of Fig. IV.8 are used. Due to their large errors, results obtained with the second-order approximation are plotted in the insert. Reproduced with permission from Ref. 343.

At higher densities, very large errors can result from the second-order terms which rapidly increase with pressure, as shown by the insert in Fig. IV.9. For very closely spaced lines, as in the considered CO_2 Q-branch, the squares of the terms $W_{\ell\ell'}/(\sigma_\ell - \sigma_{\ell'})$ in Eq. (IV.31) are large, leading to values of $n_p^2 g_\ell$ getting quickly greater than unity with increasing n_p. These elements show that the interest of the second-order approach is not clear and that using Eq. (IV.1) is much less hazardous.

IV.2.2b Strong overlapping

Entire bands.   In order to study the high density (HD) behavior of the spectral shape, it is convenient to use the symmetrized expression of Eq. (IV.11), ie:

$$I(\sigma, n_b, T) = \frac{1}{\pi} \langle\langle u_0(T) | [\Sigma - L_a - in_p{}^s W(T)]^{-1} | u_0(T) \rangle\rangle, \qquad (IV.32)$$

where $|u_0\rangle\rangle$ is given by Eq. (IV.13). When W verifies the sum rule in Eq. (IV.17), one can easily show that $||u_0(T)\rangle\rangle$ is eigenvector of ^sW with the associated eigenvalue $a_0 = \Delta\gamma - i\Delta\delta$ [zero within the rigid rotor approximation, cf Eq. (IV.14)]. The idea,[348,349] somehow the counterpart of the low density development, is to consider that the eigenvectors $|u_m\rangle\rangle + |\Delta u_m\rangle\rangle$ of $[L_a + in_p{}^s W]$ are not very different, at elevated density n_p, from those $|u_m\rangle\rangle$ of ^sW (which form an orthogonal basis with associated eigenvalues a_m). Introducing them into Eq. (IV.32) and keeping only the first terms lead to (with simplified notations):

$$I^{HD}(\sigma, n_p, T) = \frac{1}{\pi} \frac{1}{\sigma - A_0(n_p, T)} + \frac{1}{\pi} \sum_{m \neq 0} \frac{\langle\langle u_0 | \Delta u_m \rangle\rangle \langle\langle \Delta u_m | u_0 \rangle\rangle}{\sigma - A_m(n_p, T)}, \qquad (IV.33)$$

where $A_m(n_p)$ is the eigenvalue associated with $|u_m\rangle\rangle + |\Delta u_m\rangle\rangle$. The perturbation terms are obtained in a way similar to that used for low densities, leading to:

$$\langle\langle u_0|[\Sigma - L_a - in_p{}^s W]^{-1}|u_0\rangle\rangle = \frac{1}{\sigma - A_0(n_p, T)} - \frac{1}{n_p{}^2}\sum_{m\neq 0}\frac{S_m(T)}{\sigma - A_m(n_p, T)}, \quad (IV.34)$$

with

$$A_0(n_p, T) = \langle\langle u_0|L_a|u_0\rangle\rangle + in_p a_0(T)$$
$$+ \frac{i}{n_p}\sum_{m\neq 0}\frac{\langle\langle u_0|L_a|u_m\rangle\rangle\langle\langle u_m|L_a|u_0\rangle\rangle}{a_m(T) - a_0(T)}\left(1 + \frac{a_0(T)}{a_0(T) - a_m(T)}\right),$$
$$A_{m\neq 0}(n_p, T) = \langle\langle u_m|L_a|u_m\rangle\rangle + in_p a_m(T),$$
$$S_m(T) = \frac{\langle\langle u_0|L_a|u_m\rangle\rangle\langle\langle u_m|L_a|u_0\rangle\rangle}{[a_m(T) - a_0(T)]^2}.$$
$$(IV.35)$$

As density increases, the second term in Eq. (IV.34) vanishes (as $n_p{}^{-2}$), only leaving the first term which corresponds to a *single* (complex) *Lorentzian equivalent* line. Its position, including a pressure shift, is $\text{Re}\{A_0(n_p,T)\}$, which reduces to $\bar{\sigma} + n_p\bar{\delta}$ when the imaginary part of W is diagonal. The unperturbed position $\bar{\sigma}$ and collisional shifting coefficient $\bar{\delta}$ of the equivalent line are then intensity-weighted averages of the individual line positions σ_ℓ and shifts δ_ℓ, ie:

$$\bar{\sigma}(T) = \langle\langle u_0|L_a|u_m\rangle\rangle = \sum_\ell \rho_\ell(T)d_\ell{}^2 \sigma_\ell \bigg/ \sum_\ell \rho_\ell(T)d_\ell{}^2, \quad (IV.36)$$

$$\bar{\delta}(T) = -\text{Im}\left\{\langle\langle u_0|{}^s W|u_0\rangle\rangle = \sum_\ell \rho_\ell(T)d_\ell{}^2 \delta_\ell \bigg/ \sum_\ell \rho_\ell(T)d_\ell{}^2\right\}. \quad (IV.37)$$

The width (HWHM) $\text{Im}\{A_0(n_p,T)\}$ of the equivalent line, when imaginary elements of W are disregarded so that all $a_m(T)$ are real, is given by:

$$\bar{\Gamma}(n_p, T) = [n_p\Delta\gamma(T)] + \frac{\gamma^{(2)}(T)}{n_p}, \quad (IV.38)$$

with $\gamma^{(2)}(T) = \sum_{m\neq 0}\frac{\langle\langle u_0|L_a|u_m\rangle\rangle\langle\langle u_m|L_a|u_0\rangle\rangle}{a_m(T) - \Delta\gamma(T)}\left(1 + \frac{\Delta\gamma(T)}{\Delta\gamma(T) - a_m(T)}\right). \quad (IV.39)$

This high density behavior of the effective width [Eq. (IV.38)] of a cluster of strongly overlapping lines is corroborated by the measurements for an isotropic Raman Q branch of CO_2 plotted in Fig. IV.10. Below about 10 Am, $\bar{\Gamma}$ is dominated by the $\gamma^{(2)}/n_p$ contribution, translating the narrowing of the structure observed in Figs. IV.3b and IV.6b. As the density increases, the contribution of $n_p\Delta\gamma$ in Eq. (IV.38), due here to a vibrational dephasing,[350] becomes significant and the Q branch broadens linearly with

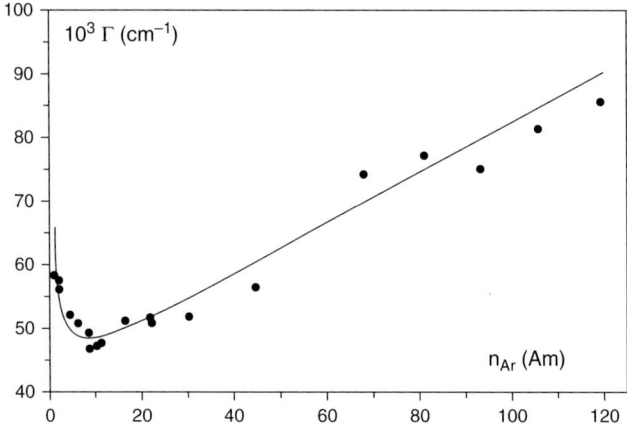

Fig. IV.10: Half width at half maximum of the Ar-broadened isotropic Raman Q branch of the $2\nu_2$ band of CO_2 at room temperature vs Ar density. (•) are experimental values. Courtesy of B. Lavorel, see also Ref. 350. The line gives the best fit using the expression $\Gamma = \Delta\sigma_Q + n_p\Delta\gamma + \gamma^{(2)}/n_p$.

density. Recall that the value of $\bar{\Gamma}$ at very low density, when collisional effects are small, is due to the spectral spreading $\Delta\sigma_Q$ of the lines composing the Q branch. Its value is related, as in Fig. IV.4, to the intensity-weighted root mean square of the line positions.[333] Finally, Figs. IV.3a and IV.6a demonstrate that the band tends to look like a single Lorentzian line at elevated density, corroborating Eq. (IV.34).

Isolated clusters of lines. The preceding results are valid when an *entire* band, including *all* transitions of *all* branches, is considered. When restricted to a few collisionally coupled transitions spectrally *isolated* from the others of the band, the situation is different. Consider a narrow cluster of lines which strongly overlap with each other but remain separated from those outside the cluster, ie. $\forall\,\ell \in$ cluster:

$$\Gamma_\ell = n_p\gamma_\ell \gg |\sigma_\ell - \sigma_{\ell'}|\ \forall\,\ell' \in \text{cluster, and, } \Gamma_\ell \text{ and } \Gamma_{\ell'} \ll |\sigma_\ell - \sigma_{\ell'}|\ \forall\,\ell' \notin \text{cluster.}$$

This is encountered in narrow infrared CO_2 Q branches, in the manifolds in the R branch of the ν_3 band of methane, or in transitions associated with the hyperfine structure due to nuclear spin, for instance. Starting from Eqs. (IV.11) and (IV.13) in which the sum is restricted to the cluster lines, one can introduce the mean line position given by Eq. (IV.36) and the following average of the relaxation matrix elements (see also Ref. 46):

$$\bar{w} \equiv \langle\langle u_0|^s W|u_0\rangle\rangle = \frac{\sum_{\ell,\ell'\in\text{Cluster}}\sqrt{\rho_\ell(T)\rho_{\ell'}(T)}d_\ell d_{\ell'}\langle\langle\ell'|^s W|\ell\rangle\rangle}{\sum_{\ell\in\text{Cluster}}\rho_\ell(T)d_\ell^2}$$

$$= \frac{\sum_{\ell,\ell'\in\text{Cluster}}\rho_\ell(T)d_\ell d_{\ell'}\langle\langle\ell'|W|\ell\rangle\rangle}{\sum_{\ell\in\text{Cluster}}\rho_\ell(T)d_\ell^2}. \tag{IV.40}$$

The operator in Eq. (IV.32) can then be expanded, leading to:

$$\langle\langle u_0|[\Sigma - L_a - in_p{}^sW]^{-1}|u_0\rangle\rangle = \frac{1}{\sigma - \bar{\sigma} - in_p\bar{w}}\left\{\sum_{m=0}^{\infty}\frac{\langle\langle u_0|[\Delta L_a + in_p\Delta W]^m|u_0\rangle\rangle}{[\sigma - \bar{\sigma} - in_p\bar{w}]^m}\right\}, \quad (IV.41)$$

with $\Delta L_a \equiv L_a - \bar{\sigma} \times I_d$ and $\Delta W \equiv {}^sW - \bar{w} \times I_d$, I_d being the identity matrix. It is easy to show that the m = 1 term is zero so that the spectral shape of the cluster corresponds, to first order, to a *single* (complex) *Lorentzian equivalent* line (illustrated by Fig. IV.9). It results from the m = 0 term and contains all the intensity of the cluster. Its zero pressure position and pressure-shifting and -broadening coefficients are given by:

$$\bar{\sigma}_{Cluster}(T) = \langle\langle u_0|L_a|u_0\rangle\rangle = \sum_{\ell\in Cluster}\rho_\ell(T)d_\ell^2\sigma_\ell \Big/ \sum_\ell \rho_\ell(T)d_\ell^2, \quad (IV.42)$$

$$\bar{\delta}_{Cluster}(T) = -\text{Im}\left\{\langle\langle u_0|{}^sW|u_0\rangle\rangle = \sum_{\ell,\ell'\in Cluster}\rho_\ell(T)d_{\ell'}d_\ell W_{\ell'\ell}(T) \Big/ \sum_{\ell\in Cluster}\rho_\ell(T)d_\ell^2\right\}, \quad (IV.43)$$

$$\bar{\gamma}_{Cluster}(T) = \text{Re}\left\{\langle\langle u_0|{}^sW|u_0\rangle\rangle = \sum_{\ell,\ell'\in Cluster}\rho_\ell(T)d_{\ell'}d_\ell W_{\ell'\ell}(T) \Big/ \sum_{\ell\in Cluster}\rho_\ell(T)d_\ell^2\right\}. \quad (IV.44)$$

The cluster then *broadens* with density with a broadening coefficient $\bar{\gamma}_{Cluster}$, as can be observed in Fig. IV.4b. Due to the off-diagonal real elements of W, this broadening is smaller than that, $\bar{\gamma}_{Cluster}^{Isolated}$, predicted by the isolated lines model *ie*:

$$\begin{aligned}\bar{\gamma}_{Cluster} &= \text{Re}\left\{\sum_{\ell,\ell'\in Cluster}\rho_\ell d_{\ell'}d_\ell W_{\ell'\ell} \Big/ \sum_{\ell\in Cluster}\rho_\ell d_\ell^2\right\} \\ &= \bar{\gamma}_{Cluster}^{Isolated} + \text{Re}\left\{\sum_{\ell\neq\ell'\in Cluster}\rho_\ell d_{\ell'}d_\ell W_{\ell'\ell} \Big/ \sum_{\ell\in Cluster}\rho_\ell d_\ell^2\right\}, \\ &\leq \bar{\gamma}_{Cluster}^{Isolated} = \text{Re}\left\{\sum_{\ell\in Cluster}\rho_\ell d_\ell^2 W_{\ell\ell} \Big/ \sum_{\ell\in Cluster}\rho_\ell d_\ell^2\right\} \\ &= \sum_{\ell\in Cluster}\rho_\ell d_\ell^2 \gamma_\ell \Big/ \sum_{\ell\in Cluster}\rho_\ell d_\ell^2.\end{aligned} \quad (IV.45)$$

The observed structure is thus narrower than predicted when line mixing is neglected, as confirmed by Figs. IV.4 and IV.5.

IV.2.2c Wing regions

Entire bands. Consider a single band and wavenumbers σ *far away* from *all* lines, such that $|\sigma - \sigma_\ell| \gg |\sigma_{\ell'} - \sigma_\ell|$ and $|\sigma - \sigma_\ell| \gg n_p|\langle\langle \ell'|W|\ell\rangle\rangle|$, whatever ℓ and ℓ'. Introducing a reference wavenumber σ_0, close to the band center and whose value will be given later, one can expand the $\Sigma - L_a - in_p W$ operator in the wing through:

$$(\Sigma - L_a - in_p W)^{-1} = (\sigma - \sigma_0)^{-1} \sum_{m=0}^{\infty} \left(\frac{L_a - \sigma_0 I_d + in_p W}{\sigma - \sigma_0} \right)^m. \tag{IV.46}$$

The spectral shape in the wing is then expanded as:

$$I^{\text{Wing}}(\sigma, n_p, T) = \frac{1}{\pi} \left[\sum_\ell \rho_\ell d_\ell^2 \right]^{-1} \times \sum_{m=0}^{\infty} \frac{A'_m}{(\sigma - \sigma_0)^{m+1}}. \tag{IV.47}$$

The expressions of the A'_m parameters can be deduced from Eq. (IV.46), by only keeping the terms proportional to n_p (thus disregarding A'_0) since absorption in the far wing is proportional to the perturber density within the binary collision regime [see Eq. (II.57)]. With simplified notations, and after division by the density (introducing $A_m = A'_m/n_p$), the expansion coefficients are given by:

$$A_1 = \sum_{\ell,\ell'} \rho_\ell d_\ell d_{\ell'} W_{\ell'\ell}, \quad A_2 = \sum_{\ell,\ell'} \rho_\ell d_\ell d_{\ell'} [(\sigma_\ell - \sigma_0) + (\sigma_{\ell'} - \sigma_0)] W_{\ell'\ell},$$
$$A_3 = \sum_{\ell,\ell'} \rho_\ell d_\ell d_{\ell'} \left[(\sigma_\ell - \sigma_0)^2 + (\sigma_{\ell'} - \sigma_0)^2 + (\sigma_\ell - \sigma_0)(\sigma_{\ell'} - \sigma_0) \right] W_{\ell'\ell}. \tag{IV.48}$$

When W verifies Eqs. (IV.6) and (IV.14), it is easy to show that $A_1 = A_2 = 0$, and that the first non-vanishing term reduces to:

$$A_3 = \sum_{\ell,\ell'} \rho_{\ell'} d_\ell d_{\ell'} \sigma_{\ell'} W_{\ell'\ell} \sigma_\ell. \tag{IV.49}$$

Hence, when the rigid rotor sum rule [Eq. (IV.14)] *applies*, the spectral shape in the wing *behaves as* $1/(\sigma-\sigma_0)^4$ [Eq. (IV.47) with m = 3], as already shown in Refs. 351–353. Note that more refined expansions, valid at closer distances in the wings, can be found in Ref. 352. Far away from all lines, a breakdown of the sum rule [Eq. (IV.17)] leads to a non-zero term in $1/(\sigma-\sigma_0)^2$ whereas verifying Eq. (IV.14) but not the detailed balance relation leads to a behavior in $1/(\sigma-\sigma_0)^3$. Finally, the value of σ_0 can be chosen in order to set the A_4 term to zero, leading to:[352]

$$\sigma_0 = \frac{\sum_{\ell,\ell'} \rho_{\ell'} d_\ell d_{\ell'} \sigma_{\ell'} W_{\ell'\ell} \sigma_\ell (\sigma_\ell + \sigma_{\ell'})}{4 A_3}, \tag{IV.50}$$

which proves to be close to the center of gravity of the band. Far wing behaviors, in the high frequency side of the CO_2 ν_3 band, are given in Fig. IV.11. This band is a good candidate for the rigid rotor model. Indeed, it is negligibly affected by Coriolis and centrifugal distortion effects,[354] and the CO_2-perturber interaction potentials in the upper and lower vibrational states are very similar, as shown by the small values[355] of the pressure induced shifts in this band. (typically three times smaller[122] that those plotted in Fig. III.6). Figure IV.11 demonstrates the validity of the $1/(\sigma-\sigma_0)^4$ law in the case of interactions with helium. This perturber is favorable since collisions are very short (high relative speed and short ranged potential), making the impact approximation adapted for the description of the wings (cf Apps. II.C1 and II.B4). This is not the case for pure CO_2 since the long-range quadrupole-quadrupole interaction and slower relative speed of collisions lead to much longer interaction times. This limits the validity of the impact approximation to the near wings. The influence of the finite duration of collision is then significant, the relaxation matrix depends on σ and the $1/(\sigma-\sigma_0)^4$ law breaks down. Note that dividing the measured values by the $1/(\sigma-\sigma_0)^4$ law gives a qualitative "idea" of the dependence on σ of the elements of the true relaxation matrix $W^T(\sigma)$ with respect to their impact values associated with W^1. Figure IV.11 then indicates that $\text{Re}[W^T_{\ell\ell'}(\sigma)]/\text{Re}[W^1_{\ell\ell'}]$ is, on the average, greater than unity at close and intermediate distances and then progressively decreases down to zero in the far wings, in agreement with the results of Refs. 122,339,356.

A development in the band wings can also be made using the first-order expression of Eq. (IV.22), leading to the following expansion parameters ($A'_m = n_p A_m$) for Eq. (IV.47):

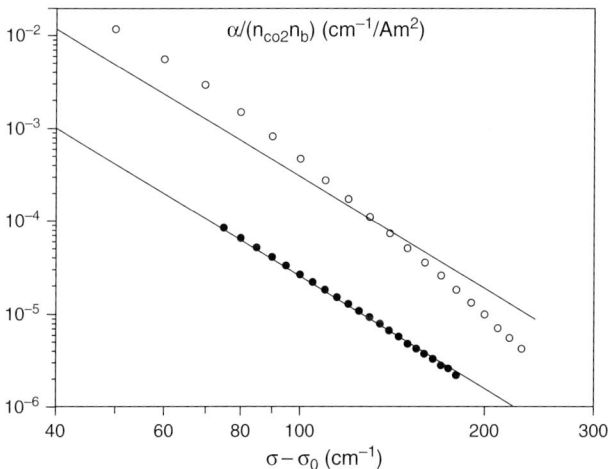

Fig. IV.11: Density (bi)normalized room temperature absorptions in the high frequency wing of the ν_3 band of CO_2 self- (o) and He (•) -broadened. The symbols give experimental values from Refs. 352,357,358. The lines correspond to properly scaled $1/(\sigma-\sigma_0)^4$ laws where $\sigma_0 = 2350 \text{ cm}^{-1}$ is the band center.

$$A_0 = \sum_\ell \rho_\ell d_\ell^2 Y_\ell, \quad A_1 = \sum_\ell \rho_\ell d_\ell^2 [W_{\ell\ell} + Y_\ell(\sigma_\ell - \sigma_0)],$$

$$A_2 = \sum_\ell \rho_\ell d_\ell^2 \Big[2W_{\ell\ell}(\sigma_\ell - \sigma_0) + Y_\ell(\sigma_\ell - \sigma_0)^2 \Big], \quad \text{(IV.51)}$$

$$A_3 = \sum_\ell \rho_\ell d_\ell^2 \Big[3W_{\ell\ell}(\sigma_\ell - \sigma_0)^2 + Y_\ell(\sigma_\ell - \sigma_0)^3 \Big].$$

One can show, using the definition of Y_ℓ [Eq. (IV.24)] and the detailed balance relation for W [Eq. (IV.6)], that A_3 is the first non-zero term *when* the rigid rotor sum rule applies. Hence the $1/(\sigma-\sigma_0)^4$ behavior is, of course, obtained again (A_1 being the first non-vanishing term if the sum rule does not apply). The expressions in Eq. (IV.51) are interesting for reasons further discussed in § IV.2.3a and IV.4.3. They will help explaining the problems connected with the storage of Y's in databases and they can be used to constraint fits of Y's from measured spectra.

Isolated clusters of lines. Similar developments can be made in the wings of a cluster of lines isolated from the others. In this case, A_1 does not vanish and the wings corresponds to those of a single Lorentzian equivalent feature whose parameters are given by Eqs. (IV.42)–(IV.44). They thus behave as $1/(\sigma-\sigma_0)^2$ but are lower than the purely isolated lines prediction due to the influence of line couplings [Eq. (IV.45)]. This is confirmed in Fig. IV.12 by the CH_4 R(8) manifold of the v_3 band.

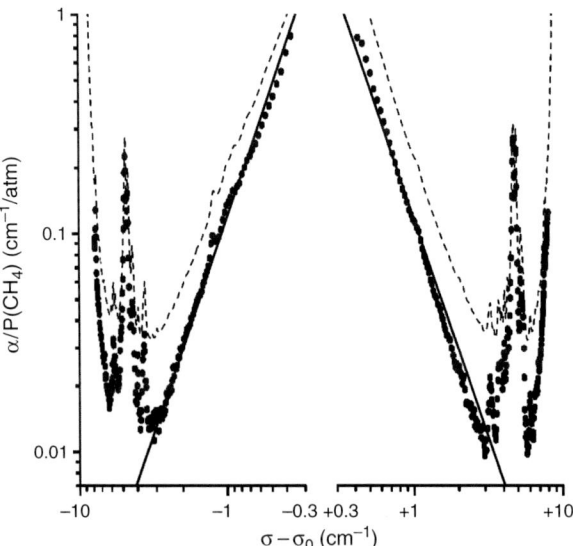

Fig. IV.12: Room temperature absorptions in the low and high frequency wings of the R(8) manifold of the v_3 band of CH_4 broadened by N_2 at 296 K and 0.985 atm. (•) are measured values, (——) corresponds to the best fit using a $1/(\sigma-\sigma_0)^2$ dependence where $\sigma_0 = 3104.4\,cm^{-1}$ is the manifold central position, and (– – –) is the purely isolated lines prediction. After Ref. 359.

IV.2.3 COMPUTATIONAL ASPECTS AND RECOMMENDATIONS

IV.2.3a Using the first-order expression

Recall that, when expanded in the wing of an *entire* band, the spectral shape is given by Eq. (IV.47), with parameters $A'_m = n_p A_m$ given by Eq. (IV.51), and that, within the impact and rigid rotor approximations, A_3 is the first non-zero term. In this case, the Y's must verify relations with high accuracy. In order to quantify the problem, consider the wing of the ν_3 band of CO_2 in He where the $1/(\sigma-\sigma_0)^4$ behavior is valid (Fig. IV.11). This implies that, for accurate predictions, one should have for $m = 0, 1, 2$:

$$A_m/(\sigma - \sigma_0)^{m+1} \ll I^{\text{Meas}}(\sigma - \sigma_0, n_p)/n_p, \qquad (IV.52)$$

where $I^{\text{Meas}}(\sigma - \sigma_0, n_p)$ is the true (measured) absorption profile. The Lorentzian normalized profile in the wing is easily derived and given by the expression:

$$I^{\text{Lor/wing}}(\sigma) = \frac{n_p A_1^0}{\pi(\sigma - \sigma_0)^2}, \qquad (IV.53)$$

where A_1^0 is the intensity-weighted average line-broadening coefficient, *ie*:

$$A_1^0 = \sum_\ell \rho_\ell d_\ell^2 \gamma_\ell \Big/ \sum_\ell \rho_\ell d_\ell^2. \qquad (IV.54)$$

Recall that the absorption, far away from the band center, can be typically three orders of magnitude smaller than the Lorentzian prediction (Figs. IV.7 and IV.5), leading to:

$$I^{\text{Meas}}(\sigma - \sigma_0) \approx 10^{-3} n_p A_1^0/(\sigma - \sigma_0)^2. \qquad (IV.55)$$

Introducing this, for the typical value $(\sigma - \sigma_0) = 100\,\text{cm}^{-1}$, into Eq. (IV.52) leads to

$$A_m \ll 10^{-3}(100)^{m-1} A_1^0, \; m = 0, 1, \text{ and } 2. \qquad (IV.56)$$

Considering that $\ll$ means errors smaller than 1%, Eq. (IV.56) becomes, for $m = 1$:

$$A_1 < 10^{-5} A_1^0, \text{ ie } \Big|1 + \sum_\ell \rho_\ell d_\ell^2 \text{Re}(Y_\ell)(\sigma_\ell - \sigma_0) \Big/ \sum_\ell \rho_\ell d_\ell^2 \gamma_\ell\Big| \leq 10^{-5}. \qquad (IV.57)$$

For the CO_2 ν_3 band, the real parts of the Y_ℓ's are typically ten times smaller than the broadening coefficients γ_ℓ,[360] and the intensity-weighted average of $|\sigma_\ell - \sigma_0|$ is a few $10\,\text{cm}^{-1}$. These numbers and Eq. (IV.57) show that, for a 1% accuracy of spectra calculations in the wings, the Y's should be accurate within better than typically 0.001%. Their storage in databases, including a proper parameterization of their temperature dependences, is thus *not* a simple problem. Even though they are described using the numbers of parameters and digits needed to obtain the required accuracy, there is still a maintenance problem. If, for some reason, the spectroscopic parameters σ_ℓ, ρ_ℓ, and/or d_ℓ are changed, the Y's must also be updated since the former set may be no longer compatible [in the sense of Eq. (IV.56)] with these new values.

A similar analysis can be made in the wings of a cluster of lines spectrally isolated from others, as discussed in Refs. 345,346. In this case, Eq. (IV.14) breaks down, since the sum is restricted to the cluster transitions [Eq. (IV.44)], but A_0 is close to zero and A_1 gives the first significant contribution. The small value of A_0 comes from the fact that the cluster is spectrally dense and far away from other transitions. Indeed, starting from Eqs. (IV.24) and (IV.51), one can write:

$$A_0 = \sum_{\ell \in \text{Cluster}} \rho_\ell d_\ell^2 Y_\ell = 2 \sum_{\ell \in \text{Cluster}} \left[\sum_{\ell' \in \text{Cluster}} \rho_\ell d_\ell d_{\ell'} \frac{W_{\ell'\ell}}{\sigma_\ell - \sigma_{\ell'}} \right]$$
$$+ 2 \sum_{\ell \in \text{Cluster}} \left[\sum_{\ell' \notin \text{Cluster}} \rho_\ell d_\ell d_{\ell'} \frac{W_{\ell'\ell}}{\sigma_\ell - \sigma_{\ell'}} \right]. \qquad (IV.58)$$

Due to Eq. (IV.6), the first term is zero. The second one is small since $|\sigma_\ell - \sigma_{\ell'}|$ is much larger than $|W_{\ell'\ell}|$ for $\ell \in$ cluster and $\ell' \notin$ cluster due to spectral separation. One can then demonstrate that 1% accuracy predictions in the cluster wings require that:

$$\left| 1 - \sum_{\ell \in \text{Cluster}} \rho_\ell d_\ell^2 [\text{Re}(Y_\ell)(\sigma - \sigma_\ell) + \text{Re}(W_{\ell\ell})] \Big/ \sum_{\ell,\ell' \in \text{Cluster}} \rho_\ell d_\ell d_{\ell'} \text{Re}(W_{\ell'\ell}) \right| \leq 10^{-2}. \qquad (IV.59)$$

The denominator is directly related to the effective broadening coefficient of the cluster with pressure [Eq. (IV.44)], which is, in infrared Q branches and methane manifolds, typically a part (1/2 to 1/4) of the average individual line widths (Figs. IV.4b and IV.5). Since the transitions are closely spaced, the Y's can be up to two orders of magnitude greater than the broadening coefficients γ_ℓ.[330,361] Hence, for a 1% accuracy of calculations one wavenumber away from the cluster, the relative accuracy on the Y's must be of the order of 0.01%. This explains why four parameters, with up to nine significant digits for each, had to be used[346] for laws parameterizing CO_2 line-coupling coefficients of Q transitions and their temperature dependences.

The preceding constraints also lead to difficulties in computations when line wing truncations are used. If one decides, in order to save computer time, that a line is disregarded if centered further away than a chosen distance from the current wavenumber, errors will arise once one (or more) transition is rejected. Indeed, these lines are then omitted in the sum giving the spectral profile and A_0, for instance, becomes non-zero, introducing more or less large deviations from the correct result. This prohibits line wing truncation, except if great caution is taken, and makes the accounting for line mixing costly since *all* lines must be kept for *all* wavenumbers or disregarded, by truncation, *all* at once.

Finally recall that the first-order approximation leads to inaccurate results when the overlapping of the pressure-broadened transitions is important (Figs. IV.8 and IV.9). This limits its use to pressures below a value which *depends* on the spectral structure, for a given temperature and radiator-perturber system. For instance, it is valid at 0.1 atm for the infrared CO_2 Q branch of the v_1-v_2 band but not for the very dense Q branch of the $v_1 + v_2 - 2v_2$ band.[343] General criteria are difficult to derive but some, which are applicable to Q branches of linear molecules, can be found in Ref. 343.

IV.2.3b Using the full relaxation matrix expression

For computations of the spectral profile using Eq. (IV.1), it is convenient to write it under a matrix form, as it is treated with computers, ie:

$$I(\sigma, n_p) = \frac{1}{\pi \sum_j \rho_j d_j^2} \sum_j \sum_{j'} \rho_j d_j d_{j'} (\sigma I_d - L_a - i n_p W)^{-1}_{j'j}, \quad (IV.60)$$

where j and j' are line numbers within the (arbitrary) ordering chosen. The diagonalization of $L_a + i n_p W$ leads to the density-dependent matrix $V(n_p)$ whose columns are the normalized eigenvectors and to the diagonal matrix $D(n_p)$ of the associated eigenvalues. One can then use the following mathematical identity:[362]

$$[L_a + i n_p W] = V(n_p) D(n_p) V(n_p)^{-1}, \quad (IV.61)$$

leading to : $(\sigma I_d - L_a - i n_p W)^{-1}_{j'j} = \sum_m V(n_p)_{j'm} \frac{1}{\sigma - D(n_p)_{mm}} V(n_p)^{-1}_{mj}. \quad (IV.62)$

Introducing this identity into Eq. (IV.60) and rearranging the sums then leads to:

$$I(\sigma, n_p) = \frac{1}{\pi \sum_j \rho_j d_j^2} \sum_m \frac{B_m(n_p)}{\sigma - D_{mm}(n_p)}, \quad (IV.63)$$

with $\quad B_m(n_p) = \sum_j \sum_{j'} \rho_j d_j d_{j'} V(n_p)_{j'm} V(n_p)^{-1}_{mj}. \quad (IV.64)$

Eq. (IV.63) corresponds to a sum of individual contributions of *equivalent* lines "m" with (complex) Rosenkranz profiles [Eq. (IV.22)] since it can be rewritten as:

$$I(\sigma, n_p) = \frac{1}{\pi \sum_j \rho_j d_j^2} \sum_m S_m^e(n_p) \frac{1 + i n_p Y_m^e(n_p)[\sigma - \sigma_m^e(n_p)]}{\sigma - \sigma_m^e(n_p) - i n_p \gamma_m^e(n_p)}, \quad (IV.65)$$

where we have introduced the intensity S_m^e, the spectral position σ_m^e, the HWHM γ_m^e, and the line-coupling coefficient Y_m^e of the equivalent line m. These parameters are all *real*, temperature and perturber *density-dependent*, and given by:

$$S_m^e(n_p) = \text{Re}[B_m(n_p)], \quad \sigma_m^e(n_p) = \text{Re}[D_{mm}(n_p)],$$
$$\gamma_m^e(n_p) = \frac{1}{n_p}\text{Im}[D_{mm}(n_p)], \quad Y_m^e(n_p) = \frac{1}{n_p}\text{Im}[B_m(n_p)]/\text{Re}[B_m(n_p)]. \quad \text{(IV.66)}$$

Once the numerical diagonalization leading to the matrices V and D, the inversion to get V^{-1} and the calculation of the equivalent line parameters of Eq. (IV.66) have been performed, computations using Eq. (IV.65) are much more efficient than a direct treatment of Eq. (IV.60) which requires a matrix inversion for each wavenumber. This is of interest when the number N_L of coupled lines is large since the CPU time required for the processing of Eq. (IV.65) is proportional to N_L whereas that involved by the use of Eq. (IV.60) typically behaves as N_L^3. Note that including Doppler effects within this approach is straightforward since Eqs. (IV.26)–(IV.28) can be used, only replacing the true line parameters by those of the equivalent lines.

As discussed in § IV.2.3b, one must be aware that *truncation* of the wings cannot be used, except with care. Rejecting equivalent lines centered further away from the current wavenumber than a pre-chosen distance removes some terms off the sum in Eq. (IV.65). This leads to errors since a number of sum rules are no longer verified. Hence, line truncation is hazardous if not properly made and it is preferable to keep all equivalent lines at all times of the spectra calculations or remove them *all at once*.

Finally, some problems may be encountered due to the finite precision of computer diagonalization and inversion routines. It can happen at low densities since the $L_a + in_p W$ matrix has widely spread diagonal elements much larger than the off-diagonal terms. For an entire band $\text{Re}[(L_a + in_p W)_{jj}]$ takes values between the positions σ_{min} and σ_{max} of the lowest and highest frequency transitions, while the off-diagonal terms $\text{Re}(W_{j'j})$ are typically of the order of a few $0.01 \text{ cm}^{-1}/\text{atm}$. A first way to limit the problem is to separate the matrix into sub-matrices when it is block-diagonal. A second trick is to replace L_a by $\Delta L_a = L_a - \bar{\sigma} \times I_d$, where $\bar{\sigma} \approx (\sigma_{max} + \sigma_{min})/2$, simultaneously replacing σ by $\sigma - \bar{\sigma}$ in Eq. (IV.65). Even when this is done, problems may be encountered since the diagonal elements are spread over $\pm(\sigma_{max} - \sigma_{min})/2$, ie typically a hundred cm^{-1} for a vibrational band. This value can be several orders of magnitude greater than the off-diagonal terms. There are thus errors in the determination of the V, D, and V^{-1} matrices due to the finite computer precision and the subsequent error propagation problems. These numerical uncertainties, although relatively small, can translate into significant errors in calculations of the wings through the breakdown of the sum rules discussed above. Tests have shown that the results may be unphysical (negative absorption coefficients) and that they depend on the computer, the compiler and the diagonalization and inversion subroutines used. The best is to use the highest precision allowed (double or quadruple precision). If problems remain, it is then preferable to compute the Y's [Eq. (IV.24)] and to use the first-order approximation [Eq. (IV.22)] since it is generally valid in these low density situations.

IV.3 CONSTRUCTING THE IMPACT RELAXATION MATRIX

Many models have been proposed for the construction of the *thermally averaged* relaxation matrix W(T) within the *impact and binary collision approximations*. They range from very simple empirical approaches for the treatment of complex molecular collisional pairs, to first-principles fully quantum calculations starting from the intermolecular potential energy surface (PES) for simple radiators and perturbers. The following of this section gives a review of the most widely used approaches, from the simplest to the most elaborated ones. In most cases, except for direct calculations starting from the PES, the models enable the construction of the *real part* of W *only*. Nevertheless, for strongly coupled lines, the imaginary part of the off-diagonal elements are small when compared with the real part, as indicated by direct calculations from the interaction potential[363–365] Furthermore, available spectra calculations indicate that they induce only slight differences with respect to predictions neglecting them.[148] Note that a summary of the literature providing comparisons between measured spectra or time-dependent response signals influenced by line mixing and direct predictions using various models is given in Sec. IV.5.

Recall the *important* properties of W mentioned before. Firstly, the diagonal elements give the individual pressure-broadening (γ_ℓ) and -shifting (δ_ℓ) coefficients, *ie*:

$$\langle\langle\ell|W(T)|\ell\rangle\rangle = \gamma_\ell(T) - i\delta_\ell(T). \tag{IV.67}$$

The second one is that W verifies the detailed balance principle (App. II.E3):

$$\langle\langle\ell'|W(T)|\ell\rangle\rangle \rho_\ell(T) = \langle\langle\ell|W(T)|\ell'\rangle\rangle \rho_{\ell'}(T), \tag{IV.68}$$

where ρ_ℓ is the relative population of the lower level of line ℓ. Finally, for a *rigid rotor* and interactions *only* dependent on rotational states, W verifies the sum rule:[338]

$$\sum_{\ell'} d_{\ell'} \langle\langle\ell'|W(T)|\ell\rangle\rangle = 0 \Leftrightarrow -\frac{1}{d_\ell}\sum_{\ell'\neq\ell} d_{\ell'}\langle\langle\ell'|W(T)|\ell\rangle\rangle = \gamma_\ell(T) - i\delta_\ell(T), \tag{IV.69}$$

where d_ℓ is the reduced matrix element of the operator (dipole, quadrupole, polarizability, susceptibility, *etc*) giving rise to the optical transition ℓ. When Eq. (IV.69) is not valid, it may be replaced by the following approximate one:

$$\frac{1}{d_\ell}\sum_{\ell'} d_{\ell'}\langle\langle\ell'|W(T)|\ell\rangle\rangle \approx \Delta\gamma(T) - i\Delta\delta(T)$$
$$\Leftrightarrow -\frac{1}{d_\ell}\sum_{\ell'\neq\ell} d_{\ell'}\langle\langle\ell'|W(T)|\ell\rangle\rangle \approx [\gamma_\ell(T) - i\delta_\ell(T)] - [\Delta\gamma(T) - i\Delta\delta(T)]. \tag{IV.70}$$

Non-zero values of $\Delta\gamma$ and $\Delta\delta$ may result from two phenomena. The first one is the eventual difference between the intermolecular potentials in the upper and lower states of the optical transitions, giving rise, for vibrational bands, to "vibrational" broadening and shifting (App. IV.A). The second one is couplings between internal states of the radiator (vibration-rotation, Fermi and Coriolis resonances, etc) affecting the *free* radiator wave functions. In this case, limiting the sum over ℓ' in Eq. (IV.69) to the lines within a single vibrational band leads to the breakdown of the sum rule, whereas the latter is exactly verified if all lines associated with all spectroscopically coupled levels are included in the sum (App. IV.B).

IV.3.1 SIMPLE EMPIRICAL (CLASSICAL) APPROACHES

IV.3.1a The hard collision model

The simplest approach, initially developed for isotropic Raman Q branches, is based on the hard (or strong) collision approximation. The latter assumes that each collision completely thermalizes the radiator states. The *real* coupling $W_{\ell'\ell}$ between lines ℓ and ℓ' is thus proportional to the relative population $\rho_{\ell'}$ of the lower level of line ℓ' [the equivalent, for populations, of the assumption made for velocities in Eq. (III.28)]. This leads to simple expressions of the spectral shape, which are *analytical* if a classical approximation of rotational states is made. Introduced in the 80's,[338,366–368] it enabled predictions of line-mixing effects at a time when the absence of tractable sophisticated approaches and the limited computer power available put strong constraints on the problem. These limitations do not hold anymore for "simple" radiator-perturber collisional pairs, but this simple approach can still be of interest for the modeling of spectra of very complex molecules. Note that various formulations of the hard collision approach have been proposed that will not be detailed. Although notations are different, the models described here are essentially similar to the Adjustable Branch Coupling approach of Refs. 369,370.

Isotropic Raman Q branches.   Within the hard collision model, the off-diagonal *real* elements $W_{\ell'\neq\ell}$ are proportional[366–368] to the relative population $\rho_{\ell'}$ of the initial level of line ℓ', ie:

$$\langle\langle \ell'|\text{Re}[W(T)]|\ell\rangle\rangle = -\rho_{\ell'}(T) \times \bar{\gamma}(T), \quad \ell \neq \ell'. \tag{IV.71}$$

The individual line-broadening values are then obtained from Eq. (IV.69), leading, for isotropic Raman Q branches with unperturbed lower and upper states ($d_\ell = 1$), to:

$$\gamma_\ell(T) = \langle\langle \ell|\text{Re}[W(T)]|\ell\rangle\rangle = [1 - \rho_\ell(T)] \times \bar{\gamma}(T). \tag{IV.72}$$

Although Eq. (IV.72) gives a rough representation of the line-to-line dependence of γ_ℓ, it can be used to determine $\bar{\gamma}(T)$ from measured line widths. In order to go further, let us introduce $\Delta^s W = {}^s W - \bar{\gamma} \times I_d$, where ${}^s W$ is given by Eq. (IV.12). Within the present model, $\Delta^s W$ is real and symmetric, and its eigenvectors form an orthogonal basis. One can easily show that $|u_0\rangle\rangle$ [*cf* Eq. (IV.13)] is an eigenvector of $\Delta^s W$ with the eigenvalue $-\bar{\gamma}$

Collisional line mixing

and that any other vector $|u_m\rangle\rangle$ orthogonal to $||u_0\rangle\rangle$ verifies $\Delta^s W||u_m\rangle\rangle = 0$. The spectral profile [Eq. (IV.11)] then takes the simple form:

$$I(\sigma, n_p, T) = \frac{I^{Lor}(\sigma, n_p, T)}{1 + i\pi n_p \bar{\gamma}(T) \times I^{Lor}(\sigma, n_p, T)}, \quad (IV.73)$$

with

$$I^{Lor}(\sigma, n_p, T) \equiv \frac{1}{\pi} \sum_\ell \frac{\rho_\ell(T)}{\sigma - \sigma_\ell - in_p\bar{\gamma}(T)}. \quad (IV.74)$$

In the case of a breakdown of the sum rule [Eq. (IV.70)], the preceding results can be generalized by replacing $\bar{\gamma}$ by $\bar{\gamma} - \Delta\gamma$ in Eq. (IV.73) [but *not* in Eq. (IV.74)]. Calculations can be further simplified[368] for linear molecules when the Q(J) line positions are approximated by $\sigma_{Q(J)} = \sigma_0 + \Delta B J(J+1)$ where $\Delta B = B' - B''$ is the difference between rotational constants in the upper and lower states. One can then use a classical approach, writing:[(3)]

$$\rho_J(T) = (B''/k_B T)(2J+1) \exp[-B''J(J+1)/k_B T], \quad (IV.75)$$

and replacing the sum over ℓ (*ie* J) by an integral. This leads to:

$$I^{Lor}(\sigma, n_p, T) = \frac{1}{\pi} \int_0^{+\infty} \frac{\exp(-u)}{\sigma - \sigma_0 - \bar{\sigma}(T)u - in_p\bar{\gamma}(T)} du, \quad (IV.76)$$

where $\bar{\sigma}(T) = \sum \rho_J(T)[\sigma_{Q(J)} - \sigma_0] \approx \Delta B k_B T / B''$ is the weighted average of the relative rotational line positions. Changing from u to $y = u - [\sigma - \sigma_0 - in_p\bar{\gamma}(T)]/\bar{\sigma}(T)$ leads to:

$$I^{Lor}(\sigma, n_p, T) = \frac{-1}{\pi\bar{\sigma}(T)} \exp(z) \int_z^\infty \frac{e^{-y}}{y} dy \equiv \frac{-e_1(z)}{\pi\bar{\sigma}(T)} = \frac{-\exp(z)E_1(z)}{\pi\bar{\sigma}(T)},$$

with

$$z = \frac{in_p\bar{\gamma}(T) + \sigma_0 - \bar{\sigma}(T)}{\bar{\sigma}(T)}. \quad (IV.77)$$

The final expression of the spectral profile is then:

$$I(\sigma, n_p, T) = \frac{1}{\pi} \frac{1}{in_p\bar{\gamma}(T) - \bar{\sigma}(T)/e_1[(in_p\bar{\gamma}(T) + \sigma_0 - \bar{\sigma}(T))/\bar{\sigma}(T)]}. \quad (IV.78)$$

Efficient computer routines are available for the calculation of the complex exponential function $E_1(z) = \exp(-z)e_1(z)$ for which expansions are also given.[182] Note that $\partial e_1(z)/\partial z = e_1(z) - 1/z$, which can be useful for fits of measured spectra. Equations (IV.73) and (IV.78) *do not* require the construction and treatment (inversion or diagonalization, *cf* § IV.2.3b) of *any* matrix. They have been applied to Raman Q branch spectra (*eg* Refs. 367,368), demonstrating their ability to take into account the narrowing with respect to the isolated line model. This was of interest at a time where more

---

[(3)] For simplicity, we omit degeneracy factors such as those associated with the nuclear spin but this equation can be easily generalized by introducing the proper weights.

physically based approaches were not tractable due to computer limitations, but this approach, which has a limited accuracy, particularly at low temperature,[368] has been seldom used in the last twenty years for the modeling of Raman spectra.

Infrared and anisotropic Raman Q branches. Extensions of Eqs. (IV.73) and (IV.78) to infrared (or CARS) Q branches were proposed in Refs. 369,371, introducing an effective shifting coefficient $\bar{\delta}$ and a parameter $\Delta\gamma_Q$. The latter accounts for the fact that, when restricted to Q lines [thus disregarding Q-(P or O) and Q-(R or S) couplings], Eq. (IV.70) must be used. Assuming constant d_ℓ values (a good approximation for bands of linear molecules not too perturbed by resonances), the spectral shape is:

$$I(\sigma, n_p, T) = \frac{1}{\pi} \frac{1}{in_p[\bar{\gamma}(T) - \Delta\gamma_Q(T)] - \bar{\sigma}(T)/e_1[(in_p\bar{\gamma}(T) + \sigma_0 n_p \bar{\delta}(T) - \sigma)/\bar{\sigma}(T)]}. \quad (IV.79)$$

This model is of little interest for simple molecular systems for which more elaborated approaches can be used. However, it can be applied to very complex spectra of heavy radiators for which the latter are not tractable. In fact, even for nonlinear molecules, satisfactory results can be obtained when some *ad hoc* branch averaged parameters are used. This is exemplified by the $2\nu_6$ infrared Q-branch of CHClF$_2$ (HCFC-22) for which fits of measured spectra were made[372] using Eq. (IV.79) and $\sigma_0, \bar{\sigma}(T), \bar{\delta}(T), \bar{\gamma}(T)$, and $\Delta\gamma_Q(T)$ as adjustable parameters together with the integrated intensity $S_Q(T)$ of the branch. The typical results of Fig. IV.13 show that the model quite well reproduces the observed profiles.

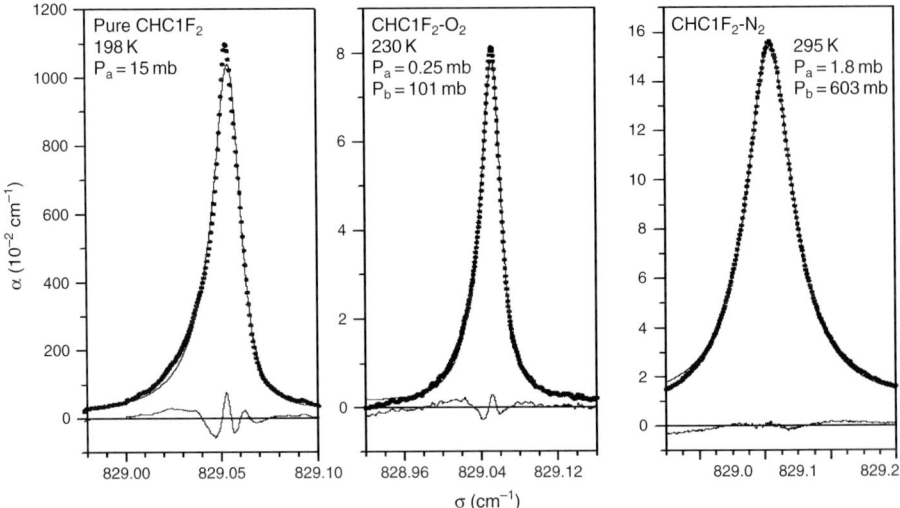

Fig. IV.13: Absorptions in the $2\nu_6$ Q-branch of ^{12}CH^{35}ClF$_2$ (HCFC-22) for various conditions. Experimental (•) and fitted (——) absorption coefficients for three mixture conditions (the lower curves are the observed-fitted residuals). Reproduced with permission from Ref. 372.

Collisional line mixing 177

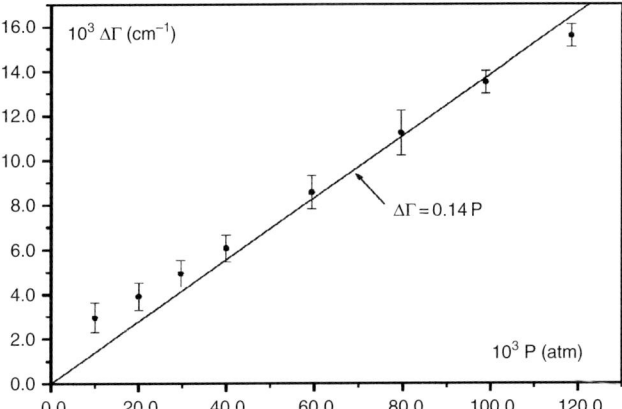

Fig. IV.14: Pure HCFC-22 line-coupling parameter $\Delta\Gamma_Q = P\Delta\gamma_Q$ within the Q branch of Fig. IV.13 deduced from fits of measured spectra for various pressures at room temperature. Reproduced with permission from Ref. 372.

Furthermore, the deduced parameters are consistent[371–373] with previous determinations and show the expected dependences on pressure as illustrated in Fig. IV.14. In this case, line mixing has a large influence since the considered Q branch broadening coefficient $\Delta\gamma_Q = 0.14 \text{ cm}^{-1}/\text{atm}$ is much smaller than what would be obtained ($\bar{\gamma} = 0.58 \text{ cm}^{-1}/\text{atm}$) neglecting this effect.[372]

Multi-branches spectra. A second extension considers all three P, Q, R (or O, Q, S) branches and the collisional couplings within each of them as well as between them.[349] Simplifying approximations are made for the couplings, the intensities, and the spectral positions of the lines, in order to reduce the number of parameters. This leads to an expression of the spectral shape depending on 8 (7 if only P and R branches are present, 6 for an isolated Q-branch) pressure- and wavenumber-independent quantities. These are the total band intensity S_{Tot}, the central wavenumber of the band σ_0, the average positions $\bar{\sigma}_P(T)$ and $\bar{\sigma}_R(T)$ of P and R lines, the average broadening and shifting coefficients $\bar{\gamma}(T)$ and $\bar{\delta}(T)$, and the mean couplings $\Delta\gamma_{RQ}(T)$ between R and Q lines and $\Delta\gamma_{RP}(T)$ between R and P lines. All the other quantities are deduced from this set assuming the symmetry of the band. The resulting expression of the profile for a band including the three branches is:[349]

$$I(\sigma, n_p, T) = \frac{1}{\pi} \frac{1}{in_p\bar{\gamma} + A(\sigma, n_p, T)^{-1}}, \text{ with} \quad (IV.80)$$

$$A(\sigma, n_p) = \frac{\frac{l_P + 2l_Q + l_R}{4} + in_p(\bar{\gamma} - \Delta\gamma_{RQ})\frac{l_P + l_R}{2} l_Q}{1 + in_p(\bar{\gamma} - \Delta\gamma_{RQ})\frac{l_P + 2l_Q + l_R}{4}}$$

$$- \frac{in_p(\bar{\gamma} - \frac{\Delta\gamma_{RQ} + \Delta\gamma_{RP}}{2})\left\{\frac{l_R - l_P}{2\sqrt{2}}[1 + in_p(\bar{\gamma} - \Delta\gamma_{RQ})l_Q]\right\}^2}{B(\sigma, n_p, T)}, \text{ and} \quad \text{(IV.81)}$$

$$B(\sigma, n_p, T) = \left[1 + in_p(\bar{\gamma} - \Delta\gamma_{RQ})\frac{l_P + 2l_Q + l_R}{4}\right]^2$$

$$\times \left\{1 + in_p\left(\bar{\gamma} - \frac{\Delta\gamma_{RQ} + \Delta\gamma_{RP}}{2}\right)\left[\frac{l_P + l_R}{2} - \frac{\frac{in_b(\bar{\gamma} - \Delta\gamma_{RQ})}{4}\left(\frac{l_R - l_P}{2\sqrt{2}}\right)^2}{1 + in_p(\bar{\gamma} - \Delta\gamma_{RQ})\frac{l_P + 2l_Q + l_R}{4}}\right]\right\}, \quad \text{(IV.82)}$$

where l_X refers to the "Lorentzian-type" profile in the X branch, defined by:

$$l_X(\sigma, n_p, T) \equiv \sum_{\ell \in X} \frac{\rho_\ell(T) d_\ell^2}{\sigma - \sigma_\ell - n_p\bar{\delta}(T) - in_p\bar{\gamma}(T)} \bigg/ \sum_{\ell \in X} \rho_\ell(T) d_\ell^2. \quad \text{(IV.83)}$$

For a band involving P and R branches only, the spectral shape is still given by Eq. (IV.80) but A must be replaced by:

$$D(\sigma, n_p, T) = \frac{l_P + l_R}{2} - \frac{in_p(\bar{\gamma} - \Delta\gamma_{RP})\left[\frac{l_P - l_R}{2}\right]^2}{1 + in_p(\bar{\gamma} - \Delta\gamma_{RP})\frac{l_P + l_R}{2}}. \quad \text{(IV.84)}$$

Since Eq. (IV.83) still requires the knowledge of the individual line parameters a classical model can be used, leading, for a linear molecule, to:[349]

$$l_P(\sigma, n_p, T) = E\left(\sigma - \sigma_0 - n_p\bar{\delta}(T) - in_p\bar{\gamma}(T), \frac{\bar{\sigma}_P(T) + \bar{\sigma}_R(T)}{2}, -\frac{\bar{\sigma}_P(T) - \bar{\sigma}_R(T)}{2}\right),$$

$$l_Q(\sigma, n_p, T) = E\left(\sigma - \sigma_0 - n_p\bar{\delta}(T) - in_p\bar{\gamma}(T), \frac{\bar{\sigma}_P(T) + \bar{\sigma}_R(T)}{2}, 0\right), \quad \text{(IV.85)}$$

$$l_R(\sigma, n_p, T) = E\left(\sigma - \sigma_0 - n_p\bar{\delta}(T) - in_p\bar{\gamma}(T), \frac{\bar{\sigma}_P(T) + \bar{\sigma}_R(T)}{2}, \frac{\bar{\sigma}_R(T) - \bar{\sigma}_P(T)}{2}\right),$$

where $E(z, a, b)$ is the complex exponential integral defined by:

$$E(z, a, b) = \int_0^{+\infty} \frac{2y e^{-y^2}}{z - ay^2 - by\sqrt{4/\pi}} dy. \quad \text{(IV.86)}$$

This approach was first validated using high pressure CO_2 spectra,[349,374] but it is of more interest, using *ad hoc empirical* parameters, for complex spectra of non-linear molecules (O_3 and $CHClF_2$).[375] Some of the approximations made in Ref. 349 have been released,[375] introducing specific intensities S_P, S_Q, and S_R for the three branches and an average Q branch position $\bar{\sigma}_Q$. Spectra can then be adjusted using Eqs. (6)–(10) of Ref. 375 and $S_P, S_Q, S_R, \sigma_0, \bar{\sigma}_P, \bar{\sigma}_Q, \bar{\sigma}_R, \bar{\gamma}, \bar{\delta}, \Delta\gamma_{RP}, \Delta\gamma_{RQ}$ as adjustable parameters. The values obtained from fits of O_3 and HCFC-22 spectra are consistent with those of other sources and the collisional quantities show reasonably linear pressure dependences.[375] A comparison between measured spectra of ozone and those calculated using a single set of parameters is displayed in Fig. IV.15, showing good agreements at all pressures. Again, important line-mixing effects affect these spectra as discussed in Ref. 375.

Note that, for species such as CFC's and HCFC's, this approach is, to our knowledge, the only one permitting spectra predictions while explicitly expressing the dependence of the absorption shape on the perturber pressure. The alternative is the use of strictly empirical absorption cross sections[376,377] obtained from measured spectra.

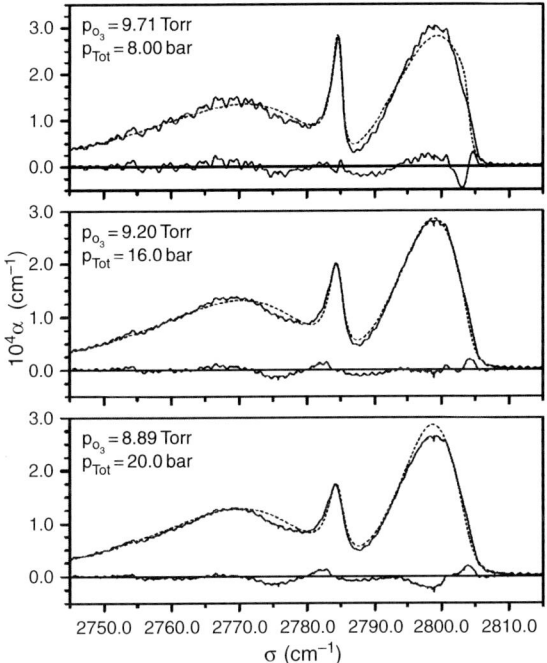

Fig. IV.15: Absorptions in the $\nu_1 + \nu_2 + \nu_3$ O_3 band perturbed by N_2 at room temperature (298 K) for three pressures: (——) experimental values ; (- - -) calculated values (see text). The lowest line is the $\alpha^{Meas} - \alpha^{Calc}$ deviation. Reproduced with permission from Ref. 375.

IV.3.1b The ovaloid sphere model

A less empirical alternative to the preceding classical models, but only written for line mixing between rovibrational components of dipole (k = 1) stretching absorption bands of linear molecules, can be found in Refs. 271,351,378. In these studies, the relaxation matrix real element connecting lines with initial rotational levels J and J' belonging to branches q and q' (q = −1 and +1 for P and R lines, respectively, so that $J_i = J$, $J_f = J + q$) is given by:

$$\langle\langle J', q'|Re(W)|J, q\rangle\rangle = N(J)[\delta_{q,q'}\delta_{J,J'} - \int_0^{2\pi} f(J', J, \alpha)D^1_{qq'}(-\pi/2, \alpha, +\pi/2)d\alpha], \quad (IV.87)$$

where $D^1_{qq'}$ is a Wigner rotation matrix.[65] $f(J',J,\alpha)$ defines the probability of a change of the rotational angular momentum from $\vec{J}$ to $\vec{J}'$ ($J = \|\vec{J}\|$, $J' = \|\vec{J}'\|$, α being the angle between them). Its expression, derived from the ovaloid sphere model, is:[271,351]

$$f(J', J, \alpha) = N(J', J, \alpha)/N(J), \text{ with,}$$

$$N(J', J, \alpha) = A\frac{B''J'}{2k_BT}\exp\left\{-\frac{B''}{2k_BT}\left[(J'^2 - J^2) + \frac{Z}{2} + \frac{(J'^2 - J^2)^2}{2Z}\right]\right\}, \quad (IV.88)$$

$$\text{where } Z = \frac{1+G}{1-G}\|\vec{J}' - \vec{J}\|^2$$

B'' being the rotational constant. As explained in Ref. 351, G and A can be deduced from parameters describing the intermolecular potential at short distances. They can also be considered as adjustable quantities and obtained from fits of measured data. Note that $G \approx 1$ corresponds to the soft collision model, for which the rotational angular momentum is insensitive to collisions [$f(J', J, \alpha) = \delta_{J',J} \times \delta_{\alpha,0}$]. On the opposite, G = 0 and G = −1 correspond to strong perturbations, G = −1 being associated with anticorrelated collisions for which $\vec{J}' = -\vec{J}$. The value of N(J), which governs the broadenings of the P(J) and R(J) lines, is obtained from Eq. (IV.88) and the relation:

$$N(J) = \int_0^\infty dJ' \int_0^{2\pi} N(J', J, \alpha)d\alpha. \quad (IV.89)$$

The relaxation matrix defined by Eqs. (IV.87) and (IV.88) verifies the classical analog of Eq. (IV.68) and Eq. (IV.89) ensures that Eq. (IV.69) is satisfied (assuming that the matrix elements of the tensor coupling radiation and matter are independent on J and q).

The present model has been little applied although interesting results were obtained in Refs. 351,379. These work shows that the potential parameters deduced from fits of measured spectra are consistent with values from other sources. Furthermore, satisfactory modeling of non-Lorentzian behaviors are obtained in the near wing of the fundamental CO band, in the region of the $3\nu_3$ band head of CO_2, and in the central part of the 2–0 CO band at high pressure.

IV.3.2 STATISTICALLY BASED ENERGY GAP FITTING LAWS

IV.3.2a Introduction

Energy gap (EG) *fitting laws*[380] are a rather simple way to construct the *real* part of the relaxation matrix W. The idea is to express the *real* elements of W in terms of state-to-state collisional transfer quantities $K(i' \leftarrow i)$ between the initial levels [and eventually $K(f' \leftarrow f)$ between the final levels] of the lines $\ell = f \leftarrow i$ and $\ell' = f' \leftarrow i'$. These (in cm^{-1}/Am or cm^{-1}/atm) are related to the thermally averaged collisional cross sections $<\sigma(i' \leftarrow i)>$ by the relation:

$$K(i' \leftarrow i) = \frac{n_0 <v_r>}{2\pi c} <\sigma(i' \leftarrow i)>, \qquad (IV.90)$$

where $<v_r>$ is the mean relative speed and n_0 is the number of perturbers per unit volume at the density of 1 Am (or 1 atm if pressure units are used). In the statistical fitting law approach, the $K(i' \leftarrow i)$ are represented as functions of the *energy gap* $\Delta E_{i'i} = |E_i - E_{i'}|$, ie for a downward transition $(E_i > E_{i'})$:

$$K(i' \leftarrow i, T) = f_{EG}(E_i - E_{i'}, T) \text{ for } E_i \geq E_{i'}, \qquad (IV.91)$$

where f_{EG} is an *analytical* function depending on a set of temperature-dependent parameters $a_1, a_2, \ldots, a_n$. The reciprocal upward state-to-state quantity is deduced from Eq. (IV.91) and the detailed balance relation, *ie*:

$$K(i \leftarrow i', T) = \frac{\rho_i(T)}{\rho_{i'}(T)} K(i' \leftarrow i, T). \qquad (IV.92)$$

There is a large variety of such semi-empirical laws, as reviewed in Ref. 380, the most commonly used being recalled in the Table below. The model parameters a_i are often obtained from fits of line broadening data, as explained later.

A first *limitation* of these fitting laws comes from the fact that Eq. (IV.91) depends on the energies *only* and thus cannot always be used straightforwardly without loss of physical significance. Whereas it can describe some collisional transfers, it may be wrong for others. As a first example, consider collisional state-to-state transfers between rotational levels within the vibrational bending state $(v_2 = 1)$ of CO_2 where both odd and even J levels are present. The molecule being almost symmetric, the even $\leftrightarrow$ even and odd $\leftrightarrow$ odd changes of the rotational angular momentum J have large cross sections whereas those associated with even $\leftrightarrow$ odd J changes are much smaller. The $<\sigma_{J' \leftarrow J}>$ (and $K_{J' \leftarrow J}$) thus show oscillations as J' varies.[381,382] Their relative amplitudes depend on J and cannot be modeled using a single energy gap law. Assumptions must then be made, such as the neglect of even $\leftrightarrow$ odd transfers or the use of a specific law to represent them. Another example of the need to take into account propensity rules while using fitting laws is given by symmetric-top molecules whose cross sections $\sigma_{J',K' \leftarrow J,K}$ strongly depend on J,K and J',K'.[363] Again, a priori assumptions are necessary, making the application of energy gap laws to non-linear molecules not straightforward. Finally, expressions only dependent on the energies do not apply to groups of sub-levels which practically have the same energy. It is the case, for instance, of hyperfine structures due to the nuclear spin within a given rotational level J. Since $\Delta E \approx 0$, all state-to-state rates predicted by Eq. (IV.91) are equal, which is obviously not the case (*eg* Refs. 383,384).

A second difficulty arises when line mixing is considered, since this requires switching from the level space (that of the K's) to the line space (that of W). This is not a problem for spectroscopically unperturbed isotropic Raman Q branches since it is exact to write, in the case of a linear molecule for instance:

$$\langle\langle Q(J')|\text{Re}[\mathbf{W}^{\text{Iso-Raman}}]|Q(J)\rangle\rangle = -K(J' \leftarrow J), J' \neq J. \tag{IV.93}$$

On the other hand, it cannot be done straightforwardly for spectra which involve different branches. Specific laws depending on the branches to which the lines belong must then be introduced. One will have to use, a minima, an approximation like:

$$\langle\langle \ell'|\text{Re}[\mathbf{W}]|\ell\rangle\rangle = -A_{XY}K(i_{\ell'} \leftarrow i_\ell), i_\ell \neq i_{\ell'}, \tag{IV.94}$$

where i_ℓ and $i_{\ell'}$ designate the initial levels of line ℓ and ℓ', respectively, and A_{XY} depends on the branches [(X,Y) = P, Q, R or (X,Y) = O, Q, S] to which ℓ and ℓ' belong. These parameters, which scale the amounts of coupling within (X = Y) and in between (X ≠ Y) branches, depend on the vibrational symmetry of the band and on the collision partner, as discussed later. Note that the A_{XY}'s are phenomenological and can only be determined from measured spectra or from more physically based models. Also recall that different laws may be required according to the types of levels (*eg* even/odd J).

For spectra calculations using the preceding equations, one needs to know the fitting law parameters a_i (Table IV.1), which is straightforward if state-to-state data are available. In the absence of such knowledge, they are generally obtained from fits of measured line-broadening data, using expressions such as:

$$\langle\langle \ell|\text{Re}[\mathbf{W}]|\ell\rangle\rangle = \gamma_\ell = \sum_{i_{\ell'} \neq i_\ell} K(i_{\ell'} \leftarrow i_\ell), \tag{IV.95}$$

Table IV.1: Commonly used energy gap fitting laws

Name		Expression
Power Gap Law	PGL	$f_{\text{PGL}}(\Delta E) = a_1|\Delta E|^{-a_2}$
Exponential Gap Law	EGL	$f_{\text{EGL}}(\Delta E) = a_1 e^{-a_2|\Delta E|}$
Modified Exponential Gap Law	MEGL	$f_{\text{MEGL}}(\Delta E) = a_1 \left[\frac{1+a_2 E_i}{1+a_3 E_i}\right]^2 e^{-a_4|\Delta E|}$
Exponential Power Gap Law	EPGL	$f_{\text{EPGL}}(\Delta E) = a_1|\Delta E|^{-a_2} e^{-a_3|\Delta E|}$

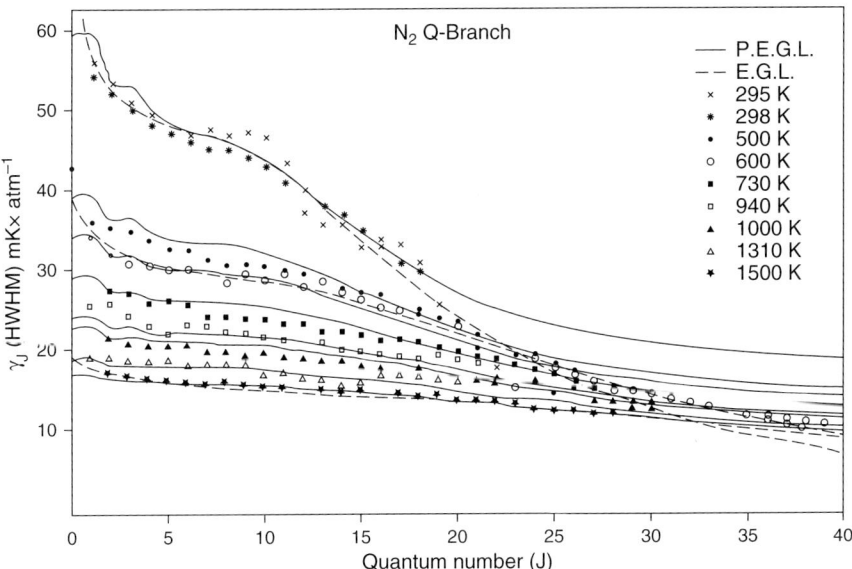

Fig. IV.16: Measured (symbols) and calculated (———: PGL law, – – –: EGL law)) self-broadened half-widths of N_2 Raman Q(J) lines vs J and temperature. Reproduced with permission from Ref. 341.

or, alternatively, introducing the upper states:

$$\gamma_\ell = \frac{1}{2}\left[\sum_{i_{\ell'} \neq i_\ell} K(i_{\ell'} \leftarrow i_\ell) + \sum_{f_{\ell'} \neq f_\ell} K(f_{\ell'} \leftarrow f_\ell)\right]. \tag{IV.96}$$

These equations are rigorous for spectroscopically unperturbed isotropic Raman Q lines *only*. They must be used with care for P and R lines since they neglect elastic collisions which interrupt the phase of the radiator without quenching it. As detailed in Ref. 385, some systems involve large elastic contributions invalidating the use of Eq. (IV.96). For linear molecules, Eqs. (IV.91), (IV.92), and (IV.95), and the energy gap laws of Table IV.1, generally lead to accurate descriptions of the line-to-line variations of the broadening coefficient, as illustrated by Fig. IV.16 and Ref. 386.

IV.3.2b Q branches of linear molecules

Isotropic Raman Q branches of linear molecules permit a straightforward application of fitting laws since Eqs. (IV.93) and (IV.95) are then valid (provided that the Q branch spectroscopy is not perturbed by internal interactions such as Coriolis). An example in the case of N_2, plotted in Fig. IV.17, demonstrates the quality of the exponential law (EGL) whereas the polynomial law (PEGL, identical to PGL in Table IV.1) underestimates line-coupling effects. This is due to the fact that the PGL underestimates the decrease of $K(J' \leftarrow J)$ with $|J - J'|$. The couplings between closely spaced Q(J) and Q(J') lines are thus too small, leading to too broad Q branch profiles.

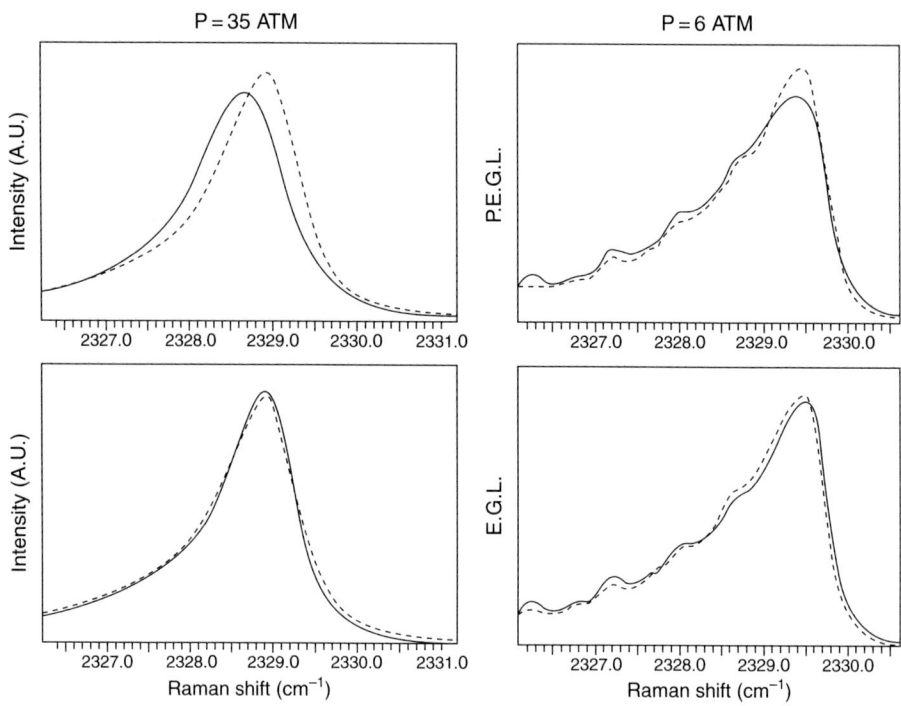

Fig. IV.17: Measured (- - - - -) and calculated (line) pure N_2 Raman Q branch spectra at room temperature and total pressures of 35 atm (left) and 6 atm (right). The PGL (top) and EGL (bottom) laws have been used. Reproduced with permission from Ref. 341.

An extension can be made to infrared or anisotropic Raman Q branches of linear molecules in the absence of even/odd J transfers. This was first done by Strow et al[387] for the $(11^10, 03^10)_{II} \leftarrow 00^00$ infrared Q branch of CO_2 (which contains only even J lines). The authors used the EPGL law (cf Table IV.1) and wrote:

and
$$\langle\langle Q(J')|\text{Re}[W]|Q(J)\rangle\rangle = -K(J' \leftarrow J), J' \neq J, \qquad (IV.97)$$

$$\gamma_{Q(J)} = \frac{1}{2}\left[\sum_{J' \neq J}^{L} 2K(J' \leftarrow J) + \sum_{J' \neq J}^{U} K(J' \leftarrow J)\right], \qquad (IV.98)$$

where the sums Σ^L and Σ^U are made within the lower and upper vibrational states, respectively. In the considered band, the lower state contains only even J levels whereas all $J > 0$ values exist in the upper state. Σ^U is thus close to $2 \times \Sigma^L$ and these equations are equivalent to Eqs. (IV.94) and (IV.96), implicitly assuming $A_{QQ} = 1/2$. A similar approach was used in Ref. 388 where A_{QQ} parameters have been adjusted on measured infrared Q branch spectra of HCN and C_2H_2. Problems remained, the fact that the adjusted values of A_{QQ} depend on pressure and that the fits of measured spectra were not perfect showing the limits of the model, as discussed in Ref. 388.

More complicated are Q branches containing lines of both even and odd J, as those of CO_2 which have a non-zero vibrational angular momentum in the lower and upper states of the transitions (eg $\Delta \leftrightarrow \pi$ and $\pi \leftrightarrow \pi$ bands). In these cases, the interaction potential is dominated by even contributions. Rotational changes involving odd ΔJ values are thus much less probable than those for even ΔJ and the $K(J' \leftarrow J)$'s show more or less pronounced oscillations with J'.[381] Constructing the relaxation matrix then requires assumptions concerning the respective amounts of coupling between Q lines of identical or different J parities. The simplest approach[389,390] neglects parity-changing transfers and writes the relaxation matrix as:

$$\langle\langle Q(J')|\text{Re}[W]|Q(J)\rangle\rangle = -\frac{1}{2}K(J' \leftarrow J), \quad \text{for } J' \neq J \text{ and } (J-J') \text{ even},$$
$$\langle\langle Q(J')|\text{Re}[W]|Q(J)\rangle\rangle = 0, \quad \text{for } (J-J') \text{ odd},$$
(IV.99)

the half widths being given by Eq. (IV.96). This corresponds to Eqs. (IV.94) with $A_{QQ} = 1/2$ and $A_{QQ} = 0$ for even and odd values of ΔJ, respectively. A more refined analysis was made in Ref. 391 for a pure N_2O infrared Q branch. The authors used the EPGL law and three collisional selection rules for the transfers between lines involving levels of e or f symmetry in the upper state: (1) no $e \leftrightarrow f$ transfers and all ΔJ allowed, (2) all transfers allowed, (3) ΔJ even for $e \leftrightarrow e$ and $f \leftrightarrow f$ exchanges and ΔJ odd for $f \leftrightarrow e$ transfers. From the equations given in Ref. 391, all these approaches can be translated into relations similar to Eqs. (IV.94) and (IV.96) introducing proper parameters A_{XY} depending on the symmetries and parities of the rotational levels. The comparisons with measurements indicate[391] that model (3) gives the best results, but no further refined adjustments of the relative amounts of the various transfers was made. The most flexible energy gap fitting law approach for such collisional line-mixing problems was proposed in Ref. 346. Empirical parameters β_L and β_U, for the lower and upper vibrational states, were introduced for the modeling of $\Delta \leftrightarrow \pi$ CO_2 Q branches in the 15 μm region. The coupling elements are expressed as:[346]

$$\langle\langle Q(J')|\text{Re}[W]|Q(J)\rangle\rangle = -\beta_L \beta_U \times K(J' \leftarrow J) \quad \text{for } J' \neq J \text{ and } (J-J') \text{ even}$$
$$\langle\langle Q(J')|\text{Re}[W]|Q(J)\rangle\rangle = -(1-\beta_L)(1-\beta_U) \times K(J' \leftarrow J') \quad \text{for } (J-J') \text{ odd},$$
(IV.100)

the half-widths being given by:

$$\gamma_{Q(J)} = \frac{1}{2}\left\{\sum_{J' \neq J, = \text{parity}}^{L}\beta_L K(J' \leftarrow J) + \sum_{J' \neq J, \neq \text{parity}}^{L}(1-\beta_L)K(J' \leftarrow J) \right.$$
$$\left. + \sum_{J' \neq J, = \text{parity}}^{U}\beta_U K(J' \leftarrow J) + \sum_{J' \neq J, \neq \text{parity}}^{U}(1-\beta_U)K(J' \leftarrow J)\right\}.$$
(IV.101)

Since the sums for identical and different J and J' parities are almost equal, this approach is similar to Eqs. (IV.94) and (IV.96) with $A_{QQ} \approx \beta_L \beta_U$ for $|J'-J|$ even and $A_{QQ} \approx (1-\beta_L)(1-\beta_U)$ for $|J'-J|$ odd. Using values of β_L and β_U *adjusted* on measured spectra,

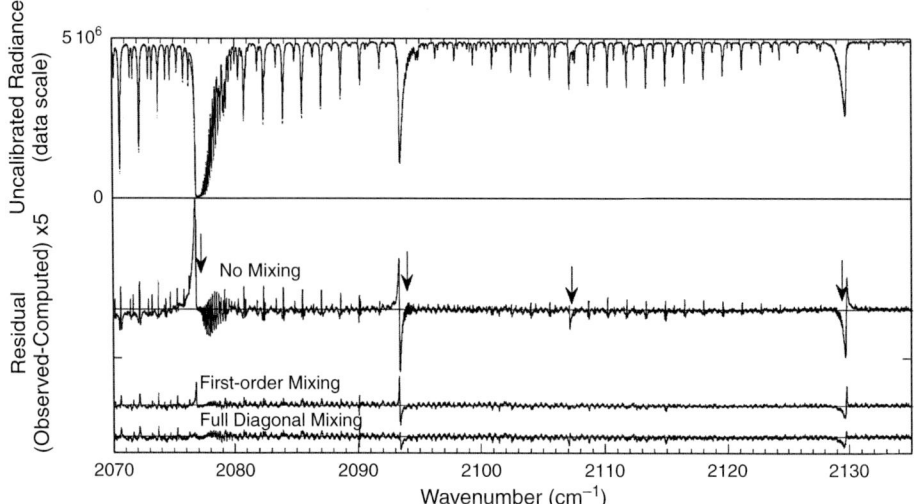

Fig. IV.18: Transmission for 3% CO_2 in 0.46 atm of N_2 at 295 K. Measured values are displayed in the upper plot. The lower part (where the Q branches are indicated by arrows) displays meas-calc deviations obtained with three models: one neglecting line mixing, the others accounting for this process using the EPGL law and Eqs. (IV.100) and (IV.101), the absorption coefficient being calculated with the first-order approximation [Eq. (IV.22)] and with a full relaxation matrix [Eq. (IV.1)]. Reproduced with permission from Ref. 392.

this model leads to accurate modeling of line-mixing effects in π-Δ CO_2 Q branches.[346] Its quality is also demonstrated, in Fig. IV.18, by more recent results[392] in the 5 μm region.

The limits of this approach for $\Delta \leftrightarrow \pi$ CO_2 Q branches are discussed in Ref. 393. It is shown that ad hoc values of β_L and β_U do enable correct modeling of the relative *average* couplings between Q(J) lines with different and identical J parities. However, the model does not well describe the significant J- and J'-dependent oscillations of the relaxation matrix elements $<< Q(J')|Re(W)|Q(J) >>$. At moderate J values, Eqs. (IV.100) lead to an almost J'-independent ratio of successive elements of W, ie:

$$\frac{\langle\langle Q(J+2\Delta J)|Re[W]|Q(J)\rangle\rangle}{\langle\langle Q(J+2\Delta J+1)|Re[W]|Q(J)\rangle\rangle} \approx \frac{\beta_L \beta_U}{(1-\beta_L)(1-\beta_U)}, \qquad (IV.102)$$

which is a rough approximation as shown in Ref. 393 and Fig. IV.19.

The *empirical* nature of Eqs. (IV.100) and (IV.101) is further demonstrated by the fact that parameters β determined from measured spectra in a given band *cannot* be used for all other transitions involving the same types of vibrational levels. Indeed, the values of β_π and β_Δ adjusted from $\Delta \leftrightarrow \pi$ Q branch spectra are of the same order since $\beta_\pi \approx 0.6$ and $\beta_\Delta \approx 0.8$.[346,393] According to Eq. (IV.100), the couplings between lines of identical J parities are thus governed by $\beta_\Delta \beta_\pi \approx 0.48$ in $\Delta \leftrightarrow \pi$ Q branches and $\beta_\pi \beta_\pi \approx 0.36$ in $\pi \leftrightarrow \pi$ Q branches [the values for different parities being $(1-\beta_\pi)(1-\beta_\Delta) \approx 0.08$ and

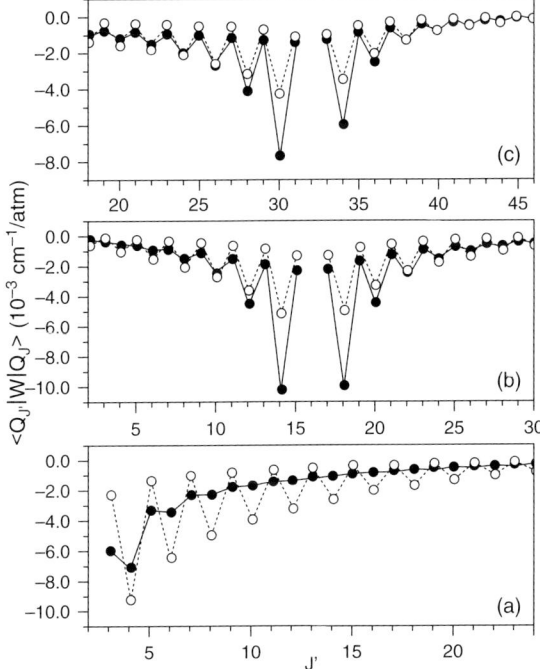

Fig. IV.19: Relaxation matrix real elements $\langle\langle Q(J')|\text{Re}[W]|Q(J)\rangle\rangle$ for CO_2–N_2 at room temperature in the $11^10_{II} \leftarrow 02^20_I$ band. (–•–) and (- - o - -) are obtained with the ECS scaling model (see § IV.3.3) and the fitting law approach of Ref. 346. (a), (b), and (c) are for J = 2, 16, and 32, respectively. Reproduced with permission from Ref. 393.

$(1-\beta_\pi)(1-\beta_\pi) \approx 0.16$]. The proposed model thus predicts comparable couplings within $\pi \leftrightarrow \pi$ and $\Delta \leftrightarrow \pi$ Q branches. This is in strong disagreement with the very different behaviors of Q branches of parallel and perpendicular bands, $\pi \leftrightarrow \pi$ Q lines being much less coupled one to the other than those of $\Delta \leftrightarrow \pi$ vibrational transitions.[335] Different values of β_π should thus be used for these two types of bands, demonstrating the limited physical basis of this approach. Note that the parameters β and A_{QQ} also depend on the collision partner. Interactions with helium lead to much more efficient Q-Q couplings than those with N_2 or Ar in perpendicular bands, this difference being significantly smaller for parallel transitions.[335,394,395]

IV.3.2c Other species and mutibranch spectra

The extension of the preceding approaches to multi-branches spectra can be done introducing ad hoc branch-dependent parameters A_{XY} (eg X = P,Q,R). To our knowledge, this has been seldom done using energy gap laws. A study, only accounting for line mixing *within* CO_2 branches and disregarding transfers between branches, can be found in Ref. 396. Less crude is the model of Ref. 397 where fitting laws and ad hoc scalings of intra-branch couplings are applied to stretching bands of CO_2. Although this is done in the time domain, the model used is equivalent to Eqs. (IV.94) with $A_{RR} = A_{PP}$ and $A_{RP} = A_{PR} = 1 - A_{PP}$.

In summary, energy gap fitting laws are easy to implement but largely *empirical* and they must be used with *care* for complex systems and for extrapolations. They do not explicitly include the influence of angular momenta and thus need adjustments for *each* type of situation. As discussed below, a more physically based alternative is offered by the scaling laws. Nevertheless, these can be difficult to implement for complex molecular systems, so that the fitting laws may be of practical interest.[398–400] Besides the prediction of line couplings, fitting laws can be used to model broadening coefficients (Fig. IV.16). If measured values do not cover the needs, Eq. (IV.98) provides a way to interpolate and extrapolate which is more physically based than the use of polynomials in J, for instance.

IV.3.3 DYNAMICALLY BASED SCALING LAWS

IV.3.3a Introduction

Dynamically based scaling laws have been developed, in the late 70's, in order to model rotational relaxation of linear molecules induced by collisions with atoms.[401,402] This was first done using the Infinite Order Sudden (IOS) approximation (*cf* § IV.3.5d), thus assuming instantaneous collisions and neglecting the energy spacing between rotational levels. This approach was improved a little later by imposing the respect of the detailed balance relation and introducing "adiabaticity corrections" in order to (approximately) include the effects of the finite duration of collisions and of the energy differences between the internal levels involved.[403,404] This so-called Energy Corrected Sudden (ECS) approximation accounts for the rotation of the molecule during collisions through a "scaling length" representing the effective range of molecule-atom interactions. Extensions to other types of levels and of radiators as listed in § IV.3.3c. Most IOS (and ECS) approaches are, strictly speaking, *limited to* collisions with atoms but the case of interactions between linear molecules has been the subject of theoretical developments.[405,406] Nevertheless, it is worth mentioning that applying molecule-atom approaches to molecule-molecule collisions has given *satisfactory* results in many cases (see references in Table IV.9). This is mainly due to the (often) *partly empirical* nature of these approaches where fits of measured spectra are used for the determination of the model parameters. It also results from the fact that perturbers with small rotational constants (CO_2, O_2, N_2), thus somehow "atom like", have been mostly studied.

The basic idea of scaling laws is to separate spectroscopic (related to the radiator internal states) and dynamical (related to the radiator-perturber interactions) effects.[401,402] This enables one to write the relaxation matrix elements (and state-to-state rotational transfer rates) as functions of angular momenta coupling terms (Wigner symbols) and of a limited set of collisional quantities. For linear molecules these quantities, denoted Q_L, can be directly calculated from the intermolecular potential or modeled analytically and deduced from fits of measured data. In order to give a more precise picture before going into details, consider a linear radiator colliding with atoms and line mixing among rovibrational transitions $v_f J_f \leftarrow v_i J_i$ between two stretching vibrational states. In this case, the IOS expression for the *real* part of the relaxation matrix is (*cf* § IV.3.5d):[402]

$$\mathrm{Re}\left[\langle\langle J'_f J'_i|W^{IOS}|J_f J_i\rangle\rangle\right] = \delta_{J_i,J'_i}\delta_{J_f,J'_f}Q_0 - (-1)^k(2J'_i+1)\sqrt{(2J_f+1)(2J'_f+1)}$$
$$\times \sum_{L\neq 0}\begin{pmatrix}J'_i & L & J_i\\0 & 0 & 0\end{pmatrix}\begin{pmatrix}J'_f & L & J_f\\0 & 0 & 0\end{pmatrix}\begin{Bmatrix}J_i & J_f & k\\J'_f & J'_i & L\end{Bmatrix}\times(2L+1)Q_L, \quad \text{(IV.103)}$$

where k is the order of the tensor coupling radiation and matter (k = 0, 1, 2 for isotropic Raman, dipole absorption, and anisotropic Raman, respectively). In this expression, where we have taken the *convention* $\mathrm{Re}(Q_L) \geq 0$, the spectroscopic characteristics are contained in the (..:) and {:::} 3J and 6J symbols,[65] which take into account the couplings of the various angular momenta involved, while the collisional dynamics are represented by the Q_L's. The latter (in cm^{-1}/atm or cm^{-1}/Am) are related[(4)] to the thermally averaged cross section $\langle\sigma_{0\leftarrow L}\rangle$ for the rotational de-excitation from level J = L to level J = 0 [(2L + 1)Q_L being that for the excitation from J = 0 to J = L] by:

$$Q_L(T) = \frac{n_0(T)}{2\pi c}\langle v_r(T)\rangle\langle\sigma_{0\leftarrow L}(T, v_r)\rangle, \quad \text{(IV.104)}$$

where $n_0(T)$ is the number of perturbers per unit volume at one atmosphere pressure (or one amagat density depending on the choice of units), and $\langle v_r\rangle$ is the mean relative speed. Equation (IV.103) applies not only to the off-diagonal elements of W, but *also* to the diagonal ones. Furthermore, since[364,407,(5)]:

$$Q_0 = \sum_{L\neq 0}(2L+1)Q_L, \quad \text{(IV.105)}$$

and, for a *rigid* rotor, the reduced matrix elements of the optical transition moment being given, for a stretching vibrational transition, by:[65]

$$d_{J_f J_i} = (-1)^{J_f}\sqrt{2J_f+1}\times\begin{pmatrix}J_i & k & J_f\\0 & 0 & 0\end{pmatrix}, \quad \text{(IV.106)}$$

one can analytically check that Eq. (IV.69) is *exactly* verified by Eqs. (IV.103), (IV.105), and (IV.106).[408]

The main *disadvantages* of the IOS approach are twofold. The first is that the detailed balance relation [Eq. (IV.68)] is not respected. Since the internal energy structure of the radiator is disregarded, the W matrix verifies the relation:

$$\langle\langle J'_f J'_i|W^{IOS}|J_f J_i\rangle\rangle = \frac{(2J'_i+1)}{(2J_i+1)}\langle\langle J_f J_i|W^{IOS}|J'_f J'_i\rangle\rangle, \quad \text{(IV.107)}$$

[(4)] Setting k = 0, $J_i = J_f = L$ and $J'_i = J'_f = 0$ in Eq. (IV.103) gives the relaxation matrix element $\langle\langle Q(0)|\mathrm{Re}[W]|Q(L)\rangle\rangle$ coupling the isotropic Raman Q(J = L) and Q(J = 0) lines. Using the corresponding values of the 3J and 6J symbols,[65] it is straightforwardly shown that $\langle\langle Q(0)|\mathrm{Re}[W]|Q(L)\rangle\rangle = -Q_L$. According to Eq. (IV.93), one thus obtains Eq. (IV.104).

[(5)] In some studies, such as Ref. 364, notations and sign conventions may be different [*eg*, (2L + 1)Q_L being replaced by Q_L which is thus associated with the excitation from level J = 0 to level J = L].

which is only an approximation of the following correct one:

$$\langle\langle J'_f J'_i | W | J_f J_i \rangle\rangle = \frac{(2J'_i + 1)}{(2J_i + 1)} e^{(E_{J_i} - E_{J'_i})/k_B T} \langle\langle J_f J_i | W | J'_f J'_i \rangle\rangle. \qquad (IV.108)$$

The second problem also results from the neglect of the energy differences between rotational levels and appears when computing line-broadening coefficients (real diagonal element of W) from Eq. (IV.103). Indeed, the calculated values slightly decrease with J_i for small J_i values and quickly become constant as J_i increases. This is generally not observed, as shown in Figs. IV.16 and IV.20, except for molecules with a small rotational constant perturbed by helium at significant relative kinetic energies. In this case, the rotational energy jumps and the collision duration are small, making the IOS a satisfactory approximation (Fig. IV.20).

In order to correct these inaccuracies, two modifications have been proposed, leading to the ECS approach. The first consists in applying Eq. (IV.103) to downward transfers *only* ($J_i \geq J'_i$), the upward ones being deduced from these using Eq. (IV.68). The second improvement, which takes into account the finite collision duration and energy spacing between rotational levels, introduces a $\Omega(J,\ell_c)/\Omega(L,\ell_c)$ factor in front of $Q_{L\neq 0}$ in Eq. (IV.103). These "adiabaticity corrections",[403] detailed later, depend on the rotational energy spacing and on an effective collision duration τ_c (through a *scaling length*

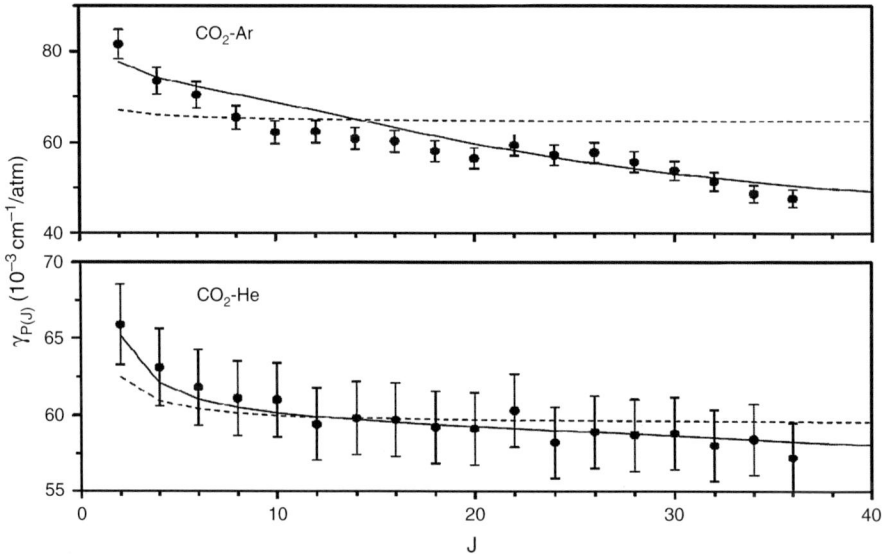

Fig. IV.20: Room temperature Ar and He pressure broadening coefficients (HWHM) of P(J) lines of the $3\nu_3$ CO_2 infrared band. (•) are measured values[168] with associated uncertainties. (—) and (– – –) are calculated values using the ECS and IOS scaling approaches, respectively. After Refs. 394,409.

ℓ_c with $\tau_c = \ell_c/\langle v_r \rangle$). In a stretching band of a linear molecule, the final ECS expression of the real *off-diagonal* elements, is:

$$\text{Re}\left[\langle\langle J'_f J'_i | W^{ECS} | J_f J_i \rangle\rangle\right] = -(-1)^k \frac{\rho_{J_>}}{\rho_J} \Omega(J_>, \ell_c)(2J_< + 1)\sqrt{(2J_f + 1)(2J'_f + 1)}$$

$$\times \sum_{L \neq 0} \begin{pmatrix} J'_i & L & J_i \\ 0 & 0 & 0 \end{pmatrix} \begin{pmatrix} J'_f & L & J_f \\ 0 & 0 & 0 \end{pmatrix} \begin{Bmatrix} J_i & J_f & k \\ J'_f & J'_i & L \end{Bmatrix} \times (2L+1) \frac{Q_L}{\Omega(L, \ell_c)},$$

(IV.109)

where $J_<$ (resp. $J_>$) is the lower (resp. upper) value of J_i and J'_i. Contrary to Eq. (IV.103), this equation *does not* apply to the diagonal terms which must be computed from the off-diagonal ones by using Eq. (IV.69). As shown by Fig. IV.20, the ECS approach enables a much better representation of the line-to-line variations of the broadening coefficients than the purely IOS model.

Besides the scaling length ℓ_c, the ECS approach for linear radiators is based on the collisional parameters $Q_{L \neq 0}$ [or on the $\langle \sigma_{0 \leftarrow (L \neq 0)}(v_r) \rangle$]. For non-linear species, as discussed later, these depend on other quantum numbers, eg $Q_{L,M,M'}$.[410] For simple molecule-atom systems, and when an intermolecular potential is available, these quantities can be directly calculated from first principles as described in § IV.3.5 and Refs. 365,408,411–413. An often used alternative is to model the $Q_{L \neq 0}$ analytically using energy gap fitting laws (Table IV.1 and Ref. 380), the associated parameters a_i being deduced from fits of experimental data. This is generally done using line-broadening coefficients, eventually simultaneously adjusting spectra or time-dependent response signals, through Eqs. (IV.69) and (IV.109) (together with a modeling of the adiabaticity correction terms Ω). As illustrated in Fig. IV.20 and references in Table IV.9, this generally leads to good agreement between experimental and calculated half widths. Furthermore, as shown by Refs. 411,413, such *empirical* values of the $Q_{L \neq 0}$'s are in satisfactory agreement with direct predictions from the interaction potential for CO_2–Ar and CO–Ar. It is important to mention that fits of *only* line-broadening data for the determination of the $Q_{L \neq 0}$'s can lead to *ambiguous* results. There are different sets leading to satisfactory predictions of line widths but to different and, for some of them, inaccurate line couplings and thus spectra predictions. Hence, only the *simultaneous* use of measured spectral or time-dependent responses in regions significantly affected by line mixing *and* of individual line broadenings can remove the ambiguity. Examples for CO_2 are given in Ref. 394 where measured values of γ_ℓ and of the shape of a Q branch have been used, and in Ref. 414 where the Q_L's have been obtained from simultaneous fits of broadening coefficients and of measured absorption in the near wing of the $3\nu_3$ band. For O_2, a discussion of this problem can be found in Ref. 415.

Due to its approximate nature, the ECS model does not lead to "perfect" predictions of all quantities and compromises must be made. For practical reasons, one often favors the modeling of the spectral shape under (pressure) conditions where line-mixing effects are important. The price paid is that the calculated widths of individual transitions are approximate, affecting the quality of spectra calculations at lower pressures. A way to correct this has been proposed,[353,416] also applicable to relaxation matrices constructed with other approaches, which consists in three steps. One first imposes accurate values to the real diagonal elements of W, setting them equal to the best broadening coefficients available. The second step is the construction of the off-diagonal elements using the ECS model [*eg* Eq. (IV.109)], hence leading to a relaxation matrix approximately verifying Eq. (IV.69). This is then corrected using a "renormalization", treating the W matrix column by column. The elements $W_{j'j}$ and $W_{jj'}$ (for $j' \neq j$) are progressively scaled, starting from the first column $j = 1$, through the changes:

$$W_{j'j} \rightarrow -W_{j'j}\frac{S_>(j)}{S_<(j)} \text{ and } W_{jj'} \rightarrow -W_{jj'}\frac{S_>(j)}{S_<(j)}, \text{ for } j' > j, \qquad \text{(IV.110)}$$

where $S_>$ and $S_<$ are recalculated at each column change ($j \rightarrow j+1$) and given by:

$$S_<(j) \equiv \sum_{j'>j} d_{j'} W_{j'j} \text{ and } S_>(j) \equiv \sum_{j'\leq j} d_{j'} W_{j'j}. \qquad \text{(IV.111)}$$

This ensures that W has accurate values on its diagonal and *exactly* verifies both the detailed balance relation and the sum rule, with the exception of the last column. This has no consequences if a sufficiently large matrix is treated in which the columns correspond, from left to right, to lines of decreasing integrated intensity, since the last ones have negligible effects on spectra calculations. Note that, in order to avoid truncation problems, W must anyway include all transitions of significant lower state relative population.

With respect to energy gap fitting laws, ECS approaches have the advantage of intrinsically including the proper couplings of angular momenta and the symmetries of the levels. It is thus adapted for the prediction of line couplings between lines of *any* type (different vibrational bands, various branches, odd-even J, *etc*), permitting calculation *without* introduction of ad hoc parameters for each type of band[335,343] or isotopomer.[417] Furthermore, the scaling laws are of rather simple implementation and have proved to enable accurate predictions in many cases (see following and Table IV.9). The remainder of this section gives a list of various ECS approaches according to the type of radiator and to the spectral structures considered. Finally note that, as energy gap fitting laws (Fig. IV.16), the ECS approach is a practical tool for interpolations and extrapolations (in J and temperature) of line broadening coefficients starting from a limited set of measured values (*eg* Refs. 386,418 and Fig. IV.20).

IV.3.3b *Rovibrational components of linear molecules*

A sophisticated version of the ECS approach for line couplings between rovibrational components of linear molecules was proposed by Bonamy *et al*[407,419,420] The relaxation matrix is expressed as a linear combination of dyadic collisional terms $\widetilde{\Gamma}_{JJ'}^{\ell\ell'}$. For lines belonging to a vibrational band, with vibrational angular momenta of the bending mode ℓ_f and ℓ_i in the upper and lower states, the *real* element of the relaxation matrix coupling the transitions $J_f \leftarrow J_i$ and $J'_f \leftarrow J'_i$ is given by[407,419]:[6]

---

[6] With respect to the notations of these papers, and for consistency with the choices made in this book, some signs have been changed and the starting and ending levels of collisional exchanges have been permuted.

$$\langle\langle J_f'J_i'|\text{Re}[\mathbf{W}]|J_fJ_i\rangle\rangle = \sum_{\ell=\ell_f-k}^{\ell_f+k} d_{J_fJ_i}^{k\ell\ell_f} \times d_{J_f'J_i'}^{k\ell\ell_f} \times \widetilde{\Gamma}_{J_i'J_i}^{\ell_i\ell}, \qquad (\text{IV.112})$$

with k = 0 for isotropic Raman, k = 1 for dipole transitions, k = 2 for anisotropic Raman spectra, *etc.* $d_{J'J}^{k\ell\ell'}$ is the associated reduced matrix element for the rovibrational transition $(J_f = J', \ell_f = \ell') \leftarrow (J_i = J, \ell_i = \ell)$, given (for a *rigid rotor*) by:

$$d_{J'J}^{k\ell\ell'} = (-1)^{J'+\ell'}\sqrt{2J'+1}\begin{pmatrix} J & k & J' \\ \ell & \ell'-\ell & -\ell' \end{pmatrix}, \qquad (\text{IV.113})$$

where (:::) is a 3J symbol.[65] The dyadic collisional quantities are expressed as:

$$\widetilde{\Gamma}_{J'J}^{\ell'\ell} = \frac{\delta_{J,J'}}{\tau_J^{\ell'\ell}} - \Delta_{J'J}^{\ell'\ell}, \qquad (\text{IV.114})$$

where the $\Delta_{J'J}^{\ell'\ell}$'s are constructed using the ECS approach, *ie*:

$$\Delta_{J'J}^{\ell'\ell} = (-1)^{\ell+\ell'}\frac{\rho_{J_>}}{\rho_J}\Omega_{J_>}(2J_< + 1)\sum_{L\neq 0}(2L+1)\begin{pmatrix} J & L & J' \\ \ell & 0 & -\ell \end{pmatrix}\begin{pmatrix} J & L & J' \\ \ell' & 0 & -\ell' \end{pmatrix}\frac{Q_L}{\Omega_L}. \qquad (\text{IV.115})$$

This equation implicitly assumes that the Q_L's depend on L *only* and *not* on the vibrational angular momenta. This is reasonable when the radiator-perturber interaction varies little with vibration and for moderate bending energies, since the molecule remains quasi-linear, as demonstrated in Ref. 365. As in Eq. (IV.109), $J_<$ (resp. $J_>$) is the lower (resp. upper) value of (J, J'), and detailed balance is verified through the term $(2J_< + 1)\rho_{J_>}/\rho_J$. The temperature- and perturber-dependent factors Ω_J and Ω_L are the adiabaticity corrections introduced by DePristo *et al*[403] They improve the IOS approximation which corresponds to $\Omega_J = \Omega_L = 1$ and $(2J_< + 1)\rho_{J_>}/\rho_J = (2J'+1)$. Ω_J is defined in terms of a scaling length ℓ_c, of the mean relative speed $<v_r>$, and of the frequency spacing $\omega_{J,J_{Next}} = (E_J - E_{J_{Next}})/\hbar$ between the level J and the next adjacent inferior rotational level J_{Next} to which it is coupled by collisions. Hence $J_{Next} = J - 1$ if the interaction potential is efficient in inducing odd changes of the rotational angular momentum whereas $J_{Next} = J - 2$ if not (for O_2, N_2, and quasi-symmetric molecules such as N_2O and bending CO_2). The adiabaticity corrections are given by:[403]

$$\Omega_J = [1 + (\omega_{J,J_{Next}} \times \ell_c/<v_r>)^2/24]^{-2}, \qquad (\text{IV.116})$$

or, alternatively:[421]

$$\Omega_J = [1 + (\omega_{J,J_{Next}} \times \ell_c/<v_r>)^2/12]^{-1}. \qquad (\text{IV.117})$$

The Q_L's ($= -\widetilde{\Gamma}_{0L}^{00}$, positive in our sign convention) are related to the cross sections of the collisional transfer from rotational level J = L to J = 0 [Eq. (IV.104)]. They are, with ℓ_c, the basic parameters of the model. For simple systems, their values can be calculated from first principles starting from the interaction potential. When this is not possible, they can be modeled analytically (cf Table IV.1) as functions of the gap $E_L - E_0 = E_L$. For this, the following Exponential Power law, leading to the ECS-EP approach, has been widely used:

$$Q_{L \neq 0}(T, p) = \frac{A(T, p)}{E_L^{\lambda(T,p)}} \exp\left[-\beta(T, p) \frac{E_L}{k_B T}\right], \tag{IV.118}$$

where A, λ, and β are the temperature and perturber-dependent parameters of the model. Remember the limitations in the use of energy gap laws discussed before since they again apply to Eq. (IV.118). Indeed, consider the expansion of the intermolecular potential on Legendre polynomials, ie $V(\vec{R}) = \sum V_L(R) P_L(\cos\theta)$. When the components V_L are significant for both even and odd L values, use of a single analytical law for all the Q_L's is wrong. This does not represent the oscillations of the Q_L from odd to even values of L. Hence, in the absence of detailed calculations of these quantities, Eq. (IV.118) is *limited* to quasi-symmetric linear radiators for which it is reasonable to assume that *only* even L values contribute significantly. This approximation, obviously valid for radiators such as O_2 and N_2, also leads to accurate results for N_2O, CO, C_2H_2, and for bending modes of CO_2 as shown in the references of Table IV.9. Note that an alternative to Eqs. (IV.116)–(IV.118) has been proposed[422] for the description of the femtosecond CARS response in pure N_2. The basic cross sections are written as:

$$Q_{L \neq 0}(T, p) = A(T, p) \exp(-\beta(T, p)\sqrt{E_L}), \tag{IV.119}$$

while the adiabaticity correction $\Omega_{J_>}/\Omega_L$ is replaced, for the coupling between the Q(J) and Q(J') lines, by a J, J', and L dependent expression:

$$\Omega_{J_>}/\Omega_L \rightarrow \exp\left\{-4[(\omega_{JJ'} - \omega_{L0}) \times \ell_c/<v_r>]/\sqrt{L(L+1)}\right\} \tag{IV.120}$$

In Eq. (IV.114), the terms $\widetilde{\Gamma}_{JJ}^{\ell'\ell}$ for $J' = J$ include contributions $\tau_J^{\ell'\ell}$, directly related[407] to the relaxation times $\tau(\vec{J})$, $\tau(\vec{J}^{[2]})$, etc, of the rotational angular momentum $\vec{J}$ and of its associated higher order tensors. Those for $\ell = \ell'$ can be calculated from the off-diagonal elements since Eq. (IV.69) translates into:[407]

$$\sum_{J'} \widetilde{\Gamma}_{J'J}^{\ell\ell} = 0 \Leftrightarrow 1/\tau_J^{\ell\ell} = \sum_{J'} \Delta_{J'J}^{\ell\ell}. \tag{IV.121}$$

Hence, once the Q_L's and ℓ_c are known, all $\tau_J^{\ell'=\ell}$ and $\Delta_{J'J}^{\ell\ell}$ terms can be calculated, but *not* the $\tau_J^{\ell' \neq \ell}$. The model contains *no* information on these quantities which must be obtained from other sources. Their values can be calculated from the interaction potential,[423,424] or deduced from specific experiments (Nuclear Magnetic Resonance, depolarized Rayleigh scattering, the viscomagnetic effect),[407,425] or obtained from fits of measured spectra in bands of various symmetries.[426,427,591] If one needs to model line-mixing effects for numerous vibrational symmetries, the number of unknown parameters quickly becomes important. This is shown in Table IV.2 which summarizes the $\tau_J^{\ell'\ell}$ involved in

Table IV.2: Angular momentum relaxation terms involved in the construction of the relaxation matrix for dipole radiation (k = 1) according to the vibrational symmetry (angular momenta ℓ_f and ℓ_i) of the band

Band symmetry	ℓ_f, ℓ_i	$\tau_J^{\ell\ell'}$ involved
$\Sigma \leftarrow \pi$	0, 1	(τ_J^{11})
$\Sigma \leftarrow \Sigma$	0, 0	(τ_J^{00}), τ_J^{01}
$\pi \leftarrow \Sigma$	1, 0	(τ_J^{00}), τ_J^{01}, τ_J^{02}
$\pi \leftarrow \pi$	1, 1	(τ_J^{11}), τ_J^{10}, τ_J^{12}
$\Delta \leftarrow \pi$	2, 1	(τ_J^{11}), τ_J^{12}, τ_J^{13}
$\pi \leftarrow \Delta$	1, 2	(τ_J^{22}), τ_J^{20}, τ_J^{21}

the construction of relaxation matrices for various vibrational angular momenta and k = 1. Only the values between parentheses ($\ell = \ell'$) can be calculated from Eq. (IV.121).

The preceding model and above mentioned problem can be *greatly* simplified by assuming that $\tau_J^{\ell'\ell} = \tau_J^{\ell\ell}$, an approximation which is satisfactory, at least for CO_2.[409,591] Within this assumption, *all* values can be calculated from Eq. (IV.121) once the parameters ℓ_c and Q_L are known. Furthermore, using some 3J and 6J algebra, Eqs. (IV.112)–(IV.115) and (IV.121) then lead to the more compact following expression for the *off-diagonal* elements:

$$\text{Re}[\langle\langle J_f' J_i'|W|J_f J_i\rangle\rangle] = -(-1)^{\ell_i+\ell_f+k}\frac{\rho_{J_>}}{\rho_J}\Omega_{J_>}(2J_< + 1))\sqrt{(2J_f+1)(2J_f'+1)}$$
$$\times \sum_L \begin{pmatrix} J_i' & L & J_i \\ \ell_i & 0 & -\ell_i \end{pmatrix}\begin{pmatrix} J_f' & L & J_f \\ -\ell_f & 0 & \ell_f \end{pmatrix}\begin{Bmatrix} J_i & J_f & k \\ J_f' & J_i' & L \end{Bmatrix} \times (2L+1)\frac{Q_L}{\Omega_L},$$
(IV.122)

where {:::} is a 6 J symbol.[65] For $\ell_i = \ell_f = 0$, this equation is identical to Eq. (IV.109) and, within the purely IOS limit, it becomes:

$$\text{Re}[\langle\langle J_i' J_f'|W^{IOS}|J_i J_f\rangle\rangle] = \delta_{J_i',J_i}\delta_{J_f',J_f}Q_0 + (-1)^{\ell_i+\ell_f+k+1}(2J_i'+1))\sqrt{(2J_f+1)(2J_f'+1)}$$
$$\times \sum_L \begin{pmatrix} J_i' & L & J_i \\ \ell_i & 0 & -\ell_i \end{pmatrix}\begin{pmatrix} J_f' & L & J_f \\ -\ell_f & 0 & \ell_f \end{pmatrix}\begin{Bmatrix} J_i & J_f & k \\ J_f' & J_i' & L \end{Bmatrix} \times (2L+1)Q_L,$$
(IV.123)

which is identical (within the above mentioned sign changes) to those given in Refs. 365,402. One can again check, using Eq. (IV.105) and the transition moment in Eq. (IV.113), that Eq. (IV.69) is rigorously verified within the IOS framework.[408]

As summarized in Tables IV.9 and IV.12, the ECS approaches [Eqs. (IV.116) or (IV.117) and (IV.118) together with (IV.112)–(IV.115) or (IV.122)] have been applied to line mixing for many linear radiators under a large variety of conditions. Numerous studies are devoted to infrared or Raman Q branches in the spectral domain, showing that the ECS model correctly predicts the narrowing of these structures with respect to results obtained assuming collisionally isolated lines. Some studies also consider entire bands, as exemplified in Fig. IV.21. In this case, not only the Q branch shape is accurately modeled but also the P and R branches and the sub-Lorentzian behavior of the wings. The agreement in the far wings may seem surprising due to the expected breakdown of the impact approximation at such distances from the absorption maximum (App. II.B4). It is due here to the dominant contribution of local absorption features, such as the $\nu_1-\nu_2$ and $\nu_1+\nu_2-\nu_1$ vibrational transitions near 720 and 790 cm^{-1}.

An example in the time domain is displayed in Fig. IV.22. In this case, line mixing mostly affects the response at short time delays between the pump and probe. This corresponds, in the spectral domain, to the large effects affecting the weak absorption regions in the wings and in the troughs between transitions.

The quality of the ECS approach for the prediction of infrared and Raman spectra and time domain responses has positive consequences for practical applications. It is the case, as discussed in Sec. VII.4, for remote sensing studies of atmospheric properties and local soundings of gas media by Raman techniques.

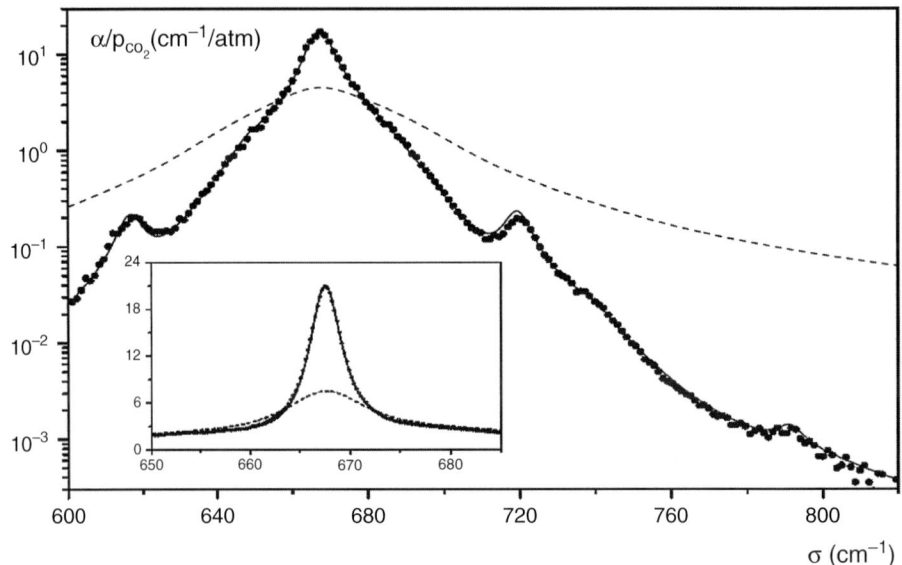

Fig. IV.21: Absorption by CO_2 diluted in N_2 in the 12–17 μm region. (•) are laboratory measurements, while (—) and (– – –) have been calculated with (ECS) and without (purely Lorentzian lines) the inclusion of line mixing. The larger plot corresponds to 255 K and 159 atm and the ν_2 band Q branch at 296 K and 70 atm is displayed in the insert. After Ref. 353.

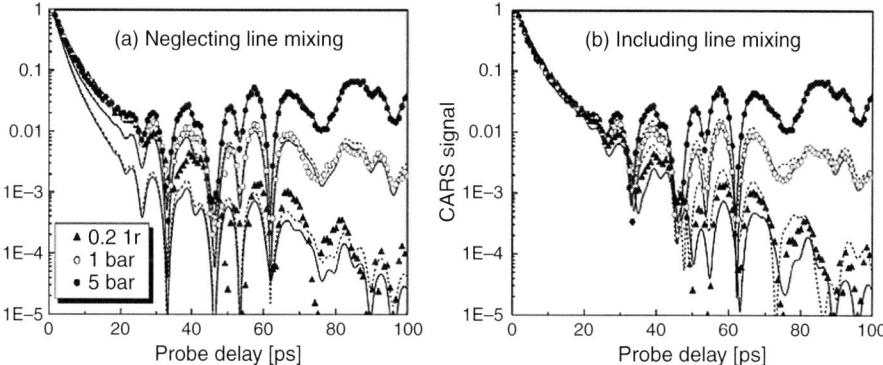

Fig. IV.22: Femtosecond CARS transients (log scale, *vs* the pump-probe delay) for pure N_2 in the Q branch of the fundamental band at room temperature and three pressures. Laboratory measurements (symbols) are compared with calculations (lines) with [(ECS, in (b)] and without [purely Lorentzian lines, in (a)] the inclusion of line mixing. Reproduced with permission from Ref. 110.

IV.3.3c *Other spectra and systems*

While collisional couplings between rovibrational transitions of linear molecules have been extensively studied, few studies are devoted to other situations. These are summarized below and references are also given in Table IV.9.

Hyperfine rotational components of linear molecules. The IOS expression of the relaxation matrix describing line couplings within nuclear spin hyperfine components of a given rotational or rovibrational transition of a linear molecule can be derived from Refs. 412,428. This has been done in Ref. 384 which provides all needed information [Eqs. (1)–(6)]. This IOS/ECS approach was first applied to hyperfine components of microwave lines of HCN broadened by He[412] using Q_L values deduced from an electron gas potential energy surface. The results show that all hyperfine components of the R(J = 0) line have identical widths and are not collisionally coupled whereas this is not the case for the ones within the R(J = 1) manifold. In this last case, assuming the same widths and no line mixing leads to errors of about 10% but, in the absence of experiments, no comparison with measurements was made. Thanks to the high spectral resolution and signal-to-noise ratio of a laser spectrometer, the quality of this IOS/ECS approach was demonstrated recently.[384] In this work, the R(0) and R(1) lines of the v = 1 ← 0 transitions of HI diluted in He are studied at room temperature. Measured absorptions are fitted in order to deduce the Q_0 and Q_2 parameters, which are the only ones required to construct the W matrix for the considered manifolds. It is shown that the values determined from spectra at various pressures are consistent, and that they enable calculation of the absorption shape under all studied conditions with an accuracy within the (small) noise level. This study confirms the predictions of Ref. 412 since it proves that the three hyperfine components within R(0) do have the *same* widths and are *not* collisionally coupled, whereas this is *not* the case for the nine transitions composing the R(1) manifold.

Electronic spectra of linear molecules. Assuming that the intermolecular potential is independent of the electronic spin and within the pure Hund's case b, the IOS expression for the relaxation matrix elements can be derived from Ref. 428. The extension including ECS corrections was then made in Ref. 429 where the information needed for calculations can be found [Eqs. (10)–(16)] (see Ref. 415 for model improvements). To our knowledge, the only studies in which this approach has been used for the modeling of spectra influenced by line mixing are those of Refs. 415,429. These papers present comparisons between measured values of the absorption in the A-band of O_2 and predictions using the ECS model. Good agreement with both laboratory and atmospheric measurements is obtained provided that Q_L parameters fitted on both line-broadening data *and* spectra are used and that ad hoc empirical corrections to the sum rule are introduced.[415] These are attributed to the fact that the A-band is not a pure Hund's case b and to the expected influence of the electronic state on intermolecular interactions.

Symmetric-top molecules. IOS expressions for line couplings between rovibrational components of symmetric-top molecules are given in Ref. 410. These include the effects of the (eventual, *eg* NH_3) inversion motion and are based on dynamical factors $Q(L,M,M')$, which now depend on three numbers. The M-diagonal values $Q(L,M,M)$ can be identified with the state-to-state rotational cross section $\sigma(J=0,K=0 \leftarrow J=L,K=M)$.[410,430] Adiabaticity corrections $\Omega_{J,K}$ derived from first- and second-order approaches, leading to the ECS model for symmetric-top molecules, have been proposed by Richard and DePristo.[430] Their inclusion in the expressions of the relaxation matrix elements was then made in Eqs. (10)-(16) of Ref. 408. For NH_3 in mixtures with He and H_2,[408,431] the $Q(L,M,M')$'s were calculated from an intermolecular potential, and the predicted spectra are in good agreement with measurements (particularly for NH_3–He) under conditions where strong line-mixing effects are observed, as illustrated in Fig. IV.23. For CH_3F in He, in the absence of any reliable potential energy surface, Thibault *et al*[432] assumed that only the $Q(L,M,M')$ with $M=M'=0$ are significant. These were modeled using the power energy gap law (Table IV.1) with parameters obtained from fits of measured line-broadening data. The model was then applied to the CH_3F ν_3 band shape at room temperature. Comparisons with measured Q branch profiles at moderate density and with all P, Q, R branches at high pressure demonstrate[432] the quality of the approach.

Asymmetric-top molecules. The expression of line couplings between transitions of an asymmetric-top molecule starting from the symmetric-top expression is quite straight-forwardly obtained.[410] It is simply based on the projection of the wavefunctions $|J\tau m\rangle$ of the asymmetric-top on those $|JKm\rangle$ for a symmetric-top, as obtained from the construction and diagonalization of the free radiator internal states Hamiltonian. The IOS relaxation matrix is then given by Eq. (37) of Ref. 410. It could be corrected in order to obtain its ECS equivalent, introducing detailed balance corrections and adiabaticity factors following the case of symmetric-top molecules discussed above. To our knowledge, this has never been applied to line mixing, but it would be of great interest for the modeling of this process in the case of water vapor (see App. V.A). This might

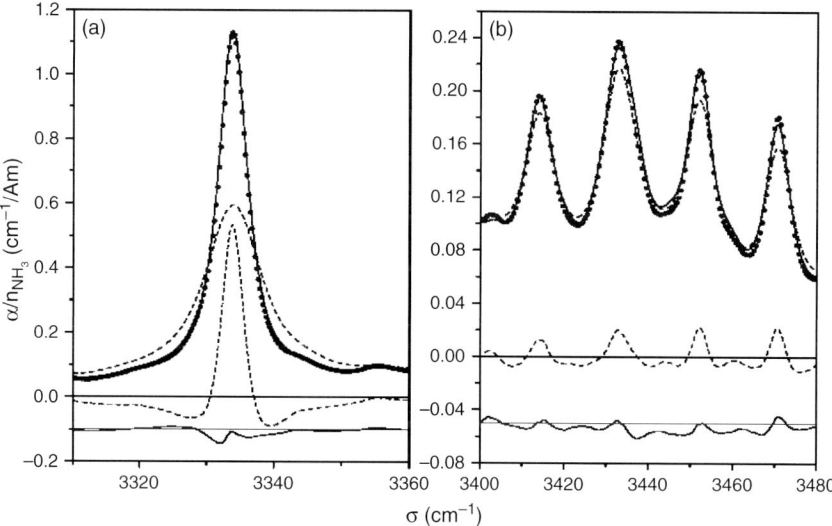

Fig. IV.23: Absorptions in the ν_1 band of NH_3 diluted in 125 Am of He at room temperature in the region of the Q branch (a) and of the R(3) to R(6) manifolds (b). (•) are measured values, while (—) and (– – –) have been calculated with and without the inclusion of line mixing, respectively (meas-calc deviations are given by the two lower traces). After Ref. 408.

enable to calculate the near wing behavior and thus to make a junction between the core regions, where isolated line profiles (chapter III) are generally valid, and the far wings, for which quasistatic approaches (Sec. V.3) have been developed.

Diatom-diatom case. Infinite Order Sudden expressions for collisions between two diatomic molecules have been proposed in Refs. 405,406. State-to-state rotational relaxation is considered so that the expressions are only suitable for the modeling of line-mixing effects in isotropic Raman Q branches. Nevertheless, to our knowledge, this model has only been tested comparing predicted and measured line-broadening coefficients,[405,433,434] regardless of the spectral consequences of collisional couplings.

IV.3.4 SEMI-CLASSICAL MODELS

IV.3.4a Introduction

In *semi-classical* approaches, trajectories are described classically and parameterized by the initial relative speed v_r and impact parameter b, while the internal degrees of freedom are described through quantum mechanics. The relaxation matrix elements are then given, with simplified notations, by a thermal average over v_r and all initial perturber internal states p, *ie*:

$$\langle\langle f'i'|W|fi\rangle\rangle = \frac{n_0 \langle v_r \rangle}{2\pi c}\int_0^\infty xe^{-x}\sum_p \rho_p \sigma^{(k)}_{f'i',fi,p}(x)dx \qquad (IV.124)$$

where x is the reduced initial relative kinetic energy $x = m^* v_r^2/(2k_B T)$ and ρ_p is the relative population of the perturber internal state p. The *spectroscopic* cross section $\sigma_{f'i',fi,p'}^{(k)}$ which *depends on* the order k of the tensor coupling radiation and matter, is obtained from the integral over all values of b of the "interruption function" S, ie:

$$\sigma_{f'i',fi}^{(k)}(x) = \int_0^\infty 2\pi b S_{f'i',fi,p}^{(k)}(b,x) db. \qquad (IV.125)$$

The S's are expressed, within the impact approximation, in terms of the evolution operator $U(+\infty, -\infty, b, x)$ for a *complete collision centered at time t = 0*, ie the *scattering* (or diffusion) operator S. For a linear molecule colliding with an atom, for instance, S is given, for lines within the same vibrational band, by [see Ref. 147 and Eq. (II.47)]:

$$S_{f'i',fi}^{(k)}(b,x) = \sum_{m_i,m_i',m_f,m_f',q} (-1)^{J_f+J_f'+m_f+m_f'} \begin{pmatrix} J_i' & k & J_f' \\ m_i' & q & -m_f' \end{pmatrix} \begin{pmatrix} J_i & k & J_f \\ m_i & q & -m_f \end{pmatrix} \frac{\sqrt{2J_i'+1}}{\sqrt{2J_i+1}}$$

$$\left\{ \delta_{J_i,J_i'} \delta_{J_f,J_f'} \delta_{m_i,m_i'} \delta_{m_f,m_f'} - \langle v_i J_i' m_i' | U(+\infty, -\infty, b, x) | v_i J_i m_i \rangle \right. \qquad (IV.126)$$

$$\left. \times \langle v_f J_f' m_f' | U(+\infty, -\infty, b, x) | v_f J_f m_f \rangle^* \right\}.$$

Note that, within this semi-classical frame, W does not verify the detailed balance relation. It only fulfills the approximate one of Eq. (IV.107) due to the neglect of exchanges between translational and internal degrees of freedom. The same relation is verified by both the real and imaginary parts of W.

The semi-classical models presented thereafter have been *largely* used for the calculation of the broadening- and shifting-coefficients (*diagonal* part of W), but *few* references consider the off-diagonal (line coupling) terms. Hence, after a presentation of the models, the studies devoted to line mixing are all discussed in § IV.3.4f.

IV.3.4b The Neilsen and Gordon approach

The most sophisticated impact semi-classical approach, *limited* to interactions between linear molecules (no bending) and atoms, was proposed by Neilsen and Gordon[147] (NG). The collision is centered at time zero and the classical path is divided into a series of points $k = -n, +n$, beginning and ending at distances outside the range of the intermolecular potential. Each is defined by its time t_k and the intermolecular vector $\vec{R}_k$, parameterized by its magnitude R_k and the angle β_k with respect to the quantification axis z as described in Fig. IV.24. For given values of the initial relative speed v_r and impact parameter b, R_k and β_k can be computed from the *isotropic* part of the interaction potential using classical mechanics.

The evolution operator from time t_{-n} to time t_{+n} ($t_{-n} = -t_{+n}$) can be expressed as:[147]

$$U(t_n, t_{-n}) = \exp\left[-\frac{1}{i\hbar} H_a t_n\right] \times D^{-1}(\beta_n) T_n D(\Delta\beta_n) T_{n-1} D(\Delta\beta_{n-1}) \cdot \times \ldots$$

$$\times D(\Delta\beta_{-n+2}) T_{-n+2} D(\Delta\beta_{-n+2}) T_{-n+1} D(\beta_{-n}) \times \exp\left[\frac{1}{i\hbar} H_a t_{-n}\right], \qquad (IV.127)$$

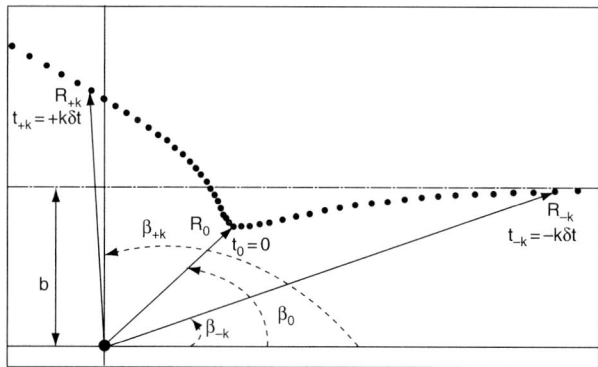

Fig. IV.24: Classical trajectory description and definition of notations.

where $D(\beta)$ is the rotation matrix for angle β and (*cf* Fig. IV.24):

$$\beta_k = \beta\left(\frac{t_k + t_{k-1}}{2}\right) \text{ and } \Delta\beta_k = \beta_k - \beta_{k-1}, \text{ with } \beta_{-n}=0 \text{ and } \beta_{+n}=\sum \Delta\beta_k. \quad \text{(IV.128)}$$

Provided that $t_{k+1} - t_k$ is chosen small enough, the total Hamiltonian $H(t) = H_a + V(t)$ (H_a being the radiator internal states Hamiltonian and V the radiator-perturber interaction potential) varies little between two successive times and T_k can be approximated by:

$$T_k = \exp\left[\frac{1}{i\hbar}(t_k - t_{k-1})H\left(\frac{t_k + t_{k-1}}{2}\right)\right]. \quad \text{(IV.129)}$$

Thanks to the introduction of the rotations D, the matrix elements of T are calculated using the intermolecular axis as the quantification axis. Due to the cylindrical symmetry of the interaction potential, $H(t)$ is then diagonal with respect to the rotation quantum number m (projection of the total angular momentum $\vec{J}$). If $V[\vec{R}(t)]$ is expanded on Legendre polynomials, *ie* $V(R,\beta) = \sum_L V_L(R)P_L(\cos\beta)$, one has:[147]

$$\begin{aligned}\langle J'm'|H(t)|Jm\rangle &= \langle J'm'|H_a + V(t)|Jm\rangle \\ &= \delta_{J,J'}\delta_{m,m'}E_J + (-1)^m \delta_{m,m'} \sum_L \sqrt{(2J+1)(2J'+1)} \\ &\quad \times \begin{pmatrix} J' & L & J \\ 0 & 0 & 0 \end{pmatrix}\begin{pmatrix} J' & L & J \\ m & 0 & -m \end{pmatrix}V_L[R(t)]. \end{aligned} \quad \text{(IV.130)}$$

The main problem of the NG approach is the computer time required. Indeed, since the operator U constructed step-by-step using Eq. (IV.127) is non-diagonal in J and m, the number of its elements for J ranging from 0 to J_{max} is $(J_{max}+1)^4$. The time required for the matrix products in Eq. (IV.127) thus typically scales as $\approx J_{max}^6$ remembering that these multiplications are made for each initial kinetic energy, impact parameter, and time step. Calculations are lengthy but (today) tractable up to $J_{max} \approx 20$ but not at all for $J_{max} \approx 60$ (this requiring a computation time about a thousand times greater, the U matrix being complex with one million elements). This constraint on J_{max} restricts the use of the NG approach to practical calculations of the diagonal terms of W *only*, for the following reasons. If one wants to test a calculated relaxation matrix by comparisons between predicted and measured spectra influenced by line mixing, W must include all lines with significant initial level population in order to avoid truncation errors. For a radiator with rotational constant B'', a criterion corresponding to a cut-off below 1% of the maximum relative population leads to:

$$J_{max} \approx 2.5\sqrt{k_B T/B''}, \quad (IV.131)$$

eg $J_{max} \approx 60$ for CO_2, ≈ 30 for N_2, and less than 15 for HCl at room temperature. The problem then appears when considering the validity of the model. Indeed, for the *impact* calculation of the coupling between lines R(J) and R(J') to be correct, one must have $|\omega_{R(J)} - \omega_{R(J')}|\tau_c \ll 1$ where τ_c is the duration of efficient collisions (see App. II.C1). The latter can be estimated as $\tau_c \approx \ell_c/<v_r>$, from a typical potential efficient range ℓ_c and the mean relative velocity $<v_r> = (8k_B T/\pi m^*)^{1/2}$. The preceding impact criterion for the coupling of the R(J) and R(J + ΔJ) lines becomes:

$$|\omega_{R(J)} - \omega_{R(J+\Delta J)}|\tau_c \ll 1 \Leftrightarrow 2\pi c(2B''\Delta J)\ell_c \sqrt{\frac{\pi m^*}{8k_B T}} \ll 1, \quad (IV.132)$$

ie:

$$B''(cm^{-1}) \ll \left[2\pi c(2\Delta J)\ell_c \sqrt{\frac{\pi m^*}{8k_B T}}\right]^{-1} \approx \frac{4}{\ell_c(\text{Å})\Delta J}\sqrt{\frac{T(K)}{m^*(g)}}. \quad (IV.133)$$

If we consider the values $\ell_c \approx 3$ Å, $T \approx 300$ K and $m^* \approx 30$ g, representative of many collisional systems, correct description of the R(J)–R(J') coupling for $|J-J'| = 8$ leads, through Eq. (IV.133), to $B'' \ll 0.5\,\text{cm}^{-1}$. According to Eq. (IV.131), this imposes the unacceptable constraint $J_{max} \gg 50$. Even in the favorable case of collisions with He ($\ell_c \approx 1$ Å and $m^* \approx 4$ g) the result is $B'' \ll 5\,\text{cm}^{-1}$, leading to $J_{max} \gg 18$ which is almost not tractable with today's computers. Hence, the feasibility of calculations without significant truncation errors is incompatible with the respect of the impact approximation. For molecules with a small rotational constant colliding with He, the impact model is valid but a full relaxation matrix ($J_{max} \approx 50$) cannot be processed. For radiators with a large rotational constant, a full matrix can be calculated but the impact model breaks down. Whereas the diagonal elements of W are then correctly predicted (thanks to $|\omega_{R(J)} - \omega_{R(J)}| = 0$), the *off-diagonal* elements are strongly underestimated within the impact frame (see Fig. VIII.3 and the discussion in Ref. 435). These constraints (today) limit the use of the NG approach to the prediction of the diagonal terms of the relaxation matrix for light radiators. In this case, it leads to good agreement[19,131] with measurements and purely quantum calculations. In particular, thanks to the qualities of both the interaction potential energy surface[23] and NG approach, subtle effects on line widths and shifts of HF in Ar due to rotation-vibration couplings have been analyzed.[19]

IV.3.4c The Smith, Giraud and Cooper approach

A first simplification of the NG approach is obtained through the *peaking* approximation (PA) in which one considers that the intermolecular axis $\vec{R}(t)$ does *not* rotate during the efficient part of the collision. In the Neilsen-Gordon algorithm, the PA consists in dropping all the rotation matrices D(...). The product of the T_k operators in Eq. (IV.127) can then be re-combined, leading to:

$$U^{PA}(+\infty, -\infty, b, x) = S^{PA}(b, x) = \exp_t\left[\frac{-i}{\hbar}\int_{-\infty}^{+\infty} e^{+iH_a t/\hbar} V(t) e^{-iH_a t/\hbar} dt\right], \quad (IV.134)$$

where $\exp_t$ is a *time-ordered* exponential. The PA makes the time evolution (scattering) matrix diagonal in m, resulting in much smaller computer times and storage requirements. Rigorously exact for b = 0, this approximation mostly affects low J lines and is reasonable for small impact parameters and high relative kinetic energies.[19,436]

A further approximation neglects the non-commutation of the interaction potential at different times. The time ordering in Eq. (IV.134) can then be disregarded, leading to:

$$U^{SGC}(+\infty, -\infty, b, x) = S^{SGC}(b, x) = \exp\left[\frac{-i}{\hbar}\int_{-\infty}^{+\infty} e^{+iH_a t/\hbar} V(t) e^{-iH_a t/\hbar} dt\right]. \quad (IV.135)$$

This equation essentially corresponds to the approach of Smith, Giraud, and Cooper[385] (SGC), although further simplifications were made by these authors which were released later on.[19,131,437] When compared to the PA, the commutative assumption introduces new errors affecting low J transitions.[19]

In summary, with respect to the NG approach, the PA only neglects the rotation of the intermolecular axis, while the SGC formalism neglects both the rotation of $\vec{R}(t)$ and the time ordering. Thanks to these assumptions, the SGC approach is of tractable use even when couplings between many lines is considered. Nevertheless, likely due to these approximations and to its limitation to linear radiators colliding with atoms, the SGC model has been little used.

IV.3.4d The Anderson, Tsao and Curnutte approach

The preceding semi-classical formalisms are limited to linear molecules colliding with atoms. In order to treat more complex collisional pairs, a simplified model (ATC) was proposed by Anderson,[40] Tsao and Curnutte.[438] Since detailed but lengthy developments can be found in Ref. 438, only the main features of the model are recalled here using simplified notations. Starting from a more general expression similar to Eq. (IV.126), the relaxation matrix elements are expressed in terms of the scattering matrix $S(b,v_r)$ by:

$$S^{(k)}_{f'i',fi,p}(b,x) = \sum_{\beta_i,\beta_i',\beta_f,\beta_f',q} C(i, f, i', f', k, \beta_i, \beta_i', \beta_f, \beta_f', q) \quad (IV.136)$$

$$\times \left\{ \delta_{i,i'} \delta_{f,f'} \delta_{\beta_i,\beta_i'} \delta_{\beta_f,\beta_f'} - \langle i'\beta_i' p | S(b,x) | i\beta_i p \rangle \langle f'\beta_f' p | S(b,x) | f\beta_f p \rangle^* \right\}.$$

In this expression, i and f denote the quantum numbers defining the lower and upper levels of the radiator optical transitions and the β's designate all other (sub) quantum numbers. Similarly, p designates the internal states of the perturber. The quantity C depends on quantum numbers of the radiator lines of interest and on the order (k) of the tensor coupling radiation and matter.[438] Its expression for rotational transitions of a linear molecule is the first line of Eq. (IV.126). In order to go further, Eq. (IV.134) is used and the exponential is expanded to *second order*, leading to:

$$S^{(k)}_{f'i',fi,p}(b,x) = i[S_{1,i,i',p}(b,x) - S_{1,f,f',p}(b,x)] \qquad (IV.137)$$
$$+ [S_{2,i,i',p}(b,x) + S_{2,f,f',p}(b,x) - S_{2,i,i',f,f',p}(b,x)],$$

with

$$S_{1,i,i',p}(b,x) = \sum_{\beta_i,\beta'_i,\beta_f,\beta'_f,q} C(i,f,i',f',k,\beta_i,\beta'_i,\beta_f,\beta'_f,q)\langle i'\beta'_i p|P(b,x)|i\beta_i p\rangle \delta_{\beta_f,\beta'_f}\delta_{f,f'},$$

$$S_{2,i,i',p}(b,x) = \sum_{\beta_i,\beta'_i,\beta_f,\beta'_f,q} C(i,f,i',f',k,\beta_i,\beta'_i,\beta_f,\beta'_f,q)\langle i'\beta'_i p|P(b,x)^2|i\beta_i p\rangle \delta_{\beta_f,\beta'_f}\delta_{f,f'},$$

$$S_{2,i,i',f,f',p}(b,x) = \sum_{\beta_i,\beta'_i,\beta_f,\beta'_f,q} C(i,f,i',f',k,\beta_i,\beta'_i,\beta_f,\beta'_f,q) \qquad (IV.138)$$
$$\times \{\langle f'\beta'_f p|P(b,x)|f\beta_f p\rangle^* \times \langle i'\beta'_i p|P(b,x)|i\beta_i p\rangle\},$$

where the operator P(b,x) is given by:

$$P(b,x) = \frac{1}{\hbar}\int_{-\infty}^{+\infty} e^{i(H_a+H_p)t/\hbar} V(t,b,x) e^{-i(H_a+H_p)t/\hbar} dt, \qquad (IV.139)$$

H_p being the Hamiltonian associated with the free perturber internal states. *Straight-line trajectories at constant relative speed* are then assumed and the anisotropic interaction potential is (essentially) *limited* to electrostatic contributions. For the electric multipole θ_a of the radiator interacting with the multipole θ_p of the perturber, each of these terms takes the typical form:[438]

$$V_{\theta_a\theta_p}(R,\Omega_a,\Omega_p) = \frac{\theta_a\theta_p}{R^{q_a+q_p+1}} D^{q_a,q_p}(\Omega_a,\Omega_p). \qquad (IV.140)$$

In this expression q_a and q_p are the orders of the multipole moments ($q=1$ for dipole, $q=2$ for quadrupole, *etc*) and $D^{q_a,q_p}(\Omega_a,\Omega_p)$ is associated with the dependence of the interaction on the set of angles Ω_a and Ω_p defining the relative orientations of the two molecules. The matrix elements in Eq. (IV.138) can then be expressed in a compact form, since the integration over time is analytical, thanks to the trajectory model for which $R(t) = [b^2 + (v_r t)^2]^{1/2}$, leading to the so-called "resonance functions".[438,439]

A problem of the straight trajectories is that, if straightforwardly applied, it leads to un-physical situations where the molecules go one through the other (for b=0 for instance). Furthermore, $S^{(k)}_{f'i',fi,p}(b,x)$ can take unacceptable high values, due to the expansion limited to second order. In order to correct this, a *"cut-off"* procedure is used.[40,438] It consists in replacing $S_{fi,fi,p}^{(k)}(b,x)$ by unity for impact parameter values lower than b_0, where b_0 is defined by:

$$b_0(p,x) = \text{Max}[\sigma_0, b_0'(p,x)], \ b_0' \text{ being solution of}:$$
$$S_{2,i,i,p}(b_0') + S_{2,f,f,p}(b_0') + S_{2,i,i,f,f,p}(b_0') = 1, \tag{IV.141}$$

where σ_0 is a rigid sphere radius which can be deduced from the isotropic part of the interaction potential. One of the limits of the ATC approach, resulting from Eq. (IV.141) and from taking only electrostatic interactions into account, is obvious. For weakly polar systems, the S_2 terms can be small (zero for atomic perturbers) so that b'_0 is lower than σ_0. In this case the broadening coefficients are governed by the value of σ_0 and given by $\gamma_{fi} \approx (n_0 <v_r> /2\pi c) \times \pi\sigma_0^2$. They are thus almost independent on the line, in disagreement with measured values for many systems (eg Fig. IV.20).

With a *few* exceptions discussed in § IV.3.4f, the ATC approach has mostly been applied to the prediction of the broadening and shifting coefficients [i = i', f = f' in Eqs. (IV.136)-(IV.138)]. For this problem, it has been a *main* step which enabled, as early as the mid sixties, calculations of the self-broadening of linear molecules[440] but also of complex species such as water vapor.[441] For completeness, recall that (similar) alternatives to the ATC model have been proposed.[442–444] As expected, these approaches lead to good predictions for strongly polar systems where interactions at long distances are dominant, validating the cut-off procedure and the straight trajectory model. On the contrary, due to the simplified representations of the interactions and of dynamics, inaccurate results are obtained when close collisions are the main source of the broadening.[445,446] This is why, after having been largely used for about twenty years, ATC-like models have been progressively abandoned once the more refined but still tractable approach described below was proposed.

IV.3.4e *The Robert and Bonamy approach*

Robert and Bonamy (RB) have proposed[322] an extension of the ATC approach for linear radiators providing a more correct modeling of the dynamics and interactions at short distances. This greatly enlarges the applicability of the model, enabling correct predictions of the broadening and shifting coefficients for weakly polar systems, collisions with atoms, and/or high temperatures. The improvements are threefold. The first one is the inclusion, in the interaction, of an *atom-atom* Lennard-Jones potential, *ie*:[321,322]

$$V_{at-at}(R, \Omega_a, \Omega_p) = \sum_{i \in a} \sum_{j \in p} \left[\frac{e_{ij}}{r_{ij}(R, \Omega_a, \Omega_p)^{12}} - \frac{d_{ij}}{r_{ij}(R, \Omega_a, \Omega_p)^{6}} \right], \tag{IV.142}$$

where the sums extends over all atoms i and j composing the radiator (a) and perturber (p), respectively. e_{ij} and d_{ij} are the atom-atom parameters and $r_{ij}(R, \Omega_a, \Omega_p)$ is the distance between i and j for a configuration defined by the intermolecular distance R and the orientations Ω_a and Ω_p of the collision partners. These distances are expanded (to fourth order) as a series of powers of 1/R,[322] thus expressing the interaction potential as a sum of terms having a form similar to Eq. (IV.140). The second change *removes* the ATC cut-off procedure. This is done by using a cumulant expansion and schematically corresponds to an exponentiation of all S terms in Eq. (IV.137), as previously done for molecular scattering S-matrix elements. Note that various ways to include higher order terms are discussed in Ref. 447. The last improvement concerns the description of the *dynamics at short distances*, achieved through a parabolic description of the trajectories determined from the isotropic part $V_0(R)$ of the interaction potential. This leads to the introduction[322] of the true instantaneous relative speed $v_c(R_c,v_r)$ and of an "equivalent straight trajectory" speed $v'_c(R_c,v_r)$ which describes the relative motion near the point of closest approach $R(t=0)=R_c(b,v_r)$. These quantities can be expressed [analytically if $V_0(R)$ is a 6–12 Lennard-Jones] from the initial relative speed v_r and impact parameter b. The final RB expression (with simplified notations) of the broadening and shifting cross section is:

$$\sigma_{fi,fi}^{(k)}(v_r) = \sum_p \rho_p(T) \int_{R_0(v_r)}^{+\infty} 2\pi R_c \left[\frac{v'_c(R_c,v_r)}{v_r}\right]^2$$
$$\times \{1 - \exp[-iS_{1,p}(R_c,v_c) - S_{2,p}(R_c,v_c)]\}dR_c, \qquad (IV.143)$$

with

$$iS_{1,p} + S_{2,p} \equiv i(S_{1,i,i,p} - S_{1,f,f,p}) + (S_{2,i,i,p} + S_{2,f,f,p} + S_{2,i,i,f,f,p}),$$

where $R_0(v_r)$ is the turning point for b = 0 and all the S functions depend on R_c and v_c. They include various terms resulting from the different contributions to the interaction potential and involve the relevant resonance functions.[322]

The RB approach, initially developed for linear molecules,[322] was extended to asymmetric-top (and thus symmetric-top) species.[448,449] In this case, the potential was later on expanded from 4th to 8th order[450] and a fully complex expression was derived.[451] The case of spherical-top molecules was also treated.[452–454] Finally, exact trajectories have been implemented,[455,456] following the work of Ref. 457.

With a *few* exceptions described in § IV.3.4f, most applications of the RB formalism are limited to the pressure-broadening and, eventually, -shifting coefficients. When the radiator-perturber interaction potential is well known, the calculated values compare well with measurements (within typically 10%, often much better), except in situations where collisionally induced rotational transfers involve significant energy jumps with respect to the kinetic energy. In these cases (high J and/or low T), the semi-classical assumption breaks down, since rotation-translation exchanges are neglected, and inaccurate results are obtained (Fig. IV.25). These discrepancies may also be further amplified by a poor knowledge of radiator-perturber forces at very short distances.

The bibliography providing RB-type calculations of pressure-broadening and -shifting coefficients is much too vast to be exhaustively reviewed here. Table IV.3 gives a few references as starting points for bibliographical searches.

When off-diagonal elements of W are concerned, the only studies based on the RB model use state-to-state cross sections. Similarly to what is done with fitting laws (§ IV.3.2a), the *real* element coupling lines ℓ and ℓ' is written as [Eq. (IV.94)]:

$$\text{Re}\langle\langle\ell'|W|\ell\rangle\rangle = -A_{XY}K(i_{\ell'} \leftarrow i_\ell), \text{ for } \ell \neq \ell' \qquad (IV.144)$$

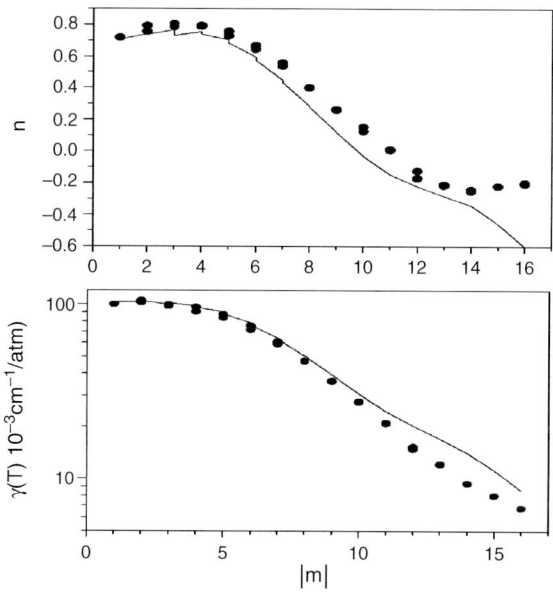

Fig. IV.25: Measured (•) and calculated (———) room temperature broadening coefficients $\gamma(296\,\mathrm{k})$ and temperature dependence coefficient n [such as $\gamma(T) = \gamma(296\,\mathrm{K})(296/T)^n$] for the N_2 broadened H_2O lines $J \pm 1_{1,J} \leftarrow J_{0,J}$ and $J \pm 1_{0,J} \leftarrow J_{1,J}$ of the ν_2 band. Results are plotted vs $|m|$ with $m = J_f$ and $m = -J_i$ for R and P lines, respectively. After Ref. 159.

Table IV.3: Some applications of the RB approach to pressure broadening and shifting coefficients. p (rows) and a (columns) designate the perturber and radiator

↓(p) (a)→	Linear	Symmetric-top	Asymmetric-top	Spherical-top
Atom	455,458,459	460,461	462,469	453
Linear	160,463–467	468,469	159,470–473	452,454
Symmetric-top		328,474,475		
Asymmetric-top	386,476		477,478	
Spherical-top			479	

where i_ℓ and $i_{\ell'}$ designate the initial levels of lines ℓ and ℓ', respectively. A_{XY} depends on the branches $[(X,Y) = P,Q,R$ or $(X,Y) = O,Q,S]$ to which ℓ and ℓ' belong and $K(i_{\ell'} \leftarrow i_\ell)$ is associated [Eq. (IV.90)] with the cross section for the collisional transfer from i_ℓ to $i_{\ell'}$. Within the RB approach, *some* of these can be calculated following Refs. 480,481. In order to do this, one sets $k = 0$, $f = i$, $f' = i'$ and the state-to-state quantity is given by:

$$K(i' \leftarrow i) = \frac{n_0 <v_r>}{2\pi c} \int_{R_0}^{\infty} dR_c\, 2\pi R_c \left(\frac{v'_c}{<v_r>}\right)^2 S'_{2,i' \leftarrow i}(R_c, <v_r>). \qquad (IV.145)$$

$S'_{2,i'\leftarrow i}$ is obtained[481] from the contribution of $i' \leftarrow i$ transfers to the total value of $S_{2,i}$ after renormalization in order to verify the sum rule of Eq. (IV.95), ie:

$$S'_{2,i'\leftarrow i}(r,v) = \frac{\bar{S}_{2,i'\leftarrow i}(r,v)}{\sum_{i'} \bar{S}_{2,i'\leftarrow i}(r,v)} \{1 - \exp[-S_{2,i}(r,v)]\} \qquad (IV.146)$$

with $\bar{S}_{2,i'\leftarrow i}(r,v) = \text{Min}[S_{2,i'\leftarrow i}(r,v), 1]$. Due to the limited expansion of the atom-atom potential and to the second-order expansion, this enables calculations of the "first" state-to-state quantities *only* (from i to a limited number of close by i' levels). However, these are generally the largest ones, likely relatively well calculated within this approach. An advantage of this model is that the collisional propensity rules are intrinsically contained in the calculated values (contrary to energy gap fitting laws, § IV.3.2b).

IV.3.4f Applications to line-mixing problems

As mentioned in § IV.3.4b, the NG approach is either (today) too computer costly ("heavy" radiators) or not suitable due to the breakdown of the impact assumption ("light" radiators), for the calculation of a full relaxation matrix.

The SGC approach (§ IV.3.4b) has been little applied to line-mixing problems, although it needs much less computer efforts. To our knowledge, the only direct application of this formalism is Ref. 482. In this work, rotational state-to-state cross sections (directly related to line-coupling elements between isotropic Raman Q lines) have been calculated for N_2–Ar. Due to lack of experiments, no comparison with measurements was made, leaving the question of the model quality opened. For completeness, recall that an approach similar to the SGC was used by Lam[483] to calculate line-broadening and -mixing effects on O_2 and H_2O microwave absorption. Very crude simplifying approximations were made for the description of the interaction potential (reduced, for O_2, to a single term) and a spherically averaged representation of the collision partner (isotropic perturber approximation) was used. For O_2, significant line-mixing effects in the 60 GHz band were predicted, but comparisons with measurements made later-on[347] showed the limits of the model. For H_2O, the calculations also indicate that line mixing has influence but they fail in modeling the observed absorption around 100 GHz.

More successful has been the use of ATC- and RB-like approaches for the calculation of relaxation matrix real elements and of their spectral consequences. In a series of studies, Buffa, Tarrini, and co-workers have investigated line broadening and line mixing for various types of spectral components. The general theoretical frame is given in Refs. 484–486. The approach is then reduced, through the use of a second-order expansion and of electrostatic interactions only, to a model similar to that described in § IV.3.4d. A number of interesting results concerning line mixing have been obtained. In Ref. 487, line couplings among doublets of self-broadened CHF_2Cl microwave lines has been evidenced experimentally and analyzed theoretically. It is shown that line mixing leads to a significant reduction of the doublet width (as expected from Fig. IV.2) when the two transitions overlap significantly. This result is well predicted by semi-classical calculations with the strong dipole-dipole interaction only. For hyperfine components, variations of the broadening from one transition to another (also mentioned in § IV.3.3c) were observed and explained by calculations but no line-mixing effects were evidenced.[484] Finally, Stark components have been analyzed in detail in Refs. 486,488. In these studies, the various M components of some rotational $J',K' \leftarrow J,K$ lines of self-broadened CH_3F are spectrally separated by an external electric field. Low pressure experiments then enable the determination of the individual broadening and shifting coefficients. These show dependences on M, well described[486] by ATC-like calculations based on the dipole-dipole interaction. Furthermore, when unresolved (no electric field and/or strong overlapping due to pressure) the clusters have effective broadening coefficients significantly smaller than the average value of those of the individual M components. This result, due to line mixing [cf Fig. IV.2 and Eq. (IV.45)], is analyzed in Ref. 488 with the above mentioned semi-classical approach. The comparisons between experimental and calculated signals then demonstrate the quality of the model and the large influence of line mixing, as illustrated in Fig. IV.26.

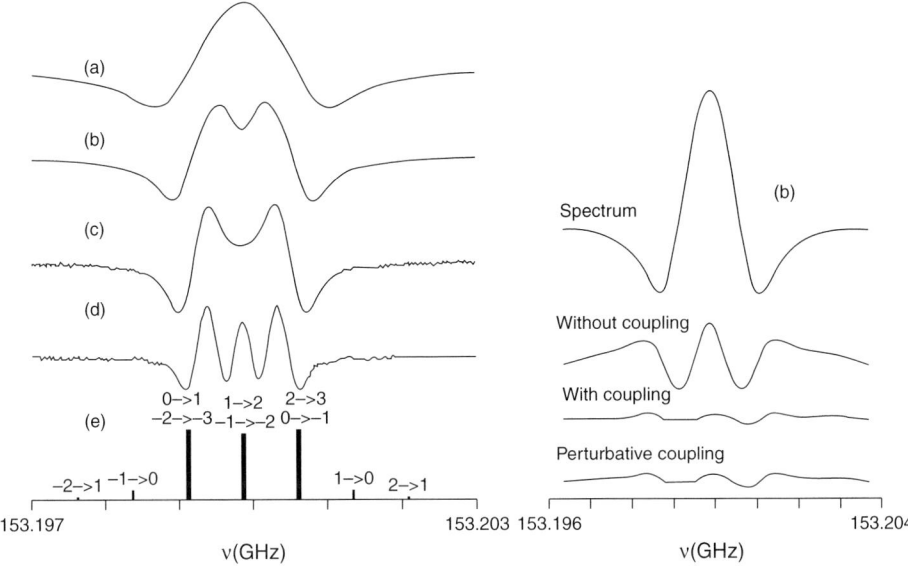

Fig. IV.26: The self-broadened $3,2 \leftarrow 2,2$ rotational transition of CH_3F as observed by a frequency modulation second harmonic absorption technique in the presence of an external electric field. The left panel displays measured signals for various pressures [between 4 and 24 mTorr from (d) to (a)] and modulation frequencies. The stick spectrum (e) gives the frequencies and M quantum numbers of the components. The right panel gives the measured spectrum and $2 \times$ (meas-calc) deviations for the pressure of 24 mTorr. These correspond, from bottom up, to predictions accounting for line couplings with the first-order approximation, with a full relaxation matrix, and neglecting all couplings. Reproduced with permission from Ref. 488.

The first test of the semi-empirical line-mixing approach based on the RB model [§ IV.3.4e] was made for methane infrared spectra.[359] State-to-state data calculated[489] as described above were used together with ad hoc scalings [A_{XY}, cf Eq. (IV.144)] of line couplings. These A_{XY} parameters, depending on the branch only, were adjusted on a single high pressure N_2-broadened spectrum in the v_3 band of CH_4 at room temperature. Comparisons for other densities show the consistency of the model although the evolution of the spectral shapes of high J R(J)- and P(J)-manifolds are not well reproduced. This approach was refined later[336,490] by introducing A_{XY} parameters for each R(J) and P(J) manifold. The robustness of the model was then demonstrated by the agreement between measured and calculated spectra obtained without the introduction of *any* new adjustable parameter, at various temperatures and in both the v_3 and v_4 bands.[491,492] The quality of predictions is illustrated in Fig. IV.27 which shows that line mixing affects the structures (Q branch and manifolds) and the wings. As shown in Refs. 359,493, refined propensity rules are involved in collisional transfers between methane rotational levels. These are contained in the present model as discussed in appendix A of Ref. 359. Note that the significant evolution of the effects of line mixing from one manifold to another shown in Fig. IV.5 is well reproduced[336] and that the quality of the approach has been confirmed using atmospheric spectra (Refs. 494,495 and Fig. VII.10).

This approach was also applied to Raman spectra of CH_4,[496,497] leading to good agreement with measurements (Fig. VII.9), provided that a vibrational dephasing contribution [parameters $\Delta\gamma$ and $\Delta\delta$, see Eq. (IV.70) and App. IV.A] is included and

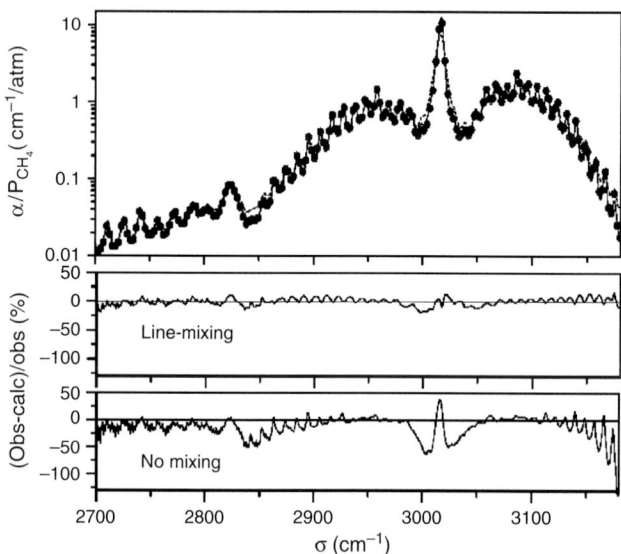

Fig. IV.27: Room temperature absorption in the v_3 band of CH_4 diluted in H_2 for a pressure of about 48 atm. (•) are measured values (not all plotted) while (———) and (- - -) have been calculated with and without line mixing. (meas-calc)/meas relative deviations (%) are plotted in the lower part. After Ref. 495.

adjusted. A second radiator for which semi-classical state-to-state data have been used for the modeling of line mixing is ammonia. A satisfactory description of line couplings within doublets associated with the inversion motion is obtained for pure NH_3, other perturbers leading to poorer results (*cf* Refs. 328,498 and those therein).

IV.3.5 QUANTUM MODELS

IV.3.5a Introduction

A variety of impact quantum approaches have been proposed to calculate the matrix elements of the scattering operator S from the knowledge of the interaction potential energy surface (see below and Refs. 499–503). From these elements one can calculate state-to-state cross sections as well as the broadening, shifting, and off-diagonal terms of the relaxation matrix for various types of spectroscopy. The most elaborated approach is the Close-Coupling (CC, § IV.3.5b) scheme introduced by Shafer and Gordon.[144] It is essentially "exact" within the *impact* approximation but requires large computational efforts, as discussed for the NG approach (§ IV.3.4b) which is its semi-classical equivalent. To treat molecular systems and high rotational states for which the CC model is too costly, simplified models have been proposed. The first is the Coupled-States (CS, § IV.3.5c) or Centrifugal Sudden approach[504,505] in which the collisional angular momentum operator describing the Coriolis coupling is approximated in order to partly decouple different partial waves. This makes the problem diagonal with respect to the orbital angular momentum for the relative motion. In this sense, the semi-classical equivalent of the CS is the peaking approximation (§ IV.3.4c). Starting from the CS model, calculations can be further simplified by neglecting the internal energy structure of the molecules, leading to the Infinite Order Sudden (IOS, § IV.3.5d) approach. This enables a factorization of the expression of the relaxation matrix elements which depend on a limited set of collisional quantities. Within the IOS frame, these can be computed at low computer cost from the radiator-perturber interaction potential. Except for very few studies discussed in § IV.3.5e, comparisons of the results of calculations with spectroscopic measurements are limited to the diagonal terms of the relaxation matrix. In other papers, state-to-state relaxation cross sections, which describe line couplings between isotropic Raman Q lines, are considered but no comparison with measured spectral shapes is made. As for semi-classical models, the bibliography associated with the use of quantum approaches for such calculations is too vast to be reviewed here. The references in Table IV.4 are limited in number but provide bibliographical paths for various collisional pairs.

Table IV.4: Some applications of quantum approaches to pressure broadening (and eventually shifting) coefficients and to rotational relaxation cross-sections. p (rows) and a (columns) designate the perturber and radiator

↓(p) (a)→	Linear	Symmetric-top	Asymmetric-top	Spherical-top
Atom	23,435,506–508	363,509,510	22,511	493,512
Linear	433,434,513,514	515,516	517,518	

In order so simplify notations, the case of a linear molecule colliding with atoms is considered here. Equations for other systems are not given but proper references are indicated or can be found in Table IV.4.

IV.3.5b The Close-Coupling approach

Within the binary and impact approximations, the relaxation matrix elements for a linear molecule colliding with atoms are given, within the Close-Coupling scheme (CC), in Refs. 144,364. With obvious notations, the cross section is given by:[7]

$$^{CC}\sigma^{(k)}_{f'i',fi}(E_r) = \left(\frac{\pi\hbar^2}{2m^*E_r}\right) \sum_{\lambda,\lambda',\widetilde{J_i},\widetilde{J_f}} (-1)^{J'_i+J_i+\lambda+\lambda'} \begin{Bmatrix} J'_i & k & J'_f \\ \widetilde{J_f} & \lambda' & \widetilde{J_i} \end{Bmatrix} \begin{Bmatrix} J_i & k & J_f \\ \widetilde{J_f} & \lambda & \widetilde{J_i} \end{Bmatrix}$$

$$\sqrt{(2J'_i+1)/(2J_i+1)(2\widetilde{J_i}+1)(2\widetilde{J_f}+1)} \tag{IV.147}$$

$$\times \left\{ \delta_{J'_iJ_i}\delta_{J'_fJ_f}\delta_{\lambda',\lambda} - \langle J'_i\lambda'|S^{(\widetilde{J_i})}(E_r+E_{J_i})|J_i\lambda\rangle \langle J'_f\lambda'|S^{(\widetilde{J_f})}(E_r+E_{J_f})|J_f\lambda\rangle^* \right\}$$

where {:::} is a 6J symbol,[65] k is the order of the tensor of the optical transitions, E_r is the initial relative kinetic energy and m* is the reduced mass. λ and λ' are the orbital angular momentum quantum numbers for the relative motion of the atom-linear molecule system before and after collision, respectively, while $\widetilde{J} = \vec{J} + \vec{\lambda}$ is the total angular momentum. $S^{(\widetilde{J})}$ designates the scattering S-matrix defined in the total angular momentum representation,[144] evaluated for the total (kinetic + internal) initial energy. These quantities can be calculated by solving a set (one for each $\widetilde{J}$ block) of close-coupling Schrödinger equations for the radial wavefunction $F_{\widetilde{J},\lambda}^{J',\lambda'}(R)$,[144,519] each of them being:

$$\left[\frac{\hbar^2}{2m^*}\frac{d^2}{dR^2} - \lambda'(\lambda'+1)\frac{\hbar^2}{2m^*R^2} + E - E_J\right] F_{\widetilde{J},\lambda}^{J',\lambda'}(R) = \sum_{J'',\lambda''} V_{\widetilde{J},J',\lambda',J'',\lambda''}(R) F_{\widetilde{J},\lambda}^{J'',\lambda''}(R), \tag{IV.148}$$

where E is the total energy, and E_J is that of the rotational level J. In this equation, J,λ and J',λ' are the initial and final quantum numbers before and after collision. The terms $V_{\widetilde{J},J',\lambda',J'',\lambda''}(R)$ are obtained from the interaction potential. When the later is expanded on Legendre polynomials [$V(R,\beta) = \sum_L V_L(R) P_L(\cos\beta)$], the matrix elements are given by:[144]

$$V_{\widetilde{J},J',\lambda',J'',\lambda''}(R) = \sum_L (-1)^{\lambda'+\lambda''+\widetilde{J}} \sqrt{(2\lambda''+1)(2\lambda'+1)(2J'+1)(2J''+1)}$$

$$\times \begin{pmatrix} \lambda' & L & \lambda'' \\ 0 & 0 & 0 \end{pmatrix} \begin{pmatrix} J' & L & J'' \\ 0 & 0 & 0 \end{pmatrix} \begin{Bmatrix} \widetilde{J} & \lambda' & J' \\ L & J'' & \lambda'' \end{Bmatrix} V_L(R). \tag{IV.149}$$

---

[7] For consistency with the "spectroscopic" notations used in this book, J and $\widetilde{J}$ denote the radiator rotation angular momentum and the total angular momentum, respectively. This is different from the notations generally used for such "collisional" problems[144,364] which correspond to the changes J → j and $\widetilde{J}$ → J.

The behavior of the radial wavefunctions $F_{J,\lambda}^{J',\lambda'}(R)$ at large distance directly gives the impact S matrix elements of Eq. (II.147) since:[144]

$$\lim_{R\to\infty} F_{J,\lambda}^{J',\lambda'}(R) = \delta_{\lambda,\lambda'} \exp[i(R\sqrt{2m^*(E-E_J)/\hbar} - \lambda\pi/2)]$$
$$- \left(\frac{E-E_J}{E-E_{J'}}\right)^{1/4} \exp[i(R\sqrt{2m^*(E-E_{J'})/\hbar} - \lambda'\pi/2)]\langle J'\lambda'|\widetilde{S^{(J)}}|J\lambda\rangle. \tag{IV.150}$$

In order to reach this limit, one must solve Eq. (IV.148). This is done numerically, starting at a value of R which lies well inside the innermost classical turning point and carrying out the outward propagation to a distance for which the potential is negligible when compared with the initial relative kinetic energy. This has to be done for many values of the orbital angular momentum quantum numbers up to a limit determined from convergence tests. For a number of systems the $\langle J'\lambda'|S^{(J)}|J\lambda\rangle$'s are readily computed by the MOLSCAT[308] and MOLCOL[309] programs. In the MOLCOL package, a rotating body-fixed (BF) frame is used and the CC equations are solved by means of the Johnson log-derivative propagator.[520] The radial solutions in the BF frame then have to be matched, at large distances where the potential has attained its threshold value, to Bessel space-fixed asymptotic radial functions through a unitary transformation. In the MOLSCAT package, the faster log-derivative-Airy propagator of Alexander and Manolopoulos[521] is often used.

For complexity and computer time constraints, but also according to the availability of accurate interaction potential energy surfaces (PES), most applications of the CC scheme have been made for (linear molecule)-atom systems, furthermore restricted to relatively low rotational quantum numbers in order to limit the number of energetically opened channels. In many cases, only state-to-state rotational relaxation rates or the diagonal elements of the relaxation operator are predicted (cf references in Table IV.4 and those therein). For line widths and shifts, very good agreement with measurements is generally obtained when an accurate PES is available, as illustrated by the broadening and shifting coefficients by Ar of lines of HF,[23] HCl,[24] and CO.[413] Theoretical developments and results within the CC frame are also available for collisions of symmetric-top, asymmetric-top, and spherical-top molecules with atoms, as well as for interactions with diatomic perturbers (see Table IV.4). Nevertheless, unresolved problems remain when collisional broadening at (very) low temperature is considered, as discussed in Refs. 506,522–524. This is illustrated for CO-He in Fig. IV.28, showing that predictions are very satisfactory in a wide temperature range, but that they increasingly overestimate the line width and shift below 10 K. A number of reasons for these discrepancies have been invoked but no definitive explanation has been found yet (see Refs. 506,525,526 and those therein).

IV.3.5c The Coupled-States approach

The Coupled-States (CS, or Centrifugal Sudden) method[504,505,527,528] is obtained from the CC equations by approximating the Coriolis coupling for the colliding pair. It assumes an effective orbital momentum eigenvalue and freezes the centrifugal potential. Setting the operator $\vec{\lambda}^2$ to an effective eigenvalue parameter $\hbar\lambda(\lambda+1)$ decouples (in the BF frame) all equations with different $\widetilde{J}_z$ projections for each total angular

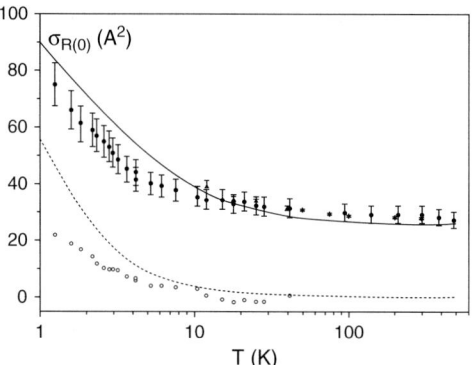

Fig. IV.28: Pressure broadening (•, ▲, ___) and shifting (o, - - -) cross sections for the R(0) line of CO in by He vs temperature. The symbols designate experimental values (• and o from Ref. 522 and ▲ from Ref. 526) while the lines are calculated results. Courtesy of F. Thibault, after Ref. 526

momentum $\widetilde{J}$. This is related to the $\widetilde{J}_z$-conserving approximation[527] in the sense that only couplings between different rotational states and not between multiplets (thus the name Coupled States) may appear in the dynamical treatment. This decoupling of different partial waves allows a factorization of the CC equations. In order to derive the CS expression of spectroscopic cross sections, the first step rotates the body-frame S matrix to the space-frame through the following transformation:

$$\langle J'\lambda'|S^{(\widetilde{J})}|J\lambda\rangle \rightarrow \sqrt{(2\lambda'+1)(2\lambda+1)} \sum_{m,m'} \begin{pmatrix} J' & \widetilde{J} & \lambda' \\ m' & -m' & 0 \end{pmatrix} \begin{pmatrix} J & \widetilde{J} & \lambda \\ m & -m & 0 \end{pmatrix} \langle J'm'|S^{(\lambda')}|Jm\rangle. \quad (IV.151)$$

The CS approximation then assumes that m is conserved (it is thus the quantum equivalent to the peaking approximations), so that:

$$\langle J'm'|S^{(\lambda')}|Jm\rangle = \delta_{m,m'}\langle J'm|S^{(\lambda')}|Jm\rangle. \quad (IV.152)$$

Introducing Eqs. (IV.151) and (IV.152) into Eq. (IV.147) and using algebra lead to the following expression of the line-coupling cross section:[364,527]

$$^{CS}\sigma^{(k)}_{f'i',fi}(E_r) = \left(\frac{\pi\hbar^2}{2m^*E}\right)(-1)^{J'_f+J_f} \sum_{\lambda,m_i,m_f} \begin{pmatrix} J'_i & k & J'_f \\ m_i & m_f-m_i & -m_f \end{pmatrix} \begin{pmatrix} J_i & k & J_f \\ m_i & m_f-m_i & -m_f \end{pmatrix}$$

$$\times \sqrt{(2J'_i+1)/(2J_i+1)}(2\lambda+1) \quad (IV.153)$$

$$\times \left\{\delta_{J'_iJ_i}\delta_{J'_fJ_f} - \langle J'_im_i|S^{(\lambda)}(E_r+E_{J_i})|J_im_i\rangle\langle J'_fm_f|S^{(\lambda)}(E_r+E_{J_f})|J_fm_f\rangle^*\right\}.$$

As for CC calculations, the S matrix elements in Eq. (IV.153) are readily computed, for a number of collisional pairs, with the MOLSCAT computer package.[308]

The advantage of the CS is that it reduces the number of equations to the number of rotational levels included in the basic set and thus significantly lowers the computational time. This makes the CS applicable to collisional quantities for (high) values of the radiator rotational quantum number and/or of the kinetic energy for which CC calculations are not tractable. Concerning accuracy, the CS is, as the PA, a good approximation for collisions with He. For heavier perturbers, it leads to poorer predictions (within typically $\pm 10\%$, eg Refs. 529,530) for low values of J and/or of the energy E_r, but it becomes more and more accurate as J and/or E_r increase. More generally, this method is well adapted for short range potentials or interactions for which the influence of the anisotropy at long range is small. The CS model thus appears as very complementary to the CC approach, permitting the extension of predictions toward higher J and E_r values (see, for instance, Refs. 364,531). As for CC calculations, CS predictions have been essentially limited to state-to-state relaxation and pressure-broadening and -shifting coefficients. For these problems, a variety of colliding pairs have been considered (see the references in Table IV.4 and those cited therein) but few applications to line mixing have been made, as discussed in § IV.3.5e.

IV.3.5d The Infinite Order Sudden approach

The Infinite Order Sudden (IOS) approximation starts from the CS equations and neglects the energy spacing between the radiator rotational levels. It thus sets the rotational constant to zero and freezes the molecular rotation. As a first result, the S matrices in Eq. (IV.153) are evaluated for the same energy E_r. For a linear molecule colliding with atoms, one can project the S operator on Legendre polynomials P_L, so that the matrix elements in Eq. (IV.152) are expressed as:[401]

$$\langle J'm|S^{(\lambda)}(E_r)|Jm\rangle = (-1)^m\sqrt{(2J+1)(2J'+1)}\sum_L \begin{pmatrix} J & J' & L \\ m & -m & 0 \end{pmatrix}\begin{pmatrix} J & J' & L \\ 0 & 0 & 0 \end{pmatrix}S_L^{(\lambda)}(E_r). \quad \text{(IV.154)}$$

Introducing this expression into Eq. (IV.153) and carrying out some algebra lead to:[401]

$$^{IOS}\sigma_{f'i',fi,p}^{(k)}(E_r) = (-1)^{k+1}(2J_i'+1)\sqrt{(2J_f+1)(2J_f'+1)} \\ \times \sum_{L\neq 0}\begin{pmatrix} J_i' & L & J_i \\ 0 & 0 & 0 \end{pmatrix}\begin{pmatrix} J_f' & L & J_f \\ 0 & 0 & 0 \end{pmatrix}\begin{Bmatrix} J_i & J_f & k \\ J_f' & J_i' & L \end{Bmatrix}\sigma(L\leftarrow 0, E_r), \quad \text{(IV.155)}$$

where $\sigma(L\leftarrow 0, E_r)$ is the cross section for the excitation from level J = 0 to level J = L at a relative kinetic energy E_r. Its value is obtained from the coefficients $S_L^{(\lambda)}(E_r)$ in the decomposition of the matrix elements of S on Legendre polynomials, ie:[401]

$$\sigma(L\leftarrow 0, E_r) = \frac{\pi\hbar^2}{2m^*E_r}\frac{1}{2L+1}\sum_\lambda (2\lambda+1)\left[\left|S_L^{(\lambda)}(E_r)\right|^2 - 1\right] \quad \text{(IV.156)}$$

A way to obtain the values of $S_L^{(\lambda)}(E_r)$ is to solve the following one-dimensional equation derived from Eq. (IV.148):

$$\left[\frac{\hbar^2}{2m^*}\frac{d^2}{dR^2} - \lambda(\lambda+1)\frac{\hbar^2}{2m^*R^2} + E_r - V(R,\theta)\right]\phi_\lambda(R,\theta,E_r) = 0, \qquad (IV.157)$$

and then to deduce $S_\lambda(\theta, E_r)$ from the asymptotic behavior of $\phi_\lambda(R,\theta,E_r)$ at large R values [Eq. (IV.150)], before projecting them on Legendre polynomials, through:

$$S_L^{(\lambda)}(E_r) = \frac{2L+1}{2}\int_{-1}^{+1} S_\lambda(\theta, E_r) P_L(\cos\theta) d(\cos\theta). \qquad (IV.158)$$

Other more simplified approaches can also be used, for which the reader can refer to Refs. 499–502.

Using Eqs. (IV.104) and (IV.105), and $\sigma(L \leftarrow 0) = (2L+1)\sigma(0 \leftarrow L)$, it is easy to show that Eq. (IV.155) leads to Eq. (IV.103). The limits of the IOS approach have been discussed in § IV.3.3 and it is well known that this approximation is practically limited to collisions of heavy radiators with He. In this case, one of its advantages is its simple implementation and low computer cost, permitting easy evaluation of $\sigma(L \leftarrow 0, E_r)$ and thus of the line-mixing relaxation matrix. An example can be found in Ref. 408, where NH_3-He is considered (Fig. IV.23). For this system, the use of more accurate approaches to compute a full relaxation matrix in the ν_2 band, where several hundred lines must be taken into account, is hardly tractable.

IV.3.5e Applications to line-mixing problems

Since applications of the IOS (and ECS) approach to line-mixing problems have been discussed in § IV.3.3 and references are given in Table IV.9, this is not further discussed here. When line couplings are considered, there are few applications of the CC and CS approaches. The first calculations have been made[364] for infrared lines of CO broadened by He. In this purely theoretical study, relaxation matrix elements are calculated using the CC, CS, and IOS approaches. Due to computer limitations, only a *few* line-coupling terms are calculated within the CC frame, for small values of the rotational quantum number J. This is not sufficient for spectra calculations but enables a test of simplified models. It is then shown that the CS approximation leads to values accurate within a few % for this favorable CO-He system. Comparisons with IOS predictions show that this approximation leads to satisfactory results for the large values of $\text{Re}(W_{\ell'\ell})$ (those for lines ℓ and ℓ' involving close values of the angular momentum J). On the opposite, due to the neglect of the internal energy structure, the IOS model overestimates the small couplings. Starting from these results, comparisons between CS and IOS predictions and experiments have been made,[148] considering absorption in the troughs between transitions and a high pressure spectrum of the 1–0 band of CO in He. The results, illustrated by Fig. IV.29, show that both approaches lead to a correct description of line-mixing effects. Ref. 148 also gives the demonstration that the off-diagonal imaginary elements of W have a *small* influence on calculated spectra. To our knowledge, this is the only quantitative justification of the widely

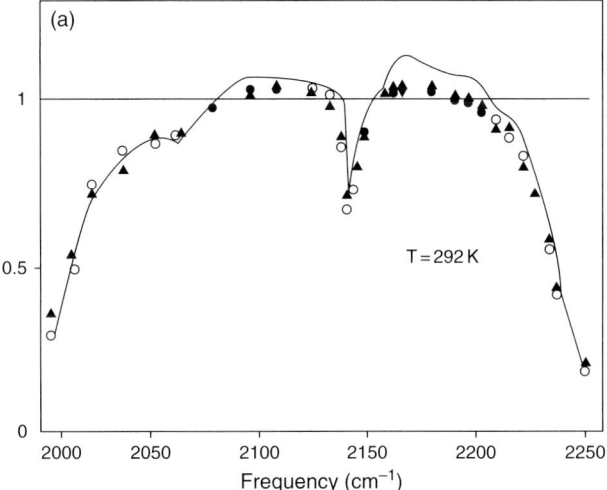

Fig. IV.29: Comparison between predictions using the CS (▲) and IOS (o) approaches and measured values (——) of the band correction function (ratio of absorption to the results of purely isolated Lorentzian lines predictions). Results are for the 1-0 band of CO broadened by He at 292 K. Reproduced with permission from Ref. 148.

used neglect of these elements. It is worth noting that the results of CC calculations obey[148] the conjugated detailed balance relation: $W_{\ell'\ell}\rho_\ell = \rho_{\ell'}W^*_{\ell\ell'}$. This is discussed in Ref. 146 where it is stated that this is wrong and associated with the use of on-the-energy-shell calculations bound to the impact approximation (see App. II.E3). Together with the inaccurate CC predictions at low temperature (Fig. IV.28), this remains an open problem.

Later on, the isotropic Raman Q branch of CO-(He and Ar) has been analyzed[413,532] using similar models. Good agreement is obtained (Fig. IV.30) with measured spectra under conditions where line mixing significantly influences the spectral shape. Nevertheless, for the lowest temperature (87 K) investigated in Ref. 413, unexplained discrepancies between experimental and predicted CO-Ar profiles are obtained.

Quantum calculations of line-mixing effects have also been made for He broadened D_2 Raman spectra.[533,534] Since the lines are largely spaced, the line-coupling process manifests[535] through an asymmetry of the line profiles, due to the first-order line-mixing coefficient Y [Eqs. (IV.22) and (IV.24)]. The calculations for depolarized spectra lead to values of the Y's which are very small, today beyond the possibility of an experimental check. On the contrary, for isotropic Raman Q lines, the asymmetry is much larger and calculated Y values for D_2–He are comparable (a few 10% lower) to the measured ones for pure D_2.

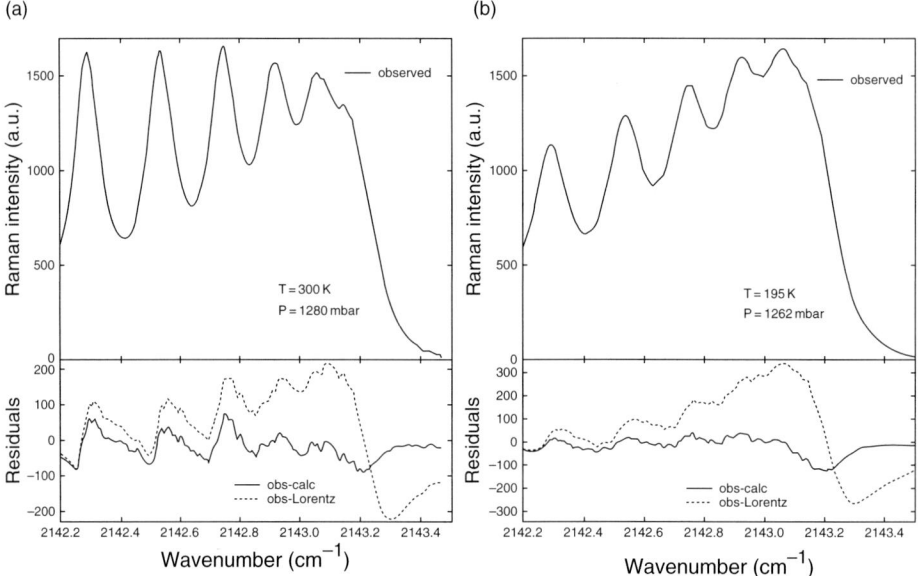

Fig. IV.30: Isotropic Raman Q branch profiles of CO in Ar at about 1.3 atm, 300 K (a), and 195 K (b). The upper plots give the measured values and the lower part shows the deviations with respect to calculation including line mixing (———, CC relaxation matrix) and disregarding (. . . .) this effect. Reproduced with permission from Ref. 413.

IV.4 DETERMINING LINE-MIXING PARAMETERS FROM EXPERIMENTS

IV.4.1 INTRODUCTION

The number of studies in which spectroscopic parameters have been determined from measured spectra is considerable. In most cases, isolated line profiles are used, disregarding, rightly or wrongly, the influence of collisional interferences. Thanks to the sensitivity and spectral resolution of modern laboratory techniques, there is an increasing number of studies in which parameters related to the line-mixing process are obtained from fits of measured data. Various approaches have been applied, depending on the experimental information available (absorption, Raman signal in the spectral or time domains, *etc*), on the fitting technique (multispectrum or spectrum-by-spectrum), and on the way line mixing is included in the fits (W matrix or first-order coefficients Y_ℓ). These empirical determinations are more and more used, due to the improving accuracy of laboratory measurements and increasing accuracy requirements of applications. These are now getting so high that purely theoretical approaches, except for very simple molecular systems, will hardly be able to quickly provide accuracies comparable with the signal-to-noise ratio of modern experimental techniques. As an example, the objective of the Orbiting Carbon Observatory,[536] which is an accuracy better than 1% on the remote sensing of atmospheric air masses and CO_2 amounts, is getting beyond the current capability of theory, but it may be attained by using *empirical* values deduced from high quality laboratory spectra.[537,538]

The procedure used to determine parameters from experimental recordings consists in the minimization of the root mean square deviation between measured and calculated spectra:

$$\text{RMS} = \left[\sum_{m=1}^{M} w_m (s_m^{\text{Exp}} - s_m^{\text{Calc}})^2 \bigg/ \sum_{m=1}^{M} w_m \right]^{1/2}. \quad \text{(IV.159)}$$

In this expression, the sum extends over measurement points which may correspond to various wavenumbers or delays for experiments in the time domain, and to different sample conditions. s^{Exp} designates the vector containing the measured quantity, s^{Calc} is its theoretical equivalent, and w are weights (*eg* the inverse of the local signal-to-noise ratio). In order to be more illustrative, consider transmission measurements for a pure gas in which the absorption coefficient is modeled using the first-order line-mixing approach (§ IV.2.2a). In this case, s^{Calc} depends, a minima, on the wavenumbers σ_m of the measurements, and on the spectroscopic parameters of the lines ℓ. These, which depend on the path length L_m, temperature T_m and total pressure P_m, are the integrated absorbance $P_m L_m S_\ell(T_m)$, the line positions $\sigma_\ell + P_m \delta_\ell(T_m)$ including the collisional shifts δ_ℓ, the pressure-broadened halfwidths $P_m \gamma_\ell(T_m)$, and the first-order line-mixing parameters $P_m Y_\ell(T_m)$. Data analysis is generally made from sets of measurements at a given temperature using two approaches.

In the widely used *spectrum-by-spectrum* fits, measurements for different conditions (*ie* $\neq$ L, P, T) are adjusted *separately*, and the sum in Eq. (IV.159) is restricted to the wavenumbers σ_m. When applied successively to the spectra, this leads to a set of values of $P_i L_i S_\ell(T)$, $\sigma_\ell + P_i \delta_\ell(T)$, $P_i \gamma_\ell(T)$, and $P_i Y_\ell(T)$ for the various recording conditions P_i and L_i at a given temperature T. In a second step, these results are fitted using linear dependences on P_i or $P_i L_i$, leading to the line parameters $S_\ell(T)$, σ_ℓ, $\delta_\ell(T)$, $\gamma_\ell(T)$, and $Y_\ell(T)$. These quantities can then be adjusted through a third fit if measurements have been made for different values of T. For the line intensity and broadening coefficients, for instance, the expressions are:

$$S_\ell(T) = S_\ell(T_0) \frac{\rho_\ell(T)[1 - \exp(-hc\sigma_\ell/k_B T)]}{\rho_\ell(T_0)[1 - \exp(-hc\sigma_\ell/k_B T_0)]} \quad \text{and}$$
$$\gamma_\ell(T) = \gamma_\ell(T_0) \times [T_0/T]^{n_\ell}. \quad \text{(IV.160)}$$

These equations can be used to determine the (T- and P-independent) parameters $S_\ell(T_0)$, $\gamma_\ell(T_0)$, and n_ℓ which are stored in spectroscopic databases.[2–5] The spectrum-by-spectrum approach is simple to implement but the eventual insufficiencies of the model used (s^{Calc}) may be difficult to detect. Indeed, the number of adjusted parameters being important (4 for each line in each spectrum in the preceding example), the quality of the fits may be quite good even though the model is inappropriate. Problems can only be detected by *closely* looking at the pressure (and/or path length) dependences of the retrieved quantities (*ie* the quality of the second fit). In order to illustrate the problem, we have calculated the absorption by two lines of identical intensities and widths, with *no* pressure shift ($\delta_1 = \delta_2 = 0$), weakly coupled through first-order line-mixing coefficients ($Y_1 = -Y_2$). Results of pressure-by-pressure fits using Lorentzian profiles, thus disregarding the collisional coupling ($Y_1 = Y_2 = 0$), are plotted in Fig. IV.31. The residuals are very small, not clearly indicating that line mixing needs to be included.

Furthermore, the insufficiency of the model used is still not obvious when looking at the retrieved parameters plotted in Fig. IV.32. These show "nice" linear variations *vs* pressure with slopes, for the width and intensity, which are very close to the true values.

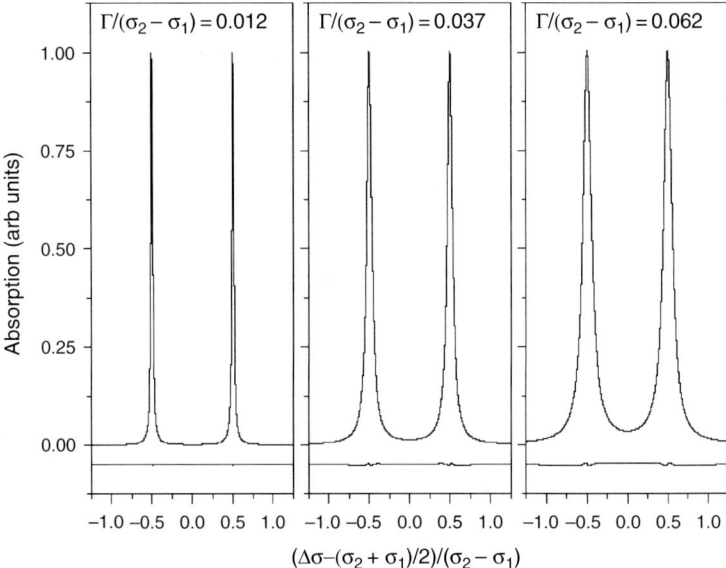

Fig. IV.31: Spectrum-by-spectrum fits of absorption by two weakly coupled lines using purely Lorentzian profiles. For each pressure, the plot displays the absorption and the observed-fitted residuals.

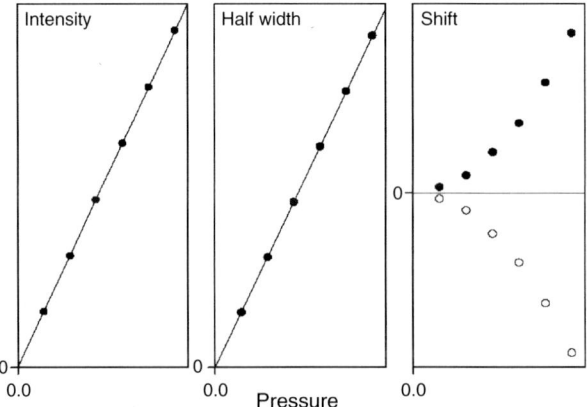

Fig. IV.32: Pressure dependences of dimensionless line parameters obtained from the spectrum-by-spectrum fits of Fig. IV.31. The symbols give the adjusted values of the intensity, widths, and shift of the two lines while the straight lines give the true values.

Collisional line mixing

The non-zero shifts show that the modification of the spectral shape due to line mixing is taken into account by wrongly shifting the Lorentzian profiles. The only clue that the model used is not the proper one is the slight non-linearity of the line positions with pressure, which is a characteristic signature of line mixing.[328,539] Indeed, for a weakly coupled line, one can show that the absorption coefficient maximum is located at:

$$\sigma_{\text{Max}}(P) = \sigma_\ell + \delta_\ell P + \frac{\gamma_\ell}{Y_\ell}\left[\sqrt{1+(Y_\ell P)^2} - 1\right] \approx \sigma_\ell + \delta_\ell P + \frac{\gamma_\ell Y_\ell}{2}P^2. \quad \text{(IV.161)}$$

This quadratic behavior of peak absorption with pressure was used[539,540] for the determination of line-coupling coefficients of He-broadened CO lines.

In the *multispectrum* fitting technique,[541] spectra are adjusted *simultaneously* and the sum in Eq. (IV.159) includes the various wavenumbers, pressures, and path lengths. Experiments for different temperatures can also be adjusted simultaneously when the effects of T on the fitted parameters is described analytically as in Eq. (IV.160). This is seldom done in the absence of physically based laws describing the evolution of the line-shifting and line-coupling coefficients with temperature. Hence, for coupled lines, the pressure- and length-*independent* spectroscopic coefficients S_ℓ, σ_ℓ, δ_ℓ, γ_ℓ, and Y_ℓ for each temperature are directly determined. When N_S spectra are adjusted, each containing N_L lines, the number of unknowns is then $5N_L$ whereas it is $4N_L N_S$ if the spectra are adjusted one-by-one. Much more severe constraints are thus imposed to the fitted quantities and the multispectrum treatment more clearly points out the eventual insufficiencies of the spectral model used. This is exemplified in Fig. IV.33 by a doublet of

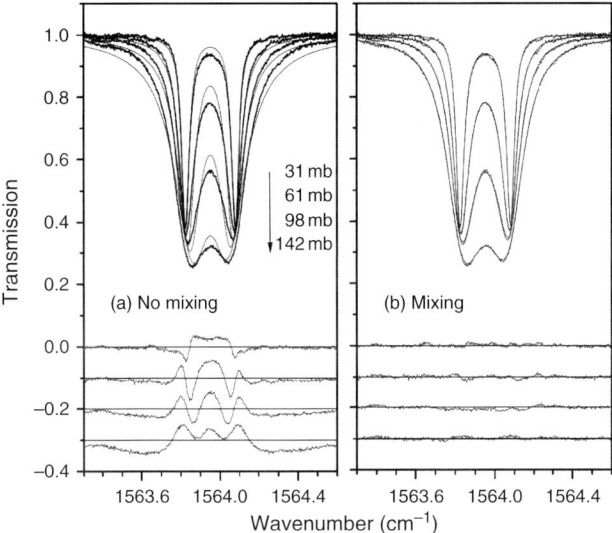

Fig. IV.33: Multispectrum fits of measured room temperature transmissions for the ^{P}P(4,4) doublet of the ν_4 band of pure NH$_3$ *vs* pressure with (b) and without (a) the inclusion of line mixing. Transmissions are displayed in the upper part of the plots, shifted observed-fitted residuals being given below. After Ref. 328.

pure NH_3 where the multispectrum fit residuals obviously indicate the defaults of the approach neglecting line mixing. Fitting the spectra one-by-one would lead to much smaller residuals (8 parameters for each pressure, *ie* 56 for the seven pressures studied in Ref. 328, whereas only 10 are adjusted in the multispectrum fit) and the need to include line mixing would not be as easily detected.

Besides line parameters, other quantities are often included in the fits. For transmission measurements, it is generally the case of the 100% transmission base line, and a description of the zero is also sometimes used. If the instrument function is not accurately known, some adjustable quantities can be introduced in order to describe it. Also note that more refined spectral profiles have been used,[490,542–544] including both line-mixing and velocity effects, as further discussed in Sec. VIII.2.

IV.4.2 RELAXATION MATRIX ELEMENTS

When the number of lines is small, one can determine the relaxation matrix (W) elements from fits of measured spectra. For N_L coupled lines, the number of unknowns in the real part of W is then $N_L(N_L+1)/2$ since one can impose the detailed balance relation connecting elements $W_{\ell\ell'}$ and $W_{\ell'\ell}$ [Eq. (IV.68)]. For a doublet, the derivation of the inverse matrix elements of Eq. (IV.1) is analytical (see Ref. 545 for a triplet) and the absorption coefficient (disregarding Doppler and velocity effects) depends on nine parameters. For a highly diluted binary mixture, the expression is:

$$\alpha(\sigma,n_a,n_p) = \frac{n_a}{\pi}\text{Im}\left\{\frac{[S_1(\sigma-\sigma_2')+S_2(\sigma-\sigma_1')]-i(S_1 n_p\gamma_2+S_2 n_p\gamma_1+2n_p\xi\sqrt{S_1 S_2})}{[(\sigma-\sigma_2')(\sigma-\sigma_1')-n_p^2\gamma_1\gamma_2+n_p^2\xi^2]-i[n_p\gamma_1(\sigma-\sigma_2')+n_p\gamma_2(\sigma-\sigma_1')]}\right\},$$
(IV.162)

with $\sigma_j' \equiv \sigma_j + n_p\delta_j$. In these expressions, S_j, σ_j, δ_j, and γ_j are respectively the integrated intensity, unperturbed position, pressure-shifting and pressure-broadening coefficients of line $j=1,2$. Line mixing is contained in the parameter ξ, related to the relaxation matrix elements and initial levels relative populations (ρ) by:

$$\xi = \sqrt{\rho_2/\rho_1}W_{12} = \sqrt{\rho_1/\rho_2}W_{21}.$$
(IV.163)

S_j, σ_j, δ_j, γ_j, and ξ can be obtained from fits of spectra recorded at various pressures, as shown in Fig. IV.33 for two NH_3 lines. Similar results can be found in Ref. 546 in the case of A+ and A− transitions of CH_3D, and in Fig. IV.34 for two couples of H_2O lines.

The situation gets complicated when couplings between many lines must be taken into account, since the number of unknowns in W increases as $\approx N_L^2/2$. The limited experimental information available and the uncertainties generally make the retrieval of the entire matrix an ill-posed problem introducing errors and instabilities in the solution. One must thus use constraints through a regularization procedure or the introduction of

Collisional line mixing

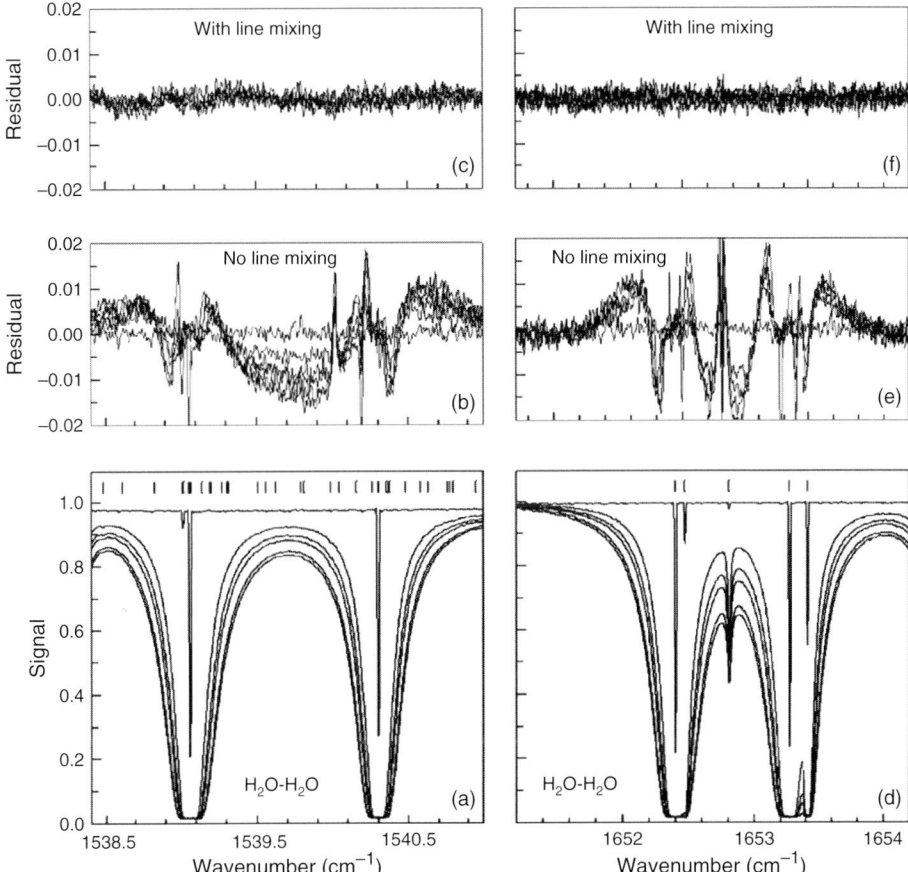

Fig. IV.34: Multispectrum fits of measured transmissions of pure water vapor at room temperature for six pressures near two couples of lines of the ν_2 band ($1_{0,1} \leftarrow 2_{1,2}$ and $2_{1,2} \leftarrow 3_{0,3}$ near 1539.4 cm^{-1}, and $3_{0,3} \leftarrow 2_{1,2}$ and $2_{1,2} \leftarrow 1_{0,1}$ near 1652.8 cm^{-1}). Reproduced with permission from Ref. 329.

some a priori information concerning the solution (some coupling elements are set to zero from physical considerations, a specific analytical form is imposed to the W matrix elements searched for, *etc*). For example, consider a cluster of closely spaced lines and a pressure for which they strongly overlap. In this case, the absorption is close [Eqs. (IV.41)–(IV.44)] to that of a single equivalent Lorentzian feature whose *effective* broadening coefficient is the average of all couplings, *ie*:

$$\gamma_{\text{Cluster}} = \sum_{\ell,\ell' \in \text{Cluster}} \rho_\ell d_\ell d_{\ell'} \text{Re}\{W_{\ell'\ell}\} / \sum_{\ell \in \text{Cluster}} \rho_\ell d_\ell^2. \qquad (\text{IV.164})$$

The spectrum thus contains information on the (weighted) average of the W elements *only* and one can hardly deduce a full matrix without introducing large a priori assumptions. Such a situation, although less extreme, was encountered in fits of relaxation matrices on measured Q branch spectra[547,548] which required a regularization procedure. Nevertheless, some ambiguity remains since the retrieved matrix depends on the "initial guess" used at the start of minimizing iterations.[547]

IV.4.3 FIRST-ORDER LINE-COUPLING COEFFICIENTS

Situations with many coupled lines are simplified when the first-order approximation [§ IV.2.2a and Eq. (IV.25) or (IV.26)] can be used. The adjusted quantities are then the positions (σ_ℓ), intensities (S_ℓ), pressure-broadening (γ_ℓ) and -shifting (δ_ℓ) coefficients, and first-order line-coupling coefficients Y_ℓ (in practice restricted to their real part) of the transitions. The number of unknowns, $5N_L$, may then be significantly smaller than that involved when a full relaxation matrix is searched for [$N_L(N_L + 7)/2$]. The interest of the first-order approach is then obvious, as illustrated by a study of a CO_2 infrared Q branch[361] where the collisional parameters γ_ℓ and Y_ℓ of 23 lines have been determined. In this case, fitting a full W matrix is not tractable since it would involve more than two hundred unknowns. A first illustration of fits using the first-order approach is displayed in Fig. IV.35. The six components of the R(6) manifold of the ν_3 band of methane have been included, leading to the determination of the line-coupling coefficients Y_ℓ. Note that only five of these were adjusted since the single line of E symmetry was assumed uncoupled (Y = 0) to the others which all have F or A symmetries.[330] This remark points out a first difficulty of carrying out fits when the spectrum is complex. Indeed, if some lines are not coupled to others and this is not imposed in the adjustments, this may lead to errors on the retrieved quantities. A second difficulty is the validity of the first-order (weak overlapping) approach which is assumed a priori. As for the check that a proper model (which are the coupled lines ?) is used, the fact that a higher order expansion is not needed can only be checked by looking at the fit residuals. Note that if significant structures in the residuals do indicate insufficiencies, their absence is *not* a definitive proof of the validity of the approach.

Direct determinations of spectral-shape parameters from measured spectra are accurate only when the amount of experimental information is sufficient to remove a large part of the correlations between the unknowns. It is the case when spectra cover a pressure range from low values of P where the transitions are well separated (giving relatively uncorrelated information on the parameters of isolated lines) to higher values where they start to overlap (providing additional information on line mixing). The use of variable path lengths is also interesting, moderate optical thicknesses providing information on core regions mostly sensitive to *isolated* line parameters, whereas thick paths emphasize absorption in the troughs and wings where the signature of line mixing is more pronounced. This is illustrated by the results of Refs. 537,538 and Fig. IV.36, where an entire band of CO_2 is treated through simultaneous fits of numerous spectra. In Ref. 538, the coupling between lines is represented using a tri-diagonal real W matrix, assuming that a line is collisionally coupled only to its nearest neighbors of the same

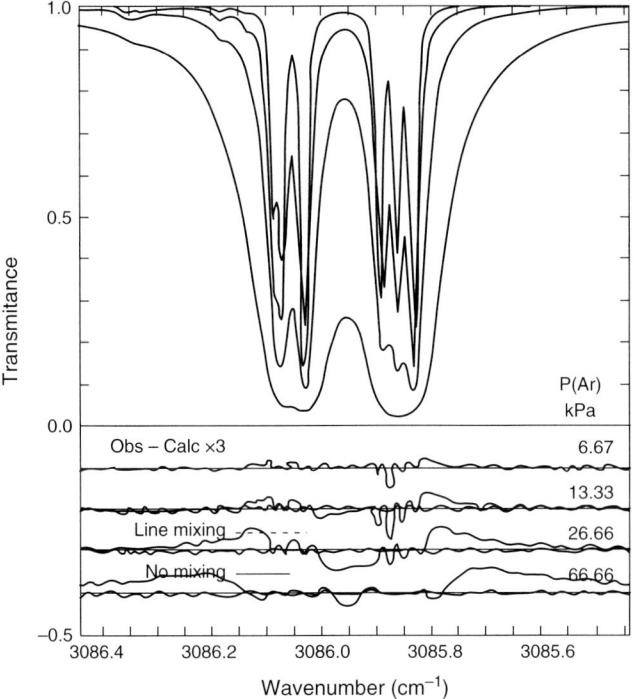

Fig. IV.35: Least squares fits of the argon-broadened R(6) manifold of the ν_3 band of CH_4 at room temperature. Measured transmissions for four total pressures are plotted in the upper part. Shifted 3 × (obs-fit) deviations are displayed below, the fits without line mixing leading to significant errors, those including line mixing being practically perfect. Reproduced with permission from Ref. 330.

band. This is a crude approximation,[364,426] and the retrieved W matrix must thus be considered as *effective*. It will nevertheless give accurate spectra predictions when applied for pressures not significantly above those of the measurements used for its determination. Furthermore, first-order line-mixing coefficients calculated from these W values using Eq. (IV.24) are likely correct.

When an *entire* band is studied, it may be of interest to constrain the Y's through relations [Eq. (IV.51) with $A_0 = A_1 = A_2 = 0$] derived from the far wing behavior in the case of a *rigid* rotor, which lead to:

$$\sum_\ell \rho_\ell d_\ell^2 Y_\ell = 0, \quad \sum_\ell \rho_\ell d_\ell^2 \sigma_\ell Y_\ell = -\sum_\ell \rho_\ell d_\ell^2 \gamma_\ell, \quad \sum_\ell \rho_\ell d_\ell^2 \sigma_\ell^2 Y_\ell = -2\sum_\ell \rho_\ell d_\ell^2 \sigma_\ell \gamma_\ell,$$

(IV.165)

where ρ_ℓ and d_ℓ are, for each line ℓ, the relative population of its lower level and the transition moment of the optical transition, respectively [related to the line integrated

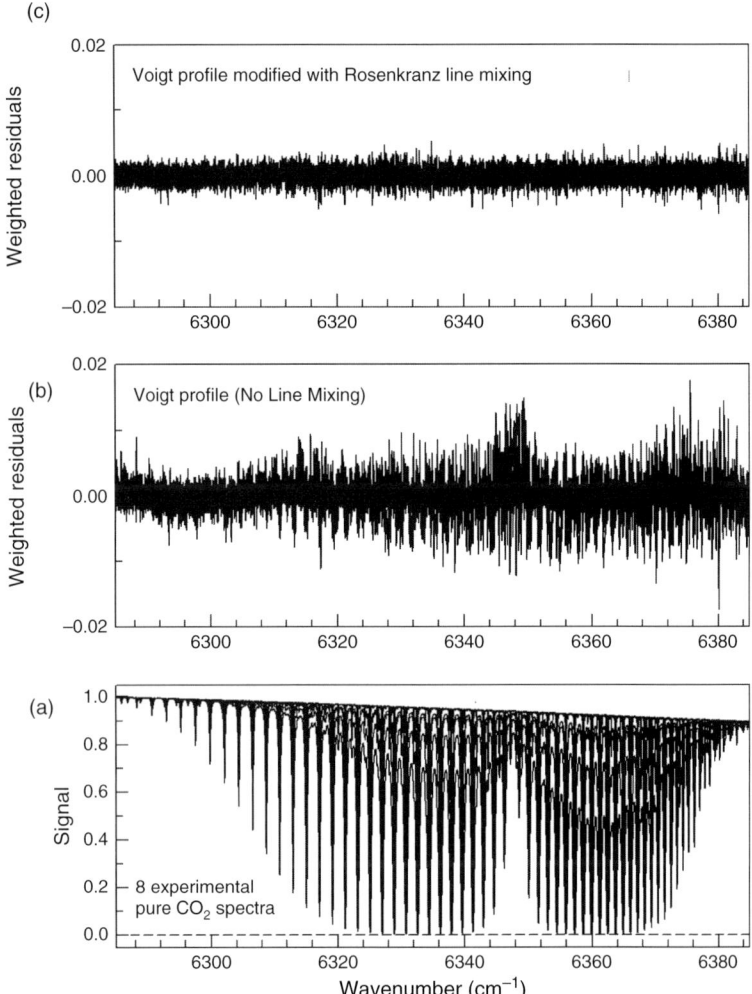

Fig. IV.36: Multispectrum fits of room temperature spectra in the $30^0 1_2 \leftarrow 00^0 0_1$ band of pure CO_2. (a): 8 observed spectra. Obs-fit residuals obtained with models disregarding line mixing and taking this process into account are displayed in panels (b) and (c) respectively. Courtesy of C. Miller, after Ref. 538.

intensity by Eq. (IV.9)]. The constraints of Eq. (IV.165) can, in principle, be included in minimization procedures for fits of the Y's on measured spectra.

When experimental information is poor, fitting line-coupling coefficients may be an ill-posed problem requiring constraints, particularly when the measurements cover a limited pressure range. An example is given in Ref. 549 where a significantly overlapping measured spectrum of the 60 GHz oxygen band is adjusted. In order to do this, numerous constraints have been put on the Y's and a Tikhonov regularization procedure was used.[549,550] This enables a fit within the noise level, but the Y's obtained must be considered as "effective". Indeed, since the measured data are few and the overlapping of the lines is strong while the number of unknowns is large, there are several solutions minimizing obs-fit deviations. The results then depend on the constraints and regularization which have been chosen a priori. Furthermore, if, due to overlapping, a higher order approach [Eq. (IV.1) or (IV.30)] was necessary, the insufficiency of the first-order expansion directly translates into errors on the Y's.

IV.4.4 MIXED THEORETICAL MODEL AND MEASURED SPECTRA FITTING APPROACHES

Many of the models presented in Sec. IV.3 are based on parameters whose determination requires the use of experimental spectra. This is true for simple empirical approaches (§ IV.3.1), but also for fitting laws (§ IV.3.2) which rely on measured absorption to scale the amounts of coupling [A_{XY}, see Eq. (IV.94)]. Even within the frame of the self-consistent ECS scaling approach (§ IV.3.3), it was shown[394,414,415] that line-broadening data are not sufficient for the unambiguous determination of the model parameters. Only the simultaneous use of measured absorptions and of line-broadening coefficients can remove this ambiguity. It is then possible, at least formally, to determine the parameters of the model from least squares fits of *all* the available experimental information. This has been seldom done, likely due to the difficulty and computer cost of this non-linear minimizing problem. The parameters are generally determined through somehow "manual" fitting attempts using few measurements (*eg* line widths and a single spectrum[359,394,398]). There have been a few studies where full fittings have been carried out. First examples can be found in Refs. 340,551 where transients of femtosecond nitrogen Raman scattering recordings have been fitted including rotational relaxation (*ie* line mixing) with the ECS approach (§ IV.3.3). The parameters used to describe the basic quantities Q_L, through an energy gap exponential power [(Eq. (IV.118)] law for instance, are then directly adjusted[340] on the measured signals. Another example is a study[384] of line mixing within the R(0) and R(1) hyperfine structures of HI diluted in He using the IOS approach (§ IV.3.3c). The relaxation matrices then depend on the single collisional parameter $Q_{L=0}$ for the R(0) manifold and on $Q_{L=0}$ and $Q_{L=2}$ for the R(1) lines. These quantities were fitted to measured spectra, showing consistent pressure dependences and leading to agreement between experimental and calculated absorptions within the small noise level.[384]

IV.5 LITERATURE REVIEW

This section presents a review of the available literature on line-mixing effects, made with three *restrictions*. *(i)* The first one is that we have only retained studies dealing with the *off*-diagonal elements of the relaxation matrix and with the first-order line-coupling coefficients Y_ℓ. The very large bibliography limited to the pressure-broadening and -shifting coefficients is not reviewed here since it is much too vast. This is the case for experimental determinations of γ_ℓ and δ_ℓ (*eg* Refs. 552,553) as well as for theoretical calculations (Tables IV.3 and IV.4). These parameters can be found (at least for self- and air-broadening and -shifting) in regularly updated molecular spectroscopic databases such as HITRAN,[2] where reference lists are provided. *(ii)* The second limitation, in order to keep the number of citations reasonable, is that, when there are several papers by a given group devoted to the same subject, we only cite representative studies. The readers are invited to follow the bibliographical tracks starting from the given references. *(iii)* Finally, only relatively recent literature is considered. Within these restrictions, the authors of this book hope that no fundamental reference was omitted and, if it was the case, they apologize to the authors of the missing papers.

IV.5.1 AVAILABLE LINE-MIXING DATA

In Table IV.5 are listed references in which values of (some) line-mixing parameters directly deduced from measurements, as explained in Sec. IV.4, are given. Table IV.6 gives studies in which computed line-mixing databases are given.

Table IV.5: Literature providing first-order (Y_ℓ) or relaxation matrix ($W_{\ell\ell'}$) couplings derived from fits of laboratory measurements as discussed in Sec. IV.4. IR, MW, R, and V in the third column refer to InfraRed, MicroWave, Raman, and Visible respectively. The stars indicate studies in which a model was used

Rad	*Pert*	*Region*	*Refs.*
C_2H_2	C_2H_2	IR	548*
CH_3D	CH_3D, N_2	IR	554
CH_4	CH_4, Ar, etc	IR	490,542,544, 555–557
		R	558
CHF_2Cl	CHF_2Cl	MW	487
CO	N_2,He	IR	540,559
CO_2	CO_2,air Ar,He	IR R	361,543,561 560*
D_2	D_2	R	535

Table IV.5: (Continued)

Rad	Pert	Region	Refs.
H_2O	H_2O, H_2, etc	IR	329
HCN	HCN	IR	548*
HD	HD	R	562
HI	He	IR	384*
N_2O	N_2O	IR	548*
NH_3	NH_3, H_2, etc	IR	328,498,563
O_2	O_2, air	MW R V	550*,549* 564 565
OCS	Ar	IR	566

Table IV.6: Literature providing self-consistent calculated line-mixing databases

Rad.	Pert.	Region	Model	Refs.
CH_4	Air, H_2	IR, ν_3 and ν_4	Semi-class	492,495
CO_2	air	Some IR Q br. All IR Q br. All IR P,Q,R br.	Fitting law Scaling law Scaling law	346 343,417 567
O_2	air	A-band	Scaling law	415

IV.5.2 COMPARISONS BETWEEN PREDICTIONS AND LABORATORY MEASUREMENTS

Below are given references in which comparisons are made between measured spectra or time-dependent response signals and the values calculated using various models described in Sec. IV.3. These include simple empirical approaches (§ IV.3.1 and Table IV.7), energy gap fitting laws (§ IV.3.2 and Table IV.8), scaling laws (§ IV.3.3 and Table IV.9), semi-classical approaches (§ IV.3.4 and Table IV.10), and quantum models (§ IV.3.5 and Table IV.11).

Table IV.7: Literature providing comparisons between laboratory spectra influenced by line mixing and values calculated using the simple models of § IV.3.1

Radiator	Region	Refs.
CH_3Cl	IR Q br	568
CH_3F	IR P,Q,R	569
CH_4	IR P,Q,R	370,374
$CHClF_2$	IR Q br. IR P,Q,R	372 375
CO	IR P,Q,R	338,351,379
CO_2	IR Q br. IR P,Q,R	371,570,571 338,349,351,360,370,374
N_2	Raman	367,368
N_2O	IR Q br.	338,371
O_3	IR P,Q,R	375
SiH_4	Raman	572

Table IV.8: Literature providing comparisons between laboratory spectra influenced by line mixing and values calculated using the fitting laws of § IV.3.2

Radiator	Region	Refs.
C_2H_2	IR Q br	388
CH_3Br	IR ^{R}Q br	400
CH_3Cl	IR ^{R}Q br	399
CO_2	IR Q br IR P,Q,R Raman	392,393,573–575 360,396,397 576
CO	Raman	577
HCN	IR Q br	388,578
HD	Raman	562,579
N_2	Raman	341,580
N_2O	IR Q br	390,391,581
NO	IR	582
NO	Raman	583
O_2	Raman	386
O_3	IR Q br	398,584

Table IV.9: Literature providing comparisons between laboratory spectra or time-dependent response signals influenced by line mixing and values calculated using the scaling laws of § IV.3.3

Radiator	Region	Refs.
CO	IR P,R Raman	585,586,413
	Raman	413
NO	IR P,R	587
O_2	A-band	415,429
	Raman	386,588
NH_3	IR P,Q,R	431
OCS	IR Q br	566
C_2H_2	IR Q br	589
CH_3F	IR Q br	432
CO_2	IR Q br	335,590
	IR P,Q,R	353,414,591
	Raman	104,592,593
N_2O	IR Q br	395
N_2	Raman	104,110,422,594–597

Table IV.10: Literature providing comparisons between laboratory spectra influenced by line mixing and values calculated using the semi-classical models of § IV.3.4

Radiator	Region	Refs.
CH_3F	MW	488
CH_4	IR	491,495,598–600
	Raman	496,497
CHF_2Cl	MW	487
H_2O	MW	483
HD	IR	136
NH_3	IR	328,498
O_2	MW	483
OCS	IR	566

Table IV.11: Literature providing comparisons between laboratory spectra influenced by line mixing and values calculated using the quantum models of § IV.3.5

Radiator	Region	Refs.
CO	IR	148
	Raman	413,532
D_2	Raman	533

IV.5.3 COMPARISONS BETWEEN PREDICTIONS AND ATMOSPHERIC MEASUREMENTS

The table below gives references in which atmospheric emission or transmission spectra influenced by line mixing are compared with the results of calculations. It is limited to studies where *direct* calculations, with *no* adjustable parameter, have been made. Cases where some quantities are fitted for remote sensing or thermometry purposes are discussed in chapter VII.

Table IV.12: Literature providing comparisons between measured *atmospheric* spectra influenced by line mixing and values calculated using models described in Sec. IV.3

Rad.	Atmosphere	Region	Model	Refs.
CH_4	Earth	ν_3 and ν_4	Semi-class	492
	Jupiter, Saturn			495
CO_2	Earth	IR Q br.	Fitting law	346,392,601
			Scaling law	343,417
	Earth	IR P,Q,R br.	Scaling law	602,603
	Mars	IR Q br.	Scaling law	604
N_2O	Earth	IR Q br.	Scaling law	334
O_2	Earth	A-band	Scaling law	415

IV.6 CONCLUSION

This chapter shows that the knowledge on impact line-mixing (and -broadening) processes has considerably progressed in the last decades.

From the theoretical point of view, much has been done since the early formal developments of Fano[44] and the application[367] of simple models to practical systems. In particular, the development of the Energy Corrected Sudden approach has enabled large scale modeling of line-mixing effects for many molecular systems of interest for applications. Furthermore, the increased feasibility of quantum calculations starting from the intermolecular potential opens interesting perspectives.

From the experimental point of view, considerable progresses have also been made, following the improvement of laboratory measurement techniques. These now enable direct determinations of collisional parameters with an accuracy progressively becoming higher than that of theoretical models. They also permit to detect weak line-mixing signatures not discernable in the past which call for new theoretical developments. In a similar way, due to the sensitivity achieved by sounding instruments, proper modeling of impact collisional spectral shapes becomes essential for practical applications, as discussed in chapter VII.

A number of specific remaining problems mentioned in this chapter should keep the research field active. Besides these, a challenge is to relax two of the main approximations made. The first is the neglect of the influence of the radiator translational motion. The latter, which results from the speed dependence of collisional parameters and from the collision-induced changes of the radiator velocity, has been extensively studied in the case of isolated transitions (chapter III). Extended studies should be undertaken for collisionally coupled lines, as discussed in Sec. VIII.2. The second limiting assumption made in this chapter is the impact approximation which restricts the applicability of resulting models to regions close to the optical resonances (Apps. II.C1 and II.B4). Whereas approaches described in chapter V have been proposed for the far wings, the treatment of the intermediate (mid-wing) regions remains a problem for which paths are discussed in Sec. VIII.3.

APPENDIX IV.A VIBRATIONAL DEPHASING

Vibrational dephasing has been mentioned before as a reason for the breakdown of the sum rule in Eq. (IV.69) leading to non-zero values of $\Delta\gamma$ and $\Delta\delta$ in Eq. (IV.70). In order to get a simple picture of the process, consider a linear molecule whose wave functions are well represented assuming that their rotational part is independent on vibration, ie $|vJm\rangle = |v\rangle \times |Jm\rangle$. Let us also restrict ourselves to isotropic Raman Q lines, collisions with atoms and to a semi-classical approach for which the relaxation matrix elements are related to the following quantity [Eqs. (IV.124)–(IV.126) with k = 0]:

$$S^{(0)}_{Q(J'),Q(J)}(b,x) = \sum_{m,m'} \frac{1}{2J+1} \left\{ \delta_{J,J'}\delta_{m,m'} - \langle v_i J'm'|S(b,x)|v_i Jm\rangle \langle v_f J'm'|S(b,x)|v_f Jm\rangle^* \right\}.$$

(IV.A1)

In this equation, the elements of the scattering matrix S, directly related to the evolution operator U for a completed collision centered at time t = 0, are given by:

$$\langle vJ'm'|S(b,x)|vJm\rangle = \langle vJ'm'|U(+\infty,-\infty,b,x)|vJm\rangle$$
$$= \langle vJ'm'|\exp_t\left\{\frac{1}{i\hbar}\int_{-\infty}^{+\infty}[H_a + V(t,b,x)]dt\right\}|vJm\rangle.$$

(IV.A2)

In this expression, $\exp_t(...)$ is a *time-ordered* exponential, H_a is the Hamiltonian describing the internal rotational states of the radiator and V is the radiator-perturber interaction potential. Let us assume that V does *not* depend on the vibrational state, *except* for its isotropic part $V_{iso}(R)$. The equilibrium value of $V_{iso-eq}(R)$ being used to calculate the classical path trajectories, the total Hamiltonian can be split into two terms. The first is independent on rotation but vibrationally-dependent, while the second depends on rotation but is independent on vibration, ie: $H(t) = H_V(t) + H_R(t)$ with $H_V(t) = \Delta V_{iso}(t) \equiv V_{iso}[R(t)] - V_{iso-eq}[R(t)]$ and $H_R(t) = H_a + V_{ani}(t)$ where $V_{ani}(t)$ contains the anisotropic part of the radiator-perturber interactions. If one neglects the vibrational changes induced by ΔV_{iso} (*ie* the terms $\langle v|\Delta V_{iso}|v'\rangle$ for $v \neq v'$, which are much smaller than the diagonal elements $\langle v|\Delta V_{iso}|v\rangle$), $H(t)$ is then *diagonal* with respect to the vibration quantum number and Eq. (IV.A2) can be rewritten as:

$$\langle vJ'm'|S(b,x)|vJm\rangle = \langle v|\exp\left[\frac{1}{i\hbar}\int_{-\infty}^{+\infty}\Delta V_{iso}(t,b,x)dt\right]|v\rangle \quad \text{(IV.A3)}$$
$$\times \langle J'm'|\exp_t\left\{\frac{1}{i\hbar}\int_{-\infty}^{+\infty}[H_a + V_{ani}(t,b,x)]dt\right\}|Jm\rangle.$$

Introducing the operator:

$$S_R = \exp_t\left\{\frac{1}{i\hbar}\int_{-\infty}^{+\infty}[H_a + V_{ani}(t,b,x)]dt\right\}, \quad \text{(IV.A4)}$$

acting on rotation *only*, and the *vibrational phase shift* η_v defined by:

$$\eta_v(b,x) = \exp\left\{\int_{-\infty}^{+\infty}\left[\frac{1}{i\hbar}[\langle v_i|\Delta V_{iso}(R(b,x,t))|v_i\rangle - \langle v_f|\Delta V_{iso}(R(b,x,t))|v_f\rangle]\right]dt\right\}, \quad \text{(IV.A5)}$$

Eq. (IV.A1) can be rewritten as:

$$S_{Q(J'),Q(J)}^{(0)}(b,x) = \sum_{m,m'}\frac{1}{2J+1}\left\{\delta_{J,J'}\delta_{m,m'} - e^{-i\eta_v(b,x)}\langle J'm'|S_R(b,x)|Jm\rangle\langle J'm'|S_R(b,x)|Jm\rangle^*\right\}. \quad \text{(IV.A6)}$$

Since S_R is a unitary matrix $S_R^{-1} = S_R^+$, where S_R^+ is the adjoint of S_R, and one has:

$$\langle Jm|S_R^{-1}S_R|Jm\rangle = \langle Jm|S_R^+S_R|Jm\rangle = \langle Jm|I_d|Jm\rangle = 1$$
$$\Leftrightarrow \sum_{J',m'}\langle Jm|S_R^+|J'm'\rangle\langle J'm'|S_R|Jm\rangle = \sum_{J',m'}\langle J'm'|S_R|Jm\rangle^*\langle J'm'|S_R|Jm\rangle = 1. \quad \text{(IV.A7)}$$

Introducing this relation into Eq. (IV.A6) and summing over all J' values, straightforwardly lead to:

$$\sum_{J'} S^{(0)}_{Q(J'),Q(J)}(b,x) = \frac{1}{2J+1} \sum_m [1 - e^{-i\eta_v(b,x)}] = 1 - e^{-i\eta_v(b,x)}. \quad \text{(IV.A8)}$$

After integrations over the impact parameter b and kinetic energies x [Eqs. (IV.124) and (IV.125)], Eq. (IV.A8) demonstrates that the sum rule does indeed correspond to Eq. (IV.70) with:

$$\Delta\gamma - i\Delta\delta = \frac{n_0 <v_r>}{2\pi c} \int_0^\infty xe^{-x} \int_0^\infty 2\pi b[1 - e^{-i\eta_v(b,x)}]dbdx. \quad \text{(IV.A9)}$$

Such non-zero values of $\Delta\gamma$ and $\Delta\delta$ are responsible for the broadening and shifting of isotropic Raman Q branches at elevated pressure, as shown in Fig. IV.3 for the shift, in Fig. IV.10 for the broadening, and illustrated again by Fig. IV.A1.

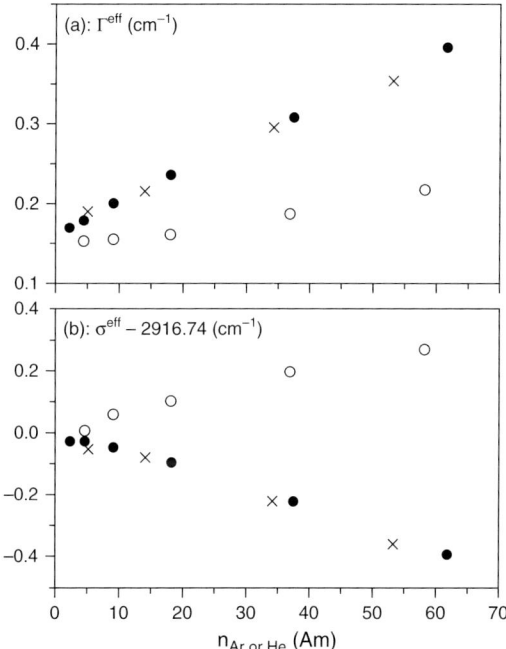

Fig. IV.A1: Measured width (a) and spectral position (b) of the v_1 isotropic Raman Q branch of CH_4 at room temperature *vs* the perturber density. (○) are values for CH_4 in He,[496] while (×)[605] and (•)[496] are for CH_4 in Ar. Reproduced with permission from Ref. 496.

Besides in isotropic Raman Q branches at elevated density, these values can also be observed for high J transitions of light rotors for which rotational contributions to the broadening and shifting are very small.[19,472] Indeed, they vanish since the inelastic rotational transfers, involving larger and larger energy jumps, become less and less probable with increasing J. $\Delta\gamma$ et $\Delta\delta$ can thus also be identified as the purely *vibrational* broadening γ_v and shifting δ_v of *individual* transitions, ie when the contribution of rotational relaxation is negligible. Indeed, setting $J = J'$ and $V_{anis} = 0$ in the preceding equations straightforwardly leads to $\gamma_{Q(J)} = \gamma_v$ and $\delta_{Q(J)} = \delta_v$. This is illustrated by Fig. IV.A2, which confirms that both $\gamma_{R(J)}$ and $\delta_{R(J)}$ tend to become constant with increasing J. Since η_v is (generally) small, the exponential in Eq. (IV.A9) can be expanded, leading to:

$$\Delta\gamma = \gamma_v = \frac{1}{2} \frac{n_0 <v_r>}{2\pi c} \int_0^\infty xe^{-x} \int_0^\infty 2\pi b \eta_v(b,x)^2 db dx$$

$$\Delta\delta = \delta_v = -\frac{n_0 <v_r>}{2\pi c} \int_0^\infty xe^{-x} \int_0^\infty 2\pi b \eta_v(b,x) db dx.$$

(IV.A10)

Since the matrix element $<v|\Delta V_{iso}|v>$ is proportional to the number v of vibrational quanta, one can show, using Eqs. (IV.A5) and (IV.A10) that δ_v is proportional to $(v_f - v_i)$ while γ_v is proportional to $(v_f - v_i)^2$. These relations are indeed well verified by the high J asymptotic experimental values of Fig. IV.A2 (also see Refs. 19,472).

Note that the preceding analysis leading to the identification of the sum rule breakdown parameters $\Delta\gamma$ and $\Delta\delta$ to the purely vibrational width and shift is only valid for isotropic Raman Q branches and when the vibrational dependence of V_{ani} can be

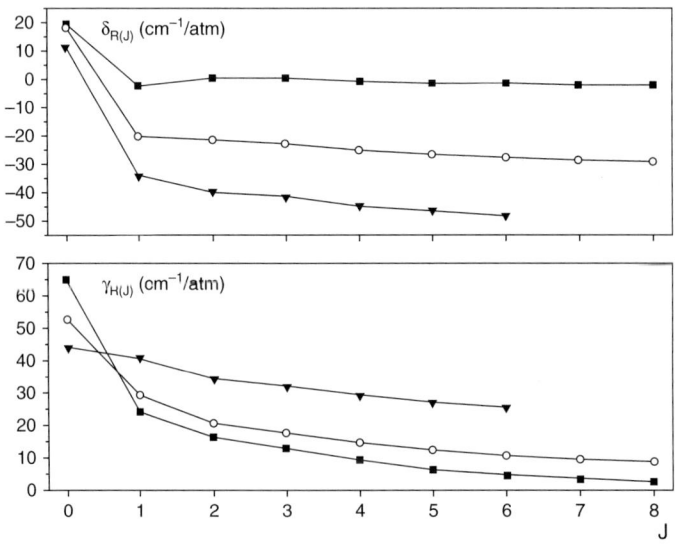

Fig. IV.A2: Room temperature measured pressure broadening (γ) and shifting (δ) coefficients for R(J) lines of HF in Ar.[214,606] ■, ○, and ▲ are results in the $0 \leftarrow 0$, $1 \leftarrow 0$, and $2 \leftarrow 0$ vibrational bands, respectively.

disregarded. When this approximation cannot be made, the situation is more complicated and gets even more complex when the spectrum involves different branches as in infrared and anisotropic Raman spectroscopies.

APPENDIX IV.B   PERTURBED WAVE FUNCTIONS

The rigid rotor (RR) sum rule in Eq. (IV.69) can be re-written as:

$$\vec{d}^{RR}.W^{RR} = \vec{0}. \tag{IV.B1}$$

where $\vec{d}$ is a (row) vector whose components d_j correspond to the various lines $j = 1, N$ within the ordering chosen, and W is the relaxation matrix, $W_{i,j}$ being the element coupling line j to line i. When the internal wave functions of the isolated radiator are perturbed, they can be obtained from the diagonalization of the associated Hamiltonian matrix H_a. The latter can be constructed, taking all proper internal couplings into account, on the basis of unperturbed rigid rotor-harmonic oscillators (RRHO) wavefunctions. For a polyad of vibrationally interacting states for instance, H_a includes all vibrational and rotational levels, the off-diagonal elements describing the anharmonicities and various couplings (*eg* Fermi, Coriolis, *etc*). The diagonalization of H_a leads to the matrix E of the normalized eigenvectors and to the diagonal one D containing the associated energies, *ie*:

$$H_a = EDE^{-1}. \tag{IV.B2}$$

The element $E_{i,j}$ is the coordinate of the jth true radiator eigenstate on the ith RRHO wavefunction. For simplicity, let us consider that only the upper states of the optical transitions are perturbed, the lower ones being well described by RRHO wave functions. Let us also assume that the radiator-perturber interaction is independent on vibration (no vibrational dephasing, *cf* App. IV.A). Changing from the basis of the true (upper) states (T) to that of the rigid rotor (RR) in Eq. (IV.B1) is then straightforward using the E matrix, leading to:

$$\vec{d}^T.W^T = (\vec{d}^{RR}.E).({}^tE.W^{RR}.E). \tag{IV.B3}$$

where ${}^tE$ designates the transpose of E. Since H_a is Hermitian, the eigenvectors are real and form an orthonormalized basis. One thus has ${}^tE = E^{-1}$, and $E.{}^tE$ is the unity matrix. Equation (IV.B3) thus reduces to:

$$\vec{d}^T.W^T = \vec{d}^{RR}.W^{RR}.E, \tag{IV.B4}$$

which is $\vec{0}$ due to Eq. (IV.B1). This demonstrates that the sum rule *also* applies to spectroscopically perturbed states provided that *all* lines are taken into account in the sum and that the dependence of the interaction on the vibrational state can be neglected. This result can be easily generalized to the case where both lower and upper states of the optical transition are perturbed.

A more specific analysis was made[591] for the v_3 and v_1+v_2 bands of CO_2, which involve upper states coupled through a Coriolis interaction.[607] This coupling is small and the true radiator eigenstates can be determined by perturbation theory, leading, with obvious notations, to:

$$|v_3 J\rangle^T = |v_3 J\rangle^{RRHO} - \varepsilon \sqrt{J(J+1)})|(v_1+v_2)J\rangle^{RRHO},$$
$$|(v_1+v_2)J\rangle^T = |(v_1+v_2)J\rangle^{RRHO} + \varepsilon \sqrt{J(J+1)})|v_3 J\rangle^{RRHO}, \quad \text{(IV.B5)}$$

where $\varepsilon \ll 1$ (but $\neq 0$) results from the Coriolis interaction. One can show[591] that the sum rule is indeed verified, provided that the lines of the *two* bands are included in the sum. The spectroscopic mixing of the rovibrational states gives rise to non-zero couplings between lines of the two bands. Indeed, even if V is *diagonal* with respect to the unperturbed vibrations, there are vibrationally off-diagonal couplings since:

$$\begin{aligned}{}^T\langle(v_1+v_2)J'|V|v_3 J\rangle^T &= -\varepsilon\sqrt{J(J+1)}^{RR}\langle(v_1+v_2)J'|V|(v_1+v_2)J\rangle^{RRHO} \\ &+ \varepsilon\sqrt{J'(J'+1)}^{RR}\langle v_3 J'|V|v_3 J\rangle^{RRHO} \quad \text{(IV.B6)} \\ &\approx \varepsilon\left(\sqrt{J'(J'+1)} - \sqrt{J(J+1)}\right)^{RR}\langle J'|V|J\rangle^{RR} \neq 0.\end{aligned}$$

Neglecting these corresponds to restricting the sum rule to within a vibrational band and leads to deviations from Eq. (IV.69). Note that the non-zero coupling elements between lines of the v_3 and v_1+v_2 bands and the perturbation of line-coupling elements within each band are very small (since $\varepsilon \ll 1$). They thus have negligible spectral consequences. Furthermore, it was demonstrated[591] that calculating the diagonal elements of W (the broadenings) from the off-diagonal ones [Eq. (IV.69)] using rigid rotor dipole matrix elements and restricting the sum to a single band is correct. This gave a theoretical justification to the procedure empirically used in Ref. 608.

APPENDIX IV.C RESONANCE BROADENING

The models for the prediction of the relaxation matrix described in Sec. IV.3 (and particularly those of § IV.3.4 and IV.3.5) disregard "resonance broadening" which occurs when collisions involve identical molecules. Strictly speaking, they are thus valid only when the radiator and perturber are different species, thus *distinguishable*. For self-broadening, the distinction between the two molecules of the colliding pair is not meaningful. Indeed, the optical excitation can be transferred through a *resonant exchange* from one molecule to the other and still be considered as *among* the radiators. Theoretical approaches have been proposed in a number of papers (see Refs. 47,52,290,406,609,610 and those cited therein) for the treatment of this process in the case of allowed lines. Most of them are of difficult application to molecular systems of practical interest, and few quantitative results are, to our knowledge, available. Let us nevertheless synthesize the conclusions of two studies in which calculations have been performed.

An extension of the Anderson-Tsao-Curnutte approach (§ IV.3.4d) is proposed in Ref. 610. It is semi-classical, limited to a second order expansion of the scattering operator and to electrostatic interactions. Tractable equations are given, in which new contributions due to resonant exchange are added to the S_2 terms of the usual[438] ATC theory. Starting from this, computations have been made for self-broadened pure rotational lines of HCl, HF, and HI.[610] These are favorable systems having a strong permanent dipole for the resonant transfer. It is found that the contribution of the resonant terms to line widths is no more than a few per cent. On the opposite, a significant blue shift is predicted which has, to our knowledge, found no experimental confirmation yet. In Ref. 406, Close-Coupling (§ IV.3.5b) and Coupled-States (§ IV.3.5c) approaches are proposed for collisions between identical particles. Calculations are made for the rotational relaxation in pure N_2. The results show that, for this problem, assuming distinguishable molecules is a good approximation. The maximum difference between results obtained including and neglecting the effects of identical particles is 19%, but this is obtained for specific couples of the collisions partners rotational states at a relatively low kinetic energy. The difference decreases rapidly with increasing energy and it is much smaller for other couples of rotational states. After the averaging leading to state-to-state rates $K(J' \leftarrow J)$, the effect of indistinguishability becomes very small, except for large $\Delta J = |J-J'|$. In this case, the identical molecules rate is significantly larger than the corresponding distinguishable molecules rate, but both are very small.

In summary, the available theoretical results indicate that resonant exchange has a small influence on collisional cross-sections. Furthermore, in situations where the optical excitation involves a vibrational change, as in infrared absorption and Raman scattering, the effect of identical molecules is expected to be negligible.[406,610] Indeed, the resonant exchange then requires a change in the vibrational state, much less efficient than pure rotational transfers. Hence, all the models presented in this chapter *can be reasonably used for pure gases* and are not practically restricted to radiators highly diluted in a bath of different species. Finally note that, in the practical applications discussed in chapter VII, self-broadening of allowed transitions often plays a marginal role. In planetary atmospheres, except for CO_2 in Venus and, to a lesser extent, H_2O on Earth, the molecules showing allowed (dipolar) absorption are highly diluted. Only the foreign broadening by the most abundant species (N_2 and O_2 for Earth, H_2 for Jupiter, N_2 for Titan, *etc*) is thus relevant, and collisions between undistinguishable partners need not to be considered. This is not the case for N_2, O_2, H_2 spectra, but these involve collision-induced absorption processes (chapter VI) for which indistinguishability can be properly taken into account.[84]

Experimental evidence of the effects of resonant exchange is not easy to obtain but it may be attainable thanks to the measurement accuracy achieved in the last decades. From this point of view, studies of purely rotational lines of hydrogen halides may bring interesting information. In particular, one could investigate the widths and shifts of the same transition of different isotopomers of a gas of natural composition, such as HCl: 76% $H^{35}Cl$ + 24% $H^{37}Cl$. Indeed, the (binary) collisions leading to the self-broadening of $H^{35}Cl$ then involve identical molecules in 76% of the cases, this fraction being 24% for self-broadened $H^{37}Cl$. Eventual relative differences in the widths of identical lines of these two isotopomers could then be analyzed with a model, such as that of Ref. 610, properly accounting for the intrinsic influence of the isotopomer (mass, dipole moment, *etc*), disregarding and including resonant exchange.

V. THE FAR WINGS
(BEYOND THE IMPACT APPROXIMATION)

V.1 INTRODUCTION

Near resonances, the shape of *isolated* transitions is often described by the Lorentzian or Voigt profile (except for very long wavelengths for which the Van Vleck-Weisskopf shape is used). This neglects refined but generally small effects resulting from changes of the radiator velocity and from the speed dependence of collisional parameters, discussed in chapter III. When the spectral separation between optical transitions is of the order of, or smaller than, their pressure-broadened widths, the lines overlap and collisional *line mixing* may lead to more complicated spectral shapes, as analyzed in chapter IV. For *moderate* distances $\Delta\sigma$ from the line centers, ie such that $|\Delta\omega| = 2\pi c|\Delta\sigma|$ is smaller than the inverse of the duration τ_c of efficient collisions (leading to values of $|\Delta\sigma|$ from a few cm^{-1} to a few tens of cm^{-1}, *cf* App. II.B4), the breakdown of the impact approximation remains small and the frequency dependence of the relaxation matrix W may be neglected (*cf* App. II.C). Consequently, *impact* models such as those presented in Secs. IV.2 and IV.3 give satisfactory predictions in the near wings. However, to calculate the absorption (or scattering) of radiation in systems involving optically thick paths, an accurate description of spectral regions far away from the line and band centers is often required. This is important for the prediction of the absorption/emission of radiation by CO_2 and H_2O in the atmospheres of the Earth (and Venus), as illustrated by Figs. I.4, VII.15, VII.16. The weak but important continua formed by the far wings of atmospheric water vapor lines in the 3–5 and 8–12 µm intervals and in the millimeter wave region are good examples of the cumulative effects of non-Lorentzian line wings (App. V.A). As discussed in § VII.5.1, the quality of their modeling conditions the results of radiative transfer calculations and the accuracy of quantities retrieved from atmospheric spectra by remote sensing experiments.

From a theoretical point of view, the non-Lorentzian behavior of the far wings is expected since line mixing (Chapt. IV) leads to the breakdown of models assuming isolated line contributions through the off-diagonal elements of the relaxation matrix W. For angular frequency detunings $|\omega-\omega_{fi}|$ from the centers ω_{fi} of the most intense $f \leftarrow i$ lines larger than the inverse of the collision duration τ_c, the frequency dependence of W, disregarded in the preceding chapter, is no more neglectable. Until relatively recently, little if any work had been done on the detailed analysis of this frequency dependence. The reason for this situation is that there is no simple method for solving such a problem, which manifests primarily in terms of "off-the-energy-shell" transition operator in a fully quantum approach (Refs. 139,143,611 and App. II.C2). An additional difficulty arises when considering distances from line centers of the order of, or greater than,

$k_BT/\hbar$. As shown in chapter II, the absorption coefficient $\alpha(\omega)$ is expressed in terms of the initial statistical equilibrium density operator $\rho(0)$ for a system including (within the binary collision approximation) one radiator (Hamiltonian H_a) and one perturber (Hamiltonian H_p) interacting through the potential operator V, ie:

$$\rho(0) = e^{-H_M/k_BT}/\text{Tr}(e^{-H_M/k_BT}), \quad \text{where } H_M = H_a + H_p + V, \tag{V.1}$$

where Tr(...) denotes the trace over the eigenstates of H_M. For detuning much smaller than $k_BT/\hbar$ (about 200 cm^{-1} at 300 K), one usually ignores *initial statistical correlations* by approximating $\rho(0)$ by:

$$\rho(0) \approx e^{-(H_a+H_p)/k_BT}/\text{Tr}\left[e^{-(H_a+H_p)/k_BT}\right]. \tag{V.2}$$

As seen in § II.3.3, this approximation leads to an absorption coefficient which no longer satisfies the fluctuation-dissipation theorem. This has minor consequences near resonances but certainly not in the very far wings. In summary, in the far wings, the theoretical challenge is to include *simultaneously* in the formalism, the following three basic mechanisms: (1) line mixing which leads to contributions of the various optical transitions to the total spectral shape that are not additive, thus invalidating the notion of isolated lines; (2) the finite duration of collisions; *and* (3) initial statistical correlations which *both* contribute to the frequency dependence of the relaxation matrix.

The first theoretical (but *formal*) treatment of a "non impact" profile was proposed by Fano who extended[44] the theory beyond the impact description of intermolecular collisions. It was then reformulated some years later by Ben-Reuven[45,46] within the resolvent approach (*cf* App. II.E). Recall that a review describing a variety of formalisms proposed before 1981 was made by Breene in his monograph.[48] Since the mid 1990's, a number of spectral-shape theories have been proposed in order to calculate the *far* wings of optical transitions of molecular systems. However, due to the complexity of many of these formalisms, their use for practical computations remains of formidable difficulty. Most of them, which nevertheless enable a better understanding of the approximations needed in order to derive tractable models, remain at a highly *formal* level. They are summarized in Table V.1, with some "founding" papers.

In the following of this chapter, we mainly concentrate on more approximate but *tractable* approaches, permitting predictions of the spectral profile for molecular systems of practical interest. Since the problem is a very difficult one, it has often been necessary to use phenomenological models based on empirical corrections to the Lorentzian (or Voigt) profile, as described in Sec. V.2. More physically based approaches are discussed in the following sections. They are obtained within the quasistatic approximation (Sec. V.3), from perturbative treatments of the evolution operators (Sec. V.4), and from extensions (Sec. V.5), beyond the impact approximation, of the dynamically based scaling laws introduced in § IV.3.3.

The far wings 243

Table V.1: Some "basic" and formal references

References	characteristics
44	« Founding »: All about the relaxation operator $W(\omega)$.
45,46	« Founding »: The resolvent approach in the frequency domain.
57	The basic approach in the time domain.
141,142	« Founding »: The differences between the time-domain and the frequency-domain formalisms.
140	How to maintain the detailed balance principle at each step of the derivation, through a rigorous analysis of initial correlations.
72	On the symmetry properties of the memory function and $W(\omega)$ relaxation matrix.
612	How to build a generalized $W(\omega)$ matrix which respects all known fundamental relations.
613	A careful examination of the validity of various approximations when going from line centers to the wings.
139,143,611	How to relate off-the-energy-shell matrix elements to their on-shell values in order to evaluate the frequency dependence of the Fano collision operator within a purely quantum approach.

V.2 EMPIRICAL MODELS

Before considering more physically based but significantly more complex approaches, some empirical methods developed since the mid 1960's are first presented. These were proposed, due to the complexity of first principle calculations of the spectral shape in the far wings, in order to fulfill the needs of applications.

V.2.1 THE χ FACTOR APPROACH

The first approach introduces a line-shape correction χ modifying the Lorentzian (or Voigt) profile in the wings. This factor is considered, in a given spectral region, as independent on the line, so that the absorption coefficient[1] α at wavenumber σ, for a highly diluted binary mixture with radiator and perturber densities n_a and n_p at temperature T, is written as:[2]

---

[1] The χ factor approach has been, to our knowledge, exclusively used for the description of dipolar absorption spectra.
[2] In the case of any mixture, the binary χ function in Eq. (V.3) must be replaced by its weighted average accounting for the species molar fractions x_i, *ie*:

$$\chi_a(\Delta\sigma, T, x_i) = \left[\sum_{\text{species i}} x_i \gamma_\ell^{a-i}(T) \chi_{a-i}(\Delta\sigma, T)\right] \Big/ \left[\sum_{\text{species i}} x_i \gamma_\ell^{a-i}(T)\right],$$

where γ_ℓ^{a-i} and χ_{a-i} are the broadening coefficient of line ℓ and the line-shape corrective factor induced, in absorbing species a, by collisions with species i.

$$\alpha(\sigma, n_a, n_p, T) = \sum_{\ell} n_a S_\ell(T) \times \text{Voigt}_\ell(\sigma, n_a, n_p, T) \times \chi_{a-p}(\sigma - \sigma_\ell, T), \quad (V.3)$$

where S_ℓ [Eq. (IV.9)] and Voigt$_\ell$ [Eqs. (III.20)-(III.22)] designate respectively the integrated intensity and normalized Voigt profile of line $\ell = f \leftarrow i$. For long wavelengths, negative resonances must be added through the terms associated with the transitions $i \leftarrow f$ centered at negative wavenumbers. In this case, and considering the wings where the Voigt profile reduces to a Lorentzian shape, Eq. (V.3) becomes:

$$\alpha(\sigma, n_a, n_p, T) = \sum_{\ell, \sigma_\ell > 0} n_a S_\ell(T) \times \left\{ \frac{\Gamma_\ell(T)\chi_{a-p}(\sigma - \sigma_\ell, T)}{\pi(\sigma - \sigma_\ell)^2} + \frac{\Gamma_\ell(T)\chi_{a-p}(\sigma + \sigma_\ell, T)}{\pi(\sigma + \sigma_\ell)^2} \right\}, \quad (V.4)$$

where $\Gamma_\ell (=n_p\gamma_\ell)$ is the pressure-broadened half widths of line ℓ. χ, which depends on the molecular system and on temperature, is a function of the spectral distance $\Delta\sigma = \sigma - \sigma_\ell$ to the transition center. While it is often considered as symmetric [$\chi(-\Delta\sigma) = \chi(+\Delta\sigma)$], more refined asymmetric profiles accounting for Eq. (II.31) have also been proposed,[614] ie $\chi(-\Delta\sigma) = \chi(+\Delta\sigma)\exp(-hc\Delta\sigma/k_B T)$. The *imposed condition* $\chi(|\Delta\sigma| \to 0) \to 1$ ensures that Voigt profiles are obtained near line centers. As shown in § V.3.3, χ can be identified in the very far wings [$|\sigma - \sigma_\ell| \gg |\sigma_{\ell'} - \sigma_\ell|$] as some band average of the wavenumber-dependent relaxation matrix elements for all the significantly contributing ℓ and ℓ' lines. The empirical approach consists in representing $\chi(\Delta\sigma)$ by analytical functions (such as piece-wise exponentials[615] or combined polynomial and exponential laws[616]) whose parameters are deduced from fits of measured spectra. This has been done for various systems and spectral regions, as summarized in table V.2.

Note that different functions χ are obtained, depending on the molecular system, as exemplified in Fig. V.1.

Table V.2: Infrared line-shape correction factors χ available in the literature

Molecules	Region	References
CO_2 in (CO_2,N_2,O_2,air)	4 µm	357,614,615,617–624
	2.3 µm	618,625
CO in (CO, N_2)	5 µm	626
H_2O in (H_2O, N_2)	>8 µm	616, see also App. V.A
	5 µm	627
CH_4 in H_2	3.3 µm	628
NH_3 in (NH_3,H_2)	5 µm	629

The far wings 245

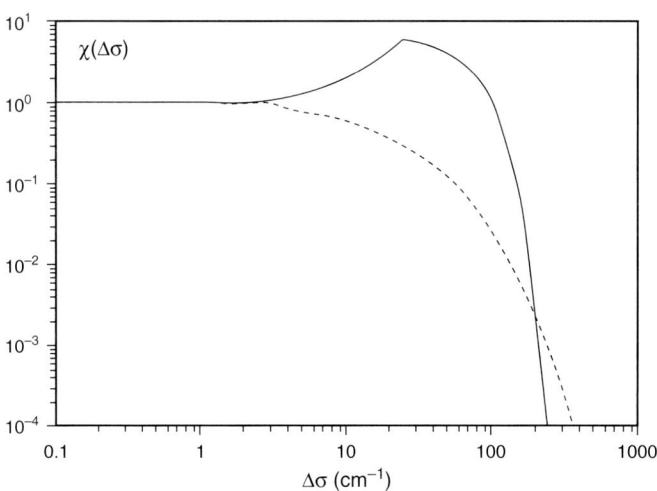

Fig. V.1: Room temperature wing correction line-shape factors χ for pure CO_2 (- - -) and H_2O in N_2 (———). After Refs. 615,616.

For radiators with closely spaced lines (*eg*, CO, CO_2) and relatively narrow vibrational bands, line mixing (chapter IV) quickly becomes efficient when going away in the wings, due to the overlapping of the line contributions. It leads to a strongly increasing sub-Lorentzian behavior (Figs. IV.7 and V.5), resulting in rapidly decreasing values of χ (*eg* CO_2 in Fig. V.1). For radiators with broad vibrational bands with widely spread lines, the influence of line mixing becomes important only far away in the wings, at distances where many line contributions overlap significantly. In the intermediate $|\Delta\sigma|$ region, the profile is dominated by the influence of the finite duration of collisions on the diagonal elements of the relaxation matrix since the overlapping and the consequences of line mixing remain weak. Since the values of $\text{Re}[W_{\ell\ell}(|\sigma-\sigma_\ell|)]$ may increase with $|\sigma-\sigma_\ell|$ (*eg* Fig. VIII.3), this may lead to a super-Lorentzian behavior (and $\chi>1$) in the mid-wings (see Refs. 627,630,631 and χ for H_2O in Fig. V.1), before line couplings become efficient and χ decreases down to zero.

This empirical approach can, in principle, lead to very accurate predictions of the wings, provided that accurate laboratory measurements and a sufficient number of adjustable parameters are used in the fits. Nevertheless, the use of χ factors raises problems. The first is that measured laboratory spectra do not generally cover all the needs of applications. Some spectral regions and/or the temperature dependence may not have been studied, for instance because optical paths attainable in the laboratory are insufficient. Examples are given by H_2O on Earth and CO_2 on Venus where the atmospheres involve integrated molecular amounts and optical thicknesses significantly larger than what can be achieved using Fourier transform laboratory experiments. Extrapolations must thus be made, which may lead to erroneous results. This extrapolation problem is illustrated by CO_2 whose χ factors derived in the 4 μm region lead to inaccurate predictions when used near 2.3 μm[625] and 15 μm.[632] This can be expected

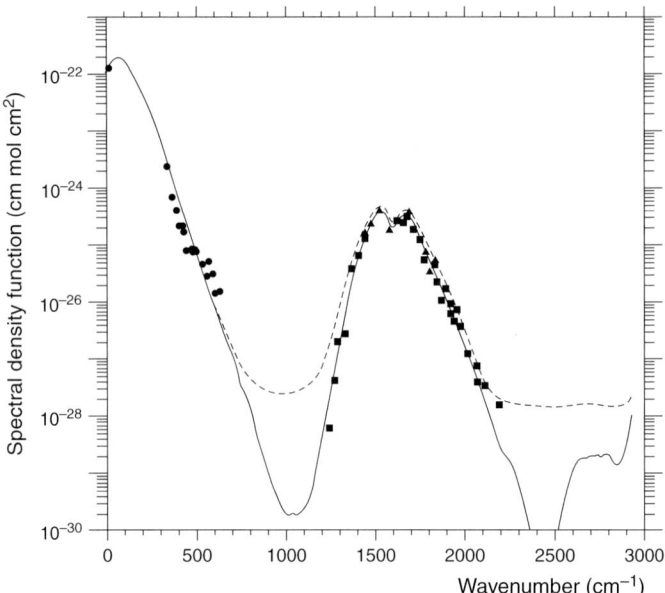

Fig. V.2: H$_2$O–N$_2$ wing absorption due to lines centered more than 25 cm^{-1} away. Symbols indicate measured values (Ref. 633 and those therein). The lines are predicted values based on empirical line shapes, following Ref. 616 and corresponding to two versions of the CKD (see § V.2.2 and App. V.A) continuum (- - -) CKD0 (1989), (____) CKD2.2 (1996).

from the fact that the shape of the wings depends on the vibrational symmetry of the considered band (eg $\Sigma \leftarrow \Sigma$ with only P and R branches for ν_3 near 4 µm and $\Pi \leftarrow \Sigma$ with P, Q, and R branches for ν_2 near 15 µm). The problem is also illustrated by the results for water vapor displayed in Fig. V.2. Indeed, whereas successive empirical line-shape corrections lead to consistent predictions in the intense absorption regions where laboratory data are available, the values extrapolated at wavenumbers in between the vibrational bands, where absorption is weak, differ very significantly.

A second difficulty comes from the empirical nature of χ factors deduced from fits of laboratory spectra. This makes them database dependent and they can thus, strictly speaking, only be used together with the spectroscopic data [σ_ℓ, S_ℓ, and Γ_ℓ, see Eq. (V.4)] used in the original fits that led to their determinations. Since changes of the parameters stored in spectroscopic databases can invalidate line-shape corrections derived using a former version, there is a maintenance difficulty.

V.2.2 THE TABULATED CONTINUA

A second practical approach (see also App. V.A), in order to include the contributions of far wings in spectra calculations, is derived by splitting the total absorption coefficient into two contributions, ie:

$$\alpha(\sigma, n_a, n_p, T) = \alpha^{\text{Close}}(\sigma, n_a, n_p, T, \Delta\sigma_c) + \alpha^{\text{Wing}}(\sigma, n_a, n_p, T, \Delta\sigma_c). \qquad (V.5)$$

The first term comes from transitions centered *nearby* the current wavenumber, *ie* those for which $|\sigma - \sigma_\ell| \leq \Delta\sigma_c$, where $\Delta\sigma_c$ is the *a priori chosen* cut-off distance. Provided that $\Delta\sigma_c$ is small enough so that $\chi(\Delta\sigma_c) \approx 1$, α^{Close} is given by the sum of the contributions of isolated transitions with Voigt profiles. More generally, it can be calculated with any appropriate impact approach. Indeed, as shown in chapter III, the Voigt line shape may be insufficient for calculations near line centers. Furthermore, assuming isolated lines can be inaccurate for clusters of closely spaced transitions when line-mixing processes (chapter IV) are efficient. The second term in Eq. (V.5), resulting from the wings, takes a simple form if $\Delta\sigma_c$ is significantly larger than the line widths Γ_ℓ ($\Delta\sigma_c \geq$ one cm^{-1} for atmospheric pressures). In this case, for $|\sigma - \sigma_\ell| \geq \Delta\sigma_c$ the Voigt shape can be approximated by $\Gamma_\ell/[\pi(\sigma - \sigma_\ell)^2]$. For an optically active species a in a gas mixture, α^{Wing} then takes the following form:

$$\alpha^{Wing}(\sigma, n_a, n_i, T, \Delta\sigma_c) = n_a \left\{ \sum_i n_i \left[\sum_{\ell, |\sigma-\sigma_\ell| \geq \Delta\sigma_c} \frac{S_\ell(T)\gamma_\ell^{a-i}(T)\chi_{a-i}(\sigma - \sigma_\ell, T)}{\pi(\sigma-\sigma_\ell)^2} \right] \right\}. \quad (V.6)$$

The quantity within brackets in Eq. (V.6), denoted C_{a-i}^0, is the so-called (density bi-normalized) "*continuum*" in the wings of the lines of species a due to collisions with species i. It depends on the chosen value of $\Delta\sigma_c$, on the wavenumber σ and on the temperature T, but *not* on the densities, leading to the simple identity:

$$\alpha^{Wing}(\sigma, n_a, n_i, T, \Delta\sigma_c) = \alpha(\sigma, n_a, n_i, T) - \alpha^{Close}(\sigma, n_a, n_i, T, \Delta\sigma_c)$$
$$= n_a \sum_i n_i C_{a-i}^0(\sigma, T, \Delta\sigma_c) = x_a n_{tot}^2 \sum_i x_i C_{a-i}^0(\sigma, T, \Delta\sigma_c), \quad (V.7)$$

where $n_{tot} \equiv \sum n_i$ is the gas total density and $x_i \equiv n_i/n_{tot}$ is the molar fraction of species i. Equation (V.7) is convenient for practical calculations since it is easy to implement and of low computer cost if tabulations of the continua are available. Values of C_{a-i}^0 can be calculated from Eq. (V.6), once the χ functions are known, and tabulated as functions of σ and T (they slowly depend on both) for each (a,i) pair. Such an approach was developed by Clough, Kneizys, and Davies,[616] leading to the widely used CKD self-broadened (a = i = H$_2$O) and foreign-broadened (a = H$_2$O, i = N$_2$) water vapor *infrared* continua.[634] These were initially generated[616] using Eq. (V.6) and χ factors adjusted on measurements. Their values have then been changed regularly, according to information brought by new laboratory and atmospheric measurements.[635,636] Some of the C_{a-i}^0's can also be obtained from experiments, by subtracting calculated values of α^{Close} from the measured absorption coefficients α^{Meas} following Eq. (V.7) (*eg* Refs. 633,637 and those therein). Finally note that, as for approaches using the factors χ, keeping the compatibility between the continua and the way α^{Close} is calculated induces a maintenance problem. If the spectroscopic data which were used in order to calculate the values of α^{Close} and subtract them from measurements to extract C_{a-i}^0 are changed, the C_{a-i}^0's must also be updated.

If one now considers the water vapor millimeter continuum, several empirical models are available. Remember that accurate modeling of this continuum absorption is essential for the accuracy of satellite retrievals using long wavelengths.[638] Here again, self and foreign continua have been adjusted from a few laboratory measurements.[639–641] Since such experiments are rather difficult, the temperature dependence of the continuum, for instance, is not very well known. The most widely used model is Liebe's Millimeterwave Propagation Model MPM89[642] and its newer version MPM93.[643] More recently, by analyzing these two models, Rosenkranz[644] concluded that the best approach, labeled ROS98 by some people, is a combination of the MPM93 self-continuum model and of the MPM89 foreign continuum. Finally, a more refined analytical parameterization has been proposed recently.[645]

V.2.3 OTHER APPROACHES

An even more simple approach has also been used, which consists in the use of Voigt profiles *truncated* at a distance $\Delta\sigma_c$ from the centers chosen in order to match measurements (*eg* Ref. 646). Although crude, this can lead to satisfactory results in regions where the density of lines is significant and for systems involving quickly decreasing sub-Lorentzian wings, so that the local contribution α^{Close} is significantly larger than α^{Wing} (*eg* Ref. 628).

Finally, an intermediate semi-empirical model accounting for line mixing was proposed in Ref. 356. In this work, an impact relaxation matrix W^I was first constructed using fitting (§ IV.3.2) and scaling (§ IV.3.3) laws. The effects of the finite duration of collisions were empirically introduced through a correction f_{corr} of impact values, *ie*:

$$W_{\ell\ell'}(\sigma) = W^I_{\ell\ell'} \times f_{corr}(\sigma - \frac{\sigma_\ell + \sigma_{\ell'}}{2}) \quad \text{with} \quad \lim_{|\Delta\sigma| \to 0}[f_{corr}(\Delta\sigma)] = 1, \quad (V.8)$$

where σ_ℓ and $\sigma_{\ell'}$ designate the spectral positions of lines ℓ and ℓ', respectively. The f_{corr} function was modeled analytically[356] and its parameters were deduced from fits of measured CO_2 spectra. Although crude, f_{corr} being assumed the same for all couples of lines, this model enables accurate predictions of the far wings, since it is adjusted to experimental spectra. Contrary to χ factor approaches, it also provides a correct modeling of impact regions affected by line mixing, thanks to the off-diagonal elements of W^I and to the unity value imposed to $f_{corr}(0)$.

V.3 FAR WINGS CALCULATIONS: THE QUASISTATIC APPROACH

Even if the basic theory of pressure broadening has been established a long time ago, and various aspects of the wings shapes have been the subject of recent studies, many studies were either overly formal (Table V.1) or, on the contrary, too empirical (Sec. V.2). In this scientific field, the major difficulty lies in the numerical application, *ie* the computation of the spectral shape starting from first principles and from a realistic radiator-perturber intermolecular potential. In our opinion, a significant step was made

The far wings 249

in the mid eighties, with the publication of Rosenkranz's *quasistatic* formalism.[647,648] Starting from the basic theory of Fano,[44] a far wing formalism is proposed, based on the quasistatic and binary collision approximations. This approach was extended and improved by other authors. Boulet et al[649] proposed a line-by-line generalization for the calculation of the frequency dependence of each of the $W_{\ell'\ell}(\omega)$ relaxation matrix elements. Later on, Ma and Tipping have eliminated many of the restrictions of the original formalism in a series of papers.[72,140,650–657] Schematically, the quasistatic approximation is the opposite of the impact one. Whereas the latter only considers completed collisions (regardless of their duration τ_c), and the restriction $|\omega-\omega_{fi}|\tau_c \ll 1$ (see App. II.C1), the former, that applies in the far wings where $|\omega-\omega_{fi}|\tau_c \gg 1$, only considers very short time scales (asymptotically a single time) within collisions.

V.3.1 GENERAL EXPRESSIONS

Let us summarize the derivation given in Ref. 656. Note that this work uses, with respect to chapter II, slightly different choices and notations. In particular, a Fourier transform in $e^{-i\omega t}$ is made and a symmetrized auto-correlation function is used. Within this frame, the absorption coefficient is then given by the following analog of Eq. (II.6):

$$\alpha(\omega) = \frac{8\pi^2}{3\hbar c} n_a \omega \sinh(\hbar\omega/2k_BT)\widetilde{F}(\omega), \quad (V.9)$$

where n_a is the number density of absorbing molecules (radiators) and sinh(…) means the hyperbolic sine. The *symmetrized* spectral density $\widetilde{F}(\omega)$ is given, equivalently to Eq. (II.34), by the following transform:

$$\widetilde{F}(\omega) = \frac{1}{\pi}\text{Re}\left\{\int_0^\infty e^{-i\omega t}\widetilde{\Phi}(t)dt\right\}, \quad (V.10)$$

of the *symmetric* auto-correlation function $\widetilde{\Phi}(t)$ given by [analog of Eq. (II.14), disregarding the effect of the radiator motion]:

$$\widetilde{\Phi}(t) = \text{Tr}\left\{\vec{d}^+ e^{-iH_M(t-i\hbar/2k_BT)/\hbar}\vec{d}\, e^{iH_M(t+i\hbar/2k_BT)/\hbar}\right\}/\text{Tr}\left\{e^{-H_M/k_BT}\right\}, \quad (V.11)$$

where $\vec{d}$ is the radiator dipole operator, Tr{…} designates a trace over the radiator and perturber variables, and d^+ is the adjoint of operator d. The idea is then to split the total Hamiltonian H_M into two parts: one commutes with the internal coordinates of the molecules while the second does not. This distinction coincides, for rovibrational states and when one neglects the vibrational dependence of the radiator-perturber interaction potential, with the division of this potential into isotropic and anisotropic contributions. Accordingly, the total Hamiltonian is decomposed as:

$$H_M = H_a + H_p + V_{iso} + V_{aniso} \equiv H_0 + V_{aniso}, \quad (V.12)$$

where H_a and H_p are the Hamiltonians of the isolated radiator and perturber, respectively, within the binary collision approximation. In order to evaluate the complex-time development operators in Eq. (V.11), one introduces the time evolution operator in the interaction representation for complex times z, ie [see Eq. (II.21)]:

$$e^{-iH_M z/\hbar} = e^{-iH_0 z/\hbar} U(z). \tag{V.13}$$

The *key* idea of the quasistatic approach is to assume, thanks to the property of the Fourier transform, that the far wings of the optical transitions [ie $\tilde{F}(\omega)$ for large values of $|\omega-\omega_{fi}|$ with respect to line centers] mainly depend on the short-time limit of the correlation function $\tilde{\Phi}(t)$. Then, following Rosenkranz,[647] $U(z)$ can be approximated, for $t = \text{Re}(z) \to 0$, by:

$$U(z) \approx 1 - \frac{i}{\hbar} \int_0^z dz' V_{\text{aniso}}(z') + \left(\frac{-i}{\hbar}\right)^2 \int_0^z dz' V_{\text{aniso}}(z') \int_0^{z'} dz'' V_{\text{aniso}}(z'') + \ldots$$

$$\approx e^{-iV_{\text{aniso}} \times z/\hbar}, \tag{V.14}$$

with

$$V_{\text{aniso}}(z) = e^{iH_0 z/\hbar} V_{\text{aniso}} e^{-iH_0 z/\hbar}. \tag{V.15}$$

This is only correct to the *first order* in z,[140] and an improved approximation has been proposed by Ma and Tipping which is valid to order z^2, while remaining tractable. Instead of using the time evolution operator $U(z)$ of Eq. (V.13), one introduces the operator $\tilde{U}(z)$ having upper and lower integral limits more symmetric, ie:

$$e^{-iH_M z/\hbar} = e^{-iH_0 z/2\hbar} \tilde{U}(z) e^{iH_0 z/2\hbar}. \tag{V.16}$$

Then $\tilde{U}(z)$ is approximated by:

$$\tilde{U}(z) \equiv 1 - \frac{i}{\hbar} \int_{-z/2}^{+z/2} dz' e^{iH_0 z'/\hbar} V_{\text{aniso}} e^{-iH_0 z'/\hbar} + \ldots \approx e^{-iV_{\text{aniso}} \times z/\hbar}. \tag{V.17}$$

Some consequences for short times have been analyzed in Ref. 649. It leads, as expected, to a formalism valid in the far wings, first defined in the frequency domain by the fact that the frequency of observation must differ from resonances by an amount much larger than the inverse of a typical collision duration τ_c for the system of interest, ie $|\omega-\omega_{fi}| \gg \tau_c^{-1}$, but with the *additional* restrictions:

$$|\omega - \omega_{fi}| \text{ and } |\omega - \omega_{f'i'}| \gg |\omega_{fi} - \omega_{f'i'}| \; ; \; |\omega_{fi} - \omega_{f'i'}| \ll k_B T/\hbar. \tag{V.18}$$

In other words, the current frequency must *also* differ from resonances by an amount that is large when compared with differences between the frequencies of strongly coupled resonances which must, at the same time, be smaller than $k_B T/\hbar$. While these two conditions are quite easily satisfied for heavy molecules with closely spaced transitions,

The far wings

they are significantly more restrictive for light absorbing species. Using Eqs. (V.16) and (V.17), one can express $\tilde{\Phi}(t)$ as[656,(3)]:

$$\tilde{\Phi}(t) = n_p \tilde{v} \, \text{Tr} \Big\{ \Big(e^{iL_a t/2} \sqrt{\rho_p} \rho_a^{1/4} d\rho_a^{1/4} \Big)^+ \rho_{iso} e^{-V_{aniso}/2k_B T} \\ \times \Big[e^{-iL_{aniso} t} \Big(e^{-iL_a t/2} \sqrt{\rho_p} \rho_a^{1/4} d\rho_a^{1/4} \Big) \Big] e^{-V_{aniso}/2k_B T} \Big\}, \quad (V.19)$$

where $\tilde{v} \equiv \text{Tr}(e^{-H_0/k_B T})/\text{Tr}(e^{-H_M/k_B T})$ [Eq. (II.25)], and the Liouville space (*cf* App. II.D) operators L_a and L_{aniso} correspond respectively to $H_a/\hbar$ and $V_{aniso}/\hbar$ in the Hilbert space. From this expression it is easy to verify that $\tilde{\Phi}(-t) = \tilde{\Phi}(t)$ which guarantees that the fluctuation-dissipation theorem is verified since $\tilde{F}(\omega) = \tilde{F}(-\omega)$ which is the equivalent of Eq. (II.31). A refined discussion devoted to the physical meaning of Eq. (V.19) can be found in Ref. 140. By introducing summations over indices i,f,i',f', where each represents all the quantum numbers specifying the internal levels of the radiator, $\tilde{\Phi}(t)$ can be written as:[656]

$$\tilde{\Phi}(t) = \sum_{if} \sum_{i'f'} e^{i(\omega_{fi} + \omega_{f'i'})t/2} n_p \tilde{v} \text{Tr}_b \Big\{ \langle f | \rho_a^{1/4} d^+ \rho_a^{1/4} \sqrt{\rho_p} | i \rangle \\ \times |\langle i | \rho_{iso} e^{-V_{aniso}/2k_B T} e^{-iV_{aniso} t/\hbar} | i' \rangle \langle i' | \sqrt{\rho_p} \rho_a^{1/4} d\rho_a^{1/4} | f' \rangle \\ \times \langle f' | e^{+iV_{aniso} t/\hbar} e^{-V_{aniso}/2k_B T} | f \rangle \Big\}. \quad (V.20)$$

In this expression Tr_b now denotes the trace over the remaining variables, including all magnetic quantum numbers. For simplicity, we now assume that the intermolecular potential does not act upon vibration. Thus, in Eq. (V.20), the vibrational quantum numbers contained in i and i' are identical as are those included in f and f', but the former may differ from the latter (in the case of a vibrational band). By choosing the Z axis along the separation between the two colliding molecules, the anisotropic interaction $V_{aniso}(R, \Omega_a, \Omega_p)$ depends on the orientation of the molecules (denoted respectively by Ω_a for the radiator and Ω_p for the perturber) and on the distance R between their centers of mass (which is considered as a parameter, since the translational motion is treated classically). In the Hilbert space associated with the internal degrees of freedom, one can introduce the eigenvectors $|\varsigma\rangle$ and eigenvalues $G_\varsigma(R)$ of $V_{aniso}(R, \Omega_a, \Omega_p)$, such that:

$$V_{aniso}(R, \Omega_a, \Omega_p)|\varsigma\rangle = G_\varsigma(R)|\varsigma\rangle. \quad (V.21)$$

[(3)] Since the three operators in H_0 [Eq. (V.12)] commute with each other, the corresponding density operator ρ_0 can be expressed as the product of three components: $\rho_0 = e^{-H_0/k_B T}/\text{Tr}[e^{-H_0/k_B T}] = \rho_a \rho_p \rho_{iso}$ with $\rho_x = e^{-H_x/k_B T}/\text{Tr}[e^{-H_x/k_B T}]$.

Then, by performing the Laplace transform of the time correlation function through Eq. (V.10) and carrying out a classical average over R, the symmetric spectral density is:

$$\widetilde{F}(\omega) = \frac{1}{\pi} \sum_{i,f} \sum_{i',f'} \widetilde{\chi}_{f'i',fi}[\omega - (\omega_{fi} + \omega_{f'i'})/2]. \qquad (V.22)$$

As an example, consider a diatomic molecule, so that $i \equiv v_i J_i$ and $f \equiv v_f J_f$ for the positive resonances ($\omega_{fi} \geq 0$). In order to simplify notations, we designate the line $v_f J_f \leftarrow v_i J_i$ by the symbol ℓ_+, and the corresponding negative resonance $v_i J_i \leftarrow v_f J_f$ by ℓ_-. Equation (V.22) can then be written as:

$$\widetilde{F}(\omega) = \frac{1}{\pi} \left\{ \sum_{\ell_+} \sum_{\ell'} \widetilde{\chi}_{\ell',\ell_+}[\omega - (\omega_{\ell'} + \omega_{\ell_+})/2] + \sum_{\ell_-} \sum_{\ell'} \widetilde{\chi}_{\ell',\ell_-}[\omega - (\omega_{\ell'} + \omega_{\ell_-})/2] \right\}. \qquad (V.23)$$

For a purely rotational band ($v_i = v_f = 0$), both sums over ℓ' in Eq. (V.23) include ℓ'_+ and ℓ'_- lines. For a vibrational band ($v_f > v_i$), since the spectral effects of couplings between negative and positive resonances are negligible due to their large spectral separation, the sum over ℓ' in the first term of Eq. (V.23) can be restricted to ℓ'_+ while that in the second term is limited to ℓ'_-. Note that the $\widetilde{\chi}$ quantities in Eq. (V.23) are *different* from those χ of § V.2.1, as discussed in § V.3.3.

V.3.2 PRACTICAL IMPLEMENTATION AND TYPICAL RESULTS

Let us briefly examine how practical computations of the spectral shape can be handled (see Ref. 656 for more details). Starting from Eq. (V.20), introducing the eigenvectors $|\varsigma\rangle$ of V_{aniso} and the eigenfunctions $|p\rangle$ of the perturber, and using Eq. (V.21), the explicit expression for $\widetilde{\chi}_{f'i',fi}(\omega)$ is:

$$\widetilde{\chi}_{f'i',fi}(\omega) = \sum_{\varsigma,\varsigma'} \widetilde{H}_{\varsigma\varsigma'}(\omega) \sum_{p,p'} \langle\varsigma'|fp\rangle\langle fp|\rho_a^{1/4} d^+ \rho_a^{1/4} \sqrt{\rho_p}|ip\rangle\langle ip|\varsigma\rangle \qquad (V.24)$$
$$\times \langle\varsigma|i'p'\rangle\langle i'p'|\sqrt{\rho_p}\rho_a^{1/4} d \rho_a^{1/4}|f'p'\rangle\langle f'p'|\varsigma\rangle.$$

In this expression, the sum over all magnetic quantum numbers is implicit, and $\widetilde{H}_{\varsigma\varsigma'}(\omega)$ is given by:

$$\widetilde{H}_{\varsigma\varsigma'}(\omega) = \widetilde{v} n_p 4\pi R_c^2 \frac{1}{|G'_{\varsigma\varsigma'}(R_c)|} e^{-V_{iso}(R_c)/k_B T} e^{-[G_{\varsigma}(R_c) - G_{\varsigma'}(R_c)]/2k_B T}, \qquad (V.25)$$

where

$$G'_{\varsigma\varsigma'}(R) = \frac{d}{dR}[G_{\varsigma}(R) - G_{\varsigma'}(R)], \qquad (V.26)$$

and R_c stands for the root of the following equation of energy conservation:

$$\hbar\omega = G_\zeta(R_c) - G_{\zeta'}(R_c). \tag{V.27}$$

In order to carry out computations, one first has to choose a complete set of basis functions in the Hilbert space. The natural choice, made for instance by Rosenkranz,[647,648] is to select the unperturbed states of the radiator + perturber molecular system, ie the $|J_a m_a J_p m_p\rangle$'s for linear molecules. Then, by constructing the matrix representing V_{aniso} in this basis and diagonalizing it, one gets its eigenvalues and eigenvectors [Eq. (V.21)]. In practice, such an approach leads to difficulties, due to computer limitations resulting from both the calculation time and memory required. Indeed, for most systems, the number and sizes of the matrices to be diagonalized are huge. One thus has to limit the number of included radiator and perturber states, hoping to still get converged values. Unfortunately, the convergence is slow[653] and accurate results cannot generally be obtained with reasonable computational efforts. Ma and Tipping have therefore proposed[653] an alternative approach based on the *coordinate representation*. Remember that the orientations of the radiator and perturber can be represented by the unit vectors $\vec{\Omega}_a$ and $\vec{\Omega}_p$, respectively. Thus, the orientation of the system a + p can be described by $\vec{\Omega}_a(\theta_a, \varphi_a) \otimes \vec{\Omega}_p(\theta_p, \varphi_p)$, which is an operator, the eigenfunctions of which are nothing but the direct product $|\delta(\vec{\Omega}_a - \vec{\Omega}_{a\varsigma})\rangle \otimes |\delta(\vec{\Omega}_p - \vec{\Omega}_{p\varsigma})\rangle$ where $\vec{\Omega}_{a\varsigma}$ and $\vec{\Omega}_{p\varsigma}$ are *specified* values.[51] These Dirac δ-functions are not only the eigenfunctions of $\vec{\Omega}_a(\theta_a, \varphi_a) \otimes \vec{\Omega}_p(\theta_p, \varphi_p)$ but also of any operator which does not contain any differential operator and depends on the orientations only. It is obvious that the anisotropic potential and the dipole moment belong to this category. Therefore, choosing:

$$|\varsigma\rangle = \left|\delta\left(\vec{\Omega}_a - \vec{\Omega}_{a\varsigma}\right)\right\rangle \otimes \left|\delta\left(\vec{\Omega}_p - \vec{\Omega}_{p\varsigma}\right)\right\rangle, \tag{V.28}$$

avoids highly time-consuming and memory-demanding diagonalizations of large matrices since V_{aniso} is *diagonal* in this coordinate representation. The main computational task is then transformed to the carrying out of multidimensional integrations over the continuous orientational variables (see Refs. 653,656 for more details). The various $\widetilde{\chi}_{f'i',fi}[\omega - (\omega_{fi} + \omega_{f'i'})/2]$ functions can be calculated from a given intermolecular potential, provided that the latter is not too complicated (ie not consisting of too many terms and parameters) so that finding the roots of Eq. (V.27) is not too costly. At this stage, a difficulty arises since it has been shown[656] that, contrary to the spectral region close to line centers, the far wings are, in this formulation, very sensitive to the parts of the interaction potential energy surface (PES) showing large angular gradients while the potential values themselves are small or negative.[655] Adjustments thus remain necessary, and they can be done starting from some realistic PES which is then optimized using fits of measured absorption data. One then has to check that this empirical potential leads to satisfactory predictions for other physical properties of the considered molecular system (virial and pressure-broadening coefficients, differential cross sections, see Ref. 656, for instance).

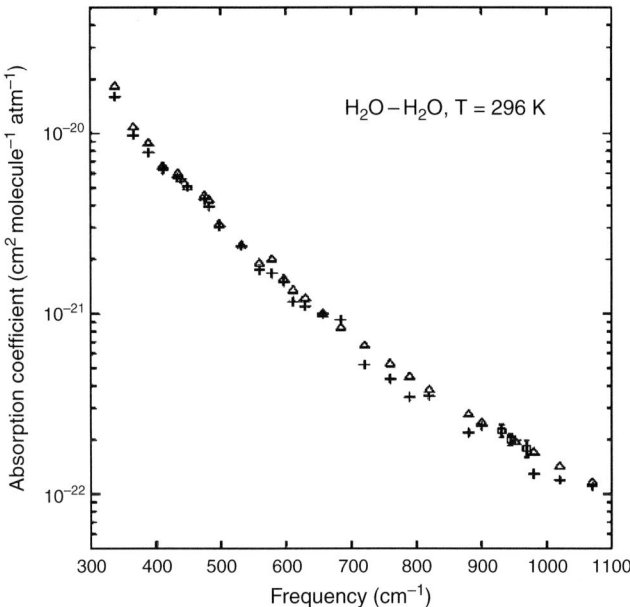

Fig. V.3: Pure water vapor density–normalized absorption coefficient (in $cm^2 molecule^{-1} atm^{-1}$) at 296 K in the 300–1100 cm^{-1} region. The calculated values are represented by (Δ). The experimental values are denoted by (+)[658] and ($\square$).[659] Reproduced with permission from Ref. 656.

The quasistatic approach presented above has been applied to calculations for H_2O and CO_2 absorption lines. These are two species whose far wings are of considerable importance for atmospheric applications (*cf* App. V.A and § VII.5.1). As an illustration of the quality of predictions, Fig. V.3 presents a comparison between experimental data and calculated absorption coefficients for pure H_2O at 296 K. In the considered 300–1100 cm^{-1} spectral region, the absorption mainly results from the high frequency wings of the pure rotational lines. This figure shows that the absorption is very satisfactorily modeled over a spectral range in which it varies by more than two orders of magnitude. A similar quality of predictions is also obtained at other temperatures (Fig. V.A4) and for H_2O broadened by N_2.[655]

Ma and Tipping have also developed a somewhat different formalism to generate the water vapor millimeter wave foreign continuum.[645,657] In this model, the active system is a pair of interacting molecules, and the bath corresponds to the other molecules. Neglecting any correlation between the active system and the bath, it is then possible to write the spectral density as a continued fraction, thanks to the Lanczos algorithm.[657] As discussed in Ref. 657, such an approach enables a direct calculation of the continuum even if it cannot reproduce the sharp features associated with "local lines". The experimental and theoretical results again agree quite well, as shown in Refs. 657,660 and Fig. V.4. Furthermore, the theoretical analysis has enabled to understand some of the limits of some widely used empirical models.[642,643] For instance, it has been shown[645] that

The far wings 255

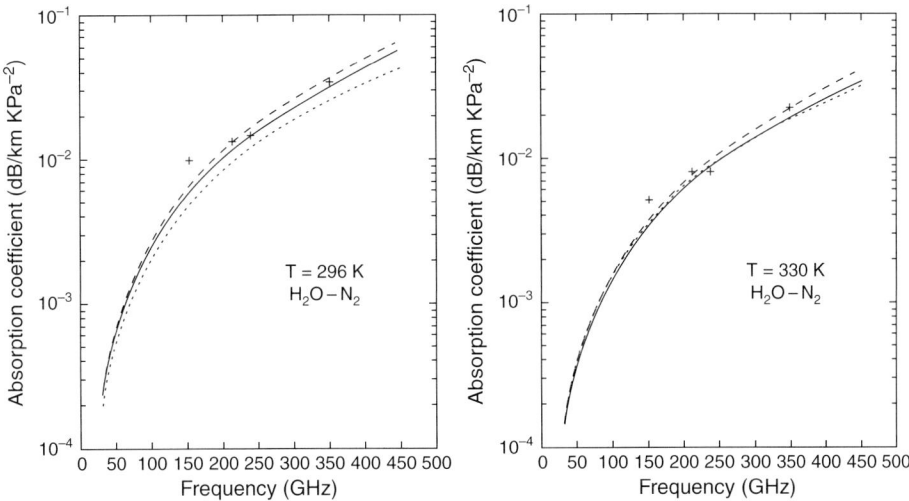

Fig. V.4: The H_2O-N_2 millimeter wave continua (in units dB/km kPa^{-2}) at 296 K (left) and 330 K (right). The calculated values are represented by the solid line. Those deduced from the measurements of Ref. 640 are represented by (+). The predictions of the MPM89[642] and MPM93[643] models are represented by the dotted and dashed lines, respectively. Reproduced with permission from Ref. 657.

the $1/T^3$ temperature law assumed in the MPM89 approach[642] wrongly ignores the T-dependence of the spectral density.

CO_2 is another species whose absorption in the far wings is important for applications including modeling of radiative heat transfer in the atmospheres of the Earth and Venus or in combustion media (*cf* § VII.5.1). Figures V.5 show that the first principles results obtained[654] from the Ma-Tipping formalism are in excellent agreement with experimental data over a wide range of temperature.

V.3.3 THE BAND AVERAGE LINE SHAPE: BACK TO THE χ FACTORS

For molecules of practical interest, there are many individual lines within each (vibrational) band and many bands in the spectrum (*eg* Fig. V.A1). In order to avoid the treatment of each of them, additional approximations have been proposed.[656] We briefly examine one of them which clarifies the physical meaning of the line shape factors χ introduced in § V.2.1. For simplicity, consider a vibrational band so that couplings between positive and negative resonances are negligible. Equation (V.19) can then be written as:

$$\widetilde{F}(\omega) = \frac{1}{\pi} \left\{ \sum_{\ell_+,\ell'_+} \widetilde{\chi}_{\ell'_+,\ell_+}[\omega - (\omega_{\ell_+} + \omega_{\ell'_+})/2] + \sum_{\ell_-,\ell'_-} \widetilde{\chi}_{\ell'_-,\ell_-}[\omega - (\omega_{\ell_-} + \omega_{\ell'_-})/2] \right\}. \quad (V.29)$$

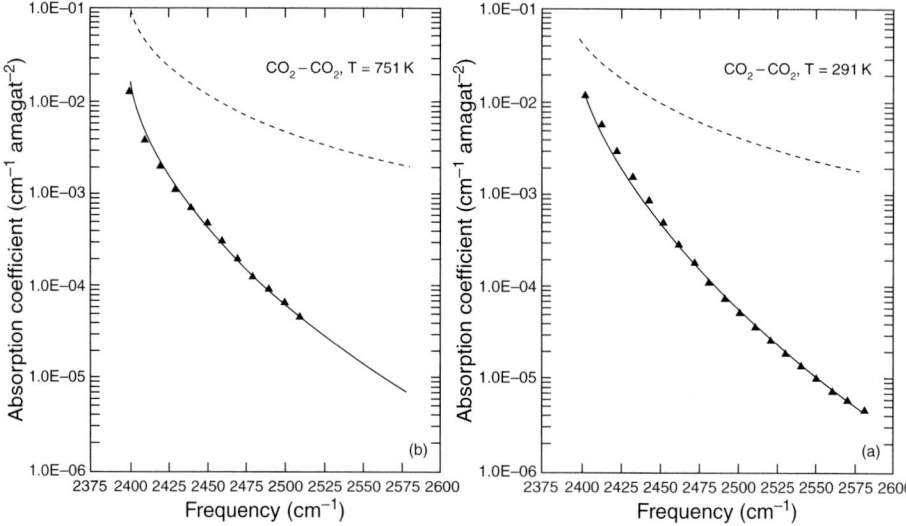

Fig. V.5: Absorption coefficients in the 2400–2580 cm^{-1} spectral region for pure CO_2. ▲ indicate experimental data.[357,358] The lines correspond to results calculated with a purely Lorentzian model (- - -) and with the quasistatic approach (——). Left: T = 751 K, right: T = 291 K. Reproduced with permission from Ref. 654.

The corresponding expression, based on the commonly used line-shape correction function χ of § V.2.1, can be derived from Eq. (V.4), leading to:

$$\widetilde{F}(\omega) = \frac{1}{\pi} \left\{ \sum_{\ell_+} \frac{\rho_{\ell_+} d_{\ell_+}^2 \Gamma_{\ell_+}}{(\omega - \omega_{\ell_+})^2} \chi(\omega - \omega_{\ell_+}) + \sum_{\ell_-} \frac{\rho_{\ell_-} d_{\ell_-}^2 \Gamma_{\ell_-}}{(\omega - \omega_{\ell_-})^2} \chi(\omega - \omega_{\ell_-}) \right\}, \qquad (V.30)$$

where ρ_ℓ, d_ℓ, and Γ_ℓ are respectively the initial level population, the reduced matrix element of the radiation-matter coupling tensor, and the collision-broadened half width (HWMH) of line ℓ. In order to identify the expressions of $\chi(\omega)$, one can perform the inverse Laplace transforms of Eqs. (V.29) and (V.30), and compare the results (in the time-domain), in the short-time limit. This corresponds to considering the very far wings, for distances much larger than the separation between transitions, leading to:

$$\chi(\omega) = \omega^2 \frac{1}{\sum_\ell \rho_\ell d_\ell^2 \Gamma_\ell} \sum_{\ell',\ell} \widetilde{\chi}_{\ell'\ell}\left(\omega - \frac{\omega_\ell + \omega_{\ell'}}{2} + \bar{\omega}\right), \qquad (V.31)$$

where $\bar{\omega}$ is some (gravity center) average frequency. This *band average* line-shape function may result in significant differences when compared to the exact calculation [Eq. (V.29)]. Nevertheless, the strongly coupled lines are generally close to each other and it may be reasonably expected that, for frequencies of interest far away from the band center, such an approximation becomes reasonable. However, its main interest is to give a physical significance to the empirical approach described in § V.2.1.

The quasistatic model is a powerful tool for the prediction of the spectral profile in the far wings of vibrational bands. Complementarily, impact approaches described in chapters III and IV are suitable for regions close to the line centers. This leaves the problem of the intermediate region unsolved so that the question "how can one calculate the *entire* spectrum ?" (*ie* from the band center, in the vicinity of the most intense optical transitions, up to the far wings) remains opened. Answers rely on the determination of the frequency dependence of the relaxation matrix over the entire frequency domain, keeping in mind that any general calculation of $W(\omega)$ must yield the impact limit values near the line centers (which is not the case for the quasistatic model described above). In the following, some advances toward such an unified treatment are described.

V.4 FROM RESONANCE TO THE FAR WING: A PERTURBATIVE TREATMENT

V.4.1 GENERAL EXPRESSIONS

The formalism developed by Borysow and Moraldi[661–664] assumes that the anisotropic part of the intermolecular potential is sufficiently weak to allow a *second order* expansion of the evolution operators appearing in the expression of the spectral density [see, for instance Eq. (II.42)]. Such a model can be applied to allowed lines but also to collision-induced spectra. However, we only consider here the allowed spectrum due to the permanent tensor coupling the active radiator to the electromagnetic field. The purely collision-induced part and the allowed-induced cross contribution are examined in chapter VI. Moreover, instead of treating the translational relative motion classically (as done in the preceding quasistatic model), Borysow and Moraldi use a quantum approach in which translation is driven by the isotropic potential only.

Recall that the spectral density for the relaxation of a tensor d is given by [analog of Eq. (II.59) but using the notations of Refs. 661,663 for consistency]:

$$F(\omega) = \frac{1}{\pi} \text{Re} \left\{ \sum_{i,f} \sum_{i',f'} \rho(J_i) d_{if} \left[i(\omega I_d - L_a) + n_p W(\omega) \right]^{-1}_{if,i'f'} d_{i'f'} \right\}, \quad (V.32)$$

where n_p is the density of perturbers and W is the density-normalized relaxation matrix. d_{if} are the matrix elements of the irreducible tensor (of rank k, with k = 1 for dipolar absorption, k = 0 or 2 for isotropic or anisotropic Raman scattering) coupling the radiator and the electromagnetic field in the line space basis, $\rho(J_i)$ is the relative population of the radiator rotational level J_i. L_a is the Liouville operator of the free active molecule which leads, in the basis used, to the diagonal matrix of the free rotational transitions frequencies ω_{fi} [Eq. (II.49)]. Starting from the results of Ref. 661, Bonamy has proposed[665] a reformulation of the relaxation matrix elements. This is done, in the "spirit" of the formalism proposed by Kouzov,[612] using the usual[40,438] classification in outer and middle terms, leading to:

$$n_p W_{if,i'f'}(\omega) = \delta_{i',i}\delta_{f',f} n_p <v_r> \Gamma^{outer}_{if}(\omega) - n_p <v_r> \Gamma^{(k)middle}_{if,i'f'}(\omega), \quad (V.33)$$

where $<v_r>$ is the mean relative speed and Γ^{outer} and Γ^{middle} have dimensions of cross sections. The outer term can be written as:

$$\Gamma^{outer}_{if} = \sum_{L_1,J'_f} (2J'_f + 1) \begin{pmatrix} J_f & L_1 & J'_f \\ 0 & 0 & 0 \end{pmatrix}^2 \widetilde{\Phi}_{L_1}(\omega - \omega_{J'_f J_i})$$

$$+ \sum_{L_1,J'_i} (2J'_i + 1) \begin{pmatrix} J_i & L_1 & J'_i \\ 0 & 0 & 0 \end{pmatrix}^2 \exp(\hbar\omega_{J'_i J_i}/k_B T)\widetilde{\Phi}_{L_1}(\omega - \omega_{J_f J'_i}), \quad (V.34)$$

where (:::) is a 3J symbol.[65] The spectral functions $\widetilde{\Phi}_L$ are thermal averages, over the perturber states J_p, of the correlation functions $\widetilde{\Phi}_{LL'}$, ie:

$$\widetilde{\Phi}_{L_1}(\Delta\omega) = \sum_{L_2} \sum_{J_p,J'_p} \rho(J_p)(2J'_p + 1) \begin{pmatrix} J_p & L_2 & J'_p \\ 0 & 0 & 0 \end{pmatrix}^2 \widetilde{\Phi}_{L_1 L_2}(\Delta\omega + \omega_{J_p J'_p}), \quad (V.35)$$

which are, themselves, averages of the cross sections over the translational degrees of freedom through the relative kinetic energy E_r:

$$\widetilde{\Phi}_{L_1 L_2}(\Delta\omega') = (k_B T)^2 \int_0^\infty E_r e^{-E_r/k_B T} \sigma_{L_1 L_2}(E_r | E_r + \hbar\Delta\omega') dE_r. \quad (V.36)$$

The middle term can also be written in terms of the same spectral functions, leading to:

$$\Gamma^{(k)middle}_{if,i'f'}(\omega) = \sum_{L_1} F^{kL_1}_{if,i'f'} \left[\widetilde{\Phi}_{L_1}(\omega - \omega_{J'_f J_i}) + \exp(\beta\hbar\omega_{J_i J'_i})\widetilde{\Phi}_{L_1}(\omega - \omega_{J_f J'_i}) \right], \quad (V.37)$$

where the symmetrized Percival-Seaton coefficient $F^{kL_1}_{if,i'f'}$ which contains *all* selection rules (collisional as well as optical ones), is given by:[666]

$$F^{kL}_{if,i'f'} = (-)^{L+k} \left[(2J_i + 1)(2J_f + 1)(2J'_i + 1)(2J'_f + 1) \right]^{1/2}$$

$$\times \begin{pmatrix} J_i & L & J'_i \\ 0 & 0 & 0 \end{pmatrix} \begin{pmatrix} J_f & L & J'_f \\ 0 & 0 & 0 \end{pmatrix} \begin{Bmatrix} J_i & J_f & k \\ J'_f & J'_i & L \end{Bmatrix}, \quad (V.38)$$

where {:::} is a 6J symbol.[65] The preceding equations are quite general and do not depend on *any* perturbative treatment. In order to go further, the anisotropic part of the intermolecular potential is assumed sufficiently small to allow a perturbative treatment of the cross sections $\sigma_{L_1L_2}(E|E')$, which are then given by:[661]

$$\sigma_{L_1L_2}(E|E') = \left(\frac{\pi\hbar^2}{2m^*E}\right) \sum_{L,\lambda,\lambda'} (2\lambda+1)(2\lambda'+1) \begin{pmatrix} \lambda & L & \lambda' \\ 0 & 0 & 0 \end{pmatrix} |\langle E\lambda|V_{L_1L_2L}(R)|E'\lambda'\rangle|^2.$$

(V.39)

Recall that the $|E\lambda\rangle$ are the eigenfunctions, which form an orthogonal basis of the Schrödinger equation for the relative translational motion, with energy E and orbital angular momentum λ, in the field of the isotropic potential $V_{000}(R)$. The intermolecular potential has been expanded as functions of the intermolecular vector $\vec{R}$ and orientations Ω_a and Ω_p of the radiator and perturber as:

$$V(\vec{R}, \Omega_a, \Omega_p) = (4\pi^2)^{3/2} \sum_{L,L_1,L_2} V_{L_1L_2L}(R) \sum_{M,M_1,M_2} (-1)^{L_1+L_2+M} \sqrt{2L+1}$$

$$\times \begin{pmatrix} L_1 & L_2 & L \\ M_1 & M_2 & M \end{pmatrix} Y_{L_1}^{M_1}(\Omega_a) Y_{L_2}^{M_2}(\Omega_p) Y_L^M(\vec{R}/R)^*$$

(V.40)

where Y_M^L is a spherical harmonic and * designates the complex conjugate.

V.4.2 ILLUSTRATIVE RESULTS

The preceding perturbative treatment was first applied to calculations of the contribution (with respect to the interaction-induced and cross components considered in Chapt. VI) resulting from the wings of the allowed component to the rototranslational Raman scattering in H_2.[661–663] As illustrated by Fig. V.6, the agreement between predictions and measurements is good in the near wings but gets poorer when going away from the transition centers. A possible explanation for this discrepancy is the important contribution of close range collisions to the wings profiles. As is well known, these are very sensitive to the anisotropic potential at short distances which is generally more poorly known. Furthermore, for such energetic collisions, a perturbative expansion may not be appropriate.

A second application of the perturbative treatment is an analysis, in Ref. 664, of the high-frequency wings of the depolarized rototranslational Raman spectra of gaseous nitrogen. Before this work, it was generally assumed[667,668] that the measured intensity was almost entirely due to the collision-induced light scattering contribution. The wrongness of this hypothesis was demonstrated since the calculated intensities in the wings, using known values of the molecular polarizabilities of N_2, are then *two orders of magnitude smaller* than the measured ones.[664,668] Using the perturbative treatment of § V.4.1 and the intermolecular potential of Ref. 669, it was shown[664] that most of the intensity at large distances (beyond 400 cm^{-1}) is due to the wings of the pressure-broadened (allowed) Raman lines, as illustrated by Fig. V.7. Here again, remaining discrepancies between measured and calculated values likely result from the perturbative treatment tied to the assumed weakness of the anisotropic part of the potential. This demonstrates the need for non-perturbative approaches which are discussed in the next section and in Sec. VIII.3.

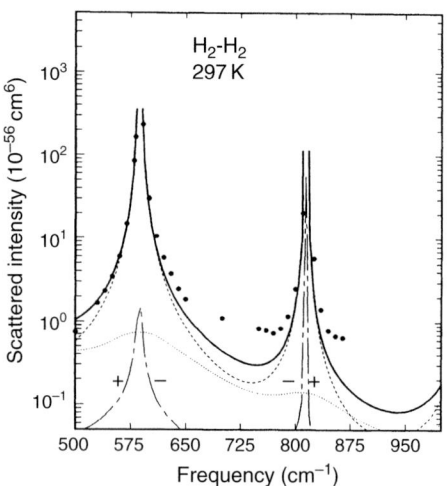

Fig. V.6: Depolarized light scattering intensity for pure H_2 at 297 K between 500 and 1000 cm^{-1}. (•) denote experimental results.[80] The total theoretical spectrum (___) is composed of the interaction-induced spectrum (....., see chapter VI), the pressure-broadened allowed spectrum (- - -), related to Eq. (V.32) through Eq. (5) of Ref. 664; and the cross term (__ _ __, cf chapter VI) whose absolute value is plotted, its sign being indicated by $+$ and $-$. Reproduced with permission from Ref. 661.

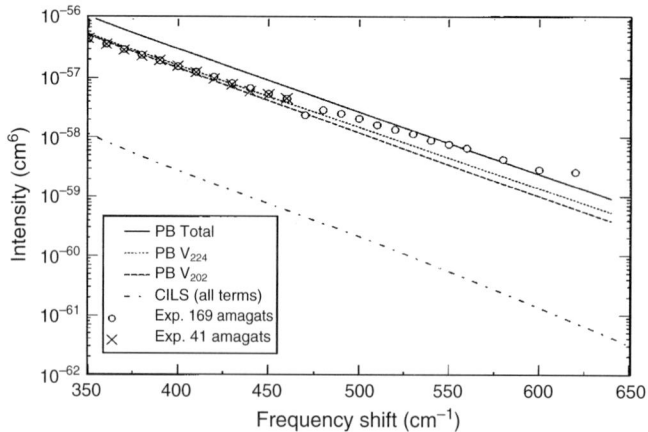

Fig. V.7: The density-normalized far wing of the depolarized Raman spectrum of pure N_2 at room temperature. (_ . _) is the collision-induced contribution while (___) is the total pressure-broadened term. The two other curves give partial contributions from specific components of the potential [cf Eq. (V.40)]. (○) and (×) are experimental results.[671] Reproduced with permission from Ref. 664.

V.5 FROM RESONANCE TO THE FAR WING: A NON-PERTURBATIVE TREATMENT

Starting from the Zwanzig-Mori formalism,[54] Kouzov has proposed[612] a non-perturbative treatment of the frequency-dependent relaxation matrix incorporating the effects of incomplete collisions and of initial correlations. It is based on a non-Markovian extension of the Energy Corrected Sudden (ECS) approximation, primarily developed within the impact limit (cf § IV.3.3b). It provides an unified treatment for the entire frequency range, from the transition centers to the far wings. Starting from Kouzov's work, Bonamy and Buldyreva[666,672] have developed a tractable approach for the modeling of the correlation function which reduces to the traditional ECS formalism in the impact limit. Note that, consistently with the use of this model, it is implicitly assumed that the radiator is a linear molecule and that the perturber is (or can be considered as) an atom, so that the interaction potential can be expanded on Legendre polynomials P_L, ie:

$$V(\vec{R}, \Omega_a) = \sum_L V_L(R) P_L[\cos\theta]. \tag{V.41}$$

V.5.1 GENERAL EXPRESSION

Starting from Eq. (V.32), but written slightly differently, since Kouzov uses[612,673] a symmetrized form of the scalar product, the spectral density is given by:

$$F(\omega) = \frac{1}{\pi} \text{Re} \left\{ \sum_{if} \sum_{i'f'} \tilde{d}_{if}^{(k)} [i(\omega I_d - L_a) + n_p \tilde{\Gamma}^{(k)}(\omega)]_{if,i'f'}^{-1} \tilde{d}_{i'f'}^{(k)} \right\}, \tag{V.42}$$

with

$$\tilde{d}_{if}^{(k)} = (-1)^{J_i} \sqrt{(2J_i + 1)(2J_f + 1)} \begin{pmatrix} J_i & k & J_f \\ 0 & 0 & 0 \end{pmatrix} N_{if}, \tag{V.43}$$

where N_{if} is given by $N_{if} = [(\tilde{\rho}_i + \tilde{\rho}_f)/2]^{1/2}$, with $\tilde{\rho}_i = \exp(-E_{Ji}/k_B T)/Z_{rot}$, where Z_{rot} is the rotational partition function. Within this metric, the expression of an *off-diagonal* element is:

$$\tilde{\Gamma}_{if,i'f'}^{(k)}(\omega) = -\frac{1 + \exp(-\hbar\omega/k_B T)}{2 N_{if} N_{i'f'}} \sum_L (2L+1) F_{if,i'f'}^{kL} \left[\tilde{\rho}_i \tilde{\Phi}_L(\omega - \omega_{f'i}) + \tilde{\rho}_{i'} \tilde{\Phi}_L^*(\omega - \omega_{fi'}) \right], \tag{V.44}$$

where $F_{if,i'f'}^{kL}$ is given by Eq. (V.38). Note that this equation ensures that the constructed matrix obeys *all* known general relations (within this metric). These are:[612] The symmetry property translating the detailed balance relation:

$$\tilde{\Gamma}_{if,i'f'}^{(k)}(\omega) = \tilde{\Gamma}_{i'f',if}^{(k)}(\omega), \tag{V.45}$$

the time reversal symmetry (*cf* Ref. 45):

$$\widetilde{\Gamma}^{(k)}_{if,i'f'}(\omega) = \widetilde{\Gamma}^{(k)}_{fi,f'i'}(-\omega)^*, \qquad (V.46)$$

and the double sided sum rule:

$$\sum_{i'f'} \widetilde{d}^{(k)}_{i'f'} \widetilde{\Gamma}^{(k)}_{i'f',if}(\omega) = \sum_{i'f'} \widetilde{\Gamma}^{(k)}_{if,i'f'}(\omega) \widetilde{d}^{(k)}_{i'f'} = 0, \; ie,$$

$$\widetilde{d}^{(k)}_{if} \widetilde{\Gamma}^{(k)}_{if,if}(\omega) = -\sum_{i'f' \neq if} \widetilde{d}^{(k)}_{i'f'} \widetilde{\Gamma}^{(k)}_{i'f',if}(\omega) = -\sum_{i'f' \neq if} \widetilde{\Gamma}^{(k)}_{if,i'f'}(\omega) \widetilde{d}^{(k)}_{i'f'}. \qquad (V.47)$$

This last equation enables the calculation of the diagonal elements from the off-diagonal ones, as widely used within conventional impact scaling approaches (§ IV.3.3). The spectral functions $\widetilde{\Phi}_L(\Delta\omega)$ [Eqs. (V.44)], tied to the Legendre components of the radiator-perturber potential [Eq. (V.41)], are the basic elements of this formalism. Note that they satisfy the following Boltzmann relation, analog of Eq. (II.31):

$$\widetilde{\Phi}_L(-\Delta\omega) = \exp(-\hbar\Delta\omega/k_B T)\widetilde{\Phi}_L(\Delta\omega). \qquad (V.48)$$

In order to simplify the problem and reduce the modeling to classical *even* correlation functions (more easy to evaluate from classical dynamics), the quantum asymmetry factor is detached by using the relation:

$$\text{Re}\left\{\widetilde{\Phi}_L(\Delta\omega)\right\} = \frac{2}{1+\exp(-\hbar\Delta\omega/k_B T)} \Phi^{class}_L(\Delta\omega), \qquad (V.49)$$

assuming, moreover, that the imaginary part is negligible. The *crucial* (but *a priori made*) approximation consists in factorizing the classical function into L-dependent and ω-dependent terms, thus assuming that they are independent from each other, *ie*:

$$\Phi^{class}_L(\Delta\omega) = \widetilde{Q}(L) \times \widetilde{\Omega}(\Delta\omega). \qquad (V.50)$$

The case of isotropic Raman ($k=0$) shows that the $\widetilde{Q}(L)$'s are proportional to the basic rate constants Q_L [*cf* Eq. (IV.104)] associated with the collisional relaxation from rotational level $J=L$ to level $J=0$, *ie*:[666]

$$\widetilde{Q}(L) = (1+\rho_L/\rho_0)\frac{Q_L}{\Omega_L}. \qquad (V.51)$$

In this expression, Ω_L is the impact (or Markovian) adiabaticity factor (*cf* Eq. (IV.116) and Ref. 403):

$$\Omega_L = [1+(\omega_{L,L_{Next}}\tau_c)^2/24]^{-2} \approx [1+(\omega_{L,L_{Next}}\tau_c)^2/12]^{-1}, \qquad (V.52)$$

where L_{Next} is the level next and lower than L to which level $J=L$ is collisionally coupled (*eg* $L_{Next}=L-2$ for CO_2). Recall that the efficient collision duration $\tau_c = \ell_c/<v_r>$ is a fitted parameter, through the interaction length ℓ_c, and that the Q_L's are often modeled by analytical laws (see Table IV.1). Within the three temperature- and perturber-dependent parameters (A, α, and β) Exponential Power law, they are given by:

$$Q_L = \frac{A}{[L(L+1)]^\alpha} \exp[-\beta L(L+1)]. \qquad (V.53)$$

Various models have been proposed[672] for the generalized adiabaticity factor $\widetilde{\Omega}(\Delta\omega)$ in Eq. (V.50), which contains most of the influence of the finite duration of collisions. Due to the exponential decay observed[668,671] in the far wings of N_2 (Fig. V.7), Bonamy and Buldyreva have proposed[666] to use the classical model introduced by Birnbaum and Cohen[674] since it leads to such a behavior. The corresponding expression for $\widetilde{\Omega}(\Delta\omega)$ is then:[666]

$$\widetilde{\Omega}(\Delta\omega) = \frac{aK_1(z)}{zK_1(a)}, \qquad (V.54)$$

where $K_1(z)$ is a modified Bessel function of second kind,[182] a and z being defined in terms of two characteristic times τ_1 and τ_2 by:

$$a = \tau_2/\tau_1 \text{ and } z = a(1 + \Delta\omega^2 \tau_1^2)^{1/2}. \qquad (V.55)$$

Note that the identification with the traditional DePristo factor [Eq. (V.52)] for small $|\Delta\omega|$ detunings requires $\tau_1 = \tau_c/\sqrt{12}$.[666]

Starting from the preceding relations, applications of the non-Markovian ECS approach consist in the following steps. The first is the determination of the ECS "impact" parameters A, α, β, and τ_c (ie τ_1) (eventually through $\ell_c = <v_r> \tau_c$). This is generally done from fits of measured impact values (line-broadening coefficients, and/or Q branch profiles, near wings, etc) as discussed in § IV.3.3. The remaining unknown parameter a (ie τ_2) is then determined by adjusting theoretical spectra in the far wings to experimental values. Note that the Birnbaum-Cohen model[674] enables an estimation of this parameter from the intermolecular potential, and thus also a check that the value deduced from the fits is acceptable.

V.5.2 ILLUSTRATIVE RESULTS

The generalized ECS model has been first applied[666] to the analysis of the N_2 depolarized Raman spectra measured by LeDuff and Teboul.[671] Typical results, obtained at room temperature for different values of the parameter a, are shown in Fig. V.8. As expected, the relatively close wing (below 300 cm^{-1}), which is dominated by contributions of nearby transitions, is practically independent of a and does not appreciably differ from the impact regime. The fit of the far wing then leads to a = 0.84, corresponding to $\tau_2 = 0.044$ ps, since $\tau_1 = 0.052$ ps. The fact that this value compares quite well with the theoretical prediction of a = 1.01 obtained from the Van der Avoird potential[669] is a first element in favor of the model. This consistency is confirmed by results obtained at 150 K

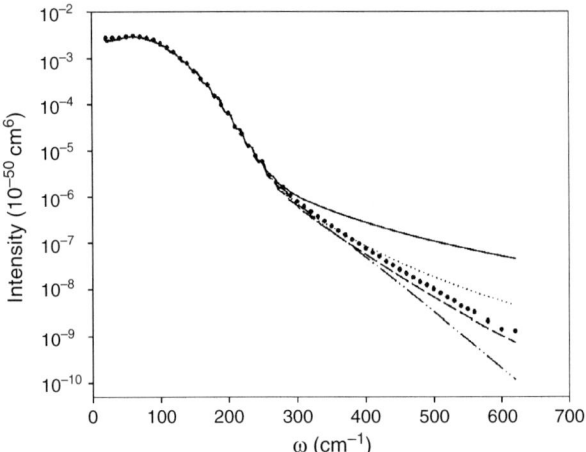

Fig. V.8: The depolarized Raman spectrum of pure N_2 for T = 295 K and a density of 169 Am The experimental results[671] are represented by (●). The lines give the theoretical values with: (___) impact theory ; present theory with (....) a = 0, (_ _ _) a = 0.8, and (_ .. _) a = 2. Reproduced with permission from Ref. 666.

where the fitted value of a = 0.8 again compares well with a = 0.89 derived from the intermolecular potential. [666]

The model was then used[675] for calculation of the anisotropic Raman spectrum of the ν_1 band of pure CO_2, including the far Stokes wing up to about 500 cm^{-1}. Figure V.9 shows that good agreement with measurements is obtained for moderate distances in the wing,

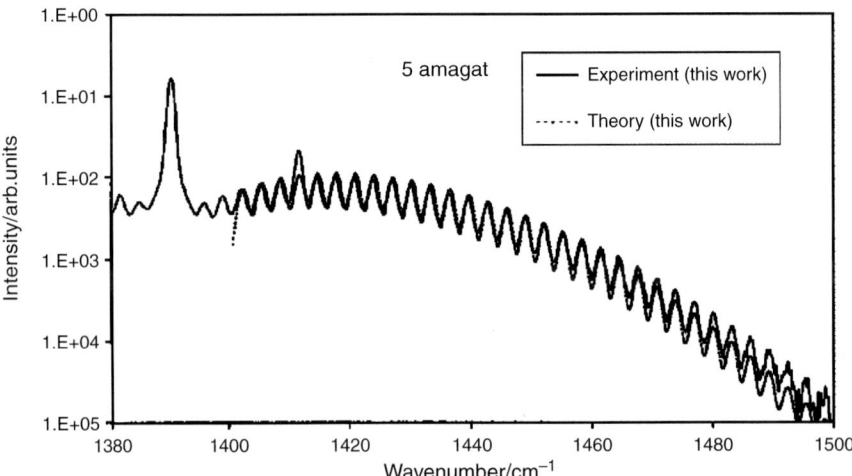

Fig. V.9: Anisotropic Raman scattering spectrum of the ν_1 band (Q and S branches) of pure CO_2 at room temperature and a density of 5 amagat. (___) are measured values, while (....) are the theoretical results. Reproduced with permission from Ref. 675.

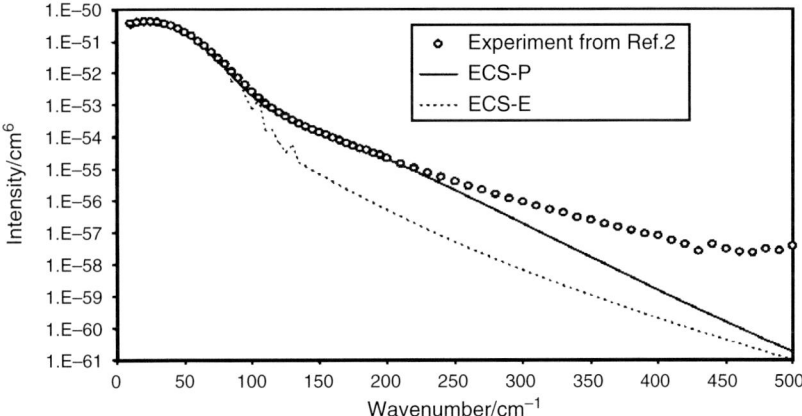

Fig. V.10: Anisotropic Raman scattering spectrum of the v_1 band of pure CO_2 at room temperature vs the frequency shift. ○ are experimental results[676]. The ECS-E ($\alpha = 0$ is imposed) and the ECS-P ($\beta = 0$ is imposed) models have been used, based on Eq. (V.53). Reproduced with permission from Ref. 675.

indicating that the impact regime is well reproduced. This could be expected since the quality of the usual (impact) ECS approach has been widely demonstrated for infrared and Raman spectra of CO_2 (§ IV.3.3b and references in table IV.9).

On the contrary, as shown in Fig. V.10, the calculated profile in the far wing is much weaker than the observed one when reasonable values of the ECS parameters are used (*ie* forbidding unrealistic values in the fits).

Two possible explanations for this discrepancy have been proposed.[675] The first one is an eventual inadequacy of the adiabaticity factor of Eq. (V.54). The second one is the underlying linear molecule-atom character of the ECS approach used, which neglects the internal degrees of freedom of the perturber. Such an approximation may be more questionable for pure CO_2 than for pure N_2, since CO_2-CO_2 collisions involve more anisotropic forces through the strong quadrupole-quadrupole interaction.

Let us mention that a second order perturbative approach, taking into account the internal degrees of freedom of the perturber (through a classical calculation of the correlation function), leads[665] to better agreement with the pure CO_2 Raman measurements of Fig. V.10 than the generalized ECS model. This corroborates the need to properly take into account the fact that the perturber is a molecule. A challenging extension of the models presented in this section would therefore be, following the study of Green[433] made within the impact limit, to generalize the non-Markovian ECS approach of Ref. 612 to pairs of linear molecules.

V.6 CONCLUSION

The understanding and modeling of (allowed) far wing profiles have greatly progressed in the last two decades. This is, in particular, due to the development of the quasistatic approach which removed (*cf* App. V.A) a large part of the ambiguity on the origin of the water vapor continua. Up to now, the Ma-Tipping formalism (Sec. V.3) has been essentially applied to H_2O, while the perturbative approach of Moraldi and Borysow (Sec. V.4) or the generalized ECS model of Kouzov, Bonamy and Buldyreva (Sec. V.5) were mainly applied to N_2. A more systematic comparison of these various models for all the available data for the far wings of various molecular systems would lead to a deeper insight into the (sometimes) different approximations made in these models. This may provide a path to improve the results obtained for the CO_2 far wing (*cf* Sec. V.5). It would also be helpful for the research of a formalism allowing a calculation of the entire profile, from resonances to the far wings. As seen in chapters III and IV, accurate and tractable models exist for the spectral shapes near the centers of the optical transitions. When put together with the quasistatic approach, this leaves the problem of the mid-wings opened. The accurate modeling of this intermediate region requires either a junction between impact and quasistatic models or the use of a self-consistent approach from resonances to the far wings. A first solution has been discussed in Sec. V.5 but it is empirical and based on parameters which must be adjusted on measured spectra. Other paths are presented in Sec. VIII.3.

Finally, we would like to stress the need for more accurate and broader scope experimental data (laboratory and atmospheric), including necessarily the investigation of the temperature dependence of the profile, for molecular systems which may not be of direct atmospheric importance. As an example, H_2O-Ar and H_2O-He are good candidates for improving the theory and experimental H_2O-rare gas continua are clearly needed. Similarly, a systematic investigation of the very far wing profiles for more "simple" radiators (HCl, HF, CO, *etc*) would be of interest, particularly in mixtures with rare gases. The situation is rather similar in Raman scattering where, as an example, a challenging continuation of the work[675] on CO_2 would be to try to obtain data for mixtures with rare gases. Such simple molecule-atom systems are "theoretically important" since they allow to avoid some hardly justified assumptions like the neglect of the perturber internal degrees inherent to the ECS approaches. Of course, new data for molecular pairs of atmospheric or planetary importance are also needed. Some of them are discussed at the end of appendix V.A.

APPENDIX V.A: THE WATER VAPOR CONTINUUM

Most planetary atmospheres have spectral regions (*windows*) of relative transparency, both in the visible, where the incoming sunlight penetrates, and in the infrared, where the emitted radiation at longer wavelengths outgoes. In the Earth, these windows are mainly determined by H_2O, CO_2, and O_2. The H_2O and CO_2 absorption bands define the atmospheric transparency regions in the infrared, while those in the millimeter wave region are due to H_2O and O_2.[677,678] In this appendix, we concentrate on the water vapor spectrum. As shown by Fig V.A1, the IR and visible windows are framed by the low and high frequency wings of the dominant H_2O vibrational bands and do not contain many strong lines. For instance, the 800–1250 cm^{-1} interval (also called the 8–12 µm window) mostly results from the positive wing of pure rotational transitions centered below 500 cm^{-1} and the negative wing of the ν_2 band centered near 1600 cm^{-1}. As discussed in chapter VII, an accurate description of these windows is essential for applications. Apart

The far wings

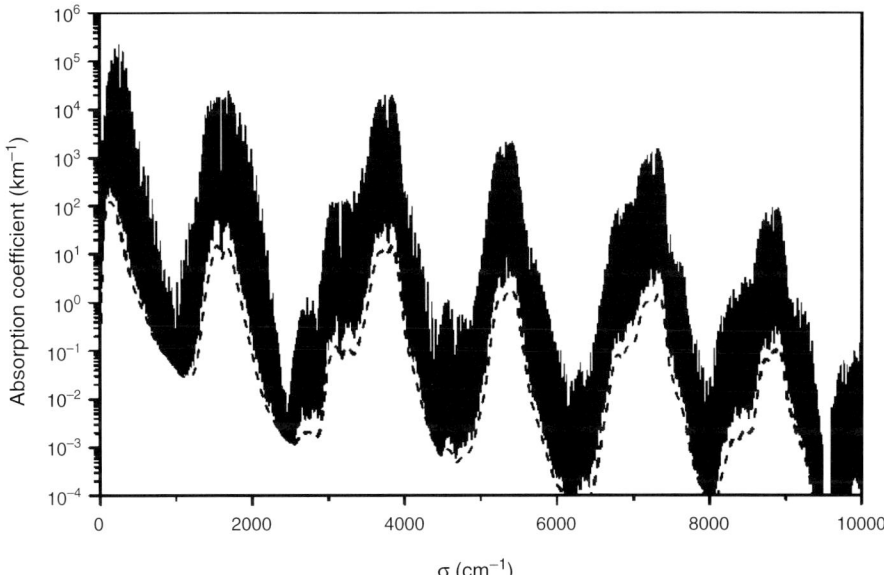

Fig. V.A1: Calculated absorption for 1% H_2O in air at 296 K and 1 atm. (___) is the total absorption, (- - -) is the continuum for a cut-off distance of 25 cm^{-1}. (cf § V.2.2)

from these regions, the in-band absorption is also of importance for satellite-borne remote sensing determinations of atmospheric humidity profiles, and retrievals again require an accurate modeling of the background absorption (eg Refs. 631,679 and § VII.5.1).

The situation in the millimeter and far infrared regions is similar. Due to the large number of strong transitions, most atmospheric observations are then limited to a few relatively transparent intervals in between the lines, as shown in Fig. V.A2. Again, the

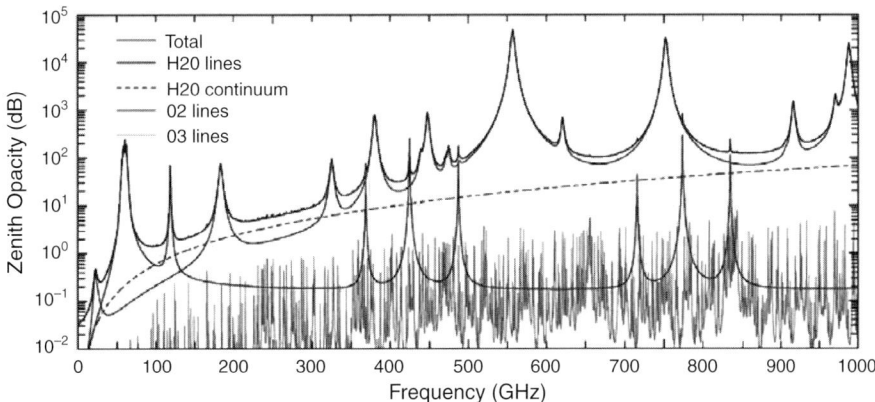

Fig. V.A2: Clear sky calculated atmospheric opacity at nadir (cf Fig. VII.2a) for a mean standard atmosphere. Reproduced with permission from Ref. 681.

"background"[680,681] carrying local absorption features must be properly modeled for accurate remote sensing.

V.A1 DEFINITION, PROPERTIES AND SEMI-EMPIRICAL MODELING OF THE H_2O CONTINUUM

The definition of the continuum (see § V.2.2) is somewhat arbitrary[616,682] since it is given by the difference between the measured ("true") absorption and that due to close-by (local) lines, calculated using a *specified model*. The latter is essentially obtained from a spectroscopic database (providing the frequencies, integrated intensities and collisional broadenings of isolated lines), from an a priori chosen upper frequency cut-off to limit the local lines definition [Eqs. (V.5)-(V.6)], and a line-shape function. Traditionally, Lorentzian or Voigt (or Van Vleck Weisskopf for the far infrared) profiles are used, extended to ± 25 cm^{-1} from the transition centers. The resulting continuum is, as shown in Figs. V.A1 and V.A2, a smooth function of wavenumber which can be parameterized independently of the details of the local absorption features. Recall that the continuum and local absorption are *completely intertwined*. This is illustrated by Fig. V.A3 which shows that different continua are obtained[631] from experiments when the local lines contribution is calculated using purely Lorentzian profiles or a super-Lorentz shape.

Knowledge of the perturber, pressure, temperature, and spectral dependences of the H_2O continuum is of considerable practical importance for atmospheric physics. Since the 1970's, many groups have contributed to the "continuum research", using a variety of experimental techniques, mostly at wavelengths greater than 10 µm. These are not reviewed here since an overview can be found in Refs. 633,684–686 and recent papers[683,687,688] provide large reference

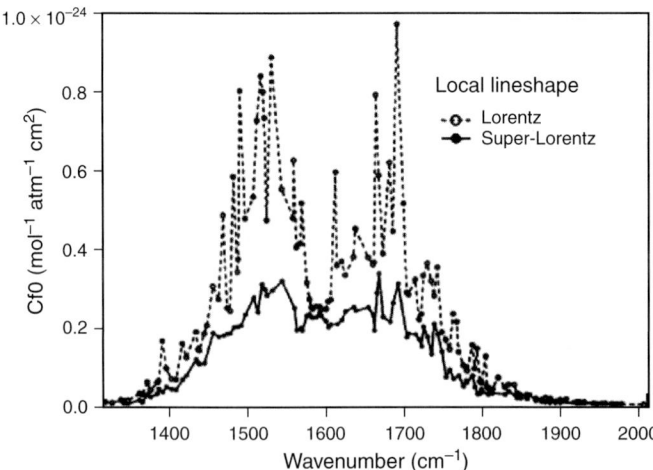

Fig. V.A3: N_2 broadened room temperature H_2O continuum absorption C_f^0 [see Eq. (V.A1)] deduced from measurement[683] in the ν_2 band using Lorentz and super-Lorentz profiles. Reproduced with permission from Ref. 631.

The far wings

lists. In the Earth atmosphere, three components are of interest: the self-broadened continuum absorption (C_s), proportional to the square of the H_2O partial pressure and the two foreign continua due to H_2O-X, (X = O_2 and N_2), proportional to the product of the partial pressures of H_2O and X.[4] These last two can be grouped into an air-broadened foreign continuum (C_f), so that Eqs. (V.5) and (V.7) become:

$$\alpha(\omega, P_{H_2O}, P_{air}, T) = \alpha^{Close}(\omega, P_{H_2O}, P_{air}, T) + P_{H_2O}^2 C_s^0(\omega, T) + P_{H_2O} P_{air} C_f^0(\omega, T). \quad (V.A1)$$

C_s^0 and C_f^0 generally show qualitatively similar temperature dependences, both decreasing with increasing T in the atmospheric range, as particularly evident for the 8–12 µm region (Fig. V.A4).[688] A more detailed summary of the properties of the continua can be found in Ref. 633.

As discussed in the next paragraph, there is still a controversy on the origin of the water vapor continuum even if some progresses have been made recently. However, the atmospheric physicists have sidestepped this problem by adopting a pragmatic approach based on semi-empirical models built from laboratory and field observations. The CKD approach (§ V.2.2) is now implemented in most atmospheric radiative transfer codes. Its most recent versions CKD_2.4,[689] and MT_CKD,[690] use rather different parameterizations, constrained to provide agreement with measurements through adjustable parameters which may have no direct physical meaning. The success of the CKD model is due to its simplicity and to the fact that it generally provides acceptable agreement with measurements. Nevertheless, recent laboratory[688,691,692] and field[686] investigations demonstrate the limits in the use of the CKD model in an "extrapolating mode" (eg Fig. V.2), for wavelengths or thermophysical conditions away from those of the experiments on which it was adjusted.

V.A2   ON THE ORIGIN OF THE WATER VAPOR CONTINUA

The nature of the water vapor continuum has been discussed in the literature for more than 50 years and still remains controversial. Note that a summary of the history of theoretical research prior to 1995 on water vapor line shapes can be found in Ref. 693. Considerable progresses have been made more recently which are summarized below after a few historical considerations. After its discovery in 1918 by Hettner,[694] the H_2O absorption in the 8–12 µm atmospheric window was attributed by Elsasser[695] to the accumulation of the far wings of the water vapor monomer (WM) lines from neighboring bands. Up to the 70's, calculations were made using Lorentzian profiles, but this approach fails to reproduce the negative temperature dependence of the continuum. For this reason, various alternative mechanisms were proposed, which are:
(1): water vapor aggregates, and more particularly water dimers (WD)[696] since larger clusters do not seem[697] to be responsible for the absorption in the 8–12 µm region;
(2): collision-induced absorption (cf Chapt. VI) arising from the transient dipole created

---

[4] Since, under the conditions of the Earth atmosphere, gases can be considered as ideal, pressures P can be used instead of number densities n, with the relation $P = nk_B T$.

during intermolecular collision;[698,699] and, (3): transitions involving metastable states (cf Chapt. VI) of the colliding pairs.[700]
Perhaps the most convincing argument in favor of the WD assumption was that it simply predicts the pressure dependence and the increase of absorption with decreasing temperature. Indeed, the absorption is proportional to the number of dimers which is proportional to $P_{H_2O}^2$ and to $\exp(-\Delta H/k_B T)$ where $\Delta H < 0$ is the change of enthalpy on dimer formation. This assumed that the other "ingredients" entering in the calculations (dimer individual line shapes, intensities, etc) do not add any other T and P dependences, a result that was not at all demonstrated until recently.[701]

It is now clearly established that the Lorentzian profile is only valid (with some restrictions discussed in Chapt. III) near the centers of collisionally isolated transitions, since it is based on the impact approximation (cf Apps. II.C1 and II.B4) and neglects both line-mixing (Chapt. IV) and finite collision duration (Chapt. V) effects. Consequently, its failure in modeling the H_2O continuum in the 8–12 μm region is not at all unexpected. Moreover, in the last decades, substantial progresses have been made (Sec. V.3) in the calculation of the far wings of the WM lines, and it has been demonstrated that models do predict the negative temperature dependences, as illustrated in Fig.V.A4.

At the same time, although absorption by WD has frequently been invoked, no rigorous calculation of their corresponding contribution existed until the recent study of Scribano and Leforestier.[701] This work presents first principle predictions of the absorption resulting from bound water dimers in the millimeter and far infrared domains, up to 944 cm^{-1}. The results demonstrate that WD may explain a significant fraction of the observed continuum, but *only* for wavenumber *below* 10 cm^{-1}, where it competes with

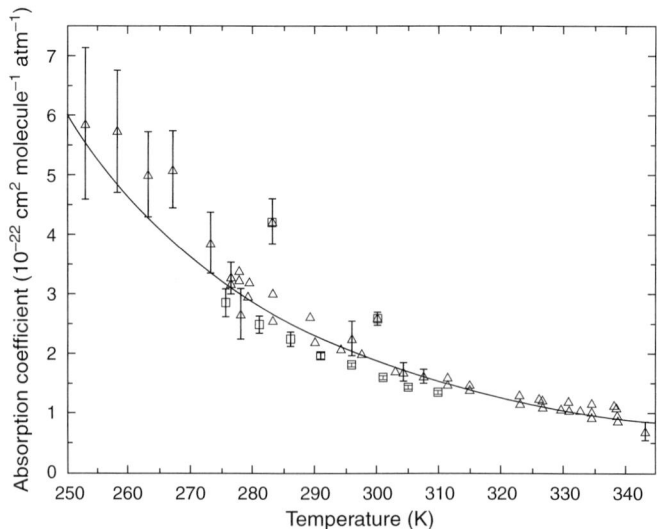

Fig. V.A4: Temperature dependence of the pure H_2O absorption coefficient at 944.2 cm^{-1}. Theoretical values are presented by the line. The symbols are laboratory measurements from Refs. 659,702,703. Courtesy of R.H. Tipping, after Ref. 704.

the WM absorption[645] due to the wings of pure rotational lines. In other words, apart from the millimeter region, the WM far wing hypothesis seems to prevail as the origin of the H_2O continuum, at least for the 8–12 μm window, the question of the contribution of dimers in the near infrared and visible regions remaining largely opened. Since the quasistatic approach (Sec. V.3) leads to good agreement with measured values for the wings of rotational lines (Figs. V.A4 and V.3), it can be easily generalized to any H_2O vibrational band, providing a powerful tool to estimate the role of WM far wings in all regions (see Figs. 3 and 4 of Ref. 693). For instance, the theoretical prediction[705] at 9466 cm^{-1} is in reasonable agreement (20%) with the measured value of Ref. 706. A second illustration is given by the results of Ref. 707, where the calculated WM far wing contribution is subtracted from measured spectra between 5000 and 5600 cm^{-1}. The remaining absorption structure is similar to that expected from a recent theoretical calculation[708] of the WD spectrum.

V.A3 THE SELF- AND N_2-BROADENED CONTINUA WITHIN THE ν_2 BAND

Up to now, we have mainly considered atmospheric windows in the far wings of adjacent bands. However, the microwindows in the troughs between lines inside a band are also of practical importance. In such regions close to strong absorption lines, the continuum is dominated by the contributions of near wings and is not as smooth as it is in between bands, as shown in Fig. V.A3. The "impact calculation" in Fig. V.A5 demonstrates that absorption in the troughs between ν_2 lines is strongly super-Lorentzian. This is qualitatively consistent with the empirical χ factor plotted in Fig. V.1 and with predictions of the quasistatic model[655,656] although the latter is not directly applicable to the in-band continuum (since strictly valid only far away in the wings).

The shape of the in-band structure obtained by subtracting the results of Lorentzian predictions from the observed spectrum has led Mlawer et al[698] to propose a new formulation of the water continuum. They arbitrarily assume that the broad residual component is due to collision-induced absorption (Chapt. VI) and/or to short life complexes of H_2O and a colliding partner. However, the magnitudes of such contributions are not at all corroborated by theoretical calculations.[709] The so-called "collision-induced component" of the MT-CKD continuum[636,690] must thus be considered as empirical, and only permitting an improved agreement with measurements through additional adjustable parameters. In other words, the insufficiency of the Lorentzian model is not a clear indication of a collision-induced signature and it may result from other processes. Among these is line mixing (Chapt. IV), since it often strongly affects absorption in the troughs between transitions. This has been demonstrated recently for H_2O,[329] a super Lorentzian (Voigt) behavior being observed between intense lines (Fig. IV.34). New experimental and theoretical developments are now needed in order to understand which specific collisional processes are involved in these asymmetric-top spectra. A possible solution to this theoretical problem may be given by scaling laws (§ IV.3.3) which have enabled accurate predictions of impact line-mixing effects for many molecular systems, but have never been applied to asymmetric-tops.

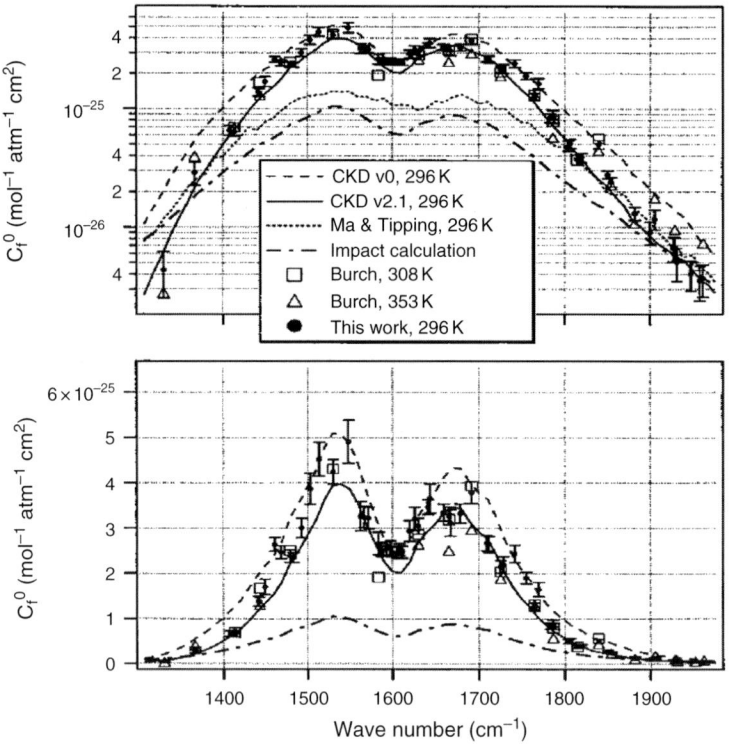

Fig. V.A5: Measured (symbols) and calculated (lines) room temperature N_2-broadened H_2O continua C_f^0 [see Eq. (V.A1)] in the ν_2 band. The same values are plotted in logarithm and linear scales in the upper and lower panels. Reproduced with permission from Ref. 683.

V.A4 CONCLUSION

The theoretical understanding of the water vapor continuum has greatly progressed in the last two decades. The role of the WM far wings is now well understood within the quasistatic approach of Ma and Tipping (Sec. V.3). It is most commonly accepted now that the wings of the allowed H_2O transitions are primary responsible for the observed continua. The problem yet to be solved is the description of the near and mid wings, which raises two major challenges. The first is, within the impact limit, the construction of a relaxation matrix describing collisional line-mixing effects (Chapt. IV) among H_2O transitions. The second is to find some interpolation scheme for the mid-wing, between the impact and quasistatic regions. Of course, other mechanisms may contribute to the continua. In fact, there is some evidence[701] that WD absorption plays a significant role in the millimeter wave while its importance in the visible is still an opened question.[710] Similarly, the contribution of CIA and of metastable states of the dimers must be further analyzed on a firm theoretical basis. Indeed, when different processes (which here have the same dependences on partial pressures) may contribute, only theory can discriminate between them by providing estimates of their relative magnitudes. One must thus still admit that the origin of the continuum in some spectral regions remains an opened question. In support for future studies, there is a strong need for new and accurate experimental data both within and in between bands. This is particularly the case for the near infrared and visible regions for which measurements would help clarifying the eventual contribution of water dimers and could be used for tests of the continuum models. Until recently this was a difficult challenge due to the weakness of background absorption at short wavelengths (see Fig. V.A1). Nevertheless, photoacoustic[711] and cavity-ring-down[712] laser techniques as well as Fourier transform measurements with very long paths,[713] now provide suitable sensitivities for such experiments. In any case, the temperature dependence is crucially needed in order to discriminate between various mechanisms eventually contributing to the continua.

VI. COLLISION-INDUCED ABSORPTION AND LIGHT SCATTERING

VI.1 INTRODUCTION

In the preceding chapters we have considered the influence of intermolecular collisions on "allowed" spectra resulting from an intrinsic permanent tensor which couples the optically active molecule to the electromagnetic field. This assumes that, during a collision, the radiator dipole or quadrupole moment (for absorption) or the polarizability (for light scattering) is not significantly modified. However, when molecules interact, transient changes in the dipole and polarizability occur at a sub-picosecond scale, leading in some cases to significant modifications of the spectral shape. Moreover, as well known since the pioneering study of Welsh and coworkers,[714] even infrared "inactive" gases (such that an isolated molecule does not absorb radiation) show broad IR bands, the so-called *Collision-Induced Absorption* (CIA) spectrum, if the gas density is high enough. This is due to the transient dipole induced during collisions involving molecules which, by themselves, possess no permanent dipole due to their symmetry. Similarly, *Collision-Induced Light Scattering* (CILS) is due to the incremental polarizability created by intermolecular interactions. The field of interaction-induced phenomena has been the subject of many reviews.[84,670,715–719] In particular, Frommhold's monograph[84] presents the experimental and theoretical foundations of CIA and CILS and provides a large bibliography. Since the reader can refer to these reviews, this chapter is restricted to illustrative examples of recent developments in the analysis of CIA/CILS spectra, with no attempt for completeness. Furthermore, we assume that the gas density is sufficiently low to neglect three and more body interactions, so that only binary collisions are considered.[84,719(Sec. A),720] Table VI.1 gives a brief summary of (recent) measurements

Table VI.1: Some illustrative measurements of CIA/CILS spectra.

CIA		CILS	
System	Refs.	System	Refs.
H_2–H_2	721,722	H_2–H_2	80,662,723
O_2–(O_2, N_2, CO_2)	724–729	N_2–N_2	668,671
N_2–(N_2, O_2)	78,727	CO_2–CO_2	675,676
CH_4–(N_2, H_2, CO_2)	730,731		

VI.2 COLLISION-INDUCED DIPOLES AND POLARIZABILITIES FOR DIATOMIC MOLECULES

A general survey of the effects of molecular interactions on gas optical properties was given by Buckingham[732] (see also Refs. 733–735). The reliable determination of collision-induced dipoles and polarizabilities raises problems very similar to those encountered in calculations of intermolecular potentials. Their discussion is beyond the scope of the present book, but readers interested by this topic can find information and bibliographical paths in Refs. 35–37,84,715,719. Schematically, induced dipoles (a similar analysis holds for the induced polarizabilities) arise from the following four mechanisms. (1) At close intermolecular distances, electronic overlap leads to a redistribution of electronic charges of the binary complex resulting in an exchange (or overlap)-induced dipole. (2) At larger distances, dispersion forces also lead to an induced dipole. (3) Simultaneously, the electric multipole field surrounding a molecule polarizes the collisional partners, leading to a multipole-induced dipole. Note that, while the multipole-induced long range components are readily obtained from a classical multipole expansion of the electric field polarizing the second molecule of the complex, the exchange and dispersion components can only be obtained from quantum chemical calculations. (4) Finally, a dipole may also result from collisional frame distortion of molecules whose unperturbed structural symmetry is inconsistent with a dipole moment. For molecules with few electrons, collision-induced electrical properties can be calculated as functions of the intermolecular separation and relative orientations by ab initio methods. A good example is the H_2–H_2 system for which accurate induced dipole surfaces have been computed.[736–738] Ab initio results for the six independent Cartesian tensor components of the collision-induced polarizability are also available.[739] For more complex systems, no ab initio result is available yet due to the formidable computer task required. For "sufficiently polarizable" species, a widely used approach assumes that the multipole induction components dominate. For N_2 pairs, the most important mechanisms are the polarization of one molecule in the quadrupolar (and to a lesser extent the hexadecapolar) field of its partner.[84] When weak dipole inductions, due to overlap and dispersion forces, exist, one has to complete the long range components by some empirical (and adjustable) short range contributions.[84,740] In summary, the spherical components of the induced operator for two colliding diatomic molecules labeled 1 and 2, are written as:[741]

$$d_\nu^{(k)}(R, r_1, r_2, \Omega, \Omega_1, \Omega_2) = \sqrt{\frac{4\pi}{2k+1}} \sum_{\lambda_1, \lambda_2, \Lambda, L} A_{\lambda_1 \lambda_2 \Lambda L}(R, r_1, r_2) Y_{L\lambda_1\lambda_2\Lambda}^{k\nu}(\Omega, \Omega_1, \Omega_2),$$

(VI.1)

with $\nu = 0, \pm 1, \ldots, \pm k$, and,

$$Y^{kv}_{L\lambda_1\lambda_2\Lambda}(\Omega, \Omega_1, \Omega_2)$$
$$= \sum_{M,M_1,M_2,M_\Lambda} (-1)^{\lambda_1+\lambda_2+L+\Lambda+M_\Lambda+v} \sqrt{(2\Lambda+1)(2k+1)}$$
$$\times \begin{pmatrix} \lambda_1 & \lambda_2 & \Lambda \\ M_1 & M_2 & -M_\Lambda \end{pmatrix} \begin{pmatrix} L & \Lambda & k \\ M & M_\Lambda & -v \end{pmatrix} Y^M_L(\Omega) Y^{M_1}_{\lambda_1}(\Omega_1) Y^{M_2}_{\lambda_2}(\Omega_2). \quad (VI.2)$$

In these expressions, k is the rank of the tensor coupling radiation and matter (k = 1 for dipole CIA, k = 2 for depolarized CILS, and k = 0 for trace scattering), (:::) is a 3J symbol,[65] and Y^M_L is a spherical harmonic. Ω_1 and Ω_2 define the orientations of molecules 1 and 2, respectively, while Ω defines the orientation of the intermolecular vector $\vec{R}$ (between the centers of mass). r_1 and r_2 are the vibrational coordinates. The coefficients A in Eq. (VI.1) satisfy some symmetry relations which are given in Ref. 84 together with their classical multipole expansion expressions.

VI.3 COLLISION-INDUCED SPECTRA IN THE ISOTROPIC APPROXIMATION

Most calculations of CIA/CILS spectra are based on the *isotropic approximation*. The corresponding formalism has been described in Sec. II.6 and is not recalled here. Furthermore, since various reviews exist[84,715–719] on applications of this approach, only illustrative results for specific molecular pairs are given below.

VI.3.1 TWO ILLUSTRATIVE EXAMPLES: H_2 AND N_2

The H_2–H_2 system is a good example of application of the isotropic approximation. Its CIA spectrum is of considerable importance in various fields of astrophysics, including the atmospheres of cold stars and planets[742,743] (see also § VII.5.2). Since hydrogen is the simplest molecule, the pure H_2 CIA is the most amenable to sophisticated calculations from first principles. Moreover, this system has a small anisotropy which may be ignored to first approximation. Extensive studies, reviewed in Ref. 84, have been made from the translational band (0–200 cm^{-1}) to the first overtone band (8000–9000 cm^{-1}). Since an accurate interaction-induced dipole moment is available,[737] which explicitly accounts for the vibrational dependence of the bond distances r_1 and r_2, it was possible to calculate [*cf* Eq. (II.86), with the difference that, in Ref. 737, the J dependence of the vibrational wavefunctions is included] the vibrational matrix elements defined by:

$$B_{\lambda_1\lambda_2\Lambda L}(R) = \langle v_1 J_1 v_2 J_2 | A_{\lambda_1\lambda_2\Lambda L}(R, r_1, r_2) | v'_1 J'_1 v'_2 J'_2 \rangle, \quad (VI.3)$$

where v_m and J_m designate vibrational and rotational quantum numbers for molecule m = 1,2, ie $v_1 = v_2 = v'_1 = v'_2 = 0$ for the rototranslational band, $v_1 = v_2 = v'_1 = 0$ with $v'_2 = 1$ for the fundamental band, and $v_1 = v_2 = v'_1 = 0$ with $v'_2 = 2$ together with the simultaneous (double) transitions defined by $v_1 = v_2 = 0$ with $v'_1 = v'_2 = 1$, for the first overtone. As mentioned in Sec. II.6, vibrationally averaged isotropic potentials are needed for the computation of the spectral profile [Eq. (II.87)]. For H_2–H_2, they are given and discussed in Table VII of Ref. 737. Once reliable isotropic potentials and induced dipole moments became available, it was demonstrated that ab initio computations of CIA lead to remarkably good agreement with measured spectra from the rotational band up to the first overtone in a wide temperature range.[737,744,745] Theoretical and experimental values agree within the estimated uncertainties of the measurements in most cases. This is illustrated by Fig. VI.1, which shows that the dominant contributions for the pure H_2 rotational band come from $\lambda_1\lambda_2\Lambda L = 0223$ and 2023 [cf Eq. (VI.1)], ie mostly quadrupole-induced components. A similar quality of predictions is obtained from 77 K to room temperature, and Fig. VI.2 shows that the theory also succeeds in the

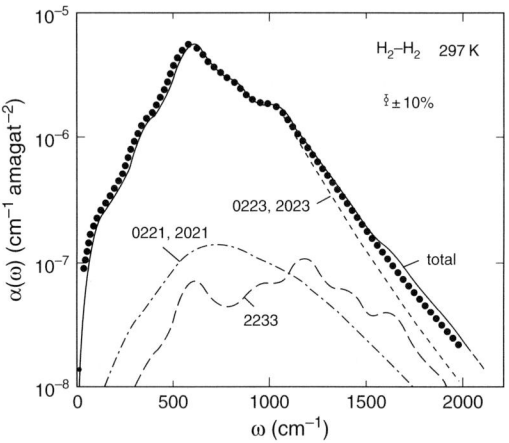

Fig. VI.1: Rototranslational spectrum of pure H_2 at room temperature. (•) denote measured values[721] while the lines correspond to calculated results, including the total spectrum (——) and some of the largest $\lambda_1\lambda_2\Lambda L$ contributions. Reproduced with permission from Ref. 737.

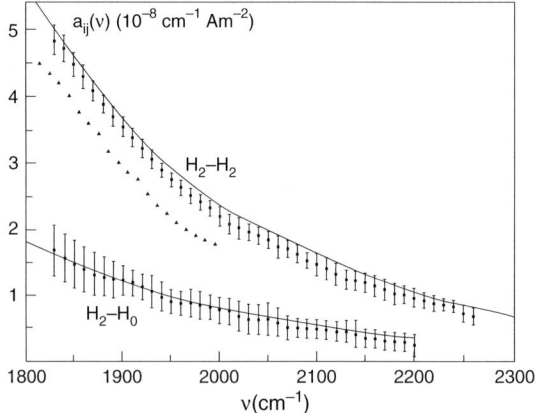

Fig. VI.2: Density (bi)normalized absorption coefficients for H_2–H_2 and H_2–He. (•) denote measured values[721,746] while the lines have been calculated from Ref. 737 for H_2 and Ref. 746b for He. Reproduced with permission from Ref. 746.

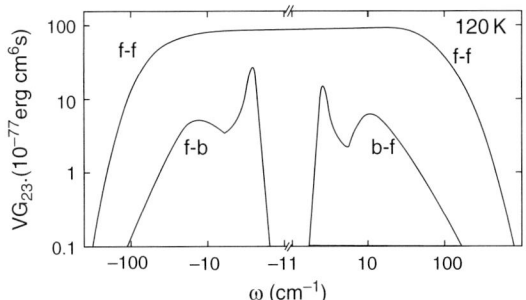

Fig. VI.3: Calculated dimer fine structures [G function of Eq. (VI.4)] central region of the bound-free (b-f) and free-bound (f-b) 0223, 2023 components compared with the free-free (f-f) component for pure H_2 CIA at 120 K. The results are plotted vs the detuning with respect to $\omega_{v'_1 J'_1 v_1 J_1} + \omega_{v'_2 J'_2 v_2 J_2}$, see Eq. (II.84). Reproduced with permission from Ref. 737.

description of the high frequency wing of the rototranslational absorption coefficients for both H_2–H_2 and H_2–He.

As discussed in Sec. II.6, CIA/CILS spectra contain free-free, as well as free-bound/bound-free and bound-bound transitions. In most cases, the spectra are dominated by the free-free component. Nevertheless, Fig. VI.3 enables the analysis of the role played by bound states. Whereas the bound-free and free-bound contributions are negligible with respect to the free-free ones in the far wings, they are much more significant near the line centers. Note that the dominant 0223 and 2023 components do not feature a bound-bound spectrum while the very weak 0221 and 2021 do, but they are totally negligible.

As shown by Fig. VI.4, the bound-free and free-bound contributions may build some "line" structures. Note that the convolution of the results with a 4.3 cm^{-1} apparatus function gives a spectrum very similar to that recorded by the Voyager mission,[747] but that the rather low resolution of this instrument "washes out" the defects of the isotropic approximation.[87,88] Better observations at lower temperatures and/or higher spectral resolutions have required more sophisticated calculations of transitions involving the $(H_2)_2$ dimer (see Sec. VI.4). In summary, the free-free transitions give the predominant profiles. They are broad with half widths of the order of a few tens of cm^{-1} since the induced dipole only exists for times of the order of the collision duration τ_c (typically 1 ps or less). When bound states are involved, the spectra exhibit more or less diffuse "structures" due to the bound-bound and/or bound-free or free-bound transitions. At low spectral resolution or high pressures these flatten and can hardly be observed.

N_2 is another important species whose CIA is needed for the modeling of some planetary atmospheres (see Ref. 748 and § VII.5.2). Calculations of the rototranslational spectra of N_2–N_2 pairs from 50 to 300 K, made within the isotropic approximation, are given in Ref. 85. This work is similar to that made for H_2–H_2 (see above and Refs. 749,750), but with a *noticeable* difference. Indeed, for N_2–N_2, no ab initio computation of the induced dipole was (and still is) available. It was thus assumed that the long range quadrupolar and hexadecapolar induction mechanisms are predominant, and a small *empirical* overlap term of quadrupolar symmetry was added. This enables to reproduce the measurements, starting from a suitable advanced isotropic potential,[751] as illustrated by Fig. VI.5.

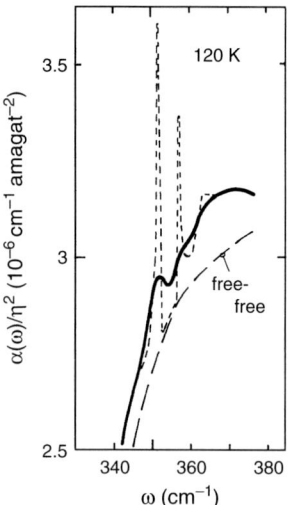

Fig. VI.4: Computed fine structures due to the hydrogen dimer (- - -) in the quadrupole-induced (0223, 2023) components, near the $S_0(0)$ line. Superimposed is the free-free component (long dash). The convolution by a 4.3 cm^{-1} slit function leads to the thick line which compares well with Jupiter observed spectrum.[747] Reproduced with permission from Ref. 737.

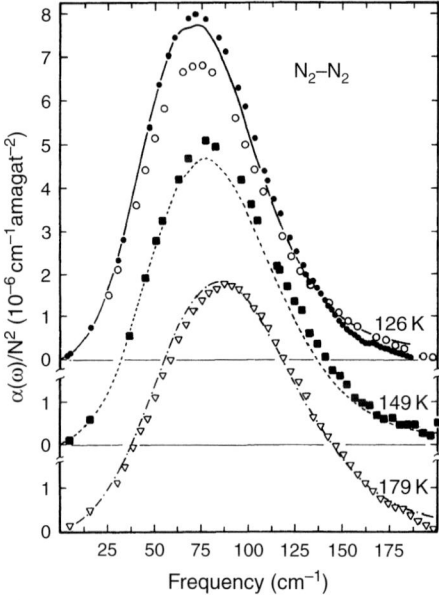

Fig. VI.5: Density (bi)normalized far infrared spectra of pure N_2 at 126, 149, and 179 K. The symbols indicate measured values,[752,753] the lines denoting calculated results. Reproduced with permission from Ref. 85.

Similar calculations of the rototranslational spectra of CH_4–(He, H_2, N_2, CH_4) can be found in Refs. 754–757. They are based on the isotropic approximation and advanced dipole surfaces.[758] However, they fail in reproducing the absorption measured at wavenumbers above 250 cm^{-1}, even though empirical (but reasonable) exchange-induced components are introduced. Additional mechanisms seem to be involved when CH_4–X pairs interact. The authors of Refs. 756,757 mention the collisional frame-distortion of the methane molecule but more work is needed for a definitive conclusion.

VI.3.2 MODELING OF THE LINE SHAPE

For practical applications, it is of interest to recast the results of quantum computations under the form of simple expressions suitable for low cost computations. According to Eqs. (II.81) and (II.84), it is sufficient to model every translational profile $G_{\lambda_1,\lambda_2,\Lambda,L}(\omega)$ (for free-free, free-bound, *etc*, components separately) at each desired temperature, according to:

$$V \times G_{\lambda_1,\lambda_2,\Lambda,L}(\omega, T) = C_{\lambda_1,\lambda_2,\Lambda,L} \times I_{\lambda_1,\lambda_2,\Lambda,L}(\omega). \tag{VI.4}$$

Here $C_{\lambda_1,\lambda_2,\Lambda,L}$ is the "strength" of the transition and $I_{\lambda_1,\lambda_2,\Lambda,L}(\omega)$ is a normalized profile such that:

$$\int_{-\infty}^{+\infty} I_{\lambda_1,\lambda_2,\Lambda,L}(\omega) d\omega = 1, \tag{VI.5}$$

for which many expressions have been proposed.[759,749] Among these, the Birnbaum and Cohen (BC) model[674] is perhaps the most elegant and one of the most successful in terms of the modeling of real quantum shapes. It is given by:

$$I_{BC}(\omega) = \frac{\tau_1}{\pi} \frac{1}{1 + (\omega\tau_1)^2} \exp(\omega\tau_0 + \tau_2/\tau_1) z K_1(z), \tag{VI.6}$$

where $K_1(\ldots)$ is a modified Bessel function,[182] with $\lim_{z\to 0}[zK_1(z)] = 1$ and $\lim_{z\to\infty}[zK_1(z)] = (\pi/2z)^{1/2} e^{-z}$], and:

$$\tau_0 = \frac{\hbar}{2k_B T} \quad \text{and} \quad z = \frac{\sqrt{\tau_0^2 + \tau_2^2}}{\tau_1} \sqrt{1 + (\omega\tau_1)^2}. \tag{VI.7}$$

Note that the parameters τ_1 and τ_2 *depend* on λ_1, λ_2, Λ, and L, and on the type (free-free, free-bound, *etc*) of transition. Together with the amplitudes $C_{\lambda_1,\lambda_2,\Lambda,L}$ they are usually obtained from adjustments of quantum-calculated spectra. As demonstrated by applications to various collisional pairs, accurate fits are obtained[85,749,760,761] even if, in some cases and particularly at the lowest temperatures, they must be improved using an extended model.[759,749] Once the τ_1, τ_2, and $C_{\lambda_1,\lambda_2,\Lambda,L}$ parameters are known, CIA can be calculated using Eqs. (VI.4)–(VI.7) with low programming and computational efforts.

The calculation of quantum profiles remains a difficult and computer time consuming problem, even when the isotropic approximation is used. The evaluation of the radial functions [Eq. (II.87)] as well as the calculation of the translational profiles [Eq. (II.85)] are not easy to handle. Thus, the efficiency of the "line-shape engineering" has suggested another procedure in order to model CIA spectra for systems which do not form bound states to a significant extent.[1] Indeed, as discussed in Ref. 83, $C_{\lambda_1,\lambda_2,\Lambda,L}$ can be identified as the zeroth-order moment of the spectral function, leading to:

$$C_{\lambda_1,\lambda_2,\Lambda,L} = \int_0^{+\infty} 4\pi R^2 e^{-V_0(R)/k_B T} B^2_{\lambda_1,\lambda_2,\Lambda,L}(R) dR, \quad (VI.8)$$

where $V_0(R)$ is the isotropic intermolecular potential. The absorption coefficient can then be written as:

$$\alpha(\omega) = n_a n_p \frac{4\pi^2 \omega}{3\hbar c}[1 - e^{-\hbar\omega/k_B T}] \sum_{\substack{\lambda_1,\lambda_2,\Lambda,L \\ v_2,J_2,v'_2,J'_2}} \sum_{v_1,J_1,v'_1,J'_1} S_{\lambda_1,\lambda_2,\Lambda,L}(v_1,J_1,v'_1,J'_1,v_2,J_2,v'_2,J'_2) \quad (VI.9)$$
$$\times I_{\lambda_1,\lambda_2,\Lambda,L}(\omega - \omega_{v'_1 J'_1 v_1 J_1} - \omega_{v'_2 J'_2 v_2 J_2}),$$

where n_a and n_p designate the densities of radiators and perturbers, respectively, and the "strengths" of the individual components are given by:

$$S_{\lambda_1,\lambda_2,\Lambda,L}(...) = \rho_{v_1 J_1} \rho_{v_2 J_2} (2J'_1 + 1)\begin{pmatrix} J_1 & \lambda_1 & J'_1 \\ 0 & 0 & 0 \end{pmatrix}^2 (2J'_2 + 1)\begin{pmatrix} J_2 & \lambda_2 & J'_2 \\ 0 & 0 & 0 \end{pmatrix}^2 C_{\lambda_1,\lambda_2,\Lambda,L}. \quad (VI.10)$$

The normalized profiles $I_{\lambda_1,\lambda_2,\Lambda,L}(\omega)$ are represented by some proper model, either previously known from theory or adjusted to measurements. Figure VI.6 gives an illustration of this approach for the first overtone band of D_2. In this case, the spectrum includes both a single vibrational transition ($v_1 = v'_1 = v_2 = 0$, $v'_2 = 2$) and a double vibrational transition ($v_1 = v_2 = 0$, $v'_1 = v'_2 = 1$).[81] The profile was then analyzed using the BC line shape for the individual components with parameters determined from fits of measured values. The observed and adjusted spectra are in very good agreement at 77, 201, and 298 K[81] as shown in Fig. VI.6 in the particular case of T = 201 K.

A second example can be found in the simultaneous fitting of collision-induced fundamental absorption bands of N_2–N_2, O_2–O_2, O_2–N_2, and N_2–O_2 pairs, where the first molecule is the one making the vibrational transition (see Ref. 727 and those therein). The normalized profile used was that obtained previously from quantum computations for the N_2 rototranslational case.[85] A similar approach is also applied in Ref. 764 for predictions of CIA in the $a^1\Delta_g(v' = 0) \leftarrow X^3\Sigma^-_g(v = 0)$ transition in O_2–CO_2, O_2–N_2, and O_2–H_2O mixtures. Finally, let us mention that a model was proposed[765,709] for

---

[1] In principle, the classification and separation of the various pair states in the phase space in free pairs, metastable (predissociating), and bound ones, is necessary for a correct assessment of their respective contributions.[762,763] Nevertheless, at low spectral resolution, high pressures and/or high temperatures, the (eventual) structures due to metastable and bound states vanish and disappear within the broad profile due to free-free transitions. Hence, Eq. (VI.9) somehow contains dimer (and metastable) intensities but not dimer structures.

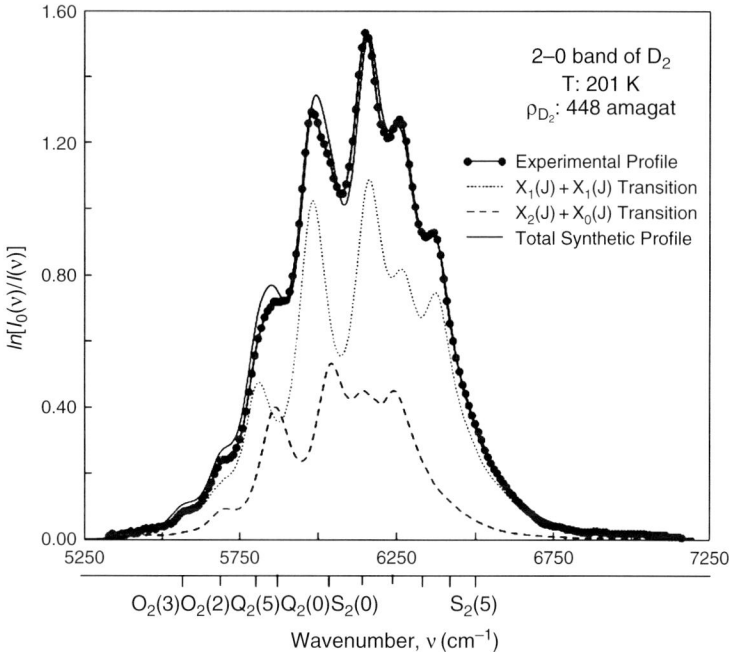

Fig. VI.6: Absorbance in the first overtone band of pure D_2 at 201 K and 448 Am. (–•–) gives the observed profile. (- - -) and (·····) are calculated results for the quadrupolar double-transition $D_2(v'=1,J'\leftarrow v=0,J)+D_2(v'=1,J'\leftarrow v=0,J)$ and for $D_2(v'=2,J'\leftarrow v=0,J)+D_2(v'=0,J'\leftarrow v=0,J)$, and (——) is their sum. Reproduced with permission from Ref. 81.

the calculation of the spectrum of a pair of molecules where one makes an allowed dipole transition while the other makes an allowed Raman transition. It has been applied to the collision-induced double transitions $CO_2(v_3=1)+X(v_2=1) \leftarrow CO_2(v_3=0)+X(v_2=0)$ for X = H_2, N_2 and O_2, and successfully compared with the experimental results of Ref. 766.

The studies mentioned in this section are based on the isotropic approximation which reduces the computational task but raises questions concerning the reliability of such calculations. In a first step, some studies[767–769] have investigated the influence of the potential anisotropy on the spectral moments of the absorption coefficient $\alpha(\omega)$, essentially considering:

$$\alpha_1 \equiv \int_0^\infty \alpha(\omega) d\omega \quad \text{and} \quad \gamma_1 \equiv \int_0^\infty \frac{\alpha(\omega)}{\omega \, \text{th}(\hbar\omega/2k_B T)} d\omega. \qquad (VI.11)$$

The spectral moments can be computed relatively easily since they only require the knowledge of static distribution functions, whereas $\alpha(\omega)$ involves all the dynamics of the process. Borysow and Moraldi have shown[767] that the effect of the anisotropy on N_2–N_2 spectral moments is relatively small (10%), somehow justifying the above mentioned isotropic studies. On the contrary, for CO_2–CO_2 pairs, the moments α_1 and γ_1 calculated within a purely isotropic approach both deviate from those deduced from experiments by about 40%.[767] In order to go further, it thus became necessary to take the potential anisotropy into account at the primary level, *ie* that of the spectral profile calculation. Some studies in this direction are discussed below.

VI.4 EFFECTS OF THE ANISOTROPY OF THE INTERACTION POTENTIAL

After the investigations of the effects of the anisotropy on low order spectral moments, calculations treating the line shape correctly became of primary importance. The goal of many of the models developed in this field was restricted to a qualitative understanding of the influence of the potential anisotropy with no direct comparisons with experiments. An example is given by the semi-classical calculations of Ref. 770 which demonstrate that very large changes of the far wings of CIA spectra can result from the anisotropy of the potential. As shown in Refs. 771,772, the anisotropy couples the translational and rotational states, thus introducing a coupling of the various dipole components. The correct method, fully accounting for the anisotropy, is based on an exact solution of the Schrödinger equation for two interacting molecules. Whereas the Close-Coupling (CC) method[519] can be used for light species such as H_2, the increasing number of coupled equations for heavier molecules requires huge computer times and storage. The first CC calculations for H_2–H_2 at low temperature were carried out by Schaefer and McKellar,[773,774] focusing on the features due to the bound $(H_2)_2$ dimers rather than on the whole spectrum. Later on, Gustafsson and coworkers have calculated CIA spectra by including a weak electromagnetic radiation field in the molecular scattering Hamiltonian and solving the Schrödinger equations numerically. Advanced anisotropic intermolecular potentials and ab initio dipole surfaces were used in order to obtain rototranslational and rotovibrational spectra at various temperatures for H_2 (or HD) colliding with He,[775,776] H,[777] H_2,[778] and Ar.[779] Below is a brief description of this formalism in the particular case of diatom-atom collisions.

The Hamiltonian describing the interaction of an atom with a rotating and vibrating diatomic molecule, in the presence of a radiation field, is given by:[775]

$$H(\vec{r}, \vec{R}, \omega) = H^{mol}(\vec{r}) - \frac{\hbar^2}{2m^*}\nabla_{\vec{R}}^2 + V(\vec{r}, \vec{R}) + V^{rad}(\vec{r}, \vec{R}, \omega) + H^{rad}(\omega). \qquad (VI.12)$$

In this expression, $V(\vec{r}, \vec{R})$ is the interaction potential for the colliding pair (reduced mass m*), $\vec{R}$ being the vector between the centers of mass and $\vec{r}$ the vibrational coordinate. The Hamiltonian of the isolated molecule is $H^{mol}(\vec{r})$, with eigenvalues E_{vJ}, while H^{rad} is that of the isolated electromagnetic field. The radiative coupling is given, within the weak radiation field approximation (assuming linearly polarized light along the angular momentum quantization z axis), by:

$$V^{rad}(\vec{r}, \vec{R}, \omega) = -\sqrt{\frac{2\pi\hbar\omega}{c}}\Phi \times d_z(\vec{r}, \vec{R}), \qquad (VI.13)$$

where Φ is the radiation flux in photon number per unit area and unit time. The standard series expansion of the induced dipole moment $d_z(\vec{r},\vec{R})$ is given by Eq. (VI.1) with $k=1$ and $v=0$. Writing the state of n photons of frequency ω as $|n;\omega\rangle$, one has:

$$H^{rad}(\omega)|n;\omega\rangle = n\hbar\omega|n;\omega\rangle, \quad (VI.14)$$

and a total state vector is then:[775]

$$\Psi_\alpha(\vec{r},\vec{R},E) \otimes |n;\omega\rangle = \sum_{\alpha',n'} v_{v',J'}(r)\frac{1}{R}F^{n''n}_{\alpha'\alpha}(R,E,\omega)\widetilde{Y}^{\widetilde{J}'M'}_{J'\ell'}(\vec{r}/r,\vec{R}/R) \otimes |n';\omega\rangle, \text{ with,}$$

$$\widetilde{Y}^{\widetilde{J}'M'}_{J'\ell'}(\vec{r}/r,\vec{R}/R) \equiv \sum_{m_1,m_2}(-1)^{J'+\ell'+M'}\sqrt{2\widetilde{J}'+1}\begin{pmatrix}J' & \ell' & \widetilde{J}' \\ m_1 & m_2 & -M'\end{pmatrix} Y_{J'm_1}(\vec{r}/r)Y_{\ell m_2}(\vec{R}/R),$$

(VI.15)

where $\alpha \equiv (v, J, \ell, \widetilde{J}, M)$, $\vec{\widetilde{J}} = \vec{J} + \vec{\ell}$ being the total angular momentum, sum of the rotational ($\vec{J}$) and translational ($\vec{\ell}$) momenta. The eigenvalue equation is:

$$[H(\vec{r},\vec{R},\omega) - E]\Psi_\alpha(\vec{r},\vec{R},E) \otimes |n;\omega\rangle = 0. \quad (VI.16)$$

By integrating five out of the six spatial variables, one obtains:[775]

$$\left[E^{mol}_{v''J''} - \frac{\hbar^2}{2m^*}\frac{d^2}{dR^2} + \frac{\hbar^2\ell''(\ell''+1)}{2m*R^2} + n''\hbar\omega - E\right]F^{n''n}_{\alpha''\alpha}(R,E,\omega)$$
$$+ \sum_{\alpha',n'}\left[V_{\alpha''\alpha'}(R)\delta_{n'',n'} + V^{rad}_{\alpha''\alpha'}(R,\omega)\delta_{n''\pm 1,n'}\right]F^{n'n}_{\alpha'\alpha}(R,E,\omega) = 0,$$

(VI.17)

which forms a coupled system of differential equations. The asymptotic boundary condition for $R\to\infty$ yields the scattering matrix element from which the transition probability between two channels (α,n) and (α',n') can be calculated.[66,519,775] In order to carry out calculation, the coupling matrix elements $V_{\alpha''\alpha'}$ and $V^{rad}_{\alpha''\alpha}$ must now be defined.

Interaction potential. For a diatom-atom system, it is appropriate to expand the potential on the basis of Legendre polynomials P_L, ie:

$$V(\vec{r},\vec{R}) = \sum_L V_L(r,R)P_L\left(\frac{\vec{r}}{r}\cdot\frac{\vec{R}}{R}\right). \quad (VI.18)$$

By neglecting non-radiative transitions between different vibrational states, the matrix elements $V_{\alpha''\alpha'}(R)$ in Eq. (VI.17) are approximated by:

$$V_{\alpha''\alpha'}(R) = \sum_L V^{v'}_L(R) \times f_L(J'',\ell'',J',\ell',\widetilde{J})\delta_{\widetilde{J}'',\widetilde{J}'}\delta_{M'',M'}\delta_{v'',v'}, \quad (VI.19)$$

$$\text{where } V^{v'}_L(R) \equiv \langle v'|V_L(r,R)|v'\rangle. \quad (VI.20)$$

Since tests show[775] that the J dependence of the vibrational matrix elements does not significantly affect results, it has been neglected in Eq. (VI.19). The $f_L(...)$'s take into account all the angular momenta couplings and are given by Eq. (24) of Ref. 775.

Radiative coupling. Using Eqs. (VI.13) and (VI.1) one obtains:

$$V^{rad}_{\alpha''\alpha'}(R,\omega) = -\sqrt{\frac{2\pi\hbar\omega}{c}}\Phi \sum_{\lambda,L} d_{\lambda L}(J'',\ell'',\widetilde{J}'',J',\ell',\widetilde{J}') B^{v''J'',v'J'}_{\lambda L}(R)\delta_{M'',M'}. \quad (VI.21)$$

The coefficients $d_{\lambda L}$, which also contain the various angular momenta couplings, are defined by Eq. (32) of Ref. 775, while the radial dipole matrix elements are given by:

$$B^{v''J'',v'J'}_{\lambda L}(R) = \langle v''J'' | A_{\lambda L}(r,R) | v'J' \rangle. \quad (VI.22)$$

Note that since the dipole function has a significant J dependence, it has been taken into account[775] in the calculations. In principle, the summations over α' and n' in Eq. (VI.17) are infinite, but since only one-photon absorption is considered, the n' summation only runs from 0 to 1. The sums over α' can also be truncated by excluding the channels which negligibly affect the radiative transition probabilities at the considered kinetic energies. Finally, the binary CIA coefficient is given by:

$$\alpha(\omega,T) = n_a n_p \lambda_0^3 \frac{1-e^{-\hbar\omega/k_BT}}{\hbar\Phi} \sum_i \rho_{0J_i}(T) \int_0^\infty e^{-E_i/k_BT} \sum_f |S_{fi}(E_i,\omega)|^2 dE_i, \quad (VI.23)$$

where $\lambda_0 = h/(2\pi m^* k_B T)^{1/2}$ is the thermal de Broglie wavelength. The subscripts $i \equiv \left(J_i \ell_i \widetilde{J}_i\right)$ and $f \equiv (J_f \ell_f \widetilde{J}_f)$ designate respectively the initial and final channels, with one or zero photon. S_{fi} is the corresponding (scattering) S matrix element, ρ_{vJ} is the relative population of level vJ, and E_i is the initial kinetic energy. Note that numerous test calculations are necessary in order to check the accuracy of various truncations and decouplings used to reduce the computer time. Results obtained with this approach are illustrated by Figs. VI.7 and VI.8. These show that, for the H_2–He rototranslational spectrum (similar results are obtained for the fundamental band[775]), accounting for the anisotropy of intermolecular forces leads to small but significant modifications of the profiles. In the region of the low J S(J) lines, it reduces the intensities by ≈10% while, in the far wing, intensities are enhanced by ≈15% (and more at higher frequencies).

Close-coupling calculations of the H_2–Ar rototranslational spectrum[779] show almost no difference with results obtained within the isotropic approximation above ≈300 cm^{-1}. For smaller frequencies, the anisotropy leads to negative corrections, typically below 10%. For H_2–H_2, the results are quite similar. Over most of the rototranslational spectrum (region of Fig. VI.7), calculations performed with and without the inclusion of the anisotropy lead to very close results, both in agreement with measured values.[778] On the opposite, the influence of the anisotropy is important for the far wing at 77 K, leading to a doubling of the absorption at the higher frequency as shown in Fig. VI.9. However, important and still unexplained discrepancies remain between measured and calculated values. They may be due to the fact that measurements in the (very weak) far wings require high pressures

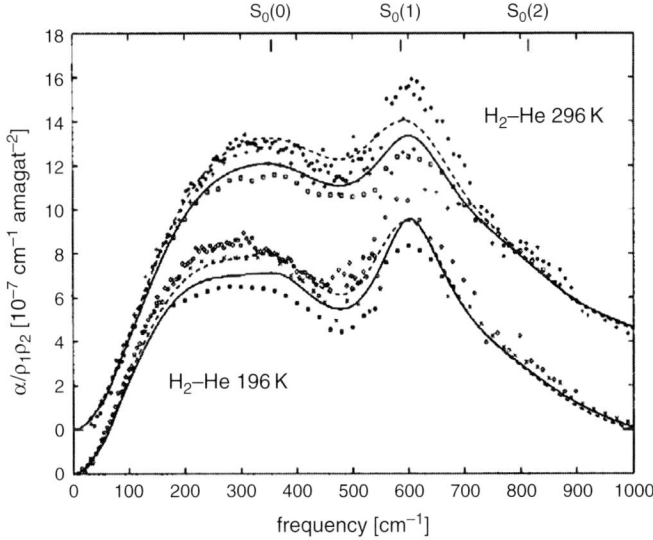

Fig. VI.7: Density (bi)normalized rototranslational CIA of H_2 in He at 196 and 296 K. Experimental values (symbols) are from Refs. 780,781. (——) and (- - -) are results calculated with and without taking the anisotropy of the interaction into account, respectively. Reproduced with permission from Ref. 775.

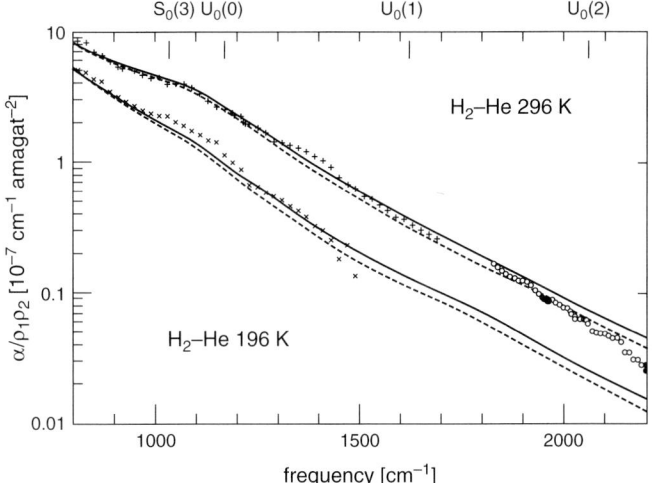

Fig. VI.8: Same as Fig. VI.7 but for the far wing of the rototranslational absorption coefficients. The experimental values are from Refs. 781,746. Reproduced with permission from Ref. 775.

for which a ternary contribution (proportional to $n_{H_2}^3$) may "pollute" the binary component. This is shown, in Fig. VI.9, by the difference between values deduced from measurements with and without a ternary contribution. Another possible explanation is uncertainties on the dipole moment and/or on intermolecular forces. Indeed, in the far

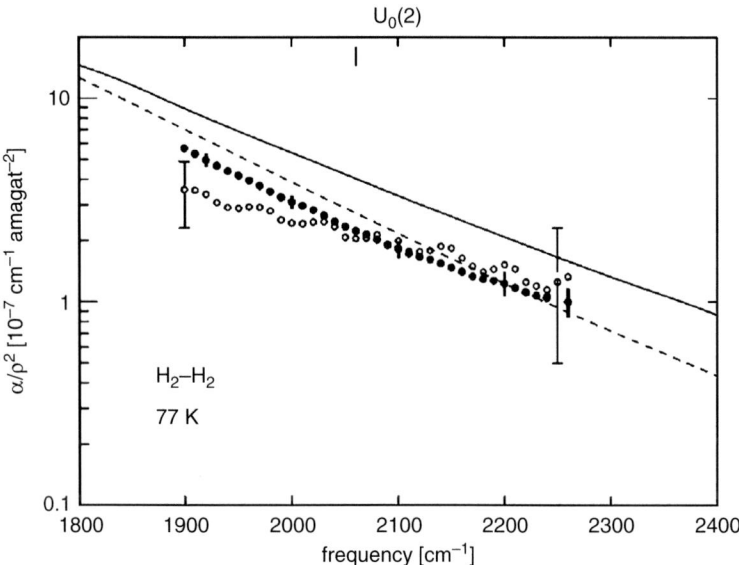

Fig. VI.9: Same as Fig. VI.8 but for pure H_2 at 77 K. Close and opened circles denote values extracted from measurements assuming a purely quadratic density dependence and including a small ternary contribution, respectively. Reproduced with permission from Ref. 778.

wings, several small dipole components contribute to the absorption, while the line shape is very sensitive to minor changes of the repulsive part of the interaction potential.

Very low temperatures favor the formation of bound states whose spectral signatures must be taken into account, particularly when high resolution spectra are analyzed. McKellar and Schaefer[774] have calculated the collision-induced spectrum of pure H_2 in the close-coupling scheme within the rigid rotor approximation. Figure VI.10 shows that the free-free contributions are not always dominant and that both sharp and relatively broader dimer features appear (see also Sec. VI.5). Some discrepancies between measured and calculated values remain, maybe due to the rigid rotor model used.

In summary, for "ordinary" density and temperature conditions, since the anisotropy of H_2–H_2 and H_2–rare gas systems is weak, calculations using isotropic and anisotropic potentials show small differences, except in the far wings (Fig. VI.9). This somehow justifies the use of the isotropic approximation for hydrogen but the question of its validity for "moderately anisotropic" pairs like N_2–X remains opened in the absence of anisotropic close-coupling calculations If one now considers strongly anisotropic systems such as CO_2–CO_2, it is obvious from the calculation of the spectral moments[767,769] that the anisotropy of the interaction potential cannot be neglected. For such molecular pairs, calculations of "exact" quantum mechanical profiles are not tractable. Nevertheless, CO_2 molecules being large and heavy, classical molecular dynamics can be used for the calculation of CIA profiles. This is the starting idea of Ref. 90, which is devoted to the far infrared CIA of pure CO_2. The procedure used to derive spectra from such computations is not detailed here but is described in studies cited in Ref. 90. Schematically, a set of classical trajectories is first chosen. For each of them, a complete time

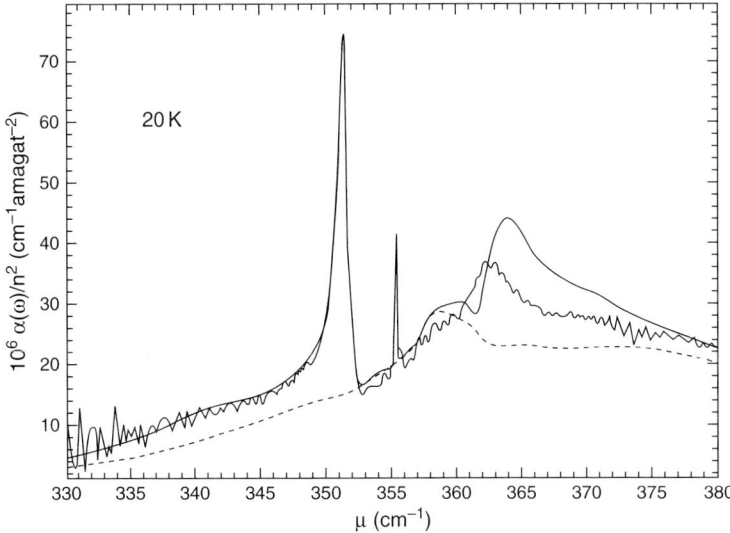

Fig. VI.10: Comparison between the theoretical and experimental ("noisy" curve) absorption spectra of para-H_2 in the $S_0(0)$ region at 20 K. The dashed curve shows the free-free contribution. Reproduced with permission from Ref. 774.

history of the induced dipole and of its correlation function is then derived. Finally, the Fourier transform of the correlation function gives the spectral shape. Note that a proper ensemble average must be made over the available phase space. Concerning the spherical components of the induced dipole [Eq. (VI.1)], the terms due to quadrupole and hexadecapole induction through the trace and the anisotropy of the polarizability were supplemented[90] by overlap components of the same symmetry considered as *adjustable* quantities. Figs. VI.11 and VI.12 illustrate the quality of the associated results,

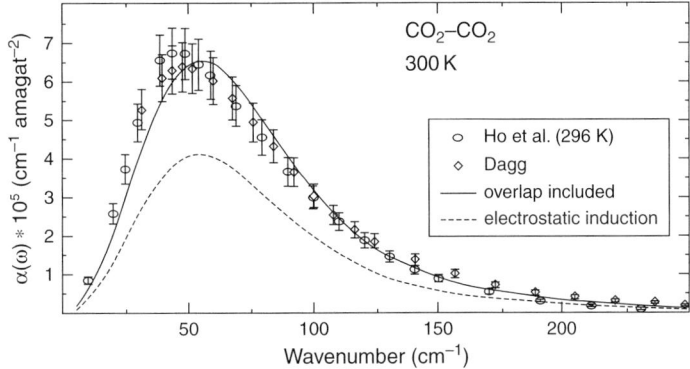

Fig. VI.11: Pure CO_2 CIA in the rototranslational band at room temperature. Symbols are measured values (from Ref. 782 and a communication to the authors by Dagg) while the lines are calculated results with only the electrostatic induction (- - -) and with the inclusion of the overlap correction (——). Reproduced with permission from Ref. 90.

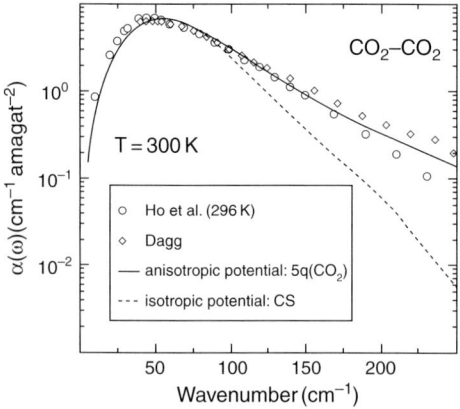

Fig. VI.12: Same as Fig. VI.11 except that calculations have been made using isotropic (- - -) and anisotropic (——) potentials. Reproduced with permission from Ref. 90.

also confirmed at lower temperatures.[90] Figure VI.11 shows the importance of the added ad hoc short range component, and Fig. VI.12 demonstrates that the anisotropy of the interaction must be taken into account, as expected from previous analyses of spectral moments.[767,769]

VI.5 THE IMPORTANCE OF BOUND AND QUASIBOUND STATES IN CIA SPECTRA

Many experimental studies of the spectroscopy of dimers and van der Waals complexes have been made using equilibrium gas samples and relatively long absorption paths, high gas pressures, and/or low temperatures. Numerous other investigations have used supersonic expansions which create a non-equilibrium environment at very low kinetic and rotational temperatures with virtually collision free conditions thus outside the scope of this book. The interpretation of observed spectra assuming that the contributions of the complex appear as sharp structures superimposed on a diffuse collision-induced background may be naive. Indeed, there is often no simple way to separate absorption due to dimers (ie bound-bound transitions) from the remaining part of the spectrum, without the use of a theoretical model since free-bound and bound-free transitions can be of significant importance. This is illustrated by Fig. VI.10, since it was shown[774] that the sharp peak near 355.4 cm^{-1} corresponds to a bound-bound dimer transition, while the peak at 351.3 cm^{-1} and the broad feature in the 360–370 cm^{-1} region are due to bound-free and free-bound transitions.

The entanglement between CIA and the spectrum of the corresponding complexes also occurs for the N_2–X and O_2–X systems (X = N_2, O_2, Ar). This is illustrated by Fig. VI.13, which corresponds to the region of the N_2 monomer vibrational frequency, and by Fig. VI.14, which was obtained for rather similar conditions but in the far infrared

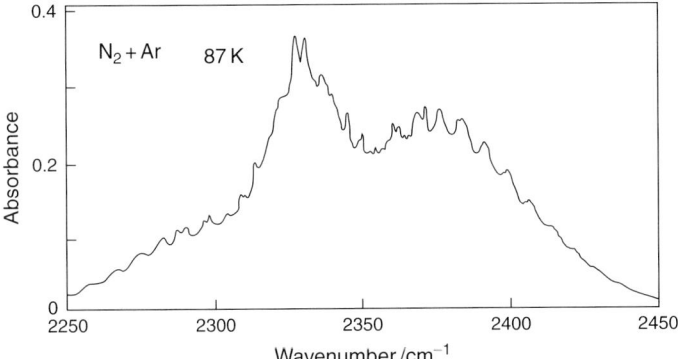

Fig. VI.13: Measured absorbance in the fundamental band of N_2 (530 Torr) in Ar (570 Torr) at 87 K for a path length of 154 m. Reproduced with permission from Ref. 89.

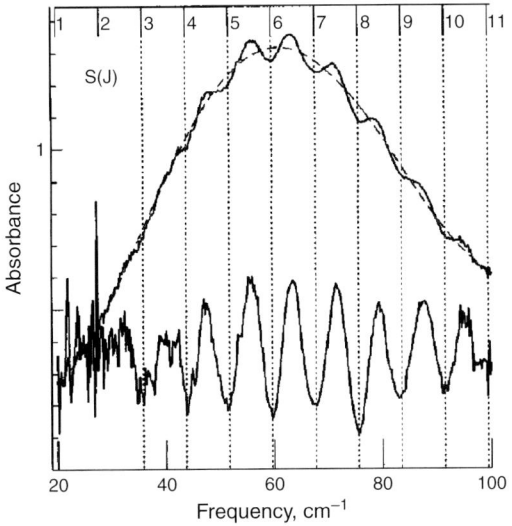

Fig. VI.14: Measured absorbance in the rototranslational band of N_2 (50%) in Ar (50%) at 89 K for a path length of 154 m and a total density of 5.2 Am. The total spectrum is shown (top curve) together with the residual ripples (lower curve, multiplied by a factor of 5 and shifted) obtained after the subtraction of a sixth order polynomial (dashed line). Reproduced with permission from Ref. 783.

region. In both cases, the overall band shape is a broad continuum due to free-free transitions but two other types of structures can be observed. The first are sharp features in the low frequency part of the spectrum below 40 cm^{-1} in Fig. VI.14, corresponding to low J values. The second are broader "ripples" becoming smooth at larger J values. In Fig. VI.14, the vertical dotted lines, which indicate the positions of S transitions of the isolated N_2 molecule, show that the absorption has minima at these frequencies rather

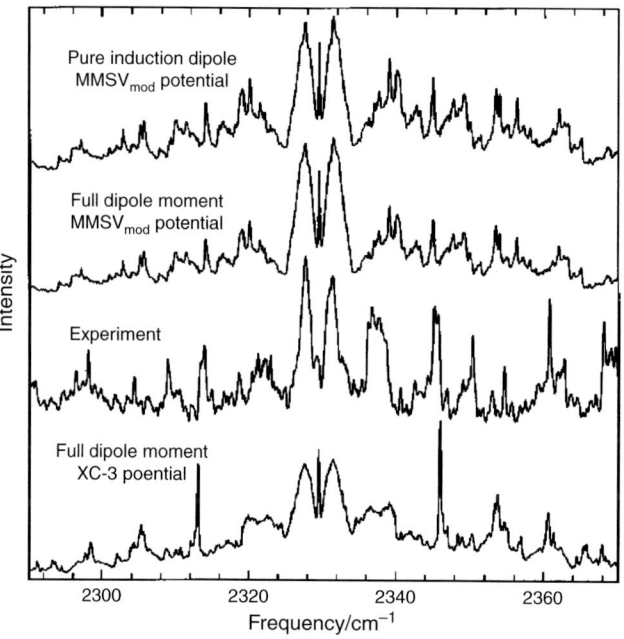

Fig. VI.15: Comparison of simulated mid-IR spectra (after removal of the broad pure CIA contribution) of N_2–Ar at T = 77 K calculated using various dipole and potential energy surfaces with the experiments (3rd from top) from Ref. 783. Reproduced with permission from Ref. 785.

than the expected maxima. The spectrum in the 2350 cm^{-1} region (Fig. VI.13) contains the same three contributions: a broad collision-induced continuum, fine structures near the band origin (cf Fig. VI.15) followed by broader ripples.

According to the analysis of Refs. 89,784,785, the narrow features observed in the low J region arise from bound-bound transitions while the ripples are due to transitions between pseudobound states of the complex [also called by some authors "rotational resonances" and by some others "rotational predissociating states", lying in the continuum, cf Sec. II.6 and Eq. (II.87)]. For higher J values, N_2 behaves as a free internal rotor within the complex. In this case, the isotropic approximation is valid, and the eigenstates can also be labeled with the quantum number ℓ for the rotation of the dimer *as a whole* about its center of mass. While $\Delta J = +2$ for the S lines, one has $\Delta \ell = \pm 1, \pm 3$. Thus, each S line has satellites consisting of N and P branches on one side and R and T branches on the other one. Since the transitions $\Delta \ell = 0$ are not allowed, there is no Q branch satellite and hence there is a minimum at the free molecule S(J) frequency. Indeed, the satellite lines are closely spaced (the rotational constant associated with the rotation of the complex is small, about 0.065 cm^{-1}) and, due to predissociation, the line widths are greater than 1 cm^{-1}. When making the sum of the individual profiles over all S transitions, the maxima of adjacent lines overlap but a minimum remains at the free molecule frequency, in agreement with the ripple part of the spectrum. Ripples have also been observed at higher temperatures (200–300 K) in the mid infrared for N_2–X and

O_2–X pairs.[77,78,725,786] They have been interpreted[(2)], [787] following Refs. 89,784, as arising from transitions between rotational resonances. Free internal rotation of N_2 in the N_2–Ar dimer is expected to occur only for high J values. For lower ones, J is no more a good quantum number because of the anisotropy of the potential and, as J decreases, one encounters *truly bound* states. Thus, at low J, the ripples disappear and sharp structures appear, which are due to bound-bound transitions (at least at low temperature; bound meaning here truly bound). The complexity of the interpretation of such experiments thus appears again. There is, today, no global calculation of the whole N_2–Ar spectrum. Therefore, in order to identify the part of the experimental spectrum which corresponds to bound-bound transitions, one has to subtract, more or less empirically, the contribution not due to bound-bound transitions. A method is to calculate them but the results may be inaccurate due to cumulative effects of uncertainties on the interaction potential and dipole moment surfaces. Once this is done, it is then possible to compare the observed N_2–Ar van der Waals spectrum with theoretical prediction, as exemplified by Fig. VI.15. For completeness, let us mention that spectral signatures of CO_2–O_2 complexes have been recently observed in the infrared spectrum of $CO_2 + O_2$ mixtures, near the O_2 fundamental vibration.[729] These results stress the importance of including bound (and metastable) states in any theoretical modeling of CIA when one of the collision partners (here CO_2) has a large electrostatic moment.

The interested reader can find more details on infrared studies of van der Waals complexes in Refs. 719(Sec. G) and 788. Our purpose here was essentially to demonstrate the entanglement between collision-induced absorption and the spectra of the corresponding complexes (the same considerations also apply to Raman spectra). As outlined by McKellar "… The absorption due to dimers is an integral part of the CIA spectrum, and there is no simple way to separate the two effects without recourse to fairly sophisticated theory…".[789]

VI.6 INTERFERENCE BETWEEN PERMANENT AND INDUCED DIPOLES (CIA) OR POLARIZABILITIES (CILS)

Up to now, in this chapter, only purely collision-induced spectra have been considered although, in most cases, permanent and induced operators coupling the molecules to the electromagnetic field simultaneously exist. Usually, in absorption spectra, the allowed lines due to the permanent dipole have intensities sufficiently large to dominate the collision-induced contribution at moderate densities. Hence, the two types of contributions need not to be simultaneously considered for the same transition. However, when they are of comparable strengths, some interesting interference features can be observed in the overall spectrum, as discovered by McKellar in the $v=1 \leftarrow v=0$ band of HD.[79] This species is quite unique since, in contrast with H_2, it has a (weak) permanent dipole moment making the intensity of the induced spectrum greater or comparable to that of the allowed one at densities of a few tens of amagat. This specific problem of comparable permanent and

---

[(2)] The interpretation of the infrared spectrum of N_2–Ar given here also applies to the spectra of other N_2 and O_2 complexes, at least in the frequency range where the ripples do not depend significantly on the anisotropy of the potential.

induced components and of their interference in the case of HD is discussed in § VI.6.2. Depolarized light scattering is another interesting situation since it involves not only the induced polarizability but also that of the isolated molecule. In the following, a review of some recent progresses in the understanding of CILS spectra is first given.

VI.6.1 DEPOLARIZED LIGHT SCATTERING SPECTRA OF H_2 AND N_2

The depolarized light scattering spectra of H_2 and N_2 have been discussed in § V.4.2, from the only point of view of the far wings of the allowed contribution due to the permanent polarizability. It was mentioned that the interpretation of the whole observed spectra requires a more refined treatment since contributions of *both* the allowed and induced components of the polarizability are involved.

Let us first consider the rototranslational Raman scattering of H_2 experimentally studied in Refs. 80,662. While the spectrum in the 10–150 cm^{-1} spectral region was quite immediately attributed[790] to the purely induced component, the interpretation of the wing up to about 1000 cm^{-1} is more complex. As shown in Fig. V.6, this part of the spectrum can be reasonably explained by the allowed far wing contribution. Nevertheless, one must analyze the importance of the purely induced contribution $\Phi_2^{II}(t)$ and of the cross terms due to $\Phi_2^{al}(t) + \Phi_2^{la}(t)$ in Eq. (II.77). Borysow and Moraldi[661] have proposed a formalism, described in Sec. V.4, for the simultaneous calculation of the three components (allowed, induced, and cross term) of the profile. It is based on a perturbative treatment of the evolution operators assuming that the anisotropic part of the potential is sufficiently weak. According to Eqs. (II.7) and (II.77), the light scattering cross section for a depolarized component (per unit volume) in a right angle experiment[80,662] results from three components, ie:

$$\frac{\partial^2 \sigma}{\partial\Omega\partial\omega} = \frac{\partial^2 \sigma^{CILS}}{\partial\Omega\partial\omega} + \frac{\partial^2 \sigma^{PB}}{\partial\Omega\partial\omega} + \frac{\partial^2 \sigma^{cross}}{\partial\Omega\partial\omega}. \qquad (VI.24)$$

The first term, which results from the Fourier transform of $\Phi_2^{II}(t)$, is the purely collision-induced (CILS) contribution, while the second, associated with the correlation function $\Phi_2^{aa}(t)$, forms the pressure-broadened (PB) allowed spectrum. Finally, the Fourier transform of the two remaining contributions $\Phi_2^{al}(t)$ and $\Phi_2^{la}(t)$ in Eq. (II.77) gives rise to the third term, *ie* the cross spectrum. Remember that, as discussed in appendix II.A, the operator d of Eq. (II.76) is expressed, in the case of depolarized CILS, in terms of the polarizability tensor and of the directions of the electric vectors in the incident and scattered radiations [Eqs. (II.A14) and (II.A15)].

The *allowed* part of the spectrum has been discussed previously. The corresponding cross section is obtained from Eqs. (II.59) and (II.A20), accounting for the fact that the rank of $\vec{d}$ is $k=2$ for the anisotropic polarizability. For a linear molecule, one has:

$$\frac{\partial^2 \sigma^{PB}}{\partial\Omega\partial\omega} = n_a \frac{\omega_{inc}\omega_{scat}^3}{c^4} \frac{\gamma^2}{15} \sum_{J_i,J_f,J'_i,J'_f} \rho_{J_i}(-1)^{J_i+J'_i} \begin{pmatrix} J_i & 2 & J_f \\ 0 & 0 & 0 \end{pmatrix} \begin{pmatrix} J'_i & 2 & J'_f \\ 0 & 0 & 0 \end{pmatrix} \sqrt{(2J_f+1)(2J'_f+1)}$$

$$\times \frac{1}{\pi}\text{Im}\{\langle\langle \beta'_f J'_f \beta'_i J'_i |[\omega - L_a - in_p W(\omega)]^{-1}|\beta_f J_f \beta_i J_i\rangle\rangle\}, \qquad (VI.25)$$

where γ is the permanent anisotropic polarizability, ω_{inc} and ω_{scat} being the angular frequencies of the incident and scattered lights, respectively. The expressions of the elements of the frequency-dependent relaxation matrix $W(\omega)$ within the approach of Ref. 661 have been given in Sec. V.4.

The *purely induced* part of the spectrum has been calculated[661,663] within the isotropic approximation described in Secs. II.6 and VI.3. Following Eqs. (VI.4), (VI.9), and (VI.10), it is given by:

$$\frac{\partial^2 \sigma^{CILS}}{\partial \Omega \partial \omega} = n_a n_p \frac{\omega_{inc} \omega_{scat}^3}{c^4} \frac{3}{20} \sum_{\lambda_1, \lambda_2, \Lambda, L} \sum_{J_1, J_1', J_2, J_2'} \rho_{J_1} \rho_{J_2} (2J_1' + 1)(2J_2' + 1)$$

$$\times \begin{pmatrix} J_1 & \lambda_1 & J_1' \\ 0 & 0 & 0 \end{pmatrix}^2 \begin{pmatrix} J_2 & \lambda_2 & J_2' \\ 0 & 0 & 0 \end{pmatrix}^2 \quad \text{(VI.26)}$$

$$\times V \times G_{\lambda_1, \lambda_2, \Lambda, L}((\omega - \omega_{J_1', J_1} - \omega_{J_2', J_2}).$$

Finally, the *cross part* results from the interference between the permanent and induced polarizabilities. Its expression has been carefully established in Ref. 663 where all needed equations can be found. Nevertheless, as shown below, its contribution to the Raman intensity turns out to be negligible in the case of H_2. Note that this is *not* a general result since the cross term plays a large role in the case of HD (§ VI.6.2). The intermolecular quantities which are needed for spectra calculations are essentially the polarizability induced in a pair of molecules and the intermolecular potential energy surface. On this basis, the rototranslational Raman scattering cross sections for pure H_2 at 300 and 50 K have been calculated and compared with measured values in Ref. 663. The results in Fig. VI.16 show that the CILS dominates for frequencies below about $250\,\text{cm}^{-1}$, corroborating the assumption made in Ref. 790. On the opposite, as already demonstrated by Fig. V.6, the $500-1000\,\text{cm}^{-1}$ range is totally dominated by the pressure-broadened allowed contribution. Light scattering in H_2 thus results from both the permanent and purely induced polarizabilities correlation functions with amounts that depend on the spectral region, and the term associated with their interference is always negligible.

Let us now consider the depolarized light scattering spectra of N_2 of Ref. 671. The scattered intensity for small frequency shifts (below about $250\,\text{cm}^{-1}$) was attributed by most authors[668,671] to the permanent polarizability through σ^{PB} in Eq. (VI.24). For larger frequency shifts, Bancewicz et al[668] first assumed that the spectrum is purely interaction induced. However, in order to match the measured data, the theoretical value of the dipole-octopole polarizability of N_2 then had to be multiplied by an unreasonable factor of about nine (see Ref. 664 for more details). This was explained in a study,[664] discussed in § V.4.2, which shows that the scattering spectrum for large frequency shifts does not arise from σ^{CILS} but from σ^{PB}. Contrary to the assumption of Ref. 668, the pressure-broadened allowed component is much larger than the collision induced one *everywhere* (for small and large frequency shifts). This is a situation very similar to that encountered in absorption spectroscopy, with the noticeable exception discussed below.

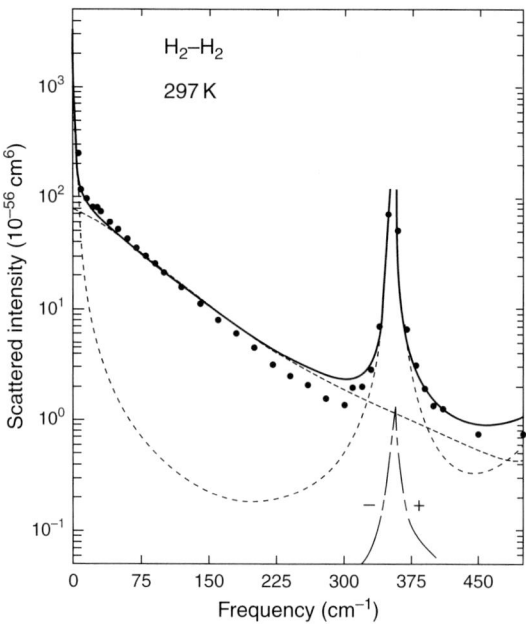

Fig. VI.16: Density (bi)normalized depolarized light scattering for pure H_2 at 297 K. (•) are experimental values.[80] The total theoretical spectrum (——) is composed of the interaction-induced spectrum (·····), the pressure-broadened allowed spectrum (- - -, see Chapt. V, related to Eq. (V.32) through Eq. (5) of Ref. 664, and the cross term (– – –) whose absolute value is plotted, its sign being indicated by $+$ and $-$. Reproduced with permission from Ref. 661.

VI.6.2 THE HD PROBLEM

Interference effects in HD absorption spectra, due to the $\Phi_2^{al}(t) + \Phi_2^{la}(t)$ contribution to the correlation function, have been first observed in Ref. 79. Since this work, extensive measurements have been made at various temperatures for pure HD and mixtures with rare gases, in both the 0←0 and 1←0 bands. The theoretical and experimental studies prior to 1985 have been reviewed by Poll[791] (see also Refs. 792,793). This interference effect, which is due to constructive and destructive interferences during a collision, is referred to as "*intracollisional interference*".

An observed spectrum of a HD-He mixture is plotted in Fig. VI.17. Sharp dipole allowed lines are then superimposed on a broad collision-induced background due to the Fourier transform of $\Phi_2^{II}(t)$ [*cf* Eq. (II.77)].

The evolution of the R(0) line profile with the perturber density n_p, as obtained after the empirical removal of the background, is shown in Fig. VI.18. While the transition has an almost symmetric Lorentzian shape at low densities, the asymmetry increases with n_p and the shape becomes essentially dispersive (with a strong negative part). This profile composed of a Lorentzian and a dispersive contribution is known, in the "HD community", as a Fano line shape. It was first introduced by Fano[795] in a different

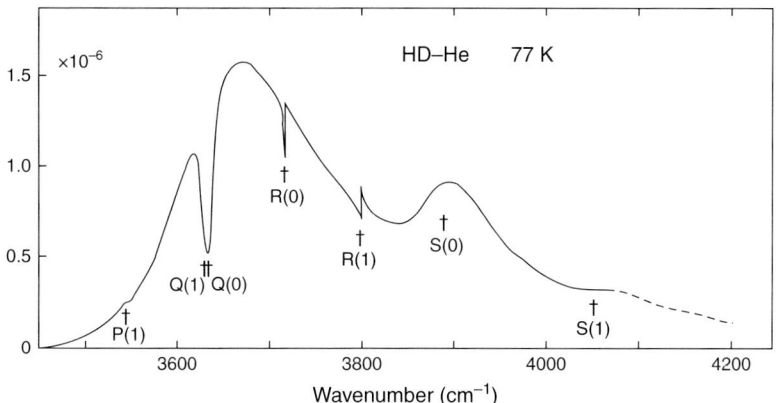

Fig. VI.17: Measured density (bi)normalized absorption (cm^{-1}/Am^2) for a HD (14.5 Am)-He (205 Am) mixture at 77 K. Reproduced with permission from Ref. 794.

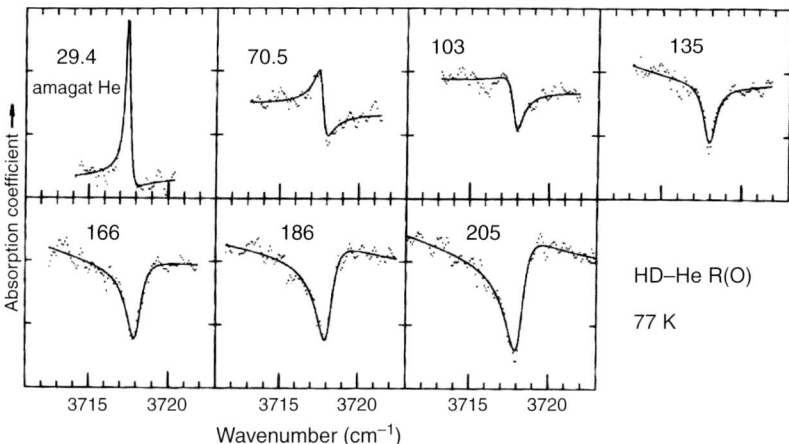

Fig. VI.18: Absorption profiles of the R(0) transition of HD in HD–He mixtures at 77 K for a HD density of 14.5 Am and He densities from 29.4 to 205 Am. The points represent the experimental spectra while the lines denote the fits using Fano line shapes. Reproduced with permission from Ref. 794.

context and appears in other processes, such as low overlapping line mixing [Eq. (IV.22)] and finite duration of collision effects [Eq. (II.B12) and Fig. III.8].

The theoretical understanding of the intracollisional interference phenomenon between the permanent and induced components of the dipole was first developed considering the integrated intensity.[796,797] The justification of the Fano line profiles was then established in a somehow heuristic manner,[798,799] some of the limits of this analysis being discussed in Refs. 800,801. Then, Gao et al[802–804] have proposed a classical path approach leading to a more general and consistent description of the process since it takes inelastic rotational changes induced by collisions into account (they are of crucial importance). However, in the pressure range of the experimental data, the intracollisional interference competes with an other mechanism, known as *"intercollisional interferences"*, which corresponds to correlation between pure induced dipoles in successive collisions [$\Phi_3^{II}(t)$, Eq. (II.80)]. This effect (§ VI.6.3), which varies, for highly diluted radiators, as $n_a n_p^2$, modifies both the shapes and the integrated intensities of the lines in the pressure range of most experiments. Then, following Ref. 798, it is customary to include it in the theoretical treatment. We give below a summary of the main results of Refs. 802,803.

Starting from the Fourier transform of the autocorrelation function $\Phi(t)$ [Eqs. (II.77)–(II.80)] given by:

$$\Phi(t) = \Phi_1^{aa}(t) + \Phi_2^{al}(t) + \Phi_2^{la}(t) + \Phi_3^{II}(t), \qquad (VI.27)$$

by making the impact and collisionally isolated lines approximations, taking the average over all orientations of the radiator, perturber, and intermolecular axis, and considering classical trajectories determined from the isotropic part of the potential, the expression of the absorption coefficient of a given "allowed" f←i optical transition is:[803]

$$\alpha(\omega, n_a, n_p) = \frac{n_a}{\pi} S_{fi} \left\{ \frac{\Gamma_{fi}}{(\omega - \sigma_{fi} - \Delta_{fi})^2 + (\Gamma_{fi})^2} (1 + \tilde{a}_{fi} n_p + \tilde{b}_{fi} n_p^2) \right. \\ \left. - \frac{(\omega - \sigma_{fi} - \Delta_{fi})}{(\omega - \sigma_{fi} - \Delta_{fi})^2 + (\Gamma_{fi})^2} (\tilde{c}_{fi} n_p + \tilde{d}_{fi} n_p^2) \right\}. \qquad (VI.28)$$

Here, S_{fi} is the unperturbed integrated intensity [Eq. (IV.9)] and $\Gamma_{fi} = n_p \gamma_{fi}$ and $\Delta_{fi} = n_p \delta_{fi}$ are respectively the usual pressure-broadened half-width and frequency shift of the transition. The profile in Eq. (VI.28) contains a dispersive term whose relative importance depends on the $\tilde{c}_{fi}$ and $\tilde{d}_{fi}$ parameters (relative to the broadening coefficient γ_{fi}). Note that the integrated absorption [area below $\alpha(\omega)$] now depends on the perturber density in contrast with the usual behavior of Lorentzian lines [as can also be observed[131] when finite duration of collision effects affect the transitions, *cf* Eq. (II.B12)]. The *real* parameters $\tilde{a}_{fi}$, $\tilde{b}_{fi}$, $\tilde{c}_{fi}$, and $\tilde{d}_{fi}$ are defined by:[803]

$$\tilde{a}_{fi} = \text{Re}[d_{fi}^{(1)} + d_{fi}^{(2)}]/d_{fi}, \quad \tilde{b}_{fi} = [\text{Re}(d_{fi}^{(1)})\text{Re}(d_{fi}^{(2)}) - \text{Im}(d_{fi}^{(1)})\text{Im}(d_{fi}^{(2)})]/d_{fi}^2, \\ \tilde{c}_{fi} = \text{Im}[d_{fi}^{(1)} + d_{fi}^{(2)}]/d_{fi}, \quad \tilde{d}_{fi} = [\text{Re}(d_{fi}^{(1)})\text{Im}(d_{fi}^{(2)}) + \text{Im}(d_{fi}^{(1)})\text{Re}(d_{fi}^{(2)})]/d_{fi}^2, \qquad (VI.29)$$

where $d_{fi} = \langle J_f \| d_a \| J_i \rangle$ is the reduced matrix element of the permanent dipole moment. $d^{(1)}$ and $d^{(2)}$ are given in terms of the collision-induced dipole $\vec{d}_I(t)$ operator by:

$$d_{fi}^{(1)} = \sum_{m_i, m_f, q} (-1)^{J_f - m_f} \begin{pmatrix} J_f & 1 & J_i \\ -m_f & q & m_i \end{pmatrix} \int_{-\infty}^{+\infty} e^{i\omega_{fi} t_0} \\ \times \left\{ \langle J_f m_f | U^+(t_0, -\infty)[\vec{d}_I(t_0)]_q U(t_0, -\infty) | J_i m_i \rangle^* \right\}_{Av} dt_0, \quad \text{(VI.30)}$$

and

$$d_{fi}^{(2)} = \sum_J e^{-(E_J - E_{J_i})/k_B T} \sum_{m_i, m_f, q} (-1)^{J_f - m_f} \begin{pmatrix} J_f & 1 & J_i \\ -m_f & q & m_i \end{pmatrix} \\ \int_{-\infty}^{+\infty} e^{-i\omega_{fi} t_0} \left\{ \langle J_f m_f | U(\infty, t_0)[\vec{d}_I(t_0)]_q U(t_0, -\infty) | J_i m_i \rangle \right. \\ \left. \times \langle J_i m_i | U(+\infty, -\infty) | J_i m_i \rangle^* \right\}_{Av} dt_0. \quad \text{(VI.31)}$$

In these equations, U(t',t) is the evolution operator for the radiator-perturber collision from time t to time t' and $[\vec{d}_I(t)]_q$ is an irreducible component of the induced-dipole operator $\vec{d}_I(t)$ in the interaction picture [cf Eq. (II.9)]. In Eqs. (VI.30) and (VI.31), $\{\ldots\}_{Av}$ denotes the average over the impact parameter and initial kinetic energy within the classical path description of trajectories. The integration over the time t_0 of closest approach has been written out explicitly in these equations. The six collisional parameters of the Fano profile in Eq. (VI.28) have expressions given in Ref. 803. They can be calculated when the intermolecular potential and induced-dipole moment surfaces are known. For extensive comparisons between measurements and predictions the reader can refer to Ref. 803, some more general statements being made now.

For the broadening coefficients, the agreement between experimental and theoretical values is satisfactory, except for the R(0) line for which semi-classical calculations overestimate elastic rates for small collision energies. If inelastic collisions are neglected (as in the previous formalism[798,799]) the parameters $\tilde{a}_{fi}$ and $\tilde{b}_{fi}$ show small dependences on J and the asymmetry parameters $\tilde{c}_{fi}$ and $\tilde{d}_{fi}$ turn out to be extremely small. The inclusion of inelastic collisions leads to results[803] which are not perfect but at least of the correct order of magnitude. A detailed analysis[792,803] has shown that the interference parameters are sensitive to distances near the turning point of the trajectories, the region where the classical-path assumption is the least valid. A fully quantum mechanical theory is thus necessary. This naturally introduces Refs. 776,805 in which a close-coupling scheme is used. It combines the radiative coupling with the diatom-atom scattering for the calculation of the intracollisional interference for He-broadened HD lines. This formalism has been discussed in Sec. VI.4, devoted to ab initio calculation of the CIA profiles for H_2-rare gas systems. We thus only describe here how the approach can be adapted in order to treat the HD case. In Eq. (VI.13), $d_z(\vec{r}, \vec{R})$, which was the induced dipole in the case of H_2, now becomes $d_z^I(\vec{r}, \vec{R}) + d_z^a(\vec{r})$, where $d_z^a(\vec{r})$ is the permanent dipole of HD. The standard expansion for the collision-induced dipole [Eq. (VI.1)] can also be used for the permanent one, which has only one non-vanishing spherical component given by:

$$d_z^a(\vec{r}) = \frac{4\pi}{\sqrt{3}} A^a(r) Y_0^1(\vec{r}/r). \quad (VI.32)$$

Spectral absorption profiles were then calculated[776,805] in a fully quantum treatment, limited to binary collisions between HD and He. This limits comparisons with experiments to low densities such that the wings of the Fano profile are linear *vs* the He density [thus neglecting the $\widetilde{a}_{fi}$, $\widetilde{b}_{fi}$, and $\widetilde{d}_{fi}$ parameters according to Eq. (VI.28)]. These theoretical absorption coefficients have then been fitted using Fano profiles and the resulting parameters are compared with those deduced from measurements in Table III of Ref. 776. Good agreement is obtained, but the accuracy of the model used has to be confirmed by similar studies for other perturbers and temperatures, and by more consistent experimental data. Indeed, since intracollisional interference effects are small, their observation requires high densities for which other effects contribute to (pollute) the spectral shape.

VI.6.3 INTERCOLLISIONAL DIPS

Intercollisional interferences corresponds to correlation between pure induced dipoles in successive collisions [$\Phi_3^{II}(t)$, Eq. (II.80)]. This effect, which varies as $n_a n_p^2$ has been mentioned as being responsible for changes of both the shapes and the

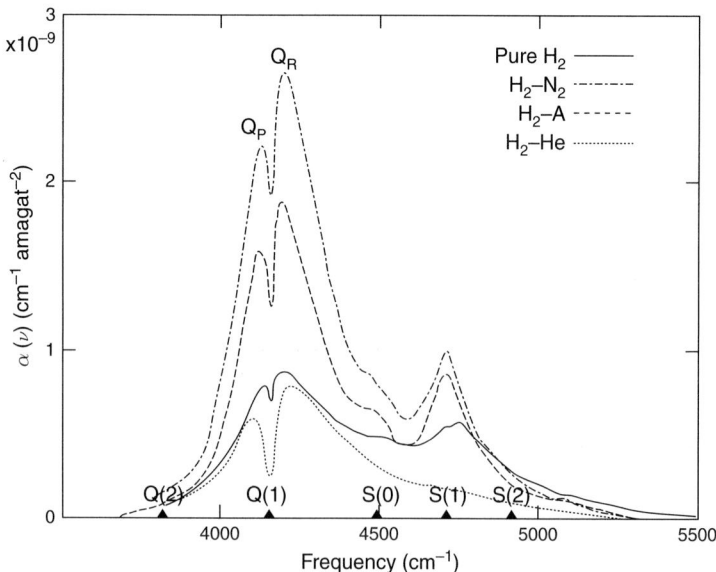

Fig. VI.19: Measured absorption in the fundamental band of H_2 for various mixtures at room temperature and about 100 atm. The dip at the position of the Q(1) line (between "Q_P" and "Q_R") is obvious. Reproduced with permission from Ref. 808.

integrated intensities of the lines of HD [through the $\widetilde{d}_{fi}$ and $\widetilde{b}_{fi}$ terms in Eq. (VI.28)]. Intercollisional interferences have also been observed in purely induced spectra.[76,806,807] Indeed, in the CIA by H_2–H_2 and H_2–He pairs, the Q branch and, to a lesser extent, the S(1) line display relatively narrow absorption minima or *dips* at the frequencies of the free H_2 molecule transitions, as shown in Fig. VI.19 and Refs. 76,808. It was demonstrated by Van Kranendonk[809] and then by Lewis,[810,811] by taking $\Phi_3^{II}(t)$ into account in the calculation of the absorption coefficient, that the dips can be modeled by (negative) Lorentzian profiles.

VI.7 CONCLUSION

For the most simple pairs, *eg* H_2–H_2, H_2–He, *etc*, the modeling of CIA/CILS spectra has reached a high degree of sophistication, thanks to the applicability of refined quantum approaches and to the availability of accurate ab initio induced dipoles or polarizabilities and intermolecular potential energy surfaces. In other words, the binary dipoles surfaces and associated spectra are well known. On this firm basis, ternary dipoles and their spectra have been the subject of intense research efforts, at least for the systems mentioned above (*cf* section A of Ref. 719 and Refs. 812–814). Indeed, as the gas density increases, ternary (and eventually higher order) features appear more or less strikingly in the collision-induced absorption or light scattering which are thus given as a power series in densities.[815] At not too high densities, the binary and ternary components can be separated experimentally, due to their differing density dependences. Ternary components have been measured in absorption,[816] as well as in light scattering.[817] Some other successful measurements of ternary spectra are discussed in Ref. 823. Such data shed some light on the nature of the interaction-induced ternary dipoles, and more particularly on the importance of the irreducible ternary component.[812] Remember that the dipole induced when three H_2 molecules collide, d(123), is the sum of the pairwise additive components d(12) + d(13) + d(23) *and* of an irreducible three-body contribution.[84] The importance of this irreducible component (discussed in Refs. 720,813,818 and those therein) has been corroborated by the first observation in 1995 of a simultaneous transition $Q_1(J_1) + Q_1(J_2) + Q_1(J_3)$ in H_2,[819] where the three interacting molecules simultaneously absorb a single photon. More work is necessary for the understanding of the irreducible ternary component, but as it appears from this short discussion, hydrogen continues to play a crucial role in elucidating new experimental and theoretical aspects of CIA/CILS.

For heavier molecules, such as N_2 and O_2, since ab initio dipoles (and polarizabilities) are not available yet, one has to represent them by a classical multipole induced-dipole approximation which neglects (or else models empirically) the exchange forces. Elaborate calculations of the induced moments at short distances are clearly needed for future research on CIA/CILS if one wants to avoid the use of ad hoc adjustments. This is a challenge for our colleagues of quantum chemistry. Among the problems of particular atmospheric importance, let us mention the interpretation of the collision-induced electronic transitions of O_2, located in the near infrared and visible regions and assigned to single and double electronic transitions in pairs of O_2 molecules[820] (for pure O_2 samples). Numerous measured data are available[724,821,822] (although new measurements of the temperature dependence are clearly needed), but, as shown in Ref. 764, the main requirement, to go further on, is a detailed and quantitative study of the possible induction mechanisms which are operative during O_2–X collisions. On the basis of a reasonable knowledge of the integrated intensities, a dynamical treatment of the collisions themselves will lead to a physically based modeling of the observed profiles. From this point of view, the challenge of primary importance for all these "intermediate" pairs (N_2–N_2, N_2–O_2, O_2–O_2, *etc*) is to go beyond the isotropic approximation, *ie* to properly take into account the influence of the anisotropy of the interaction potential in the spectral-shape calculation. From the results obtained for the spectral moments (*cf* end of section VI.3), one expects much larger spectroscopic effects of the anisotropy than those obtained for H_2–X pairs. However, close-coupling calculations like those presented in Sec. VI.4 are not yet feasible for N_2–X or O_2–X. Following a path similar to that used in quantum calculations of widths and shifts cross sections (*cf* Sec. IV.3.5c), it might be of some interest to investigate the possibility of a calculation *within the coupled-states approximation*. This requires much less computing efforts for a system of reasonable anisotropy, and, from this point of view, N_2–Ar may be a good candidate for such a test.

VII. CONSEQUENCES FOR APPLICATIONS

VII.1 INTRODUCTION

The variety of scientific fields in which the modeling of the effects of collisions on the shape of gaseous molecular spectra finds applications is almost as broad as that covered by radiative transfer calculations in gases. Except for media at very low densities, where purely Doppler profiles may be sufficient, spectra predictions need, a minima, to take into account the (Lorentz) pressure-broadening of individual transitions. Furthermore, there are many practical cases, depending on the spectral resolution, on the path lengths involved, and on the thermophysical conditions, where some other processes cannot be disregarded without significant consequences. Besides Doppler and Lorentz broadenings, refined effects on isolated line shapes, due to velocity changes and speed dependences (Chapt. III), must be taken into account at intermediate pressures for high accuracy laser soundings of gas properties. Similarly, collisional line mixing (Chapt. IV) must often be properly modeled for accurate Raman thermometry using Q branches, for instance. For large optical thicknesses, accurate descriptions of the line wings (Chapt. V) and of collision-induced absorption (Chapt. VI) may be required for both heat transfer and remote sensing applications. The media for which radiative transfer modeling is used are very diverse, with conditions ranging from the hundreds of kilometers at low pressure and temperature of some planetary atmospheres, to the few centimeters or meters at high pressure and temperature of some combustion processes. This variety of conditions is so large that all the collisional effects discussed in the preceding chapters find applications.

The practical purposes for which spectra calculations are carried out are mainly twofold. The first are predictions, using a radiative model, of spectrally integrated heat transfer properties (fluxes and dissipated powers) within the medium of interest. They are based on the transmissions and emissions of elementary paths which are then included in the computation of radiances in one-, two-, or three-dimensional systems. Examples are given by the greenhouse effect and more generally by radiative exchanges in atmospheres and also, at higher temperatures, by infrared heat transfer in combustion media. The second type of application is optical sounding, again based on spectra calculations, but where these are used for the retrieval of some characteristics of the studied medium through fits of recorded signals. This is the case of the remote sensing determination of some vertical profiles [of volume mixing ratios (vmr) and/or of pressure and temperature] in atmospheres from Fourier transform spectra. Another example is the local sounding of temperature and/or concentrations in flames or combustion exhaust gases using infrared or Raman laser techniques. In many cases the information searched for is mainly obtained from the area (*ie* the integrated intensity) of the measured spectra or time-dependent response signals. The (collisional) spectral shape is considered as known but the quality of its description *conditions* the accuracy of retrieved quantities. On the contrary, in some specific situations, the (interesting) information contained in the spectral shape *itself* has been considered. This possibility, seldom used in the past due to insufficient confidence in the models and to low spectral resolutions, is more and more exploited, thanks to the availability of accurate approaches and to the performances of modern sounding instruments.

The remainder of this chapter, devoted to applications of the knowledge of molecular spectral shapes, is divided into two parts. In Sec. VII.2, basic elements concerning radiative heat transfer and remote sensing are recalled. The consequences for applications of various manifestations of intermolecular collisions presented in chapters III–VI are then discussed in Secs. VII.3–VII.5. Examples are given of the impact that spectral shapes influenced by pressure can have on predictions of radiative characteristics of gas media and on their subsequent use for optical soundings. These fields in which the subject of this book finds applications are in constant evolution, following the improvements of the instruments and measurement methods, of the models, and of the available computer power. While applications have been, in the past, following the progresses of laboratory experimental and theoretical spectroscopy, the situation is now reversing. The quality of spectra recorded by modern sounding instruments, and the challenging atmospheric physics problems that they enable to tackle, nowadays strongly push ahead the spectroscopic needs. As an example, the Orbiting Carbon Observatory project[536] aims at measuring the CO_2 amounts in our atmosphere with an accuracy better than 1%, using satellite-borne infrared spectra. This will be remotely done using CO_2 bands near 1.6 and 2.1 µm after determination of the total pressure vertical profile from O_2 spectra near 0.76 µm. This project requires an accuracy of the spectroscopic data and models (the objective is 0.3%) for both target spectral regions and species which is not currently achieved and that will need ever more refined laboratory developments.[537] The prediction of the long term evolution of climate is also very challenging, requiring calculations of the contributions of numerous molecular species over a broad wavelength range, with results sensitive to any error on the calculated absorptions and emissions of radiation.

The authors of this book are specialists neither of radiative heat transfer, remote sensing and optical sounding, nor of atmospheric or combustion physics. This chapter is thus far from being exhaustive and is limited to examples and bibliographic indications. The aim is mostly to convince the reader that collisional effects on molecular spectra or time-dependent response signals are not only a curiosity of spectroscopy laboratories, but that they can have important consequences for other types of research.

VII.2 BASIC EQUATIONS

VII.2.1 RADIATIVE HEAT TRANSFER

Since good references are available on radiative transfer, only a few elements connected with the purpose of this chapter are given here. In particular, light scattering processes and interactions of radiation with solid surfaces are not considered. For more information on the methods, problems, and applications in the field of radiative heat transfer the reader can refer to Refs. 824–827 and those cited therein.

The basic quantities involved in radiative heat transfer calculations in gases are the spectral values, at wavenumber σ (or wavelength λ, frequency ν, angular frequency ω),

Consequences for applications

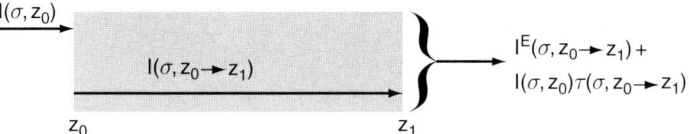

Fig. VII.1: Schematic view of gas emission and absorption.

of the *transmission* $\tau(\sigma)$ and emitted *radiance* $I^E(\sigma)$ for any path within the medium, as schematized in Fig. VII.1.

In the absence of light diffusion and assuming thermodynamic equilibrium, $\tau(\sigma)$ and $I^E(\sigma)$ are directly related to the local values of the absorption coefficient α at any point along the optical path by:

$$\tau(\sigma, z_0 \to z_1) = \exp\left\{-\int_{z_0}^{z_1} \alpha[\sigma, s(z)]\frac{ds(z)}{dz} dz\right\}, \quad \text{(VII.1)}$$

$$I^E(\sigma, z_0 \to z_1) = \int_{z_0}^{z_1} I_{BB}[T(z)]\alpha[\sigma, s(z)]\exp\left\{-\int_z^{z_1} \alpha[\sigma, s(z')]\frac{ds(z')}{dz} dz'\right\} dz. \quad \text{(VII.2)}$$

In these equations $s(z)$ is a curvilinear abscissa following the optical path from its starting point z_0 to its ending point z_1 [in the presence of gradients of the refractive index, $s(z)$ may not be a straight line]. The absorption coefficient depends on the spatial location through the local thermo-physical conditions: the total pressure (or density), the temperature, and the mixing ratios of the molecular species composing the gas mixture. I_{BB} is the black body radiance, given, in wavenumber (σ) and wavelength (λ) units, by:

$$I_{BB}(\sigma, T) = \frac{2hc^2\sigma^3}{\exp(hc\sigma/k_B T) - 1} = \frac{1.19 \, 10^{-8}\sigma(\text{cm}^{-1})^3}{\exp[1.44\sigma(\text{cm}^{-1})/T(K)] - 1} \, W/(m^2 \text{cm}^{-1} \text{sr}),$$

$$I_{BB}(\lambda, T) = \frac{1.19 \, 10^{+8}\lambda(\mu m)^{-5}}{\exp[14400.\lambda(\mu m)^{-1}/T(K)] - 1} \, W/(m^2 \mu m \, \text{sr}), \quad \text{with,} \quad \text{(VII.3)}$$

$$\int_0^\infty I_{BB}(\sigma, T)d\sigma = \int_0^\infty I_{BB}(\lambda, T)d\lambda = 5.67 \, 10^{-8} \, T(K)^4/\pi \, W/(m^2 \text{sr}).$$

The total radiance $I(\sigma, z_1)$ at the end of the path contains the emitted contribution I^E, and that, I^T, due to the transmission of the incoming radiance $I(\sigma, z_0)$ (see Fig. VII.1), ie:

$$\begin{aligned} I(\sigma, z_1) &= I^E(\sigma, z_0 \to z_1) + I^T(\sigma, z_0 \to z_1) \\ &= I^E(\sigma, z_0 \to z_1) + I(\sigma, z_0)\tau(\sigma, z_0 \to z_1). \end{aligned} \quad \text{(VII.4)}$$

The spectral quantities $\tau(\sigma)$ and $I(\sigma)$ are not those directly of interest for applications in which spectral and spatial integrations, whose ranges depend on the problem, are made.

In remote sensing, these integrations generally cover relatively narrow intervals corresponding to the spectral resolution ($\Delta\sigma$) and the solid angle ($\Delta\Omega$) defining the field of view (FOV, cf Fig. VII.2a) of the instrument, ie:

$$\bar{\tau}(\sigma, z_0 \to z_1) = \Delta\Omega^{-1} \int_{\Delta\Omega} d\Omega \int_{\sigma-\Delta\sigma_{\text{Instr}}}^{\sigma+\Delta\sigma_{\text{Instr}}} \tau(\sigma', z_0 \to z_1) F_{\text{Instr}}(\sigma - \sigma') d\sigma', \quad \text{(VII.5)}$$

$$\bar{I}(\sigma, z_0 \to z_1) = \Delta\Omega^{-1} \int_{\Delta\Omega} d\Omega \int_{\sigma-\Delta\sigma_{\text{Instr}}}^{\sigma+\Delta\sigma_{\text{Instr}}} I(\sigma', z_0 \to z_1) F_{\text{Instr}}(\sigma - \sigma') d\sigma', \quad \text{(VII.6)}$$

where $F_{\text{Instr}}(\Delta\sigma)$ is the spectral response of the instrument (instrument function, eg a sin(x)/x for unapodized FT spectra). Note that in some studies, the radiance is converted into a "*brightness temperature*" $T_{\text{bright}}(\sigma)$ such that the black-body radiance [Eq. (VII.3)] at T_{bright} is equal to the considered radiance $I(\sigma)$:

$$T_{\text{bright}}(\sigma) = \frac{hc\sigma}{k_B} \bigg/ \ln\left[\frac{2hc^2\sigma^3}{I(\sigma)} + 1\right]. \quad \text{(VII.7)}$$

In heat transfer problems, one is generally interested by total fluxes Φ (W/m^2) absorbed by solid parts and by the radiative dissipation (or net powers) Q (W/m^3) within the gas. The first is obtained by the spectral integration of the incoming radiance over the *entire* spectral range and half space ($\Delta\Omega = 2\pi$ sr).:

$$\Phi = \int_0^{+\infty} d\sigma \int_0^{2\pi} d\varphi \int_0^{\pi/2} \cos\theta \, \sin\theta \times \varepsilon_{\text{Surf}}(\sigma, \theta, \varphi) \times I(\sigma, \theta, \varphi) d\theta, \quad \text{(VII.8)}$$

where $I(\sigma,\theta,\varphi)$ is the radiance at the considered point incoming along the θ,φ (Euler) direction with respect to the normal to the surface and $\varepsilon_{\text{Surf}}(\sigma,\theta,\varphi)$ is the corresponding spectral and bi-directional absorptivity of the surface. The total radiative power per unit volume dissipated at a given point in the gas is obtained from the integration of the divergence (in x,y,z) of the local radiance over the entire space ($\Delta\Omega = 4\pi$ sr), ie:

$$Q = \int_0^{+\infty} d\sigma \int_0^{2\pi} d\varphi \int_{-\pi/2}^{+\pi/2} \text{div}[I(\sigma, \theta, \varphi)] \sin\theta d\theta. \quad \text{(VII.9)}$$

The influence of intermolecular collisions on the absorption shape obviously has consequences which depend on the spectral and spatial resolutions. Global heat transfers are mostly sensitive to effects affecting broad regions of the spectrum, since refined processes influencing narrow signatures are masked by integration over wavelength. On the opposite, remote sensing, due to higher spectral resolution, often requires accurate modeling of small spectral intervals where detailed effects occur. When heat transfer calculations in multidimensional media are made, the spectral values (of α, τ, I) are generally not calculated. Models such as those described in the preceding chapters are much too computer costly to be included in three dimensional calculations of spectrally integrated fluxes and dissipated powers. This has motivated the development of simplified approaches for the direct prediction of radiative quantities integrated over spectral intervals relatively broad with respect to the spectral scale of the variations at high resolution. Among these one finds narrow band statistical models[827,828] correlated K fictitious gas distributions,[827–830] and empirical laws constructed from line-by-line approaches.[831,832] In most cases, Lorentz (or Voigt) line shapes are used and all effects leading to the breakdown of these profiles are disregarded.

VII.2.2 REMOTE SENSING

As for radiative transfer, the techniques for optical soundings of the thermophysical properties (T, P, species amounts, *etc*) of gas media and the afferent complex minimization problems are not the central concern of this book. Only a few elements, connected with the purpose of this chapter, are given here. For more details and bibliographical paths, the readers can refer to Refs. 31–34,833–836.

Retrievals of some characteristics of gas media through the "*inversion*" of measured spectra (or response signals in the time domain) are now widely used for a number of purposes. In this problem, the data given by the instrument can be broad band spectra collected by Fourier transform or grating spectrometers with more or less high spectral resolution, a narrow wavenumber interval where the absorption or Raman signal is scanned by laser systems, or even a few discrete wavelengths as in lidar soundings. The corresponding information can be spatially localized (in situ measurements) as in CARS experiments, or integrated over long paths as in atmospheric emission remote measurements. In the first type of experiment a small part of the gas medium is sounded and the collected data depend only on localized values of the thermophysical parameters. In the second type of sounding, the measured quantities depend on the conditions all along the "line-of-sight", making the determination of local values a more difficult inversion task. Whatever the experimental technique and the measured quantity, the retrieval principle is essentially the same but more or less complex depending on the type of measurement and on the number of parameters searched for. The so-called *inversion* scheme is basically the following (see Ref. 834 for more details): Let $\vec{y}$ be the "measurement vector" containing the measured quantity (radiance, transmission, Raman signal, *etc*). Its coordinates y_m correspond to various wavelengths (or delays for experiments in the time domain) and eventually different instrument pointing characteristics (*eg* zenith angles in atmospheric limb measurements). Let F be the "*forward model*" which enables the calculation of $\vec{y}$ provided that a "*state vector*" $\vec{x} = \vec{x}^u + \vec{x}^k$ is given as input. The latter includes unknowns $\vec{x}^u$ which are searched for through the inversion procedure (mixing ratios, pressure, temperature, *etc*) and known quantities $\vec{x}^k$ which are fixed (spectroscopic parameters, instrument response, some thermophysical parameters of the medium, *etc*). The relation between $\vec{y}$ and F, in the absence of any systematic error, is:

$$\vec{y} = F(\vec{x}) + \vec{\varepsilon}, \qquad (VII.10)$$

where $\vec{\varepsilon}$ is the vector of random error due to instrumental noise. Since the remote sensing problems for which inversions of recorded signals are used are various, $\vec{x}^u$ may include different types of parameters. (i) A first and common example is the spatial profiles of total pressure, temperature, and/or mole fractions of some specific species. (ii) The populations of some internal molecular levels can also be determined from spectral signatures, giving information on deviations from thermodynamic equilibrium. (iii) Soundings using optical transitions of molecular tracers can also provide aerodynamical information such as the speed and distribution of winds or some characteristics of turbulence. (iv) Apart from the gas medium, measured spectra can also be used for the determination of some characteristics of clouds and aerosols, as well as of the temperature and spectral characteristics of (ground, sea) surfaces. In these problems, the known vector $\vec{x}^k$ then contains all the other parameters characterizing the medium which have an influence on $\vec{y}$ through F, ie those x_j^k such as:

$$\left| \left(F(\vec{x}^k + \vec{x}^u)\right)_m - \left(F(\vec{x}^{k-j} + \vec{x}^u)\right)_m \right| \geq \delta\varepsilon_m, \qquad (VII.11)$$

where $\delta\varepsilon_m$ is a fraction (chosen according to desired accuracy) of the local noise at measurement point m and $\vec{x}^{k-j}$ is identical to $\vec{x}^k$ except for the x_j^k component which is set to zero.

The retrieval algorithms are based on more or less sophisticated procedures minimizing the mean deviation between predicted and measured values (ie $\|F(\vec{x}) - \vec{y}\|$, eventually including weights).[834] The non-linearity of the forward model is generally solved by means of Newton iterations. When no regularization nor weighting is made, the evolution on $\vec{x}^u$ from for iteration i to i + 1 is obtained from:

$$(\vec{x}^u)^{i+1} = (\vec{x}^u)^i + \left({}^tKS_\varepsilon^{-1}K^i\right)^{-1}{}^tK^iS_\varepsilon^{-1}\left\{\vec{y} - F[(\vec{x}^u)^i, \vec{x}^k]\right\}, \qquad (VII.12)$$

where S_ε and K are respectively the noise covariance matrix, which weights the measurements, and the Jacobian matrix (${}^tK$ being the transpose of K), whose elements are the derivative of the forward model F with respect to the state vector $\vec{x}^u$. It is thus clear that errors in F (and/or $\vec{x}^k$) propagate and result in inaccuracies of the converged value of $\vec{x}^u$. In some problems, such as atmospheric remote sensing, use of a too thin vertical grid, on which geophysical parameters are searched for, leads to an ill-posed problem. This results from measurement noise and from the overlap of the weighting functions (due to the non-diagonal character of the Hessian matrix ${}^tKK$) which induce correlations between the parameters at various altitudes. The inversion then has to be constrained, following physical or statistical considerations, for which various approaches have been proposed. These include the use of reduced representations (ie coarser grids), or of some a priori knowledge concerning the parameters of interest through optimal estimation or Tikhonov regularization methods.[834,837–839] A second difficulty arises since one has to choose which measured information must be retained for the retrieval of the parameters of interest. In order to illustrate the problem, consider the use of Fourier transform CO_2 atmospheric spectra for the determination of the temperature profile *only*, all other atmospheric state parameters being assumed known. Spectral intervals where species other than CO_2 contribute significantly must obviously be avoided if the corresponding vmr distributions are not retrieved simultaneously.

These distributions and the associated spectroscopic information are then part of $\vec{x}^k$, leading, if not accurately known, to systematic errors on the retrieved temperatures ($\vec{x}^u$). Regions where significant uncertainties on the spectroscopic characteristics and/or spectral shapes of CO_2 lines remain must also be avoided. Finally, the parts of the spectra which are very little sensitive to temperature must be disregarded since they only bring noise to the results. Of course, it is formally possible to use *all* measured information provided that all the geophysical parameters which have an influence are adjusted in the inversion scheme. In practice, this is difficult due to the concomitant large number of unknowns and the subsequent computer power needed. Furthermore, doing this may not be interesting due to the correlations between some of the parameters and the error propagation problems that they may induce. For this reason, only more or less narrow spectral intervals are generally used, called "micro-windows" although their size gets less and less "micro" following the increase of computer power and the improvement of forward models. The vector $\vec{y}$ used in the minimization is then a sub-set of the available measurement vector and the problem is to find the best intervals according to the targets of the inversion. These windows should be the most sensitive to $\vec{x}^u$ and the less sensitive to possible errors on $\vec{x}^k$ and F. Finding the optimal ones for a given remote sensing problem is a quite difficult task as discussed in Ref. 834. In order to define some terms used in the following of this chapter, let us first recall some of the observation situations.

When spatially integrated quantities are measured for the observation of atmospheres, for instance, there are mainly three situations schematized in Fig. VII.2. In the *nadir* geometry (Fig. VII.2a), the instrument, on board various platforms (balloon, plane, satellite, probe, spacecraft), looks vertically down to the surface. It collects photons within a given solid angle called *field of view* (FOV). In *ground-based* experiments (Fig. VII.2b), the instrument is on the ground and looks up, the angle between the center of the FOV and the normal at the surface being the *zenith angle* ZA (that with respect to the tangent to the ground being the *elevation angle* EA). In *limb* observations (Fig. VII.2c), the flying instrument points with a given EA defined by the central line of the FOV and the local tangent to the surface (the ZA being defined with respect to the normal). When looking down (negative EA), the altitude of the point where the line of

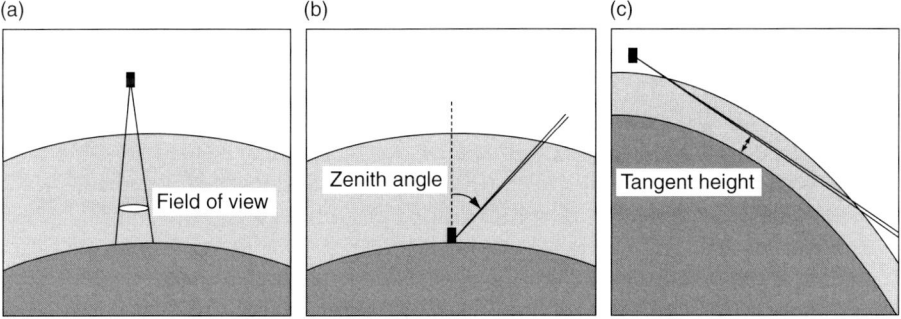

Fig. VII.2: Main geometries for spatially integrated atmospheric observations.

sight is locally parallel to the surface is called the *tangent height*. Nadir viewing provides a poor vertical resolution but generally a good horizontal resolution. On the opposite, a sequence of limb measurements for various ZA gives a good vertical resolution but a poor horizontal resolution. Ground-based measurements provide poor resolutions, both horizontally and vertically.

The photons collected by the instrument originate from the thermal emission of the atmosphere (and eventually from the ground in the case of cloud-free nadir observations) but they may also come from the sun (or the moon, a star). This is obvious in *solar occultation* measurements when the instrument directly points at the sun. It is also the case when the collected solar photons have been scattered in the atmosphere, on the ground, or by clouds or aerosols. For long wavelengths (typically $\lambda > 4\,\mu m$), and except for solar occultation measurements, the thermal emission of the atmosphere is generally dominant and the relevant quantity is the measured radiance. When directly pointing at the sun, the atmospheric transmission can be deduced from measurements after dividing by the extra-terrestrial solar radiance. The situation can be complicated at short wavelengths since one may have to include the diffusion of solar photons by molecules in the atmosphere, their scattering by aerosols and clouds, and their reflection on the surface.

When lasers are used for the sounding of gas media, the three main situations are schematized in Fig. VII.3. The simplest is that of direct absorption (Fig. VII.3a), where the attenuation of the laser intensity by the sample is measured. It is an efficient probing technique for in situ measurements over homogeneous and isothermal paths. In laser induced fluorescence, light detection and ranging, and spontaneous Raman measurements (Fig. VII.3b) the laser is shot and the scattered or induced-emitted photons are collected looking at a given point along the laser path. These techniques offer good spatial (and temporal) resolutions although some deconvolution can be required for non-homogeneous paths. Finally, two lasers, or more, can be used for double resonance or CARS (Fig. VII.3c and App. II.A), for instance. This provides the best achievable spatial resolution since only a very small volume is probed, at the crossing of the laser beams.

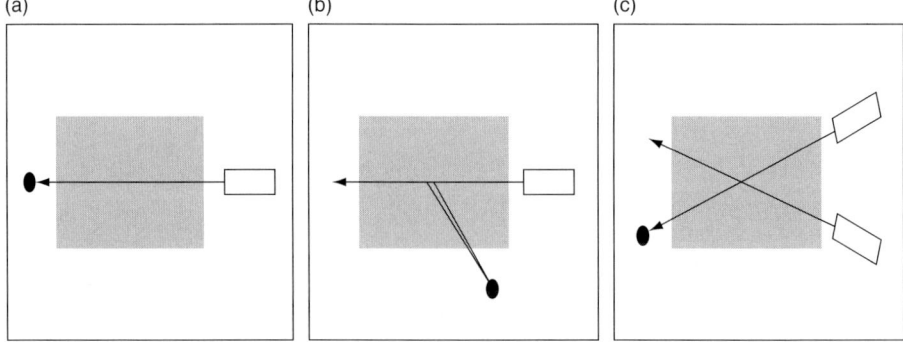

Fig. VII.3: Main geometries for laser soundings.

VII.3 ISOLATED LINES

VII.3.1 THE BASIC LORENTZ AND VOIGT PROFILES

The basic Lorentz (and Voigt) profile, parameterized by the collisional-broadening and -shifting coefficients, [cf Eqs. (III.17) and (III.21)] has been taken into account in heat transfer and remote sensing applications for a long time. Indeed, except for very low pressure where Doppler effects are dominant, the Lorentz broadening must be included in spectra calculations.

Radiative heat transfer

Even in situations where the spectral details are averaged, as in heat transfer calculations or when the spectral resolution $\Delta\sigma$ is poor when compared with the typical line width Γ, the Lorentz broadening must be taken into account. In order to evaluate its influence, consider a single transition centered at σ_0 and assume that $\Delta\sigma \gg \Gamma$. One then has to calculate the "equivalent width" Γ^{eq} of the line, defined in terms of the spectral transmission $\tau(\sigma)$ by:

$$\Gamma^{eq} \equiv \frac{1}{\Delta\sigma}\int_{\sigma_0-\Delta\sigma/2}^{\sigma_0+\Delta\sigma/2}[1-\tau(\sigma)]d\sigma \approx \frac{1}{\Delta\sigma}\int_{-\infty}^{+\infty}[1-\tau(\sigma-\sigma_0)]d(\sigma-\sigma_0). \quad (VII.13)$$

Γ^{eq} is a *key* quantity for radiative heat transfer, widely used in the development of narrow band statistical models.[827] Indeed, for values of $\Delta\sigma \gg \Gamma$ but sufficiently narrow to make the black body radiation I_{BB} constant [typically $10\,\text{cm}^{-1}$ in the thermal domain, see Eq. (VII.3)], the flux emitted by a gas column in the $[\sigma_0-\Delta\sigma/2, \sigma_0+\Delta\sigma/2]$ interval is $\Phi = \Delta\sigma \times \Gamma^{eq} \times I_{BB}(\sigma_0, T)$. For a Lorentzian line of integrated area S and half width Γ the transmission over an elementary path of length L is:

$$\tau(\sigma-\sigma_0) = \exp\left\{-\frac{SL}{\pi\Gamma}\frac{1}{1+(\sigma-\sigma_0)^2/\Gamma^2}\right\}. \quad (VII.14)$$

The equivalent width can then be rewritten as:

$$\Gamma^{eq} = \frac{\Gamma}{\Delta\sigma}\int_{-\infty}^{+\infty}\left\{1-\exp[-\text{peak}(\Gamma)/(1+u^2)]\right\}du, \quad (VII.15)$$

where $\text{peak}(\Gamma) \equiv (SL)/(\pi\Gamma)$ is the absorbance at line center. The integral in Eq. (VII.15) can be approximated by:[840]

$$\Gamma^{eq} \approx \frac{\pi\Gamma}{\Delta\sigma}\text{peak}(\Gamma)\left\{1+[\pi\text{peak}(\Gamma)/4]^{1.25}\right\}^{-0.4}. \quad (VII.16)$$

One can then study the influence of Γ on Γ^{eq} (and fluxes) from Fig. VII.4. This shows that the line broadening has a negligible influence for low peak absorbance values (peak < 0.1) where $\Gamma^{eq} \approx SL/\Delta\sigma$, but that it cannot be neglected when lines saturate (peak > 5), case for which Γ^{eq} is proportional to $\sqrt{\Gamma}$. The pressure broadening must then

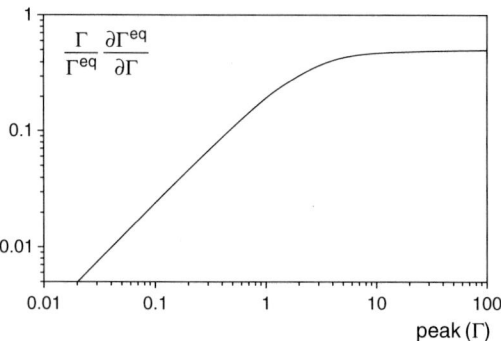

Fig. VII.4: Relative influence of the Lorentz broadening on the equivalent width of a single line as a function of the absorbance at line center.

be properly taken into account, even for calculations of quantities integrated over broad spectral ranges, since a relative error ε_Γ on Γ translates to $\varepsilon_\Gamma/2$ on Γ^{eq}.

To our knowledge, there is no quantitative analysis of the influence of the pressure-induced shifts Δ on the results of radiative transfer calculations. In fact, Δ being generally much smaller than Γ, errors on line shifts have small consequences in most cases. Nevertheless, for transitions for which Δ varies significantly with temperature and is not negligible when compared with Γ, the results of atmospheric emission computations may be erroneous if Δ is neglected. Indeed, in this case, the lines in the emission of deep and hot atmospheric layers are shifted with respect to their positions in higher and colder layers so that neglecting the shifts results in an underestimation of the outgoing radiation.

Remote sensing

The need to take into account the Lorentz broadening in remote sensing applications is obvious if the spectral resolution is high. An example is given in Fig. VII.5, which displays in situ measured spectra of a methane feature (including three lines) at different atmospheric altitudes. One can see, starting from the nearly Doppler shapes at high altitudes, the increasing broadening of the profile when getting closer to ground level (increasing total pressure). This collisional broadening must obviously be properly included in the forward model if these recordings are fitted for the determination of the CH_4 volume mixing ratio. In fact, one can show that if the absorption coefficient due to a single Lorentzian line of width Γ is least-squares fitted using a wrong broadening $\Gamma(1+\varepsilon_\Gamma)$, the resulting error on the vmr is $\varepsilon_x = \varepsilon_\Gamma/(2+\varepsilon_\Gamma)$. For moderate errors on the broadening this leads to $\varepsilon_x \approx \varepsilon_\Gamma/2$ so that a 10% overestimation on Γ leads to a +5% error on the vmr. Similarly, at poor spectral resolution where the equivalent width Γ^{eq} [Eq. (VII.16)] is measured, the effect on the vmr for moderate values of the relative uncertainty ε_Γ on Γ is given by $\varepsilon_x = -(\varepsilon_\Gamma b)/(2+b)$, with $b \equiv (\pi peak/4)^{1.25}$. For weak absorbance at line center (peak $\ll 1$), an error on the broadening does not influence the vmr. For 0.2 absorbance at the peak, for instance, a 20% error on Γ translates to less than 1% on the vmr. The error increases with the peak value (eg 10% on vmr for 20% on Γ and peak = 2) to reach $\varepsilon_x = -\varepsilon_\Gamma$ for peak $\gg 1$.

Consequences for applications 313

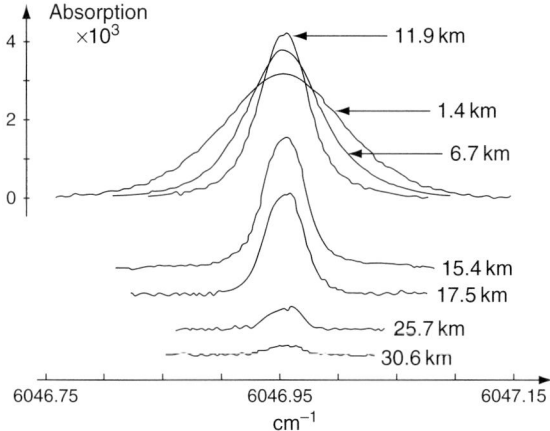

Fig. VII.5: Measured in situ absorbances by the R(3) lines of the $2\nu_3$ band of ^{12}CH$_4$ in the atmosphere as a function of altitude. Recording made by a balloon-borne tunable diode laser system. Reproduced with permission from Ref. 841.

The sensitivity of vmr determinations to Γ is discussed in more realistic situations in Ref. 27. The authors show that errors on the pressure broadening coefficient $\gamma^{O_3\text{-air}}$ significantly affect the atmospheric ozone vmr vertical profiles retrieved from limb viewing experiments (Fig. VII.2c). This is the case when high resolution spectra are used, a 60% error on $\gamma^{O_3\text{-air}}$ translating to 40% on the vmr, but also (32% error on the vmr) when the retrieval is made from broad-band radiometric measurements. The situation is similar for temperature soundings when they are made using the absorption/emission by a given species.[842–845] When the optical transitions are pressure broadened, use of incorrect line widths, and particularly of their variation from one line to another, translates into errors on the intensities and thus on temperatures, as discussed in Refs. 28,846 in the case of Raman thermometry. In Ref. 846, the Q(1) and Q(3) lines of H$_2$ are used to determine temperature. It is shown that a simultaneous change of 10% of the broadenings of *both* Q lines leads to negligible changes of the retrieved temperature. This can be expected since the two transitions are equally affected, leaving almost unchanged their retrieved intensity ratio from which T is obtained. On the contrary if *only* the Q(1) or Q(3) broadening coefficient is changed, the error on temperature reaches 20 K in the case of a 10 bar, 600 K, 5% H$_2$ + 95% N$_2$ sample.

Concerning the line-shifts Δ, besides studies in specific cases,[847–849] we have found no general analysis of the consequences of errors on Δ on quantities deduced from optical soundings. Nevertheless, this impact is expected to be small for the following reasons. The first is that an overall (line-independent) spectral shift is often adjusted so that only the line-to-line variations of Δ have an influence. The second is that the pressure-induced shifts are, in most cases, significantly smaller than the broadenings Γ (*eg* Table III.1) and than the instrument spectral resolution. Finally, fitting a measured line profile with a (wrongly) slightly shifted line only induces small errors on the area and thus on the vmr. Indeed, one can show that, for a Lorentzian line, the relative error on the vmr obtained from a least squares fit of the absorption coefficient can be approximated by $\varepsilon_{vmr} = 0.25(\varepsilon_\Delta |\Delta|/\Gamma)^2$, where ε_Δ is the relative error on Δ. Even in the (relatively) extreme situation where a shift of 10% of the broadening is completely neglected, ε_{vmr} is only of 0.25%.

If the collisional broadening is now included in all forward models, the information that it *intrinsically* contains has been seldom used in remote sensing. It is generally the area below the spectrum, and not the profile strictly speaking, which is adjusted in inversion procedures. The line-shape parameters are fixed, assuming that they are sufficiently accurately known not to introduce errors on the retrieved quantities. As mentioned above, this is discussed in Ref. 27 in the case of vmr retrievals. Another example is the remote determination of wind velocities from the Doppler shift of absorption/emission lines.[850] There again, the collisional shifts are included as fixed parameters (eventually set to zero) and the quality of their knowledge may condition the accuracy of the retrieved values. There have been, to our knowledge, very few studies where the collisional width Γ *itself* gives the desired information on the sounded medium. An example can be found in Ref. 851 where tunable diode laser transmission spectra are adjusted in order to determine temperatures. This is done from absorption by an infrared line of CO seeded in a two dimensional laminar air flow at atmospheric pressure. The half width being expressed as:

$$\Gamma = P\gamma(T_0) \times [T_0/T]^n \quad \text{with} \quad T_0 = 296K, \quad (VII.17)$$

its in situ measurement enables the determination of the temperature if the $\gamma(T_0)$ and n parameters are known. The relative uncertainty on T is then close to that on the broadening measurement (a few %[851]).

VII.3.2 MORE REFINED ISOLATED LINE PROFILES

Radiative heat transfer

As discussed in chapter III, the speed dependence of collisional parameters and the collision-induced changes of the radiator velocity lead to deviations from the Lorentz and Voigt isolated line profiles. Nevertheless, the influence of these processes does not exceed a few %, except for very light radiators with heavy collision partners. For most molecular systems of interest for practical applications [eg CO_2, CO, CH_4, NH_3, H_2O diluted in air or N_2 (Earth and Titan), in CO_2 (Venus and Mars), or in H_2 (Jupiter and Saturn)] the consequences are hence small and spectrally localized. They are thus generally neglected, this being further justified by the integration over wavelength which often reduces non-Voigt signatures to below the noise level or desired accuracy of predictions. For this reason, and due to the computer cost and the eventual lack of data and models for the molecular systems of interest, heat transfer calculations use the Voigt profile for the core regions of isolated lines, as do most remote sensing inversion programs.

Remote sensing

In spite of the weakness of velocity effects on the shapes of isolated lines, there are atmospheric studies where the consequences of departures from the Voigt profile have

Consequences for applications 315

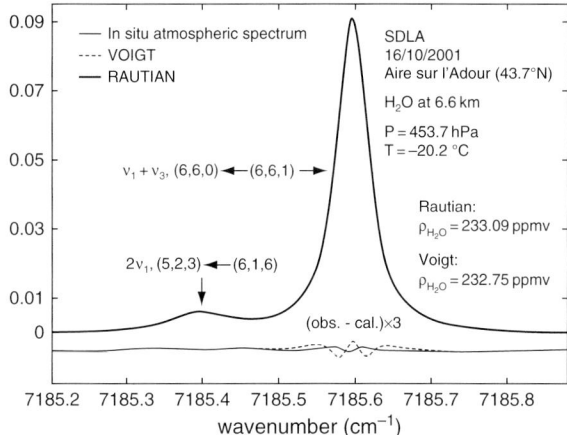

Fig. VII.6: Absorptions by H_2O lines in the atmosphere at 6.6 km altitude. Measured values have been obtained with a balloon-borne tunable diode laser system.[841] They are fitted to obtain the H_2O vmr using both Voigt and hard collision (Rautian) profiles. Reproduced with permission from Ref. 852.

been investigated, thanks to the progresses of sounding instruments. A first example is given[852] by the determination of the atmospheric humidity from in situ absorption measurements. The typical results of Fig. VII.6 show that the Voigt profile leads to fit residuals significantly above the noise level and showing the characteristic w-shaped narrowing signature (*cf* Figs. III.9 and III.12). On the opposite, the residuals are significantly reduced when a hard collision line shape (*cf* § III.3.5) is used, but the consequences on the retrieved H_2O vmr are almost negligible ($\leq 0.2\%$). This small impact of velocity changes can be expected from the weakness of the effects on the spectrum (typically $\pm 1\%$ of peak absorption) and the fact that the integrated intensity gives most of the information on the mixing ratio. As mentioned previously, even though an incorrect profile is used in the fits it gives a quite good estimate of the area below the absorption coefficient (and thus of the vmr).

A more complete analysis of the impact of deviations from the Voigt shape on remote sensing of atmospheric species amounts from isolated transitions was made in Ref. 853. In this study, vmr vertical profiles of HCl and HF are deduced from fits of ground-based transmission measurements (Fig. VII.2b). In order to quantify the impact of the line shape on the retrieved mixing ratios, the Voigt and (more realistic) soft collision (*cf* § III.3.4) models are used. The results show that, although the spectra differ by only a few percents, the differences for the vmr reach -15% at 20km and $+20\%$ at 34 km for HF (those for HCl being about twice smaller). This strong amplification of the error is due here to the fact that the ground-based spectrum provides a poor vertical resolution to the retrievals. Hence, there are significant correlations between the vmrs at various altitudes and the uncertainty on retrieved values is large (comparable to the above mentioned differences).[853] As expected, when integrated over altitude, the retrieved total column amounts are practically insensitive ($<1\%$) to the line shape, confirming the weak consequences of deviations from the Voigt profile mentioned above in the case of in situ humidity measurements.

Temperature soundings using H_2 Raman spectra give more conclusive examples showing that the use of speed-independent (homogenous) Voigt or Lorentz shapes can lead to large errors. As shown in chapter III, for this light radiator with significantly heavier perturbers, Dicke narrowing and speed inhomogeneous effects on the line profile can be very important (§ III.4.1). A detailed analysis of the consequences for Raman thermometry in the spectral domain can be found in Refs. 28,227,846. As shown by Figs. III.13 and III.20, the speed dependence of collisional parameters leads to nonlinear variations of the observed broadening vs mole fractions. Neglecting them and using purely homogeneous profiles hence leads, for the conditions of the study, to errors on retrieved temperatures which can be important as discussed in Ref. 846. On the contrary, when the $1DsdRS_{\tilde{C}}P$ approach [§ III.4.3, Eq. (III.73)] is used, good agreement is obtained between temperatures deduced from the H_2 Raman signatures and those quasi-simultaneously retrieved using the well established N_2 Raman thermometry,[28,227] as illustrated in Fig. VII.7. Note that the agreement at low density is obtained with the $3DsdRS_{\tilde{C}}P$ since the Dicke-narrowing contribution is included together with the inhomogeneous broadening. Recall that the quality of temperature determinations from N_2 Raman scattering is shown in Ref. 854 where errors on T obtained from H_2 signals were attributed to uncertainties on hydrogen line shapes.

This sensitivity of Raman soundings to the model used for the H_2 line profiles is confirmed by results in the time domain presented in Ref. 109 for H_2–N_2 mixtures at high density. This study demonstrates the thermometric consequences of *two* different effects discussed in chapter III. In this work, the temperature is deduced from the evolution, with the pump-probe delay, of the femtosecond CARS signal

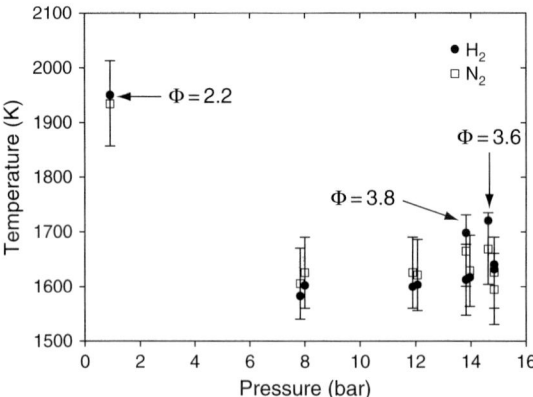

Fig. VII.7: Temperatures deduced from CARS thermometry for various stoichiometries Φ (the lowest temperature being for Φ = 4.1) and pressures in the exhaust of a H_2/air flame. (•) and (□) are values obtained using H_2 and N_2 CARS spectra respectively. Reproduced with permission from Ref. 227.

Table VII.1: Results of temperature soundings using the H_2 femtosecond CARS time response (see text). After Ref. 109.

T measured from	T(K), ΔT(K), |ΔT|/T(%)	
	5% H_2 + 95% N_2	50% H_2 + 50% N_2
Thermocouple	593	899
Homogeneous Lorentz	741, +148, 25.0	929, +30, +3.3
"Apparent" Lorentz	557, −36, −6.1	864, −35, −3.9
1DsdKS	602, +9, +1.5	885, −14, −1.6

(cf App. II.A4). The latter is adjusted using three different models, with temperature as a free parameter, leading to the results in Table VII.1. The first approach uses purely (homogeneous) Lorentzian profiles. It thus neglects the nonlinear dependence of the width of the H_2 line on the N_2 mole fraction x_{N2} in the mixture and simply writes $\gamma^{Mixture} = x_{H_2}\gamma^{H_2-H_2} + x_{N_2}\gamma^{H_2-H_2}$. This strongly underestimates the broadening for large N_2 mole fractions (see Fig. III.20). For 5% H_2 + 95% N_2, the width is about twice too large, leading to a strong overestimation of the retrieved temperature (Table VII.1). The errors are smaller for a 50%/50% mixture since the homogeneous model then underestimates the broadening significantly less (about 20%). In these results, the discrepancies are *mostly* due to the use of improper Lorentz widths Γ. The second fit ("apparent" Lorentz) still uses Lorentzian profiles, but based on the *true* width accounting for the nonlinear dependence on mole fractions. As expected, errors are significantly reduced, particularly for the 5% H_2 mixture. Remaining discrepancies are then due to the use of an improper spectral shape *only*, and, in particular, to the fact that the Lorentzian profile does not include the asymmetry due to the speed dependence of collisional parameters. Finally, when both the proper broadening and proper line profile are used (1DsdKSP, cf § III.4.4), the results are excellent. This demonstrates that accurate temperature measurements require sophisticated line shapes including *all effects* leading to the breakdown of the Voigt profile, particularly when hydrogen Raman scattering is used.

There is an increasing number of species and spectral regions for which laboratory studies bring information on deviations from the Voigt shape (cf Chapt. III). Simultaneously, the quality of spectra recorded for sounding purposes is constantly improved and the accuracy requirements on the inversion products are always stronger. More and more refined isolated line profiles will thus be included in forward models and inversion procedures and errors resulting from the use of Voigt shapes will be pointed out in an increasing number of sounding studies. A recent example can be found in Ref. 855 where it is shown that speed-dependence effects on water vapor line shapes cannot be neglected for an accurate treatment of satellite-borne atmospheric spectra.

VII.4   LINE MIXING WITHIN CLUSTERS OF LINES

Radiative heat transfer

As for isolated line refined profiles, collisional line mixing in the core region of clusters of closely spaced transitions has small consequences for radiative heat transfer calculations since the affected regions are narrow (*cf* chapter IV). Furthermore, the deviations with respect to isolated line models are often moderate and only affect the shape, the overall intensity being conserved. Hence, when total fluxes are considered, the influence of line mixing is masked by the (broad) integration over wavelength as has been shown[856] in the particular case of CO_2 Q branches in the Earth atmosphere.

Remote sensing

The need to take into account line mixing in Q branches for the Raman sounding of gas media has been known for a long time. It is often unavoidable due to the large deviations with respect to purely additive Voigt contributions, and routinely used temperature inversion programs have early included modeling of collisional line couplings. This was first done using simple models, progressively refined following the development of more accurate approaches for the construction of the relaxation matrix (*cf* Secs. IV.3 and IV.5). We have found no Raman study investigating the errors on retrieved quantities that the purely Lorentzian approach would introduce. Nevertheless, in Ref. 857, a quantitative analysis was made of the influence of different line-mixing models on temperature measurements from Raman experiments. Comparisons between the temperature measured in situ by a thermocouple and those determined from the inversion of Raman Q branch spectra of N_2, CO_2, and O_2 are presented. In this study, the relaxation matrix is constructed using the MEGL fitting law (§ IV.3.2) and the ECS scaling approach (§ IV.3.3). Large temperature differences are obtained[857] which depend on the molecule, on the pressure and temperature conditions. They reach more than 130 K, even though both approaches take line mixing into account. For thermometry using the $2\nu_2$ Q branch of pure CO_2, the errors on temperature at $T = 900$ K are $\Delta T = +136$ K and $\Delta T = +12$ K when the MEGL and ECS models are used, respectively. For pure oxygen at 1350 K, the results are different with errors $\Delta T = -52$ K (MEGL) and $\Delta T = -93$ K (ECS). This demonstrates the high sensitivity of temperature determinations to the spectral-shape model and shows that neglecting line mixing would lead to tremendous errors. These conclusions are confirmed, in the time domain, by the results of Ref. 110, in which femtosecond CARS transients for pure N_2 at room temperature are treated. It is first shown that line mixing has a large influence on the signals for short time delays between pump and probe (Fig. IV.22). These signals are then adjusted using two ECS line-mixing approaches (§ IV.3.3) in order to retrieve pressure and temperature from the measurements. Even though the ECS-P and ECS-E models used are similar in nature, significantly different results are obtained, as shown in Fig. VII.8. The errors with respect to thermocouple and pressure gauge measurements reach 20% on P (for $P = 5$ bar) and 30 K on T (for $T = 295$ K) when the ECS-P model is used whereas values retrieved with the ECS-E approach are almost perfect. This confirms the high *sensitivity*

Consequences for applications

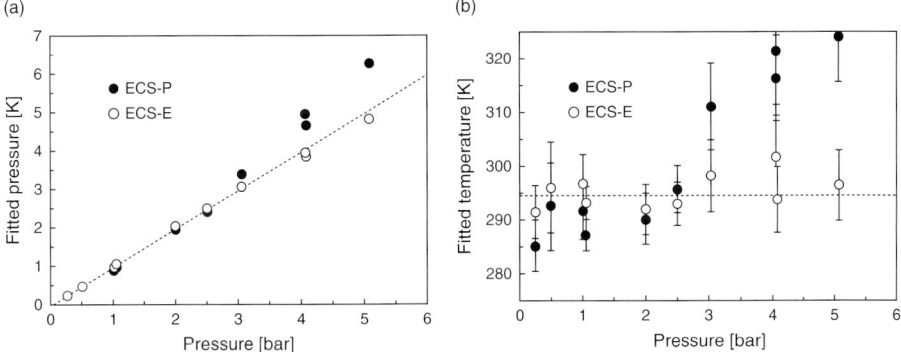

Fig. VII.8: Temperature and pressure determinations from fits of femtosecond CARS transients of pure N_2 at room temperature using two line-mixing models. The dashed line gives the true values (thermocouple and pressure gauge). Reproduced with permission from Ref. 110.

of Raman thermometry to the description of line-mixing processes and extends this statement to pressure measurements.

Another example of line-mixing effects in CARS spectra, but in the case of methane, is given in Ref. 497, even though the consequences on temperature determinations are not directly discussed. Nevertheless, the comparison between the calculated CARS intensity and that measured in the laboratory, displayed in Fig. VII.9, shows that this process must be properly modeled for temperature determinations in high pressure flames. Indeed, although the temperature is lower, the pressure is the same as that in a supercritical LOX/CH4 combustion facility where CARS spectra along the stream look very much alike.[497] It was shown later that use of proper spectral shapes enables satisfactory retrievals of flame temperatures.[858]

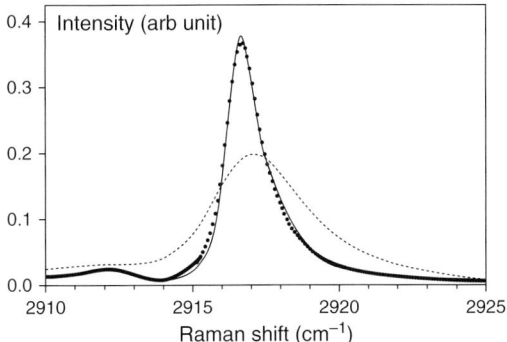

Fig. VII.9: CARS spectrum of pure CH_4 in the v_1 band at 900 K and 50 atm. (•) are measured values while (———) and (– – –) have been computed with and without the inclusion of line-mixing effects. Courtesy of E. Jourdanneau, after Ref. 497.

To our knowledge, the first study pointing out the influence of line mixing within clusters of closely spaced transitions on atmospheric spectra is that of Ref. 859 where the CO_2 v_2 infrared Q branch is considered. Later on, numerous demonstrations of this effect through comparisons of measured values with the results of direct calculations have been made, as summarized in Table IV.12. As expected from laboratory studies (eg Figs. IV.18 and IV.21), neglecting line mixing leads to the overestimation of the absorption (and thus generally of emission, but not always[603]) in the wings, to the underestimation of peak absorption, and to errors in the troughs between the lines. This is illustrated by infrared atmospheric spectra of methane and carbon dioxide in Figs. VII.10 and VII.11. Line mixing leads, with respect to predictions neglecting this effect and using isolated line shapes, to the lowering of absorption and emission in the wings. It also results in an enhancement of absorption in the central region of the affected structure, as illustrated by the trough between the CH_4 transitions in Fig. VII.10.

The spectral regions affected by line mixing have been seldom used for remote sensing until recently. This is due to tradition, since there has long been a lack of accurate models, and to the relatively increased complexity of inversion procedures when line mixing is included. It also results from the fact that the regions affected by line mixing are narrow so that disregarding them still leaves numerous other spectral windows with isolated lines to work with. This statement is illustrated by a study[862] of the consequences of line mixing in CO_2 Q branches on the remote sensing of temperature, pressure, and trace gases using spectra recorded by the MIPAS[863] limb emission (Fig. VII.2c) satellite instrument. The concern, at that time, was to check that line mixing

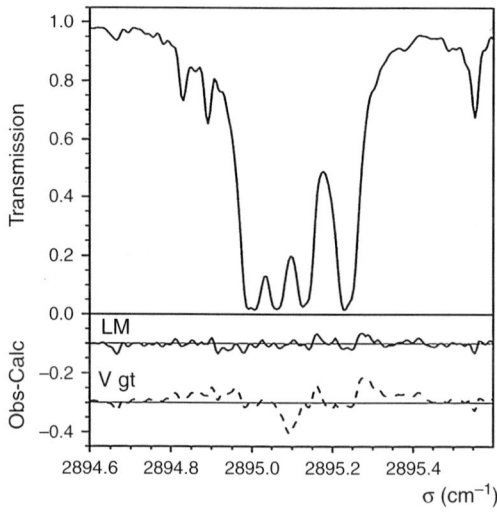

Fig. VII.10: Atmospheric transmission in the region of the P(12) manifold of the $^{12}CH_4$ v_3 band for an instrument at 29.8 km looking down to a tangent height of 15 km. Values measured with a balloon-borne instrument[860] are given in the top part of the plot while meas-calc (shifted) deviations with respect to calculations with (LM) and without (Vgt) the inclusion of line mixing are plotted below. After Ref. 494.

Consequences for applications

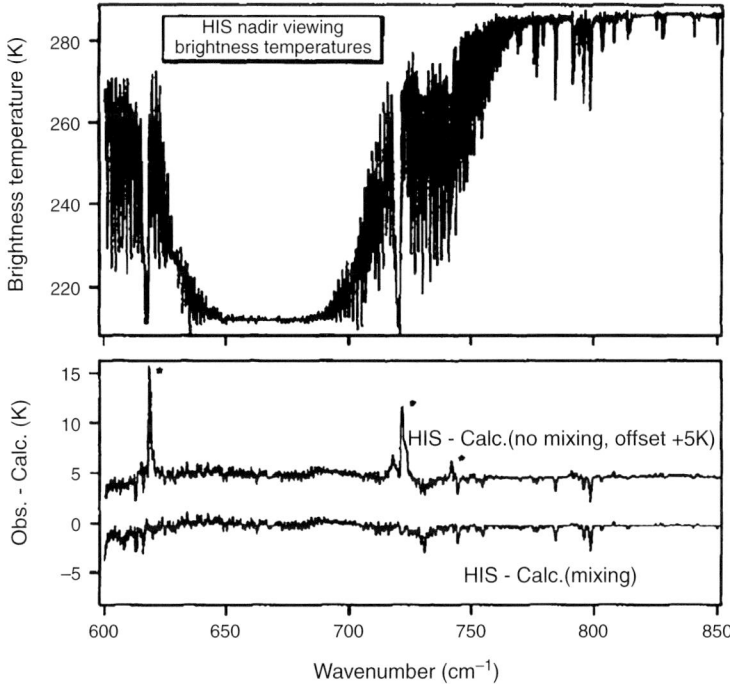

Fig. VII.11: Comparison between the atmospheric emission measured using the air plane-borne HIS instrument[861] and calculations. The upper plot is the measured brightness temperature spectrum, the lower traces give the meas-calc deviations obtained without and with the inclusion of line mixing in CO_2 Q branches (denoted by *). Reproduced with permission from Ref. 346.

has a negligible influence on the inversion products for the selected set of microwindows. However, due to its significant effects, collisional line mixing is now progressively included in atmospheric radiative transfer forward codes and inversion procedures, following the availability of models and data (cf Tables IV.5 and IV.6). There is hence an increasing number of studies demonstrating the interest of proper modeling of line mixing for atmospheric remote sensing.

A first example, concerning vmr determinations, is the retrieval of CCl_4 amounts in the atmosphere from its spectral signature in the v_3 band near 795 cm^{-1}. This feature lies in the range of the $(11^10)_I \leftarrow (10^00)_{II}$ CO_2 Q branch which is affected[343] by line mixing. The retrievals of the CCl_4 vmr from its weak spectral signature (5% absorption at most) are thus reliable only if the prediction of the CO_2 dominant contribution is very accurate. This is demonstrated in Fig. VII.12 and Ref. 602 using atmospheric transmission measured by a ground-based Fourier transform solar occultation experiment.[864] This problem is further analyzed in Ref. 865 where the errors on the CCl_4 vertical distribution due to neglecting line mixing between CO_2 lines are shown to be greater than 10%.

Staying in the field of vmr retrievals, heavy trace gases such as chlorofluorocarbons give other examples of the use of spectral windows affected by line mixing. These

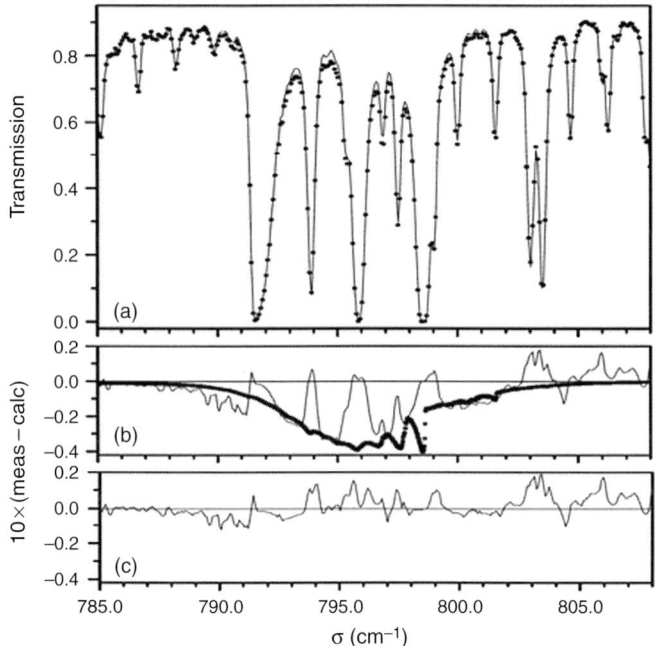

Fig. VII.12: Atmospheric transmission from ground (cf Fig. VII.2b) for an EA of 13.9° in the region of the CO_2 $v_1 + v_2 - v_1$ Q-branch and of the CCl_4 v_3 band. (a) measured (•) and computed (—, not including CCl_4) values averaged over 0.2 cm^{-1}. 10 × (meas-calc) deviations (—) when CCl_4 is not taken into account (b) and included (c) in the calculation. The absorption by CCl_4 is indicated by the symbols • in (b). Reproduced with permission from Ref. 602.

determinations generally use narrow Q branches containing many overlapping transitions, P and R branches being too broad and weak for accurate inversions. For most of these heavy molecules, the spectroscopy is still poorly known, and absorption cross sections[2,377] or strictly empirical effective line lists are thus used. Note that the retrieval is then accurate since line mixing, if present, is included in the cross sections, provided that sufficient P- and T-dependent laboratory data are available. An example, among others, is $CHClF_2$ (HCFC-22) whose mixing ratios were determined (cf Ref. 866 and those therein) using empirical modeling of the absorption in the $2v_6$ Q-branch near 828 cm^{-1}. This is thus correct, although it was demonstrated[372] that line-mixing processes strongly affect this dense structure. Due to the progresses of laboratory studies and to the increasing accuracy asked for in remote sensing studies, line mixing will be more and more included in retrievals of vmrs in atmospheres. It will likely soon be the case for methane, due to its geophysical interest and to the availability of models (see Tables IV.5 and IV.6). In fact, Fig. VII.10 shows that incorrect mixing ratios would be obtained if this P(12) manifold and Voigt line shapes were used for the vmr inversion, as confirmed recently in the case of the P(9) manifold.[557] Still for CH_4, but for the $2v_3$ R(3) manifold, the treatment of in situ measurements, such as those of Ref. 841 (see Fig. VII.4), could be improved by using the data of Ref. 556. Another species, yet to be studied, is HNO_3 since residuals remain in the fits of atmospheric spectra,[867] particularly in narrow Q branches. Similarly, problems remain[868] in the modeling of the C_2H_6 Q branch near 822 cm^{-1} and the use of proper collisional shapes would improve determinations of the abundance of this species in the atmospheres of Jovian planets.

Other examples of the positive impact of correct modeling of line mixing can be found in the remote determination of atmospheric temperatures and pressures. In a pioneering work, thanks to the confidence in the forward model,[567] Niro et al[869] have used CO_2 Q branch regions for the retrieval of P and T profiles. Contrary to the isolated P and R lines used previously (Refs. 842,870 and those therein), Q branch wings have the advantage of being *more* sensitive to total pressure. Absorption here is proportional to P^2 whereas it is independent on P at the peak of an isolated Lorentzian transition. The study[869] was carried out using a "geo-fit" retrieval[871] of a series of atmospheric limb emission spectra recorded by the MIPAS satellite instrument.[863] It shows that the use of CO_2 Q branches leads to very significant reductions (up to 70%) of the estimated errors on P and T vertical profiles. Another study based on carbon dioxide signatures is presented in Ref. 872. In this work, atmospheric temperatures are inverted from nadir recordings of the emission by CO_2 in the 15 μm region The use of a former LBLRTM[634,873] code in which line mixing is taken into account within Q branches *only* while the wings of P and R lines are modeled with an empirical correction function χ (§ V.2.1) leads to significant systematic errors on temperatures (left of Fig. VII.13).

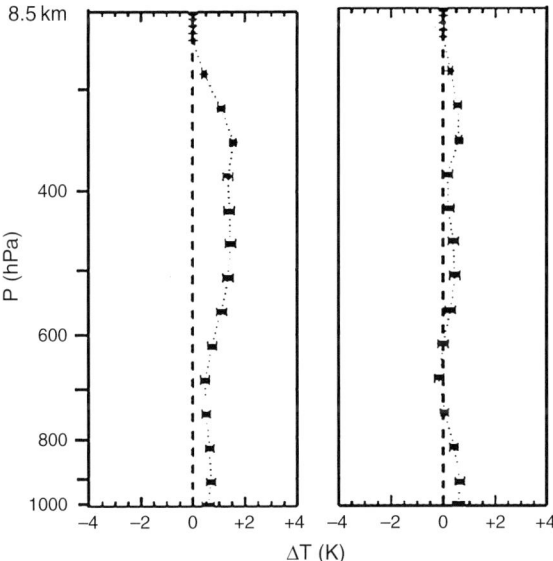

Fig. VII.13: Errors on tropospheric temperature profiles (the reference being a radiosonde) obtained from fits of airplane-borne nadir viewing CO_2 spectra in the 680–800 cm^{-1} region using two forward models. Left: line mixing within CO_2 Q branches and a χ factor for P and R lines. Right: full P+Q+R line-mixing model of Ref. 567. Courtesy of S.A. Clough, after Ref. 872.

On the other hand, with the approach of Ref. 567, which includes couplings within and between all P, Q, and R branches, the retrieved temperature profile is greatly improved (right of Fig. VII.13).

Another molecule whose spectrum permits temperature (and pressure) soundings is molecular oxygen. Like CO_2, O_2 is often used for such retrievals since it is well mixed in the atmosphere with a known mixing ratio. The vmr can then be fixed, leaving only the T + P profiles as the main unknowns. In the long wavelength domain, atmospheric temperatures are retrieved from the O_2 60 GHz band (1.7–2.3 cm^{-1}).[874] This region is significantly affected[549,550] by line couplings at atmospheric pressures and the accuracy of its modeling directly affects the accuracy of retrieved temperatures. This is discussed in Refs. 875,876 where the impact of a 10% uncertainty on the first-order line-coupling coefficients Y_ℓ (§ IV.2.1a) for cold temperatures is studied. This sensitivity analysis shows that errors of several Kelvin may be obtained. As expected, the results are the worse in the channels on the sides of the band since they are the most affected by line mixing. These results have been discussed[877] but they indicate that very large errors would be obtained if collisionally isolated line profiles were used. In the visible region, the oxygen A-band near 760 nm is used for a variety of purposes including the retrieval of some characteristics of clouds and aerosols,[878,879] as well as the determination of P + T vertical profiles[880,881] and of surface pressure.[882,883] It has been recently shown[415,884] that the use of isolated lines with Voigt profiles induces errors on the O_2 amount (*ie* the total pressure) retrieved from ground-based atmospheric transmission measurements. This is largely due to line couplings so that an accurate description of this mechanism is essential for the OCO mission[536] in which ground pressure will be determined from the O_2 A-band. This is illustrated by Fig. VII.14, where the errors on ground pressure determinations from solar occultation measurements for various

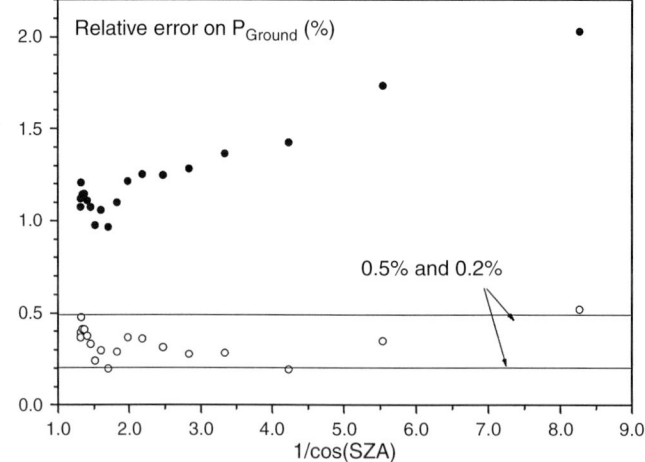

Fig. VII.14: Relative errors on surface pressure (the reference being a local pressure gauge) obtained from fits of a series (for various solar zenith angles, SZA) of ground-based solar occultation atmospheric spectra in the O_2 A-band region using two forward models with (o) and without (•) the inclusion of line mixing. After Ref. 415.

solar zenith angles are displayed. Whereas Voigt line shapes lead to significant and ZA-*dependent* errors, the inclusion of line mixing drastically improves the retrieved surface pressure and the meas-calc spectra residuals.[415] Note that the OCO mission, which aims at measuring the atmospheric carbon dioxide amounts with an extreme accuracy, will also need to take into account similar processes in the selected CO_2 bands.[537,883]

VII.5 ALLOWED BAND WINGS AND CIA

VII.5.1 ALLOWED BAND WINGS

Radiative heat transfer

Due to their spectral extension, the wings of intense allowed bands (Chapt. V) can make significant contributions to spectrally integrated heat transfer quantities in optically thick media. It is the case in atmospheres for molecules present in significant amounts, such as H_2O and CO_2 for Earth and Venus and CH_4 for Jupiter and Saturn. At smaller scales, examples are given by CO_2 in combustion processes and H_2O in high temperature and pressure systems. In such situations, the absorption in the central regions of the bands is saturated and radiative heat exchanges are sensitive to the description of the wings. Purely Lorentzian line profiles, which generally predict much too large absorption in the wings (*cf* Figs. IV.7 and V.5), often lead to a large overestimation of the atmospheric emission. This is illustrated by the intense ν_2 band of CO_2 shown in Fig. VII.15. When integrated over the spectral range of this plot, the measured

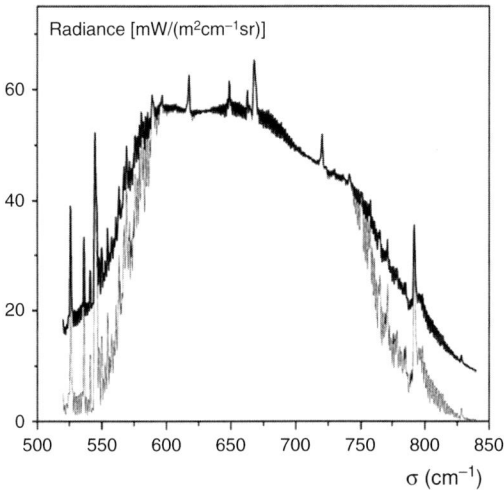

Fig. VII.15: Radiance emitted by the Earth atmosphere in the region of the ν_2 CO_2 band for a 1 cm^{-1} resolution. The thin line gives measured values obtained by a balloon-borne instrument[29] at 40 km altitude looking down to a tangent height of 10 km. The thick line corresponds to predictions using purely Voigt line shapes. After Ref. 603.

radiances lead to a flux of 10.9 W/(m²str). The associated Voigt prediction, although truncated on the sides, gives 12.8 W/(m²str). Such a 2 W/(m²str) error is unacceptable if included into a calculation of the total net flux (integrated over the entire spectrum and including the effects of ground and ocean, clouds, aerosols, and solar radiation) which must be small in order to keep the heat balance of the atmosphere. Note that, when calculations are made with the tools of Ref. 567, the error on the emitted flux becomes negligible.[603] These results are obviously further amplified on Venus where the amount of carbon dioxide is very large (nearly 100% CO_2 with a pressure of 90 atm at ground level), making the greenhouse effect considerable.[885,886] For completeness, let us mention that the wings of the intense CO_2 bands are also important for applications involving combustion gases, such as infrared signature.[887]

Water vapor is another species whose far line wings significantly contribute to radiative exchanges and the greenhouse effect in our atmosphere. This is obvious from Fig. VII.16 (also see App. V.A) which displays values of the H_2O continuum (§ V.2.2) calculated using the CKD[616] model of the LBLRTM radiative transfer computer code.[634,888] Even though only the part of the wings further away than 25 cm^{-1} from the line centers have been retained, the absorption is important over broad spectral domains. The significant participation of water vapor to atmospheric radiative exchanges mostly results from two reasons. The first is that the far wings in the thermal domain (between about 4 and 100 μm) determine the widths of the bands where the atmosphere emits and absorbs and, thus, the spectral extent of the transparency windows in between them where ground emission outgoes. In the 8 to 12 μm interval (see Fig. VII.16), for instance, most of the atmospheric absorption/emission comes from far wings of the intense H_2O lines of the rotational and ν_2 bands. These have a significant impact on radiative exchanges since the Earth surface emission peaks at the center of this interval [$I_{BB}(\lambda,300K)$, see Eq. (VII.3), has a maximum at 9.6 μm and its integral from 8 to 12 μm represents a quarter

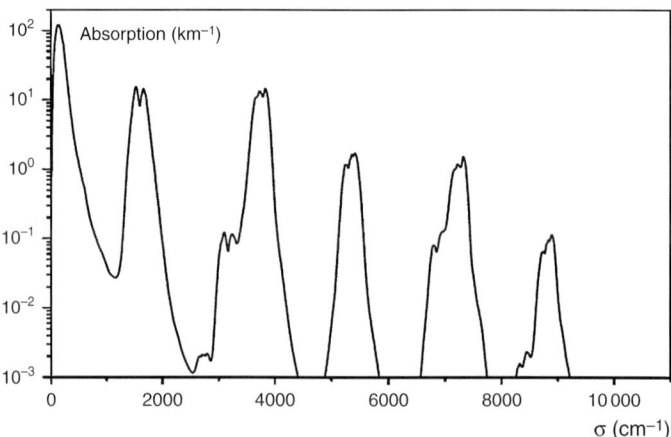

Fig. VII.16: Calculated far wing (>25 cm^{-1}) absorption by water vapor lines for 1% H_2O in air at 1 atm and 296 K. The up-to-date version of the CKD water continuum (see § V.2.2 and Ref. 634) was used.

Consequences for applications 327

of the total flux]. The second contribution of water vapor to the heat balance of the atmosphere comes from the absorption of solar photons in the short wavelength domain ($\lambda < 2\,\mu\text{m}$). In fact, the observed atmospheric attenuation of solar radiation is greater than that calculated (a still unsolved problem) and an improper modeling of H_2O line wings has been mentioned as a possible contribution to this discrepancy.[889] Further analyses of the various impacts of the water vapor absorption shape for heat transfer in our atmosphere can be found in Refs. 691,856,890,891. In particular, it was shown that a recent improvement of the CKD continuum (*cf* § V.2.2), resulting in enhanced long wavelength absorption, leads to the increase of downward surface fluxes and to the decrease of outgoing radiation.[890] This has important consequences, such as the increase of surface temperature and changes in the atmospheric temperatures leading to modifications of moisture and cloud cover.[890]

Remote sensing

Correct modeling of the far wings is also of importance for remote sensing in atmospheres, either because they affect regions containing (weak) spectral signatures due to species of interest for atmospheric physics or because they are directly used in order to retrieve geophysical parameters. This is particularly the case for soundings of deep and dense atmospheric layers whose emissions often include a significant continuum due to far wings. When vmrs are retrieved using narrow signatures, an accurate description of the background carrying out them is required, *particularly* when emission spectra are used. In order to explain this statement, consider the simple case of two adjacent layers 1 and 2 at temperatures T_1 and T_2, layer 2 being the closest to the observer. Let $A_j(\sigma)$ and C_j ($j=1,2$) be the absorbances [*ie* -ln(transmission)] in each of them due to the narrow spectral features (lines) and to the continuum, respectively. For absorption measurements (*eg* solar occultation), the experiments provide the transmission $\tau_{1+2}(\sigma)$ whose spectral expression is:

$$\begin{aligned}\tau_{1+2}(\sigma) &= \exp[-A_1(\sigma) - C_1] \times \exp[-A_2(\sigma) - C_2] \\ &= C^{te} \times \exp\{-[A_1(\sigma) + A_2(\sigma)]\}, \text{ with } C^{te} \equiv \exp[-(C_2+C_1)].\end{aligned} \quad (\text{VII.18})$$

The contribution of the far wings background through C_1 and C_2, thus only manifests itself as a multiplicative factor (Cte) which does not change the spectral profile of the narrow features. If C_1 and C_2 are not sufficiently known, Cte can easily be adjusted on measured spectra together with the absorbing species vmr. For emission measurements, the expression of the spectral radiance is:

$$\begin{aligned}I_{1+2}(\sigma) &= \{1 - \exp[-A_1(\sigma) - C_1]\}\exp[-A_2(\sigma) - C_2] \times I_{BB}(\sigma, T_1) \\ &+ \{1 - \exp[-A_2(\sigma) - C_2]\} \times I_{BB}(\sigma, T_2).\end{aligned} \quad (\text{VII.19})$$

Contrary to the case of transmission, the C_1 and C_2 terms can *strongly* change the relative variations (dynamics) of the signal due to local lines. If the continua C_j in the various layers j are badly modeled, the narrow features *themselves* can be hardly discernable. This is illustrated by Fig. 14 of Ref. 603 which shows that calculating the wings of the ν_2 CO_2 band using Lorentzian line shapes makes the signature of the ν_8 Q branch of HNO_3 completely unrecognizable. A less exaggerated situation is

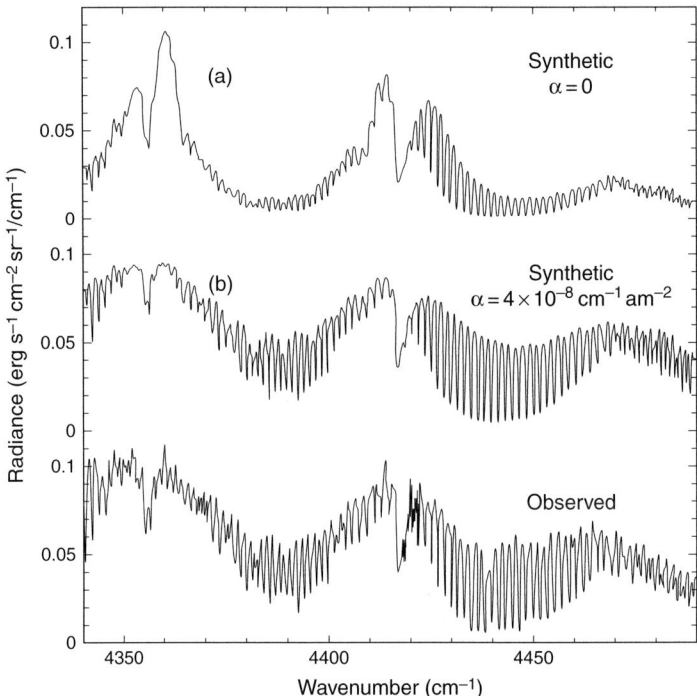

Fig. VII.17: Venus night side emission in the 2.3 µm region. The lowest curve is a measurement from the Canada-France Hawaii telescope. The two upper curves are synthetic spectra calculated with the local CO_2 lines and two altitude-independent values of the density-(bi)normalized continuum absorption C (noted α on the plots): C = 0 (top) and C = 4 10^{-8} cm^{-1}/Am^2 (middle). Reproduced with permission from Ref. 892.

illustrated by Fig. VII.17, where the Venus night side emission near 2.3 µm is displayed. As expected, the variations of the amplitude of the signal due to local lines are highly dependent on the assumed value of the background (due here to far wings of the intense transitions of carbon dioxide near 2 and 2.7 µm). Assuming no background (C = 0) leads to narrow local features with significantly underestimated amplitudes whereas better agreement with the measured spectrum is obtained using an ad hoc continuum. This is due to the fact that $C \neq 0$ increases the (background) emission of the hot and dense deep layers, thus enhancing the outgoing radiance in the troughs (emission maxima) between the narrow (cold and absorbing) lines in the high altitude layers.

Note that such adjustments, based on a single altitude-independent continuum (per squared unit density), are questionable. This can be understood by considering the simple 1 + 2 system introduced above. Indeed, its emission, calculated with $C_1 \neq C_2$, *cannot* be accurately represented by fitting a single ($C = C_1 = C_2$) background absorbance. This is shown in Fig. VII.18 where the true (•: $C_1 \neq C_2$) emission is not well fitted assuming a layer-independent continuum since the latter (——: $C_1 = C_2$), when adjusted in the wings, underestimates the peak emission. This demonstrates that vmr retrievals can be affected if they are

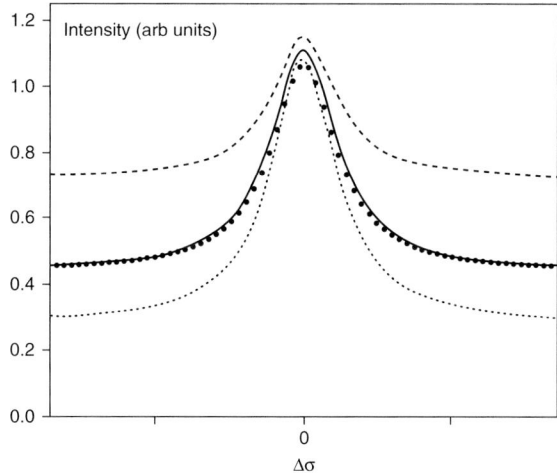

Fig. VII.18: Simulated emission at 10 μm by two layers ($T_1 = 260$ K, $T_2 = 300$ K) due to a Lorentzian local line [$A_1(\sigma)$ and $A_2(\sigma)$ peak values of 0.45 and 0.74] and a background for the values (•) $C_1 = 0.11$ $C_2 = 0.36$ (reference), (·····) $C_1 = C_2 = 0.11$, (– – –) $C_1 = C_2 = 0.36$, and (———) $C_1 = C_2 = 0.18$ (best fit). See text and Eq. (VII.19).

made in regions where there are improperly modeled and significant altitude-dependent continuum contributions. This is confirmed, in the case of H_2O on Venus, by the analysis of Refs. 892,893 and, on Earth, by those of Refs. 631,677,894. These studies show that humidity determinations are significantly sensitive to the continuum contribution due to far wings of the lines of CO_2 (Venus) and H_2O (Earth). This problem and the consequences for the determination of the vmr of trace species are further discussed in Ref. 895 where it is stated that "... a major weakness of the current gas (spectroscopic) databases is the lack of accurate information on CO_2 and H_2O continuum opacity...". Thirteen years later, the problem is not properly solved yet since "...the empirical propositions to constrain the CO_2 continuum underline the critical need for laboratory measurements...".[893]

Besides vmr retrievals, there are other remote sensing problems in which the modeling of band wings conditions the quality of results. These include previously mentioned studies of the distributions and characteristics of clouds and aerosols but also of some properties of the surface. These parameters are generally determined from spectra in "transparency" regions of the atmospheres where gaseous species are expected to make small contributions. However, these intervals in between bands contain more or less important absorption/emission by far wings and the results may suffer if the correction for their contribution is not accurate. A first example can be found in Refs. 896,897 where Earth surface temperatures are retrieved from the 8–12 μm window where the water vapor far wings can make significant contributions (cf Fig. VII.16). A similar study of some properties of the surface of Titan was made[898] in regions around 3 μm where the wings of the intense ν_3 band of methane absorb. Another field of application is the characterization of clouds on Earth, which is often made[899] using wavenumbers around 800 cm^{-1} where the high frequency wings of CO_2

lines can contribute together with the water vapor continuum. In a similar way, methane line wings must be properly modeled for studies of haze and clouds in the atmosphere of Titan.[900,901]

In all previously mentioned studies, the wings are considered as *known*, the retrieval being essentially based on the information brought by structures that they carry. On the opposite, there are a few studies where the information that the continuum *intrinsically* contains has been used. It is, in particular, the case for the determination of temperature/pressure profiles from CO_2 spectra. Indeed, the near wings, showing a continuum proportional to P^2 (or to the squared total density) and carrying high rotational quantum number lines of high sensitivity to temperature, appear as good regions for $T+P$ retrievals (see the discussions in Refs. 632,902). In spite of their interest, they have generally been disregarded due to lack of confidence in their modeling. This is exemplified by Ref. 903 where Venus atmospheric temperatures are deduced from CO_2 emission in the ν_3 band region near 4 µm. In this work, the sides of the band have been disregarded since "... The model gives untrustable results when the far wing continuum of the CO_2 band becomes dominant. This is the case for wavelengths shorter than 4.18 µm and for wavelengths longer than 4.5 µm...".[903] Note that the CO_2 15 µm band is also used for the determination of the thermal structure of Venus.[904] A pioneering study making a *direct use* of far wings is that of Ref. 905. In this study, solar occultation spectra in the high frequency side of the CO_2 ν_3 band are used for Earth atmospheric density determinations. The transmissions at 2415 cm^{-1} and 2501 cm^{-1}, dominated by the CO_2 wings and the N_2 collision-induced contribution (see Fig. VII.19),

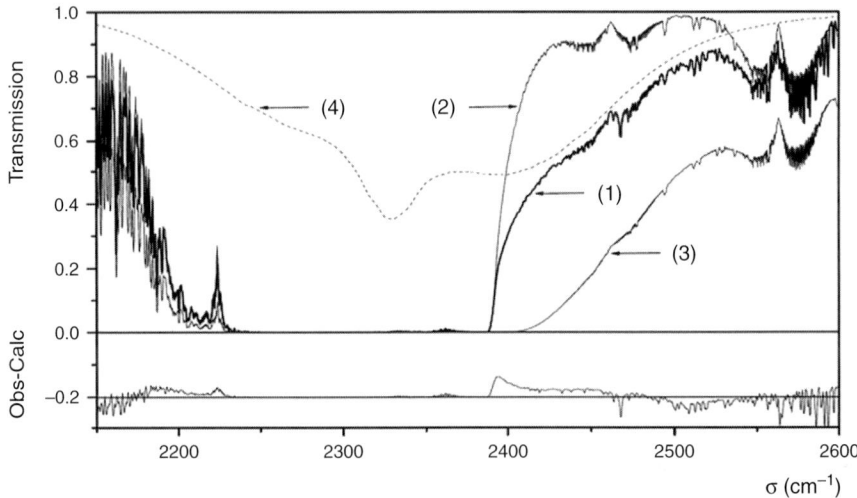

Fig. VII.19: Atmospheric transmissions (averaged over 1 cm^{-1}) for an instrument at ground level, looking up with an EA of 44.4°. Measured values (1) with a FT instrument,[864] are compared with various calculations: (2) including line mixing in CO_2 but no N_2 CIA; (3) including N_2 CIA and calculating CO_2 absorption with Voigt profiles; (4) contribution of the N_2 CIA only, calculated from the data of Ref. 78. The lowest curve gives the meas-calc residuals when both line mixing in CO_2 and the N_2 CIA are included. After Ref. 416.

are retained. The idea is to take advantage of their proportionality to the square density and of the fact that the volume mixing ratios of these two species are known. It is demonstrated[905] that this does indeed provide a high sensitivity to the vertical density distribution in the lower stratosphere and that the retrieved values are consistent with the error budget.

VII.5.2 COLLISION-INDUCED ABSORPTION

Collision-induced absorption (Chapt. VI), although due to a different physical mechanism, has characteristics and consequences for applications very similar to those of far wings of allowed transitions for three reasons. (i) The first one is that both affect broad spectral ranges and vary relatively slowly with wavenumber. (ii) The second one is that the two processes lead to absorption coefficients proportional to the square of the total density or pressure. (iii) Finally, absorptions in collision-induced bands and by far wings are both small. Values of the density-normalized CIA by non-polar gases are typically of the order of $10^{-6}\,\mathrm{cm}^{-1}\,\mathrm{Am}^{-2}$ so that significant effects are observed only for atmospheric path of several kilometers. Hence, as far wings, collision-induced absorption needs to be considered, in practice, only for molecules of significant vmr in relatively dense atmospheres (and, for smaller scales, in gases at very high pressure). It is the case, for instance, of O_2 and N_2 on Earth, of CH_4 and H_2 for Jupiter and Saturn, and of N_2, CH_4 and H_2 for Titan. Due to the similarity of the consequences of CIA and far wings, only a limited discussion is made here, all elements discussed in § VII.5.1 remaining valid.

Radiative heat transfer

Similarly to the wings of allowed bands, CIA can make significant contributions to radiative heat exchanges in atmospheres as discussed below and in Refs. 742,743,748. An example in the case of the Earth is given in Fig. VII.19 where atmospheric transmission spectra in the 4 µm region are displayed. These again demonstrate [compare curves (3) and (1)] that correct modeling of the non-Lorentzian behavior of CO_2 far wings is required, confirming the previous results in the 15 µm region (Fig. VII.15). They also show that accurate fluxes (and radiative dissipations) cannot be obtained near 4.1 µm if the N_2 CIA in the $v=1 \leftarrow v=0$ band is not taken into account. Neglecting it leads to an underestimation of $\approx 15\%$ of the solar flux absorbed by the atmosphere for the solar EA and spectral range of Fig. VII.19. Another significant contribution to the absorption of solar radiation in the Earth atmosphere is brought by the oxygen CIA bands at shorter wavelengths.[906,907] As for the far wings of water vapor lines, they were invoked as a possible explanation for the excess solar absorption problem mentioned before. Note that missing weak H_2O lines and/or improper spectroscopic parameters in databases may also contribute (*eg* Ref. 908).

Collision-induced absorption also contributes to radiative exchanges in giant planets. For Jupiter, Saturn, and Uranus, it is the case of CIA by hydrogen, since its amount is large (more than 80%). This is illustrated by Fig. VII.20, showing that the H_2 CIA is responsible for a significant part of the infrared emission by the atmosphere of Uranus. In the case of Titan, radiative transfers in the atmosphere essentially result from CIA with contributions from N_2, H_2, and CH_4. as illustrated in Fig. VII.21. Indeed, the

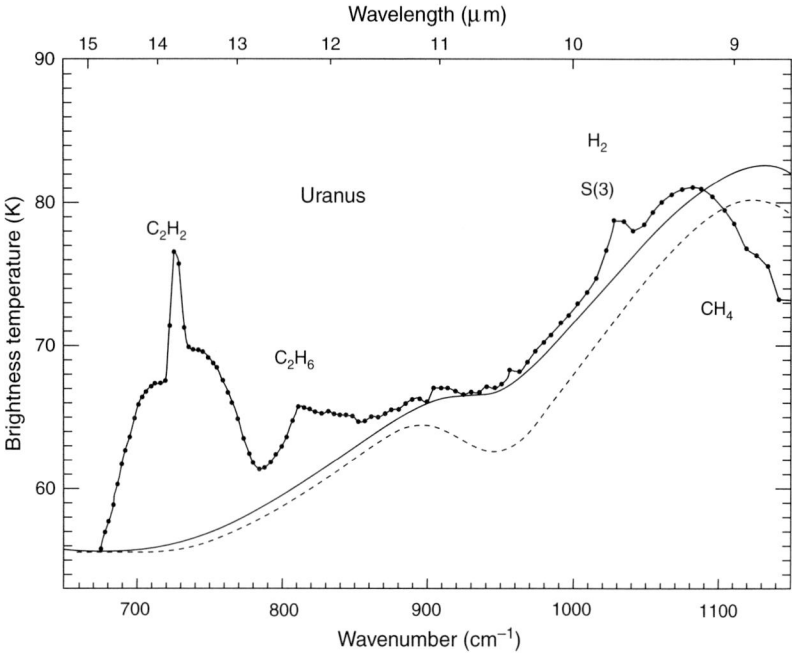

Fig. VII.20: Disk averaged emission spectrum of Uranus [in terms of brightness temperature, cf Eq. (VII.7)]. (–•–) is the observed spectrum,[909] while (——) and (- - -) show the continua due to H_2 CIA calculated with two models. Reproduced with permission from Ref. 910.

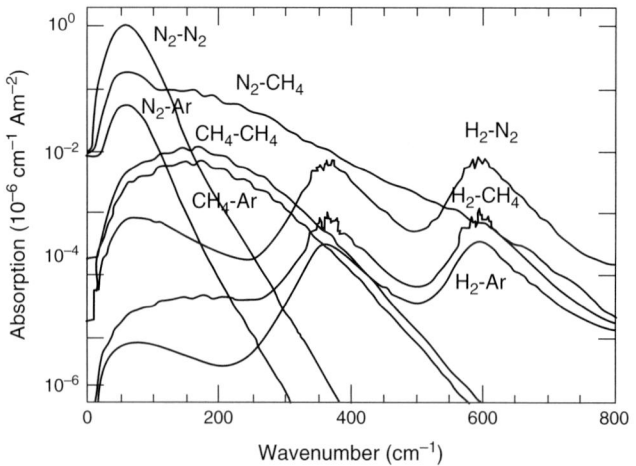

Fig. VII.21: Calculated CIA normalized by the total density for the typical Titan conditions $T = 80\,K$, and the vmrs: $x_{N2} = 0.898$, $x_{Ar} = x_{CH4} = 0.04$, $x_{H2} = 0.002$. Reproduced with permission from Ref. 911.

blackbody function [Eq. (VII.3)] at the representative temperature of 80 K peaks near 280 cm^{-1} and is completely contained in the spectral range of Fig. VII.21.

Remote sensing

The preceding demonstration that CIA can make significant contributions to atmospheric absorption and emission implies that this process must be correctly modeled for remote sensing, for the same reasons leading to the need for proper descriptions of the far wings of allowed bands. In particular, the consequences for the use, in vmr remote determinations, of narrow features carried by a broad CIA background are the same as those discussed in § VII.5.1 and they will not be further discussed. Let us only mention illustrative studies in which some CIA signatures are *directly* used for the determination of atmospheric characteristics. An application in the case of the Earth[905] was discussed above. In this work, absorptions in the region of the N_2 CIA and CO_2 v_3 band (*cf* Fig. VII.19) were used for the retrieval of density vertical profiles. Other examples are given by Refs. 911,912 where the CIA features displayed in Fig. VII.21 are used in order to deduce the volume mixing ratios of CH_4, N_2, and H_2. Finally, modeling of CIA may also be required for the determination of various properties besides those of the gaseous atmosphere itself. Examples are given by retrievals of surface temperature and the analysis of the eventual presence of clouds.[912] In particular, it has been shown recently[910] that proper modeling of the H_2 CIA obviates the former need[913] to invoke a haze contribution in order to match the observed emission of Uranus.

VII.6 CONCLUSION

Sections VII.3–VII.5 have hopefully convinced the reader that collisional effects on molecular spectra can have important consequences for applications. All the processes discussed in chapters III to VI occur in "real" media, may they affect the profile of isolated lines (chapter III), the shape of clusters of transitions (chapter IV), the far wings (chapter V), or lead to collision-induced absorption bands (chapter VI). Their proper modeling is thus required for accurate heat transfer calculations and optical soundings of gas media. However, until relatively recently, many of these effects were disregarded or very roughly modeled, most studies using isolated lines with Voigt shapes (except for CIA). The reasons for this are essentially twofold. The first one was the limited spectral resolution and signal-to-noise ratio of infrared remote sensing instruments (*eg* about 5 cm^{-1} near 12 μm for the spectrometer onboard the Venera-15 mission[914]). This masked many details of the profile, hence strongly reducing the need for an accurate description of spectral shapes. The second reason is that the qualities of radiative transfer predictions and of remote sensing products are primarily sensitive to the area below the absorption or emission. The latter being essentially determined (except for CIA) from the integrated intensities of the lines, the details of the spectral shape play only a secondary role. Significant uncertainties remained, before the 1990's, on the knowledge of these intensities, even for "key" species, and solving this spectroscopic problem was the first priority. This made the integration of sophisticated spectral profiles in forward models a subsidiary problem in view of their small impact with respect to overall uncertainties.

Note that this statement must be modulated according to the molecule and spectral region but some weak H_2O lines[915] and HNO_3 hot transitions[916] are examples for which significant corrections of some line intensities have been made recently.

Thanks to the efforts of research workers studying the spectroscopic parameters of *isolated* molecules (line positions and intensities), large progresses have been made in the last two decades, resulting in the constant improvement of the relevant databases.[2–5] Nowadays, the achieved accuracy of these parameters, the quality of sounding instruments, and the accuracy required by environmental studies, have brought the spectral-shape issue to the front stage. There are more and more situations where it is the uncertainty on the description of the spectral profile (and not any more that on its position and area) which mostly affects the optical sounding products. A first example, already mentioned, is the OCO mission[536] which will only fulfill its goals if CO_2 and O_2 spectra are predicted with an extreme accuracy,[537] including refined description of the spectral shape (see Figs. IV.36 and VII.14, and Refs. 415,538). Another demanding satellite experiment for the observation of the Earth atmosphere is the Atmospheric Chemistry Experiment[1] for which accurate line profiles are also needed.[855] Due to the importance of meteorological and climate issues, these satellite observation experiments will be followed by other dedicated missions extending the needs to other species and spectral regions. Besides laboratory research, this will likely motivate significant evolutions of molecular spectroscopic databases, as discussed in Sec. VIII.4. Another demanding field is the optical sounding of combustion processes and reactive flows,[917,918] which often involve high temperatures and pressures and complex gas mixtures. The ever more refined analysis of the processes governing the temporal and spatial variations of flames and flows leads to an increasing need for accuracy on retrieved quantities. This again induces strong exigencies for the quality of the models used to represent collisional effects on measured signals. These elements indicate that the subject of the present book will likely play a determinant role in practical applications of the near future. This conclusion is a good opportunity for a link to the next chapter, where some remaining problems are discussed.

VIII. TOWARD FUTURE RESEARCHES

VIII.1 INTRODUCTION

Unsolved or insufficiently well modeled problems concerning collisional effects on spectral shapes remain, as mentioned along the preceding chapters. Among these are completely opened theoretical questions, from formal and/or practical calculations points of view, for which almost everything remains to be done. A voluntarily extreme example is a fully quantum modeling of the near, mid, and far wings of self-broadened water vapor bands including the influence of the radiator translational motion. More reasonable, but still a difficult problem (and somehow a "curiosity" with respect to the expected limited practical consequences) is resonance exchange discussed in appendix IV.C. Similar is the case of collisional broadening and shifting below 10 K for which unexplained discrepancies remain between measured and calculated values (Fig. IV.28). There are also some situations where paths have been opened and possible solutions have been proposed but where few applications to practical systems have been made. This last chapter is devoted to cases for which results may be obtained within reasonable time scales. Three particular aspects have been retained, since they have not been discussed yet and have important consequences for applications. The first one is the inclusion of velocity effects in line-mixing models. The second point is the problem of calculating spectra from optical resonances to their far wings in a unified way. Finally, spectroscopic databases, which are essential for radiative transfer calculations and remote sensing, have to evolve in order to fulfill the increasing accuracy needs of some applications. Prospective and the estimation of the interest and applicability of new approaches being partly subjective, the following sections are not exhaustive and some specific aspects are likely missing. We nevertheless hope that no important topic, for which things are (almost) mature, was omitted in this book.

VIII.2 DICKE NARROWING IN SPEED-DEPENDENT LINE-MIXING PROFILES

VIII.2.1 MODELS OF PROFILES IN THE HARD COLLISION FRAME

Molecules may have structures of rotational energy levels such that the distances between some optical transitions are close to, or within, the Doppler line widths. The accurate description of the spectral shape of such manifolds thus requires to take into account the *combined* effects of line mixing, of the speed dependence of collisional rates, and of the Dicke narrowing (*cf* chapters III and IV). These may simultaneously contribute for pressures such that the lines overlap *and* have collisional widths comparable to the Doppler broadening. An analysis of recent extensions, for collisionally *coupled* lines, of some phenomenological models developed for isolated transitions (cf Secs. III.3 and III.4) is presented here. The main goal is to point some general characteristics and trends in the treatment of more complex spectral situations than those considered in chapters III and IV.

Starting from the semi-classical expression (Sec. II.5) for the spectral shape of Dicke-narrowed coupled lines, Ciurylo and Pine[75] have derived several useful models within the frame of the *hard* collision approximation presented below. The (N_L-dimensional column vector) spectral density for N_L collisionally coupled optical transitions can be written under the following compact form:

$$F(\omega) = \frac{1}{\pi} \text{Re} \left\{ d^+ [1 - G(\omega)(v_{VC}^H - C)]^{-1} G(\omega) \rho_a d \right\}, \quad \text{(VIII.1)}$$

where d^+ is the row vector of transition moments in the line space, v_{VC}^H is a diagonal matrix whose elements are the transition-specific hard velocity-changing collision frequencies [which are speed-independent, cf Eq. (III.28)], C is the correlated matrix for collisions simultaneously changing the radiator velocity and its phase, ρ_a is the diagonal matrix of the relative populations of the initial levels of the radiator transitions, and $G(\omega)$ is defined by:

$$G(\omega) = \langle G(\omega, \vec{v}) \rangle = \int [v_{VC}^H + i[-i\widetilde{W}(v) + L_a] - i(\omega - \vec{k}.\vec{v})I_d]^{-1} f_M(\vec{v}) d^3\vec{v}, \quad \text{(VIII.2)}$$

where $\widetilde{W}(v)$ is the ($N_L \times N_L$) speed-dependent relaxation matrix for the considered perturber density n_p [related to that W(v) introduced in Eqs. (II.47) and (II.48) and discussed in Sec. IV.3 by: $\widetilde{W}(v) = n_p W(v)^*$]. L_a is the diagonal matrix of the unperturbed transition frequencies [cf Eq. (II.49)], and $f_M(\vec{v})$ is the Maxwell-Boltzmann distribution for the radiator velocity $\vec{v}$ [Eq. (III.2)].

For *uncorrelated* hard velocity-changing (HVC) and dephasing (D) collisions (ie HVC + D collisions), C = 0 and v_{VC}^H refers only to pure HVC collisions assumed, for simplicity, to be transition-*independent*. The matrix product $G(\omega)(v_{VC}^H - C)$ in Eq. (VIII.1) thus reduces to $G(\omega) v_{VC}^H$ (where v_{VC}^H is now a scalar). In spite of this simplification, the calculation of the spectral profile remains an important task. It requires to construct and invert a $N_L \times N_L$ matrix for each velocity in order to carry out the integration over the equilibrium distribution. It must be added that the knowledge of the speed dependence of a *full* relaxation matrix $\widetilde{W}$ is, in itself, a difficult challenge. A first solution to this problem is direct calculations from the radiator-perturber potential energy surface (§ IV.3.4 and § IV.3.5). An alternative is the use of scaling laws (§ IV.3.3) in which the speed dependence of W(v) could be obtained from that of the basic parameters Q_L. As discussed in § IV.2.3b, computer time can be saved by using a diagonalizing procedure. It consists in introducing the matrix P(v) of the eigenvectors of $[L_a - i\widetilde{W}(v)]$, so that:

$$[L_a - i\widetilde{W}(v)] = P(v)[\omega_d(v) - i\Gamma_d(v)]P^{-1}(v), \quad \text{(VIII.3)}$$

where $\omega_d(v)$ and $\Gamma_d(v)$ are diagonal and real matrices, $[\omega_d(v) - i\Gamma_d(v)]$ giving the eigenvalues of $[L_a - i\widetilde{W}(v)]$. Hence $G(\omega, \vec{v})$ in Eq. (VIII.2) can be written as:

$$G(\omega, \vec{v}) = P(v) D^{-1}(\omega, \vec{v}) P^{-1}(v), \qquad \text{(VIII.4)}$$

where $D(\omega, \vec{v})$ is a diagonal matrix defined by:

$$D(\omega, \vec{v}) = v_{VC}^H + \Gamma_d(v) - i[\omega I_d - \omega_d(v) - \vec{k}.\vec{v}I_d]. \qquad \text{(VIII.5)}$$

$\omega_d(v)$ and $\Gamma_d(v)$ thus appear as the frequencies and broadenings of the *equivalent* lines associated with the eigenvectors of $[L_a - i\widetilde{W}(v)]$. This diagonalization procedure bypasses the inversion, at each frequency ω, of the matrix appearing in the definition of $G(\omega, \vec{v})$ [Eq. (VIII.2)], but still leaves $[1 - G(\omega)v_{VC}^H]^{-1}$ to be calculated in Eq. (VIII.1).

In the case of *no* collisional coupling between the spectral components (ie $\widetilde{W}_{\ell'\ell} = 0$ whatever ℓ' and ℓ with $\ell' \neq \ell$), from Eqs. (VIII.1) to (VIII.5), the leading profile is the expected sum of speed-dependent Nelkin-Ghatak (sdNG) (complex) profiles $J_{sdNG}(\omega)$ [Eq. (III.49)], weighted by the unperturbed intensities $S_\ell = \rho_\ell d_\ell^2$, ie:

$$F_{sdNG}(\omega) = \text{Re}\left[\sum_\ell S_\ell J_{sdNG}(\omega)_\ell\right]. \qquad \text{(VIII.6)}$$

In the *absence* of Dicke-narrowing (ie for $v_{VC}^H = 0$), the spectral shape becomes the *line-mixing* (labeled LM) generalization of the speed-dependent Voigt (J_{sdV}) profile (LM)sdVP [cf Eqs. (III.47) and (III.48)]:[75]

$$F_{sdV}^{LM}(\omega) = \frac{1}{\Delta\omega_D}\sum_\ell \text{Re}\left\{\int [\xi_\ell(v) + i\eta_\ell(v)] J_{sdV}[\tilde{x}_{d\ell}(v), \tilde{y}_{d\ell}(v)] f_M(\vec{v}) d^3\vec{v}\right\}, \qquad \text{(VIII.7)}$$

with

$$\begin{aligned}\xi_\ell(v) + i\eta_\ell(v) &= [d^+ P(v)]_\ell [P^{-1}(v)(\rho_a d)]_\ell, \quad \text{and} \\ \tilde{x}_{d\ell}(v) &= [\omega - \omega_{d\ell}(v)]/\Delta\omega_D \quad \text{and} \quad \tilde{y}_{d\ell}(v) = \Gamma_{d\ell}(v)/\Delta\omega_D,\end{aligned} \qquad \text{(VIII.8)}$$

$\Delta\omega_D$ being the 1/e Doppler half width given by Eq. (III.4). $J_{sdV}[\tilde{x}_\ell(v), \tilde{y}_\ell(v)]$ is the complex, *unaveraged* sdV profile ([Eq. (III.47)] without the symbol "Re" nor the v-average). In Eq. (VIII.8), the product $[d^+ P(v)]_\ell$ leads to a $(1 \times N_L)$ row vector and $[P^{-1}(v)(\rho_a d)]_\ell$ to a $(N_L \times 1)$ column vector. Their product gives the (scalar) intensity factor for the *equivalent line* ℓ.

If the Dicke narrowing is *taken into account* but *not* the speed dependence of $\widetilde{W}$, the spectral density is the full line-mixing (LM) generalization of the Nelkin-Ghatak (NG) profile [labeled (LM)NGP]:[75]

$$F_{NG}^{LM}(\omega) = \frac{1}{\Delta\omega_D}\sum_\ell \text{Re}\{(\xi_\ell + i\eta_\ell) J_{NG}(\tilde{x}_\ell, \tilde{y}_\ell + \zeta)\}, \qquad \text{(VIII.9)}$$

where ξ_ℓ, η_ℓ, $\tilde{x}_\ell$, and $\tilde{y}_\ell$ have the same definitions as above (but with *no* speed dependence), $\zeta = v_{VC}^H/\Delta\omega_D$, and $J_{NG}(\tilde{x}_\ell, \tilde{y}_\ell + \zeta)$ is the complex NG profile given by Eq. (III.32). It is worth noting that, if the HVC collision frequency is no longer considered as transition-independent in Eq. (VIII.2) but taken as *transition-specific*, the above mentioned diagonalization procedure cannot be used anymore and Eq. (VIII.9) is thus *not strictly* valid. Finally, in the particular case of no line mixing, this last equation reduces to a sum of NGP weighted by unperturbed intensities as in Eq. (VIII.6).

A last interesting case is that of weak line mixing, when a first-order perturbation[342] (§ IV.2.2a) can be used. The *first-order* spectral density, assuming that the *off-diagonal* elements of $\tilde{W}$ are *speed-independent*, is given by:[75]

$$F_{DsdNG}^{LM}(\omega) = \frac{1}{\pi}\sum_\ell \text{Re}\left\{S_\ell(1+i\tilde{Y}_\ell)\Phi_\ell(\omega)/[1-v_{VC}^H\Phi_\ell(\omega)]\right\}, \quad \text{(VIII.10)}$$

with $\Phi_\ell(\omega) = \int [v_{VC}^H + \Gamma_\ell(v) - i(\omega - \omega_\ell - \Delta_\ell(v) - \vec{k}.\vec{v}]f_M(\vec{v})d^3\vec{v}$, and (VIII.11)

$$\Gamma_\ell(v) + i\Delta_\ell(v) = \tilde{W}_{\ell\ell}(v), \quad S_\ell = \rho_\ell d_\ell^2, \quad \tilde{Y}_\ell = \frac{\eta_\ell^{(1)}}{S_\ell} = 2\sum_{\ell'\neq\ell}\frac{d_{\ell'}}{d_\ell}\frac{\tilde{W}_{\ell'\ell}}{(\omega_\ell - \omega_{\ell'})}. \quad \text{(VIII.12)}$$

In these equations, ω_ℓ are the unperturbed transition frequencies (at zero density) and $\tilde{Y}_\ell$ are the first-order line-coupling parameters [*cf* Eq. (IV.24)]. Within these approximations, the spectral density can be written under the following form:

$$F_{DsdNG}^{LM}(\omega) = \sum_\ell S_\ell \text{Re}\left\{(1+i\tilde{Y}_\ell) J_{sdNG}(\omega)_\ell\right\}, \quad \text{(VIII.13)}$$

where $J_{sdNG}(\omega)_\ell$ [Eq. (III.49)] is the complex speed-dependent NG profile for the transition ℓ. This (LM)DsdNGP is a sum of line-mixing *dispersive* (D) sdNG profiles weighted by the unperturbed intensities. In the absence of line mixing, Eq. (VIII.13) reduces to a sum of sdNGP. Recall that all the above models assume *uncorrelated* HVC/D collisions and a *transition-independent* HVC collision frequency v_{VC}^H.

For *correlated* HVCD collisions, the $N_L \times N_L$ matrix $[v_{VC}^H-C]$ [*cf* Eq. (VIII.1)] must be substituted to the scalar v_{VC}^H frequency. This matrix is now *speed dependent* and *transition specific*. For reasons explained above, the diagonalization procedure is no longer possible, and no simple analytical expression can be derived without additional assumptions. For completeness, let us mention that, in the special case of collisionally coupled doublets of transitions, an exact analytical solution can be obtained since the associated 2×2 matrix can be inverted analytically.[75,545]

Among the models described above, the (LM)NGP [Eq. (VIII.9)] and the (LM)DsdNGP [Eq. (VIII.13)] are of particular interest for fits of measured manifold spectra at low and moderate densities. They account for the main physical mechanisms underlying Dicke-narrowed line-mixing profiles and are tractable, even for a large number of transitions. Pine et al[330,490,542,563] have performed careful tests of these two models, but *empirically releasing* the transition-independent constraint on v_{VC}^H implied in the rigorous derivation[75] (*ie* taking v_{VC}^H as *transition specific*). The results are described below in order to discuss their limitations and to open paths toward further improvements.

VIII.2.2 EXPERIMENTAL TESTS IN MULTIPLET SPECTRA

Due to the importance of proper modeling of methane spectra for its spectroscopic monitoring (*cf* chapter VII) in various media including planetary atmospheres, numerous experimental and theoretical studies have been devoted to the CH_4 infrared spectral shapes (see Ref. 919 and those therein). Measurements of collisional widths of isolated lines have been made, showing, besides the usual dependence on the rotational quantum number J, a systematic influence of the tetrahedral symmetry and of the energy ordering index.[330,919] The presence of line-mixing effects at moderate pressure was first evidenced in Ref. 541 and later confirmed and analyzed using more refined fitting models (see Refs. 490,495 and those cited therein). The first test of Dicke-narrowed line-mixing profiles was made[330] using the uncorrelated *speed-independent* (LM)NGP given by Eq. (VIII.9). The determination of frequencies (ω_d) and broadenings (Γ_d) of equivalent lines requires the knowledge of *all* the W-matrix elements [*cf* Eq. (VIII.3)]. These elements not only must satisfy the detailed balance [Eq. (II.E16)], but also the following resulting sum rule:[920]

$$\sum_\ell (\xi_\ell + i\eta_\ell) = \sum_\ell \rho_\ell d_\ell^2. \qquad (VIII.14)$$

This relation, which expresses the integrated intensity conservation, indicates that the sum of the line-mixing contributions vanishes for the equivalent lines (*ie* $\sum \eta_\ell = 0$). Another sum rule holds for the line widths and shifts in Eq. (VIII.9) (through $\widetilde{x}_\ell$ and $\widetilde{y}_\ell$), that is:

$$\sum_\ell (\Gamma_{d\ell} + i\Delta_{d\ell}) = \sum_\ell (\Gamma_\ell + i\Delta_\ell). \qquad (VIII.15)$$

These relations can be used to reduce the number of independent unknowns in the fits or to test the reliability of the parameters obtained from unconstrained adjustments. Another important point concerns the *symmetry effects* in the line-mixing process. Indeed, even if spectral components within a manifold overlap significantly, they may be uncoupled due to symmetry considerations. This is the case for the A, E, and F symmetries due to nuclear spin in tetrahedral molecules, which are essentially uncoupled by collisions.[921] This leads to a block-diagonal form of the relaxation matrix for which the sum rules of Eqs. (VIII.14) and (VIII.15) pertain to each symmetry group *separately*.

For the (full) line-mixing speed-independent Dicke-narrowed (LM)NGP model, the $\widetilde{W}$ matrix was directly fitted by constraining the off-diagonal matrix elements through a relation similar to Eq. (IV.94), *ie*: $\widetilde{W}_{\ell'\ell} = -n_p F_c k_{\ell'\leftarrow \ell} P/2\pi c$,[490] where F_c is a scaling factor assumed constant for a given CH_4 J manifold (as also done in Refs 336,492). The collision rates $k_{\ell'\leftarrow\ell}$ connecting the lower levels of lines ℓ and ℓ' were calculated in the ground vibrational state of CH_4 for quasi-elastic ($\Delta J = 0$) collisional transfers.[359,489] Note that such a procedure imposing the relative amounts of line coupling within the manifolds (F_c being used to scale the absolute amounts) is needed since adjusting all the $\widetilde{W}$ elements is practically impossible (§. IV.4.2).

Before studying the adequacy of the (LM)NGP model, Pine and Gabard[490] have checked those of the speed-independent and speed-dependent approximations, by first considering the R(0) singlet and R(2) doublet (uncoupled since of E and F symmetries) lines. These are not affected by line mixing and can be treated using isolated line models. Multispectrum fits of these Ar-broadened structures using the NGP and sdNGP models [cf Eq. (VIII.6)] show significant residuals when the NGP model is used. These vanish when fits are made with the sdNGP model, confirming the influence of the speed dependence of the broadening (see § III.4.1). The situation is essentially the same for other P(J) and R(J) manifolds of the v_3 band, when *including* line mixing. This is exemplified by Fig. VIII.1 which displays results obtained when the speed-independent Dicke-narrowed full line-mixing (LM)NGP model is used [empirically considering v_{VC}^H, through ζ, cf Eq. (VIII.9), as transition-specific]. Again, the residuals are well above the noise level and characteristic of the influence of the speed dependence since they are also observed for isolated components (cf supra) and subsist at the highest pressures where Dicke narrowing effects are minimal.

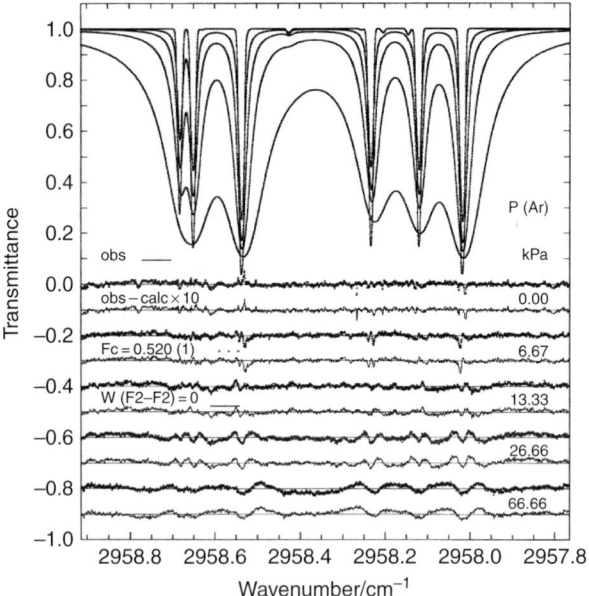

Fig. VIII.1: Multispectrum fits of the v_3 P(6) manifold of CH_4 broadened by Ar at room temperature and several pressures showing the residuals obtained with the speed-independent (LM)NGP. Besides the results obtained with the adjusted $F_c = 0.520$ value (thin line), those assuming no coupling between F2 type transitions (dark trace) are also shown. Reproduced with permission from Ref. 490.

The speed-dependent (LM)DsdNGP [Eq. (VIII.13)] was then tested under the same conditions (in particular, taking v_{VC}^H as transition specific). This model requires the knowledge of the speed dependence of the line-broadening $\Gamma(v)$ and -shifting $\Delta(v)$. In the absence of accurate ab initio data, Pine and Gabard[490] used the R^{-q} model defined through Eq. (III.57) (the potential parameter q being an additional adjusted quantity). The resulting multispectrum fits are displayed in Fig. VIII.2, for the same spectra as in Fig. VIII.1. Note that these results, calculated from Eq. (VIII.13), assume that the intensity, the first-order line-mixing coefficient, and the velocity-changing collision rate are speed-independent (in particular, the speed dependence of the off-diagonal part of the relaxation matrix is *disregarded*). In these fits, the intensities S_ℓ, the line broadenings Γ_ℓ and shifting Δ_ℓ, the collision frequencies $v_{VC_\ell}^H$, the first-order line-mixing coefficients $\widetilde{Y}_\ell$, and the potential parameter q, are adjusted. Their values for the Ar-broadened v_3 P(6) manifold can be found in Tables 4 and 6 of Ref. 490. Similar results hold for the other manifolds ($J \leq 10$) of CH$_4$–Ar, and close results are obtained for CH$_4$–N$_2$.[330,490]

Figure VIII.2 clearly evidences the need to take line mixing into account at all pressures, (LM)DsdNGP fit residuals then being almost reduced to the noise level. Adjustments of similar quality were obtained throughout the P branch for all manifolds and pressures. For the R branch, the overlapping of the manifold transitions at the highest

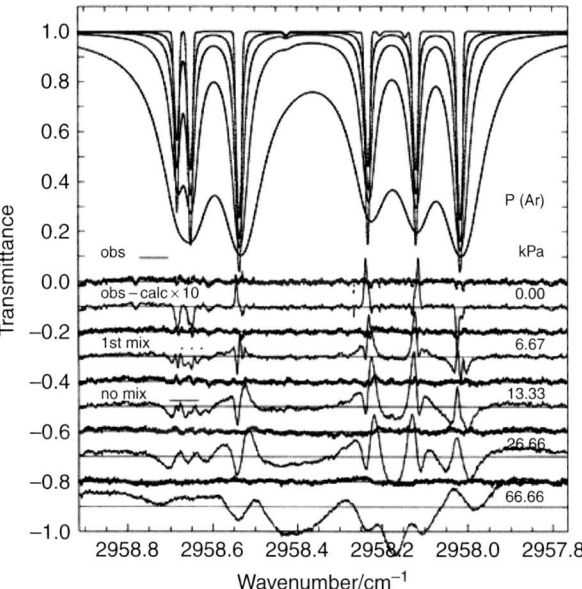

Fig. VIII.2: Same as Fig. VIII.1 except that the speed-dependences of Γ and Δ are now included. Hence the (LM)DsdNGP model is used (dark trace), which takes into account line mixing, together with the sdNGP (light trace) which assumes collisionally isolated lines. Reproduced with permission from Ref. 490.

pressures is too strong to justify the first-order line-mixing approximation and fits were restricted to pressures below 200 Torr. Note that the results of Ref. 490 evidence the strong reduction of the uncertainties on S_ℓ, Γ_ℓ, and Δ_ℓ obtained with the multispectrum fitting procedure. A second feature is the significant modification of the line-broadening and -shifting coefficients (a few percents) induced by the speed dependence. Much more drastic changes occur in the velocity changing collision frequency (a reduction by a factor of several units), which can be attributed in part to the HVC/D correlations not included in the (LM)DsdNGP and (LM)NGP (cf § III.3.6 and III.4.3). An additional fitting parameter is the potential range q (Table 6 of Ref. 490) for which q(Ar) = 4.72 and q(N_2) = 4.68 are obtained from the P(6) manifold. These values indicate similar speed dependences for the two perturbers considered and their magnitude is reasonably consistent with the type of dominant interactions in CH_4–Ar and CH_4–N_2 molecular pairs. However, the artificiality of the R^{-q} model and the restriction of the speed dependence to the line-broadening and -shifting parameters (disregarding that of the line-mixing parameters) prevent to go further in the analysis of the (LM)DsdNGP model. In order to overcome this difficulty, accurate ab initio calculated data for the speed dependence of Γ and Δ should be used, as done for HF–Ar (cf Ref. 172 and § III.4.3). The calculations should also include off-diagonal elements of the relaxation matrix in order to take into account the speed dependence of the line-coupling parameters [cf Eq. (VIII.13)].

Dicke narrowing and presumably line mixing were previously observed[922] in the Q branch of the v_3 band of CH_4 broadened by various collision partners. After this study in which spectrum by spectrum fits disregarding line mixing were made, Pine and Gabard[542] have reexamined the recorded spectra. They used multispectrum fits and the (LM)DsdNGP model, leading to adjustments within the noise level for all spectra, regardless of the pressure and perturber. In order to explore the influence of the speed-dependence [still *only* through $\Gamma(v)$ and $\Delta(v)$], similar adjustments were made using the (LM)DNGP (thus disregarding the speed-dependence), leading to the same quality of agreement. Furthermore, the speed-independent (LM)NGP full line-mixing model also accurately fits the observed spectra, except for minor and localized discrepancies. This is consistent with previous results (see § IV.3.4f) showing that an accurate description of absorption can be obtained in wide pressure ranges using adjusted (through F_c) speed-independent $\widetilde{W}$ elements. But these results for the Q branch[542] are in contrast with those for P and R manifolds[490] where the (LM)DNGP model produced systematic residuals attributed to speed-dependent broadening. This comparison thus raises the question of the real role played by the speed dependence in CH_4 spectral shapes, due to the over-simplified classical path R^{-q} model. Beyond this specific point, let us point out the main limitations of the above line-shape models for Dicke-narrowed line-mixing profiles, recalling that, in the applications, the transition-independent constraint on the velocity-changing collision frequency has been released.

Ignoring the speed dependence of the first-order line-mixing parameter in Eq. (VIII.13) might seem drastic, but it should be noted that, for the considered P and R manifolds, quasi-elastic collisions are involved, with energy differences negligible when compared to the kinetic energy. Disregarding the imaginary part of the off-diagonal elements of the relaxation matrix should be not too restrictive either, considering the small values of the line-shifts. Neglecting the HVC/D correlation is probably a more restrictive assumption. Its role in isolated line shapes has been carefully studied (cf § III.4.3 and III.4.4). However, recall that, if this correlation is taken into account, the diagonalization (§ VIII.2.1) cannot be used and there is no analytical solution for a general $N_L \times N_L$ correlated relaxation matrix. It must be added that there is currently no practical way of including speed dependence in a full $\widetilde{W}$ matrix inversion scheme. Furthermore, this speed dependence should not be restricted to the diagonal elements. One must also remember the limitations tied to the hard collision approximation, analyzed in detail for isolated lines in § III.4.3. In this case, a pertinent generalization has been introduced through (Rautian-Sobelman) RSP models incorporating both hard and soft events in the collisional process. Extensions to line-mixing profiles seems not easy, at least on firm physical bases.[75] Finally, let us mention the possibility of full ab initio methods for which directions have been formally given by Monchick.[290] However, practical applications of such approaches require a formidable computational task. Even if possible in the future, the problem coming from the difference between the accuracy nowadays attained by high quality measurements (better than 0.1%) and that (about one percent) of the best ab initio calculations (cf § III.5.4) remains.

To conclude this discussion, the speed-dependent hard collision line-mixing profiles can be considered as reasonably accurate and useful tools, in spite of the mentioned limitations of the hard collision model. However, they require particular care in the treatment of the line-broadening and -shifting speed dependences through the multi-spectrum fitting procedure. At the present level of refinement of spectral-shape models (especially for isolated lines, cf chapter III), properly modeling this speed dependence appears as the main difficulty.[21,923] A possible alternative, intermediate between the (analytical) hard collisional models and full ab initio methods could be the generalization of the Keilson-Storer memory model (cf § III.4.4) for speed-dependent line-mixing profiles. The scalar radiating dipole appearing in Eq. (III.94) for isolated lines must then be replaced by the row vector of transition moments in the line space. Concomitantly, the kinetic equation (III.93) must be generalized to N_L coupled lines, along a scheme similar to that used in Ref. 75, and including now the Keilson-Storer memory function[187] with the possibility to use the Sonine polynomials as pertinent basis. Such an approach could enable to extend the nature of the considered collisions (from hard to soft collisions, cf § III.4.4) and also to take into account the VC/D correlation effect between velocity-changing and dephasing collisions.

VIII.3 FROM RESONANCES TO THE FAR WINGS

As shown in chapter IV, models exist for the computation of spectral shapes influenced by line mixing within the impact approximation, which may be extended, following Sec. VIII.2, to take the radiator translational motion into account. Accurate results can then be obtained, eventually with sophisticated approaches starting from the radiator-perturber potential (§ IV.3.5b), but these are *limited* to spectral regions close to the centers of the most intense optical transitions. When going away from line centers, the quality of the quasistatic model (Sec. V.3), again based on first principles and knowledge of the interaction potential, has been demonstrated, but this model is limited to the far wings. If one excludes the perturbative approach (Sec. V.4) due to its approximate nature, the problem of an unified treatment of all parts of the spectrum remains opened. This requires the calculation, starting from the interaction potential, of the frequency dependence of a relaxation matrix including all collisionally coupled lines. It is a difficult problem of considerable practical importance, for which two paths are given below. The first is an extension of the SGC semi-classical model (§ IV.3.4c) beyond the impact assumption, while the second, briefly examined, starts from the generalized ECS model (Sec. V.5).

VIII.3.1 SEMI-CLASSICAL APPROACH

Starting from Eqs. (II.58)–(II.62), provided that initial correlations and the effects of the radiator translational motion can be neglected, the frequency-dependent relaxation matrix elements are given by:

$$\left\langle \langle f'i'|W(\omega)|fi\rangle \right\rangle = -\frac{n_0 <v_r>}{2\pi c}(\omega - \omega_{fi})(\omega - \omega_{f'i'})$$

$$\times \lim_{\varepsilon \to 0^+} \int_0^\infty e^{-\varepsilon t} e^{i(\omega - \omega_{fi})t} Q_{f'i',fi}(t), \quad \text{(VIII.16)}$$

where $n_0(T)$ is the number of perturbers per unit volume at one atmosphere pressure (or one amagat density depending on the choice of units). Within a semi-classical approach, the average contained in the quantities $Q_{f'i',fi}(t)$ is over the three parameters describing a binary radiator-perturber collision. These are its central time t_0 (of closest approach), its impact parameter b and its initial relative speed v_r [or reduced relative kinetic energy $x = m^* v_r^2/(2k_B T)$]. For a linear molecule colliding with atoms and dipolar (k = 1) rovibrational transitions ($v_f J_f \leftarrow v_i J_i$), from Eq. (II.62) and App. II.C1, $Q_{f'i',fi}(t)$ is:

$$Q_{f'i',fi}(t) = \int_0^\infty 2\pi b db \times \int_0^\infty xe^{-x} dx (-1)^{J_f + J_f'} \sqrt{\frac{2J_i' + 1}{2J_i + 1}}$$

$$\times \left\{ \delta_{ii'} \delta_{ff'} - \sum_{\substack{m_i, m_i' \\ m_f, m_f', q}} (-1)^{m_f + m_f'} \begin{pmatrix} J_i' & 1 & J_f' \\ m_i' & q & -m_f' \end{pmatrix} \begin{pmatrix} J_i & 1 & J_f \\ m_i & q & -m_f \end{pmatrix} \right. \quad \text{(VIII.17)}$$

$$\left. \times \int_{-\infty}^{+\infty} dt_0 \langle v_f J_f' m_f' | U(t, 0, t_0, b, x) | v_f J_f m_f \rangle^* \langle v_i J_i' m_i' | U(t, 0, t_0, b, x) | v_i J_i m_i \rangle \right\},$$

where $U(t,0,t_0,b,x)$ is the operator describing the evolution of the system from time 0 to time t for the considered (t_0,b,x) collision. Prohibitive computer time and storage prevent calculations of the $Q_{f'i',fi}(t)$ elements, for all needed sets of the t, t_0, b, and x parameters, with the NG approach (§ IV.3.4b). Although still very costly, they are feasible with an extension of the SGC approach presented in § IV.3.4c. Within this model, a change of fact that calculated values do asymptotically verify long time behaviors which can be analytically derived (*eg* App. II.C1) from Eqs. (VIII.19) and (VIII.20). However, problems remain, and, in particular, the fact that the generalized sum rule, detailed in Ref. 72 [see also Eq. (V.47)], is not well verified.

In spite of these problems, the model leads to results opening interesting perspectives, as shown in Fig. VIII.4, since two of the main deviations from the usual Lorentzian isolated lines predictions are (semi)quantitatively reproduced: the sub-Lorentzian character of the wings of R and P branches and the super-Lorentzian absorption observed near the band center. This last behavior was observed previously,[131,924,925] together with the decrease of the intensities of P and R lines with increasing perturber density. In the past, this was interpreted[926,927] as the effects of the perturbation of the free rotation of the radiator due to the interaction potential. This leads, through radiator wave functions increasingly perturbed with increasing density, to modifications of the line intensities and the rising of a "Q branch". Another explanation was also proposed,[763,925] invoking the role of molecular complexes. The results of Fig. VIII.4 and Refs. 131,135 indicate that the finite collision duration contributes to both the super-Lorentzian absorption near the band center and the variation of the R and P line intensities with pressure.

Starting from these preliminary results, two paths could be followed to improve the model. The first one, computer time consuming but straightforward, is to carry out the average over the initial relative kinetic energy. Indeed, the far wings being sensitive to short times [through the Laplace transform in Eq. (VIII.16)], they are expected to strongly depend on the radiator-perturber relative speed through the influence of

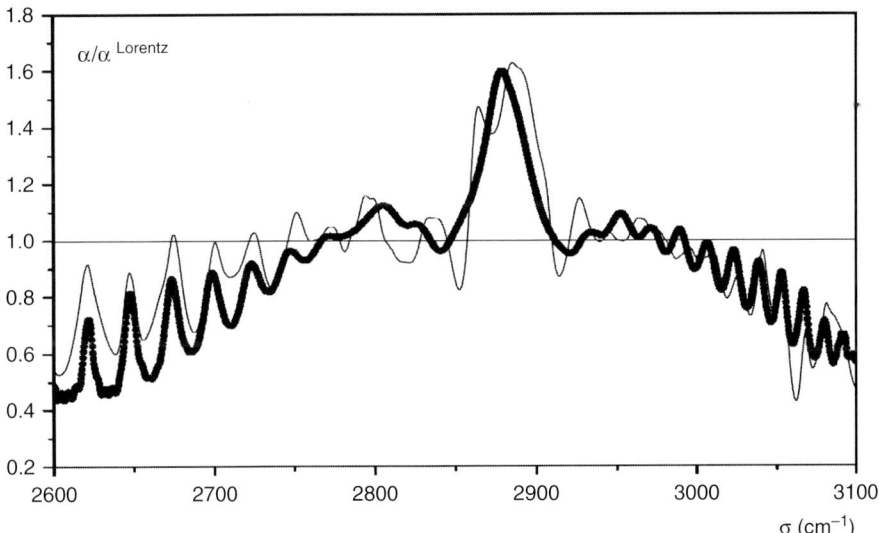

Fig. VIII.4: Measured (•) and calculated (——) absorptions divided by that calculated assuming isolated lines with purely Lorentzian profiles. The results are for the 1–0 band of HCl diluted in 350 atm of Ar at room temperature. After Ref. 135.

collisions at short distances. The second improvement would be to include the effects of initial statistical correlation through the complex time $t + i\hbar/k_B T$ [see Eq. (II.42)]. General indications on how the integration in the complex plane can be made are given in Ref. 141. Nevertheless, this is not easy since the result depend, due to the semi-classical model used, on the path followed in the complex plane.

VIII.3.2 GENERALIZED SCALING APPROACH

Section V.5 describes a model which enables a unified treatment from resonances to the far wings of collisionally coupled lines. It is based on a generalization[612,666,672] of the usual Energy Corrected Sudden (ECS) approach to non-impact situations. This is essentially obtained through the introduction of a frequency-dependent "adiabaticity" correction $\widetilde{\Omega}(\Delta\omega)$ parameterizing the influence of the finite duration of collisions on the wing. In view of the decay observed in the far wings of pure N_2 Rayleigh spectra (Fig. V.7), the choice was made[666] of an analytical expression for $\widetilde{\Omega}(\Delta\omega)$ [Eq. (V.54)] based on the classical model introduced by Birnbaum and Cohen.[674] The approach then depends on two phenomenological parameters whose values can be obtained from fits of experimental data. This has led to satisfactory predictions of N_2 spectra (Fig. V.8) but two problems remain. The first is that the far wings of the ν_1 Raman band of pure CO_2 cannot be modeled using realistic values of the ECS parameters (Fig. V.10). The second is that the chosen analytical form of the generalized adiabaticity correction automatically results in relaxation matrix elements smaller, in absolute value, than their impact equivalents. This cannot represent the "super-impact" behavior shown in Fig. VIII.3 and Refs. 339,356,655,656. These results suggest that some alternative analytical expressions should be searched for. Furthermore, the accounting for the perturber internal states would also be of interest in order to extend the model to molecule-molecule collisions. This is not easy, but paths have been opened.[405,406] Note that the previously mentioned improvements might enable to solve one important remaining problems which is (cf App. V.A) the modeling of the near and mid wings of water vapor lines.

VIII.4 TOMORROW'S SPECTROSCOPIC DATABASES

Molecular spectroscopic databases[(1)] are extremely useful to a broad scientific community since they store, in an up-to-date and *standardized* way, information scattered in numerous sources. Their main (historical) part is a table in which each row corresponds to an allowed (dipolar or quadrupolar) optical transition of a given species. Initiated, for some of them, almost four decades ago, they have been regularly updated. As an example, whereas its first edition (1973) contained about 100 000 transitions of 7 gases with a limited set of parameters, the HITRAN database[2] now provides many parameters for over 1 700 000 lines of 37 gases. Note that the consequences, for radiative

---

[(1)] There are various molecular spectroscopic databases as further described in Ref. 326. Some of them are general in terms of species and spectral coverage,[2,3] while others are more dedicated, limited to some wavelength regions,[4] or associated with a specific remote sensing instrument.[928]

transfers and remote sensing in the Earth atmosphere, of changes of the HITRAN database during the last decades are significant, as discussed in Refs. 691,856. Large efforts have been devoted to the spectroscopy of *isolated* molecules (identification, position, lower level energy, and transition moment of the lines), but the situation remains poorer for the quantities describing collisional shapes. The reason for this is that the former parameters were the *first* to be known, without which no calculation could be made, whereas the latter played only a secondary role until quite recently. The situation has changed, due to the accuracy attained for the parameters of isolated molecules and to the increased accuracy asked for by applications. There is now a strong need for the storage of extended and standardized data describing the effects of pressure on absorption/emission profiles. Although numerous results are available (*cf* chapters III-VI), only few collisional parameters are stored in databases such as the widely used ones presented in Refs. 2,3. These only provide the self- and air-broadening coefficients and the air-shifting coefficients at room temperature, and a parameter describing the temperature dependence of air-broadening. These data enable calculations for isolated lines with *speed-independent Voigt profiles*, thus disregarding almost all effects discussed in this book. The self-shifting is not given, as are the effects of temperature on the self-broadening and air-induced shift. No information is stored for collisions with many other perturbers of practical interest. Considering the requirements of remote sensing applications (see Chapt. VII), spectral-shape models more refined than the Voigt profile must now be included in radiative transfer codes. Due to the large variety of applications for which many molecular species and transitions under broad temperature, gas composition and density conditions may be of interest, extended spectroscopic databases are now needed. Some paths toward this goal have been discussed in Ref. 326. They are summarized and completed below from the point of view of the subject of this book.

VIII.4.1 ISOLATED LINES

A first and direct extension of databases would be to add the temperature dependence of self-broadening, particularly important for CO_2 and H_2O in the atmospheres of Venus and Earth, respectively. Including equivalent data for the broadenings by CO_2 (Venus), H_2 and He (Jupiter, Saturn, *etc*), and N_2 (Titan) would also be of interest. Efforts are needed to collect the available information, since many measurements have been made. These could be completed (*eg* Ref. 929) using measured values for lines of "close" quantum numbers, calculations (*cf* Tables IV.3 and IV.4), or interpolations (*eg* Figs. IV.16 and IV.20). This extension could be done by adding the proper number of columns to the existing tables. These would be two for each collision partner, providing, as already done for air-broadening, the room temperature value and a temperature exponent, *ie* for each line ℓ of each species a and each perturber p, $\gamma_\ell^{a-p}(T_0)$ and n_ℓ^{a-p} such that:

$$\gamma_\ell^{a-p}(T) = \gamma_\ell^{a-p}(T_0) \times [T_0/T]^{n_\ell^{a-p}}. \quad (VIII.22)$$

A similar work could also be done for the shifting coefficients δ (only the room temperature value of the air-induced shift is yet stored), but a problem remains to be solved. Indeed, whereas Eq. (VIII.22) satisfactory represents the temperature dependence of the broadening, the shifts can change sign with increasing temperature.[166,930] Another analytical expression must be chosen, such as a linear variation[930] with T or a function[166] of $T^{-1/2}$, (when δ is expressed in cm^{-1}/atm[(2)]), the difficulty being that it must be *standardized* and applicable to all transitions. A more flexible alternative is to store shifting coefficients on a temperature grid chosen according to available measurements and to the desired accuracy, hoping that this choice will stand for future experimental/theoretical values. Such a work could be initiated since many results are available concerning the pressure shifts.

For isolated lines, such broadening and shifting data essentially permit calculations with speed-independent Voigt profiles, thus disregarding many effects analyzed in chapter III. This is becoming insufficient for some applications requiring high accuracy, as discussed in § VII.3.2. Going beyond the usual Voigt profile first requires the choice of a line-shape model among the various ones presented in chapter III. This must be done carefully since a wrong initial choice might induce considerable database maintenance problems. Concerning the relative speed (v_r) dependence, the simplest approach, for the width cross sections, assumes a sum of power laws [Eq. (III.59)], ie:

$$\sigma_\gamma(v_r) = \sum_{i=1}^{N} a_i \times v_r^{\alpha_i}, \qquad (VIII.23)$$

where a_i and α_i depend on the transition only (but *not* on temperature). Carrying out a thermal average at a given temperature, straightforwardly leads to:

$$\gamma(T) = \frac{n_0(T)}{2\pi c}<v_r\sigma_\gamma(v_r)> = \frac{n_0(T)}{2\pi c}\sum_{i=1}^{N} a_i \times \langle v_r^{\alpha_i+1}\rangle = \sum_{i=1}^{N} A_i \times T^{(\alpha_i+1)/2}, \qquad (VIII.24)$$

with

$$A_i = a_i \frac{n_0(T)}{2\pi c} \frac{4}{\sqrt{\pi}} (2k_B/m^*)^{(\alpha_i+1)/2} \int_0^\infty x^{3+\alpha_i} e^{-x^2} dx, \qquad (VIII.25)$$

where $n_0(T)$ is the number of perturbers per unit volume at one atmosphere pressure (or one amagat when density units are used). These equations show that the parameters of a (*crude*) speed dependence with a *single* term ($N=1$, ie only a_1 and α_1) can directly be deduced from those [$\gamma(T_0)$ and n] describing the temperature dependence through Eq. (VIII.22). They are obtained from the identification of Eqs. (VIII.22) and (VIII.24), leading to:

$$\alpha_1 = 1 - 2n \text{ when cm}^{-1}/\text{atm units are used.} \qquad (VIII.26)$$

[(2)] Starting from a temperature law for a collisional parameter expressed in cm^{-1}/atm, the corresponding one, when the parameter is expressed in cm^{-1}/Am, is obtained by multiplying the former by T.

from which a_1 can be obtained using Eqs. (VIII.22), (VIII.24) and (VIII.25). Note that this assumes that the parameters a_i in Eq. (VIII.23) are *temperature-independent*. This neglects the effect of T through the populations of the perturber levels and is thus stictly valid only for atomic collision partners. A discussion on the limits of this simple approach can be found in Ref. 238. Similarly, parameters approximately describing the speed dependence of the shift can be obtained[166] from those of the analytical law describing the effects of temperature. This shows that approximate information on the speed dependence of γ is already contained (for air-broadening only) in today's databases. For better accuracy, alternative approaches must be considered. A first solution is to use more terms (N > 1) and to include the temperature dependence of the parameters a_i in Eq. (VIII.23). More refined analytical laws, such as that proposed in Ref. 275 [which has the advantage of enabling analytical developments within the Keilson-Storer memory model (§ III.4.4)] may also be considered. One could also store values of the broadening $\gamma(T,v_r)$ and shifting $\delta(T,v_r)$, for each line of each species and each perturber, on temperature and relative speed grids. These must be properly chosen in order to cover the range of temperature within which values will be used and to accurately represent the speed dependences. These data, which enable the computation of speed-dependent Voigt (or Lorentz) profiles, are insufficient when collision-induced velocity changes affect the spectral shape. Other parameters are needed, such as the frequency(ies) of the velocity-changing process(es) or the memory parameters of the KS approach (*cf* Chapt. III). In a first step, these could be considered as independent of the optical transition and of the radiator speed but they are specific of the considered model. Their storage, for each radiator-perturber pair, thus requires a dedicated file within the database and, eventually, some software to convert stored values to the desired temperature (as done for the partition functions[931]). Their dependence on temperature then has to be modeled using, for instance, simple power laws.[20] Once databases include the relevant data, the updating of radiative transfer codes is relatively easy. The line-by-line form can be kept and one only has to change the routine computing the line shape. As discussed in appendix III.A, this may require the development of efficient algorithms for low cost large scale evaluations of speed-dependent Dicke narrowed profiles.

VIII.4.2 LINE MIXING

Various studies (Sec. VII.4) have demonstrated that line-mixing effects must (sometimes) be taken into account in order to retrieve accurate information from optical soundings of gas media. There is thus a need for the inclusion of some line-coupling parameters in databases. There are essentially two ways to do this.

The first one is to store first-order line-mixing coefficients Y_ℓ [*cf* Eq. (IV.24)]. This is a priori tempting since they are associated with each transition and perturber so that the database structure could be kept. One would then only have to add the proper number of columns in the table. Nevertheless, this is not easy for reasons discussed in § IV.2.3a and Refs. 345,346. Since the Y's must verify some sum rules with a high accuracy, the constraint on the stored data are high. These must ensure a very accurate description of the temperature dependence of $Y_\ell(T)$, imposing the use of analytical laws with many parameters and digits.[346] However, if done this way, there are database maintenance problems. Indeed, any change of line-positions, intensities, or lower state energies may lead to the breakdown of the sum rules if the Y's are not simultaneously updated (which is not easy). Finally remember that the first-order approach is inadequate for strongly overlapping coupled transitions since the perturbation expansion breaks down.

In practice, it is more convenient to store the relaxation matrix elements $W_{\ell'\ell}(T)$, even though this requires a dedicated part of the database. This has been done for CO_2 Q branches in the 2004 version of HITRAN,[2] following Refs. 343,417. The relaxation matrix elements [a room temperature value and a temperature dependence exponent, as in Eq. (VIII.22)] are given in separated files. An integer flag is added in the line list, indicating that data for line mixing are available (or not) for the considered optical transition. The maintenance is then relatively easy. Indeed, if slight changes are made to the spectroscopic parameters of the isolated molecule, the $W_{\ell'\ell}$ elements (little sensitive to these parameters) need not be changed, contrary to Y_ℓ coefficients. Furthermore, a renormalization procedure can be applied [Eqs. (IV.110) and (IV.111)] in order to ensure that W exactly verifies the required properties. From the point of view of line-coupling parameters in spectroscopic databases, much remains to be done, since parameters for air-broadened CO_2 Q lines only are stored yet. These could be extended rapidly by including other available data (see tables IV.5 and IV.6). The consequences for radiative transfer codes are minor *if* the first-order approximation [Eq. (IV.22) or (IV.26)] can be used. The line-by-line approach of most codes can be kept but some software[343,567] must compute the Y's [Eq. (IV.24)] from the relaxation matrix elements. In cases of strong overlapping where Eq. (IV.1) must be used, a specific routine has to be called, based on the diagonalization procedure (*cf* § IV.2.3b) leading to the parameters of the equivalent lines. These can then be fed into the line-by-line calculation remembering that they are *pressure dependent*. As discussed in § IV.2.3, care must be taken if one wishes to save computer time through line wing truncations.

Finally, although today is likely too soon, the question of storing parameters in order to model speed-dependent line-mixing profiles will have to be considered at some point since, as described in Sec. VIII.2, there are studies evidencing such effects. A solution may be, as for the broadening and shifting, to store data representing the dependences of the relaxation matrix elements $W_{\ell'\ell}(v_r,T)$ on grids of temperature and relative radiator-perturber speed v_r.

VIII.4.3 FAR-WINGS AND COLLISION-INDUCED ABSORPTION

VIII.4.3a Far wings

Today's spectroscopic databases provide no data for the description of the far wings of allowed optical transitions. Radiative transfer and remote sensing codes builders are on their own and must collect the relevant information elsewhere. As discussed in Sec. V.2, there are essentially two practical ways to include far wings in large scale spectra calculations.

The first one uses isolated line profiles (generally Lorentzian in the wings) modified through factors χ [Eq. (V.3)]. Such corrections are available for some molecular systems (Table V.2), but the problem is then the (possibly conflicting) intricacy with the use of line-mixing data. Indeed, if a calculation including line mixing is made, in a Q branch region for instance, use of χ factors is not trivial. One has to carry out the line-mixing calculation in a limited spectral range and then turn to the χ-corrected isolated Lorentzian lines with a proper matching at the switching points. Furthermore, the available correction functions have generally been fitted to experiments at a given time, thus using a former edition of a database. Those of Ref. 615, for instance, were obtained with the 1988 version of GEISA.[3] Their use with current databases may lead to more or less important errors if significant changes of spectroscopic parameters have been made in the mean time. There is thus a maintenance problem which can be properly solved only by going back to the original experimental data and re-fitting χ factors each time databases are updated.

The alternative solution for the modeling of the far wings is the use of tabulated continua. The latter are defined [§ V.2.2 and App. V.A], for each radiator-perturber couple, as the difference between the "true" absorption and that calculated from a given *spectrally-local* model. Continua for self- and N_2-broadened water vapor lines, for instance, were constructed from a local model assuming Lorentzian lines truncated in the wings. This was first done (see Ref. 658 and those therein) using "true" values obtained from experiments in specific spectral regions. Later on, this was extended to a broad wavelength range, leading to the CKD continuum. This was done starting from reference ("true") spectra calculated using corrective factors χ adjusted on measurements.[616] Theoretical first-principles calculations of some continua have also been performed (*cf* Sec. V.3) but not made available in a database form, except for an analytical law applicable to the millimeter wave H_2O–N_2 continuum.[645] To our knowledge, the CKD water vapor continua[634,636] and the experimental ones for CO_2 and H_2O in specific regions (Refs. 614,658 and those therein) are the only ones available for practical use. The approach defining the continuum as the complement to spectrally localized predictions could be applied to other species displaying strong far wing atmospheric absorption (*eg* NH_3 and CH_4 in Jupiter and Saturn) using available χ factors (Table V.2). The values obtained for each radiator-perturber pair, which only depend on wavenumber and temperature and vary relatively slowly with wavelength, are of straightforward storage in databases and of easy use in spectra calculation codes. Nevertheless, difficulties remain. The first one is that a continuum being obtained as the complement to local contributions, it is *intrinsically* tied to the database. If modifications of the spectroscopic parameters are made, this induces changes in the calculated local absorption and adding an old continuum leads to erroneous total absorption values. The second problem is the choice of the "local" model which has, up to now, essentially been isolated lines with Voigt profiles. If line-mixing effects are included, new continua must be calculated. In this case, spectra computations require a proper switching from the line-mixing (local) approach to the continuum. This is not trivial, since wing truncations must be carefully handled (*cf* § IV.2.3).

Although it is far too early, storing wavenumber-dependent relaxation matrix elements $W_{\ell'\ell}(\sigma,T)$ in spectroscopic databases is an idea that should be kept in mind. This would ensure self-consistent calculations from the line centers to the far wings, accounting for the eventual effects of (impact) line mixing within clusters of closely spaced transitions. It would make maintenance easier, but accurate models for the construction of $W_{\ell'\ell}(\sigma,T)$ over the entire spectral range of interest remain to be built. Some paths have been opened in the preceding section, but they are far from being applicable to complex molecular pairs of practical interest.

VIII.4.3b   Collision-induced absorption

Except when interferences between the permanent and collision-induced dipoles exist (Sec. VI.6), including CIA parameters in databases is easy, thanks to two properties. The first one is that, when density (bi)normalized, the CIA depends only on the radiator-perturber couple, the wavenumber and the temperature. The second one is that the variations with wavelength and temperature are quite slow. Hence, for each radiator-perturber pair, tables can be stored (*eg* Refs. 77,78,910,932). Furthermore, their use in radiative transfer codes only needs interpolation schemes for which some laws (in T) have been proposed.[78] Note that, from the point of view of both storage in databases and radiative transfer calculations, CIA and far wings continua can be handled exactly in the same way. The case where interferences exist is more complex but of little practical interest since very seldom encountered in applications.

VIII.5   CONCLUSION

Considerable progresses have been made in the understanding of collisional effects on spectral shapes since the pioneering studies of the "fathers" of this discipline half a century ago. Together with more and more refined models, this has been made possible by four noticeable evolutions. The first one is the tremendous increase of the sensitivity, accuracy and wavelength coverage of laboratory measurement techniques. The second change is the huge evolution of the available computer power. The third one is the fact that spectral-shape theoreticians have beneficiated of the ever increased availability of refined (ab initio) intermolecular potentials provided by their colleagues of quantum chemistry. Finally, the quickly growing needs of remote sensing applications have continuously kept asking for a broader scope and better accuracy of spectral-shape models. In spite of these progresses, much remains to be done at various levels.

From the purely theoretical point of view, and even for very simple (diatom-atom) systems, many opened problems remain. Among these are the unexplained limitations of Close-Coupling calculations (§ IV.3.5) demonstrated by the inaccuracy of predicted line broadenings and shiftings at low temperature and the conjugated detailed balance relation obtained. The consequences of collision-induced radiator velocity changes and of the speed dependence of spectral-shape parameters on the far wings are also a completely opened question. Moreover, the building up of an accurate and tractable unified model for the prediction of the entire spectrum, including regions close and far away from the transitions centers, is not done yet. For more complex molecular systems of practical interest, even simplified approaches lack development. Among the important issues, let us mention the modeling of line-mixing effects on the spectrum of species such as water vapor and ozone. Another example is given by the far wings of H_2-broadened methane and ammonia lines.

From the experimental point of view, one of the important challenges is likely the increase of the optical paths attainable in the laboratory. "Conventional" Fourier transform instruments with multipass cells are practically limited to one kilometer, thus insufficient to investigate optical thicknesses involved in some atmospheres. Other techniques, such as photoacoustic, Cavity Ring Down, or Intra-Cavity Laser Absorption spectroscopies have much better sensitivities but they are often limited due to the use of lasers. This imposes limited spectral ranges or induces base-line uncertainties which may affect studies of continuum type absorptions.

For practical applications, many paths have been initiated which must be followed. The first is the inclusion of refined spectral-shape models in radiative transfer forward and retrieval codes. This is progressively done but on a kind of *ad hoc basis* which will require, at some point, some standardization. Concerning the computations of spectral profiles, accurate algorithms of high efficiency, suitable for large scale calculations must be further developed. For the input data, the solution is obviously the extension of molecular spectroscopic databases toward the storage of more and more information for the description of the spectral shapes.

The decision to write this book was taken after a PhD student came in the office of one of the authors asking for a monograph describing the various effects of pressure on molecular absorption. We hope that the present book brings a proper answer to this query and that it will be useful to the various scientific communities concerned by collisional effects on molecular spectra and their applications.

APPENDIX

ABBREVIATIONS AND ACRONYMS

Name	Meaning	See
AasdRS$_{\tilde{C}}$P	Asymmetrized asdRS$_{\tilde{C}}$P	Eq. (III.87)
asdGP	approximate speed-dependent Galatry Profile	Eq. (III.51)
asdRSP	approximate speed-dependent Rautian-Sobelman Profile	Eq. (III.73)
asdRS$_C$P	approximate correlated speed-dependent Rautian-Sobelman Profile	Eq. (III.72)
asdRS$_{\tilde{C}}$P	approximate partially correlated speed-dependent Rautian-Sobelman Profile	Eq. (III.84)
ATC	Anderson, Tsao and Curnutte	§ IV.3.4d
BC	Birnbaum and Cohen (line shape)	Eq. (VI.6) and Ref. 674
CARS	Coherent Anti-stokes Raman Spectroscopy	App. II.A
CC	Close Coupling	§ IV.3.5b
CIA	Collision-Induced Absorption	Chapt. VI
CILS	Collision-Induced Light Scattering	Chapt. VI
CKD	Clough-Kneizys-Davies (H_2O continuum)	App. V.A and Ref. 616
CKM	Collision Kernel Method	§ III.5.3
CS	Coupled States	§ IV.3.5c
DP	Doppler Profile	Eqs. (III.4), (III.23)
DP	Dicke Profile	Eq. (III.19)
DPR	DePolarized Rayleigh	App. II.A
e.m.	ElectroMagnetic	
EA	Elevation angle	Fig. VII.2
ECS	Energy Corrected Sudden	§ IV.3.3 and Sec. V.5
EGL	Exponential Gap Law	Table IV.1

Name	Meaning	See
EPGL	Exponential Power Gap Law	Table IV.1
FOV	Field Of View	Fig. VII.2
FWHM	Full Width at Half Maximum	
GALFIT	GALatry FITting program	Ref. 208
G_CP	Galatry correlated Profile	Eq. (III.36)
GEISA	Gestion et Etude des Informations Spectroscopiques Atmosphériques	Ref. 3
GHM	Generalized Hess Method	§ III.5.2
GP	Galatry Profile	Eq. (III.26)
HITRAN	HIgh-resolution TRANsmission molecular absorption database.	Ref. 2
HVC	Hard Velocity-Changing	Sec. III.3
HVCD	Hard Velocity-Changing and Dephasing	Sec. III.3
HWHM	Half Width at Half Maximum	
I_d	Identity operator or matrix	
IOS	Infinite Order Sudden	§ IV.3.3, IV.3.5d
IRS	Inverse Raman Scattering	App. II.A3
KSP	Keilson-Storer Profile	§ III.4.4
LBLRTM	Line By Line Radiative Transfer Model	Refs. 634,873
LKSP	Linearized Keilson-Storer Profile	§ III.4.4
LMNGP	Line Mixing NGP	Eq. (VIII.9)
LMsdVP	Line Mixing sdVP	Eq. (VIII.7)
LM$_D$sdNGP	Line Mixing Dispersive sdNGP	Eq. (VIII.13)
LP	Lorentz Profile	Eq. (III.17), (III.23)
LTE	Local Thermodynamic Equilibrium	App. II.B2
MDS	Molecular Dynamics Simulations	§ III.4.3
MEGL	Modified Exponential Gap Law	Table IV.1
MOLCOL	MOLecular COLlisions	Ref. 309
MOLSCAT	MOLecular SCATtering	Ref. 308

Name	Meaning	See
MPM	Millimiterwave Propagation Model	Ref. 642
NG	Neilsen and Gordon	§ IV.3.4b
NGP	Nelkin-Ghatak Profile	Eq. (III.31)
NG$_C$P	Nelkin-Ghatak Correlated Profile	Eq. (III.39)
OCO	Orbiting Carbon Observatory	Ref. 536
PA	Peaking Approximation	§ IV.3.4c
PES	Potential Energy Surface	
PGL	Power Gap Law	Table IV.1
RB	Robert and Bonamy	§ IV.3.4e
RIPS	Raman Induced Polarization Spectroscopy	App. II.A4
RSP	Rautian-Sobelman Profile	Eq. (III.67)
sC	speed Changing	Sec. III.4
sCD	speed Changing and Dephasing	Sec. III.4
sdBB	speed-dependent Billiard-Ball	§ III.5.4
sdGP	speed-dependent Galatry Profile	§ III.4.2
sdKSP	speed-dependent Keilson-Storer Profile	Eq. (III.102)
sdKS$_C$P	speed-dependent Keilson-Storer correlated Profile	Eq. (III.104)
sdKS$_{\tilde{C}}$P	speed-dependent Keilson-Storer partially correlated Profile	Eq. (III.106)
sdNGP	speed-dependent Nelkin-Ghatak Profile	Eq. (III.49)
sdNG$_C$P	speed-dependent Nelkin-Ghatak correlated Profile	Eq. (III.50)
sdVP	speed-dependent Voigt Profile	Eq. (III.48)
SGC	Smith, Giraud and Cooper	§ IV.3.4c
SRGS	Stimulated Raman Gain Spectroscopy	App. II.A
SRS	Stimulated Raman Spectroscopy	App. II.A
SVC	Soft Velocity Changing	Sec. III.3
SVCD	Soft Velocity Changing and Dephasing	Sec. III.3
VC	Velocity Changing	

Name	Meaning	See
VCD	Velocity Changing and Dephasing	
vmr	volume mixing ratio (mole fraction)	
VP	Voigt Profile	Eq. (III.21)
VVWP	Van Vleck-Weisskopf Profile	Eq. (IV.10)
WD, WM	Water Dimer, Water Monomer	App. V.A
WS	Waldmann and Snider	§ III.5.1
WSLP	Weighted Sum of Lorentzian Profiles	Eq. (III.48)
ZA	Zenith Angle	Fig. VII.2

SYMBOLS

NB: The symbols used in this book are commonly admitted and correspond to "traditions". Some may have *different* meanings depending on the context and research community (eg σ is used by spectroscopists to designate wavenumbers whereas, for collisionalists, it denotes collisional cross-sections). The following table gives the symbols frequently used along this book, but not those locally introduced and defined.

Name	Meaning
$X^*, {}^tX, X^+$	Complex conjugate, transpose, and adjoint of X
$Re(X)$, $Im(X)$, $Tr(X)$, $<X>$	Real part, imaginary part, trace, and average of X
a	Optically active molecule
A	Asymmetry parameter
b	Impact parameter
c	Speed of light
$\vec{d}$	Dipole moment vector
$\vec{k}$ k	– Radiation wave vector – Order of the radiation-matter coupling
E $\vec{E}$	– Energy (various, according to the indice) – Electric field vector
$F(\omega)$	Spectral density at angular frequency ω
$f_M(\vec{v})$	Maxwell-Boltzmann velocity distribution
h, $\hbar$	Planck's constant, $h/(2\pi)$

Appendix

Name	Meaning
H	Hamiltonian
I	– Normalized real profile – Radiance (or radiation intensity)
I or J	Normalized (complex) spectral shape
I_{BB}	Black-body radiance
J $\tilde{J}$	Rotational angular momentum quantum number Total angular momentum quantum number
k_B	Boltzmann's constant
ℓ	– Designates an optical transitions (a line) – Orbital quantum number – A typical collisional length
L_a	Liouville operator of the line positions.
m, m* m	– Mass, reduced mass – Magnetic rotational quantum number
n_s	Number density of species s
N_s	Total number of molecules of species s
P	Pressure
p	Perturber
$\vec{R}$, R	Intermolecular vector, distance
$\vec{r}$	Radiator position
S	– Integrated intensity (of a line) – Scattering (or diffusion) operator
t	Time
T	Temperature
U	Evolution operator
V	Interaction potential
V	Volume
$\vec{v}$, v v	– Velocity, speed – Vibrational quantum number
W	Relaxation matrix
Y	Collisional 1st order line-coupling coefficient

Name	Meaning
α	– Absorption coefficient – Isotropic polarizability
γ	– Broadening coefficient per unit density – Anisotropic polarizability – Keilson-Storer memory parameter
Γ	Collisional half width at half maximum
δ	– Shifting coefficient per unit density or pressure – Kronecker symbol – Dirac delta function
Δ	Collisional line shift
$\Delta\omega_D$	1/e Doppler halfwidth
λ	– Wavelength – Orbital angular momentum
ν	Frequency
χ	– Line wing corrective factor – Susceptibility
ρ	Density operator or relative population
σ	– Wavenumber – Cross section
τ	– Typical durations – Transmission
Φ	– Autocorrelation function – Radiative flux
ω	Angular frequency

UNITS AND CONVERSIONS

See also Refs. 933,934

Quantity	Unit	Related quantities / other units
Masses	g	$m(a.u.) = (1.6605\ 10^{-24}) \times m(g)$
Densities	Am	$n(molec/cm^3) = (2.6872\ 10^{19}) \times n(Am)$ $n(g/cm^3) = M(g) \times 4.460\ 10^{-5} \times n(Am)$ Ideal gas: $n(atm) = [T(K)/\ 273.15] \times n(Am)$

Appendix

Quantity	Unit	Related quantities / other units
Frequencies	s^{-1}	Hz, MHz, GHz, THz
Times	s	ms, μs, ns, ps, fs
Lengths	m or Å	R(cm) = 10^{-2} × R(m) R(Å) = 10^{-10} × R(m) R(a.u.) = 5.291 10^{-11} × R(m)
Pressures	atm	P(Pa) = (1.01325 10^5) × P(atm) P(bar) = 1.01325 × P(atm) P(Torr) = 760 × P(atm) Ideal gas: P(Am) = [273.15/T(K)] × P(atm)
Radiances	W/(m^2 s^{-1}sr)	I(Jy) = (10^{-26}) × I[W/(m^2 s^{-1}sr)] I[W/(m^2 cm^{-1}.sr)] = (2.998 10^{+10}) × I[W/(m^2 s^{-1}sr)]
Temperatures	K	T(°C) = T(K) + 273.15 T(°F) = 1.8 × [T(°C) + 32]
Wavenumber	cm^{-1}	σ(mk) = 10^3σ(cm^{-1}) λ(μm) = 10^4/σ(cm^{-1}) ν(s^{-1}) = c × σ = (2.99792458 10^{10}) × σ(cm^{-1}) ω(rad/s) = 2πc × σ = (1.88365157 10^{11}) × σ(cm^{-1})
Line intensities	cm/molec	S(cm^{-2}/Am) = (2.6872 10^{19}) × S(cm/molec) S(cm^{-2}/atm) = (7.34 10^{21}) × S(cm/molec)/T(K)
Dipole	D	d(C.m) = 10^{-21} × d(D)/c = (3.3356 10^{-30}) × d(D)
Quadrupole	D.Å	Q(C.m^2) = (3.3356 10^{-40}) × Q(D.Å)
Energies	cm^{-1}	E(K) = 1.439 × E(cm^{-1}) E(eV) = (8.065 10^4) × E(cm^{-1}) E(J) = (1.986 10^{-23}) × E(cm^{-1}) E(cal) = (8.309 10^{-23}) × E(cm^{-1}) E(erg) = (1.986 10^{-30}) × E(cm^{-1}) E(BTU) = (2.095 10^{-20}) × E(cm^{-1}) E(a.u.) = (2.195 10^{+5}) × E(cm^{-1})

REFERENCES

1. Bernath PF, McElroy CT, Abrals MC, Boone CD, *et al.* Atmospheric Chemistry Experiment (ACE): mission overview. Geophys Res Lett **32**, L15S01 (2005)
2. Rothman LS, Jacquemart D, Barbe A, Benner DC, *et al.* The HITRAN 2004 molecular spectroscopic database. JQSRT **96**, 139 (2005)
3. Jacquinet-Husson N, Scott NA, Chédin A, Garceran K, *et al.* The 2003 edition of the GEISA/IASI spectroscopic database. JQSRT **95**, 429 (2005)
4. Pickett HM, Poynter RL, Cohen EA, Delitsky ML, *et al.* Submillimeter, millimeter, and microwave spectral line catalog. JQSRT **60**, 883 (1998)
5. Müller HSP, Schlöder F, Stutzki J, Winnewisser G. The Cologne database for molecular spectroscopy, CDMS: a useful tool for astronomers and spectroscopists. J Mol Struct **742**, 215 (2005)
6. Townes CH, Schawlow AL. Microwave Spectroscopy. Dover, (1955)
7. Papousek D, Aliev MR. Molecular Vibration-Rotation Spectra. Elsevier, Amsterdam (1982)
8. Applied Laser Spectroscopy, Techniques, Instrumentation and Applications. Andrews DL editor, VCH Publishing, NY (1992)
9. Demtröder W. Laser Spectroscopy: Basic Concepts and Instrumentation. 2nd Edition, Springer-Verlag, NY, (1996)
10. Bloembergen N. Nonlinear Optics. 4th edition, World Scientific, Singapore (1996)
11. Mukamel S. Principles of Nonlinear Optical Spectroscopy. Oxford University Press, new edition (1999)
12. Davis SP, Abrams MC, Brault JW. Fourier Transform Spectrometry. Academic press (2001)
13. Herzberg G. Molecular Spectra and Molecular Structure. II. Infrared and Raman Spectra of Polyatomic Molecules. Van Nostrand Reinhold, NY, (1945)
14. Herzberg G. Molecular Spectra and Molecular Structure. I. Spectra of diatomic molecules. 2nd edition, Van Nostrand Reinhold, NY, (1950)
15. Herzberg G. Molecular Spectra and Molecular Structure. III. Electronic Spectra of Polyatomic Molecules. Van Nostrand Reinhold, NY, (1966)
16. Bernath PF. Spectra of Atoms and Molecules. Oxford University Press, NY, (1995)
17. Hollas JM. High Resolution Spectroscopy. 2nd edition, John Wiley & Sons, Sussex, (1998)
18. Computational Molecular Spectroscopy. Jensen P and Bunker PR Eds., John Wiley & Sons (2000)
19. Hartmann JM, Boulet C. Line shape parameters for HF in a bath of argon as a test of classical path model. J Chem Phys **113**, 9000 (2000)
20. Chaussard F, Saint-Loup R, Berger H, Joubert P, *et al.* Speed-dependent line profile. A test of a unified model from the Doppler to the collisional regime for molecule-molecule collisions. J Chem Phys **113**, 4951 (2000)
21. Lisak D, Hodges JT, Ciurylo R. Comparison of semiclassical line shape models to rovibrational H_2O spectra measured by frequency-stabilized cavity ring-down spectroscopy. Phys Rev **A 73**, 012507 (2006)
22. Green S. Pressure broadening data as a test of a recently proposed $Ar-H_2O$ interaction potential. J Chem Phys **95**, 3888 (1991)
23. Green S, Hutson J. Spectral line shape parameters for HF in a bath of Ar are accurately predicted by a potential inferred from spectra of the van der Waals dimer. J Chem Phys **100**, 891 (1994)

24. Roche CF, Hutson JM, Dickinson AS. Calculations of line width and shift cross sections for HCl in Ar. JQSRT **53**, 153 (1995)
25. Thibault F, Calil B, Boissoles J, Launay JM. Experimental and theoretical CO_2–He pressure broadening cross sections. Phys Chem Chem Phys **2**, 5404 (2000)
26. Korona T, Moszynsky R, Thibault F, Launay JM, et al. Spectroscopic, collisional, and thermodynamic properties of the $HeCO_2$ complex from an ab initio potential: theoretical predictions and confrontation with experimental data. J Chem Phys **115**, 3074 (2001)
27. Smith MA, Gordley L. Sensitivity of ozone retrievals in limb-viewing experiments to errors in line-width parameters. JQSRT **29**, 413 (1983)
28. Hussong J, Luckerath R, Stricker W, Bruet X, et al. Hydrogen CARS thermometry in a high-pressure H_2-air flame. Test of H_2 temperature accuracy and influence of line width by comparison with N_2 CARS as reference. Appl Phys **B 73**, 165 (2001)
29. Johnson DG, Jucks KW, Traub WA, Chance KV. Smithsonian stratospheric far-infrared spectrometer and data reduction system. J Geophys Res **D 100**, 3091 (1995)
30. Smith KM, Newnham DA, Williams RG. Collision-induced absorption of solar radiation in the atmosphere by oxygen at 1.27 µm: field observations and model calculations. J Geophys Res **D 106**, 7541 (2001)
31. Advances in Non-Linear Spectroscopy (Advances in Spectroscopy Vol. 15). Clark RJH and Hester RE Eds., John Wiley & Sons, NY, (1988)
32. Beer R. Remote Sensing by Fourier Transform Spectrometry. John Wiley & Sons, NY, (1992)
33. Spectroscopy of the Earth Atmosphere and Interstellar Medium. Edited by K.N. Rao and A Weber, Academic Press, NY (1992)
34. Eckbreth AC. Laser Diagnostics for Combustion Temperature and Species. Taylor & Francis (1996)
35. Hirschfelder JO, Curtis CF, Bird RB. Molecular Theory of Gases and Liquids. Wiley, NY, (1966)
36. Maitland GC, Rigby M, Smith EB, Wakeham WA. Intermolecular Forces. International Series of Monographs on Chemistry, Clarendon Press, Oxford (1996)
37. Stone AJ. The Theory of Intermolecular forces. International Series of Monographs on Chemistry, Clarendon Press, Oxford (1981)
38. Weidemüller M, Zimmermann C. Interactions of Ultracold Gases: From Atoms to Molecules. Wiley, NY, (2003)
39. Weiner J. Cold and Ultracold Collisions in Quantum Microscopic and Mesoscopic Systems. Cambridge university press, Cambridge (2004)
40. Anderson PW. Pressure broadening in the microwave and infra-red regions. Phys Rev **76**, 647 (1949)
41. Baranger M. Problem of overlapping lines in the theory of pressure broadening. Phys Rev **111**, 481 (1958)
42. Kolb AC, Griem H. Theory of line broadening in multiplet spectra. Phys Rev **111**, 514 (1958)
43. Baranger M. General impact theory of pressure broadening. Phys Rev **112**, 855 (1958)
44. Fano U. Pressure broadening as a prototype of relaxation. Phys Rev **131**, 259 (1963)
45. Ben-Reuven A. Symmetry considerations in pressure-broadening theory. Phys Rev **141**, 34 (1966)
46. Ben-Reuven A. Impact broadening of microwave spectra. Phys Rev **145**, 7 (1966)
47. Ben-Reuven A. Resonance broadening of spectral lines. Phys Rev **A 4**, 2115 (1971)
48. Breene RG Jr. Theories of Spectral Line Shape. Wiley, NY (1981)
49. Robert D, Galatry L. Infrared absorption of diatomic polar molecules in liquid solutions. J Chem Phys **55**, 2347 (1971)
50. Landau LD, Lipschitz EM. Statistical Physics. Pergamon (1969)

51. Cohen-Tannoudji C, Diu B, Laloë F. Quantum Mechanics. Wiley-Interscience (2006)
52. Liu WK. Symmetrized Liouville basis for indistinguishable particles. Application to spectral linewidth. J Chem Phys **72**, 4869 (1980)
53. Cohen-Tannoudji C, Dupont-Roc J, Grynberg G. Photons and Atoms. Introduction to Quantum Electrodynamics. Wiley, NY (1997)
54. Forster D. Hydrodynamic Fluctuations, Broken Symmetry, and Correlation Functions. Perseus books (1990)
55. Smith EW, Cooper J, Chappell WR, Dillon T. An impact theory for Doppler and pressure broadening-I. General theory. JQSRT **11**, 1547 (1971)
56. Rautian SG, Sobel'man JI. The effect of collisions on the Doppler broadening of spectral lines. Sov Phys Usp **9**, 701 (1967)
57. Davies RW, Tipping RH, Clough SA. Dipole autocorrelation function for molecular pressure broadening: a quantum theory which satisfies the fluctuation-dissipation theorem. Phys Rev A **26**, 3378 (1982)
58. Callen HB, Welton TA. Irreversibility and generalized noise. Phys Rev **83**, 34 (1951)
59. Kubo R. Statistical-mechanical theory of irreversible processes. I. General theory and simple applications to magnetic and conduction problems. J Phys Soc Japan **12**, 570 (1957)
60. Huber DL, Van Vleck JH. The role of Boltzmann factors in line shape. Rev Mod Phys **38**, 187 (1966)
61. Van Vleck JH, Huber DL. Absorption, emission, and linebreadths: a semihistorical perspective. Rev Mod Phys **49**, 939 (1977)
62. Albers J, Deutch JM. On the rate equation description of spectral lines. Chem Phys **1**, 89 (1973)
63. Smith EW, Cooper J, Roszman LJ. An analysis of the unified and scalar additivity theories of spectral line broadening. JQSRT **13**, 1523 (1973)
64. Zare RN. Angular Momentum. J Wiley and sons (1997)
65. Edmonds AR. Angular Momentum in Quantum Mechanics. Princeton University Press, Princeton, New Jersey (1974)
66. Roman P. Advanced Quantum Theory. Addison-Wesley (1965)
67. Fitz DE, Marcus RA. Semiclassical theory of molecular spectral line shapes in gases. J Chem Phys **59**, 4380 (1973)
68. Gordon RG. Semiclassical theory of spectra and relaxation in molecular gases. J Chem Phys **45**, 1649 (1966)
69. Gordon RG. Correlation functions for molecular motion. Adv Magnet Reson **3**, 1 (1968)
70. Smith EW, Vidal CR, Cooper J. Classical path methods in line broadening. I. The classical path approximation. J of Res of the Nat Bur of Standards **73A**, 389 (1969)
71. Smith EW, Vidal CR, Cooper J. Classical path methods in line broadening. II. Application to the Lyman series of hydrogen. J of Res of the Nat Bur of Standards **73A**, 405 (1969)
72. Ma Q, Tipping RH, Birnbaum G, Boulet C. Sum rules and the symmetry of the memory function in spectral line shape theories. JQSRT **59**, 259 (1998)
73. Zwanzig R. Ensemble method in the theory of irreversibility. J Chem Phys **33**, 1338 (1960)
74. Smith EW, Cooper J, Chappell WR, Dillon T. An impact theory for Doppler and pressure broadening-II. Atomic and molecular systems. JQSRT **11**, 1567 (1971)
75. Ciurylo R, Pine AS. Speed-dependent line mixing profiles. JQSRT **67**, 375 (2000)
76. Bouanich JP, Brodbeck C, Nguyen-Van-Thanh, Drossart P. Collision-induced absorption by H_2–H_2 and H_2–He pairs in the fundamental band. An experimental study. JQSRT **44**, 393 (1990)
77. Thibault F, Le Doucen R, Menoux V, Rosenmann L, et al. Infrared collision induced absorption by O_2 near 6.4 μm for atmospheric applications: measurements and empirical modeling. Appl Opt **36**, 563 (1997)

78. Lafferty WJ, Solodov AM, Weber A, Olson WB, et al. Infrared collision-induced absorption by N_2 near 4.3 μm for atmospheric applications: measurements and empirical modeling. Appl Opt **35**, 5911 (1996)
79. McKellar ARW. Intensities and the Fano line shape in the infrared spectrum of HD. Can J Phys **51**, 389 (1973)
80. Bafile U, Ulivi L, Zoppi M, Barocchi F. Depolarized-light-scattering spectrum of normal gaseous hydrogen at low density and a temperature of 297 K. Phys Rev **A 37**, 4133 (1988)
81. Abu-Kharma M, Gillard RG, Reddy SP. Collision-induced absorption by D_2 pairs in the first overtone band at 77, 201 and 298 K. Eur Phys J **D 37**, 59 (2006)
82. Camy-Peyret C, Flaud JM, Delbouille L, Roland G, et al. Quadrupole transitions of the 1-0 band of N_2 observed in a high resolution atmospheric apectrum. J Phys Lettres **42**, L279 (1981)
83. Poll JD, Hunt JL. On the moments of the pressure-induced spectra of gases. Can J Phys **54**, 461 (1976)
84. Frommhold L. Collision Induced Absorption in Gases. Cambridge Monographs on Atomic, Molecular, and Chemical Physics. Cambridge University Press, Cambridge, (2006)
85. Borysow A, Frommhold L. Collision induced rototranslational absorption spectra of N_2–N_2 pairs for temperatures from 50 to 300 K. Astrophys J **311**, 1043 (1986). [*erratum*: Astrophys J **320**, 437 (1987)]
86. Brocks G, van der Avoird A. Contribution of bound dimers, $(N_2)_2$, to the interaction induced infrared spectrum of nitrogen. In Phenomena Induced by Intermolecular Interaction (p. 699), Edited by G. Birnbaum, Plenum, NY, (1985)
87. Frommhold L, Samuelson R, Birnbaum G. Hydrogen dimer structures in the far infrared spectra of Jupiter and Saturn. Astrophys J **283**, L79 (1984)
88. Schaefer J. Theoretical studies of the faint features in the $S_0(0)$ line of H_2 observed in the Voyager IRIS mission. Astron Astrophys **182**, L40 (1987)
89. McKellar ARW. Infrared spectra of the $(N_2)_2$ and N_2–Ar van der Waals molecules. J Chem Phys **88**, 4190 (1988)
90. Gruszka M, Borysow A. Computer simulation of the far infrared collision-induced absorption spectra of gaseous CO_2. Mol Phys **93**, 1007 (1998)
91. Springer Handbook of Atomic, Molecular and Optical Physics. Edited by G.W.F. Drake, Springer, 2006
91b. Placzek G. Rayleigh-Streuung und Raman-Effek. Marx: Handbuch der Radiologie, p. 365, 2nd Ed., Vol. 6, Part II, Braunschweig, 1934
92. Gordon RG. Relations between Raman spectroscopy and nuclear spin relaxation. J Chem Phys **42**, 3658 (1965)
93. Gordon RG. Molecular motion and the moment analysis of molecular spectra. II. The rotational Raman effect. J Chem Phys **40**, 1973 (1964)
94. Cummins HZ, Gammon RW. Rayleigh and Brillouin scattering in liquids: The Landau-Placzek ratio. J Chem Phys **42**, 2785 (1965)
95. Bissonnette E, McCourt F. A study of the single-moment, two-moment and multilevel descriptions of depolarized Rayleigh light scattering in molecular gases. Mol Phys **70**, 65 (1990)
96. Cooper VG, May AD, Hara EH, Knapp HFP. Dicke narrowing and collisional broadening. of the $S_0(0)$ and $S_2(1)$ Raman lines of H_2. Can J Phys **46**, 2019 (1968)
97. Keijser RAJ, Van den Hout KD, Knaap HFP. Lineshape of the depolarized Rayleigh line in gases of linear molecules. Phys Lett **42**, 109 (1972)
98. Boyd RW. Nonlinear Optics. 2^{tnd}edition, Academic press, San Diego (2002)
99. Druet SAJ, Taran JPE. CARS spectroscopy. Prog Quantum Electron **7**, 1 (1981)
100. Eckbreth AC, Hall RJ. CARS concentration sensitivity with and without resonant background suppression. Comb Sci Technol **25**, 175 (1981)

101. Lang T, Motzkus M. Single shot femtosecond CARS thermometry. J Opt Soc Am B **19**, 340 (2002)
102. Morgen M, Price W, Hunziker L, Ludowise P, *et al*. Femtosecond Raman-induced polarization spectroscopy studies of rotational coherence in O_2, N_2 and CO_2. Chem Phys Lett **209**, 1 (1993)
103. Morgen M, Price W, Ludowise P, Chen Y. Tensor analysis of femtosecond Raman induced polarization spectroscopy: application to the study of rotational coherence. J Chem Phys **102**, 8780 (1995)
104. Lavorel B, Faucher O, Morgen M, Chaux R. Analysis of femtosecond Raman induced polarisation spectroscopy (RIPS) in N_2 and CO_2 by fitting and scaling laws. J Raman Spectrosc **31**, 77 (2000)
105. Tran H. Diagnostics par spectroscopie femtoseconde des caractéristiques physiques d'un milieu en combustion. Thesis, Université de Franche Comté, Besançon, France (2004)
106. Lang T, Kompa KL, Motzkus M. Femtosecond CARS on H_2. Chem Phys Lett **310**, 65 (1999)
107. Lang T, Motzkus M, Frey HM, Beaud P. High resolution femtosecond coherent anti-Stokes Raman scattering: determination of rotational constants, molecular anharmonicity, collisional line shifts, and temperature. J Chem Phys **115**, 5418 (2001)
108. Beaud P, Frey HM, Lang T, Motzkus M. Flame thermometry by femtosecond CARS. Chem Phys Lett **344**, 407 (2001)
109. Tran H, Joubert P, Bonamy L, Lavorel B, *et al*. Femtosecond time resolved CARS spectroscopy: experiment and modelization of speed memory effects on H_2–N_2 mixtures in the collision regime. J Chem Phys **122**, 194317 (2005)
110. Lavorel B, Tran H, Hertz E, Faucher O, *et al*. Femtosecond Raman time-resolved molecular spectroscopy. CR Phys Acad Sci Paris **5**, 215 (2004)
111. Shoemaker RL. Coherent transient infrared spectroscopy. In Laser and Coherence Spectroscopy, J.I. Steinfeld Ed., Plenum Press, NY (1978)
112. Dutier G, Yarovitski A, Saltiel S, Papoyan A, *et al*. Collapse and revival of a Dicke-type coherent narrowing in a sub-micron thick vapor cell transmission spectroscopy. Euro Phys Lett **63**, 35 (2003)
113. Sarkisyan D, Varzhapetyan T, Sarkisyan A, Malakyan Y, *et al*. Spectroscopy in an extremely thin vapor cell: Comparing the cell-length dependence in fluorescence and in absorption techniques. Phys Rev A **69**, 065802 (2004)
114. Brumer PW, Shapiro M. Principles of the Quantum Control of Molecular Processes. Wiley-VCH, Berlin (2003)
115. López-Puertas M, Taylor FW. Non-LTE Radiative Transfer in the Atmosphere. World Scientific Publishing Company, Series on atmospheric, oceanic and planetary physics, vol. **3**, (2001)
116. Funke B, López-Puertas M, Stiller G, von Clarmann T, *et al*. Non-LTE state distribution of nitric oxide and its impact on the retrieval of the stratospheric daytime no profile from MIPAS limb sounding instruments. Adv Space Res **26**, 947 (2000)
117. Kutepov AA, Oelhaf H, Fischer H. Non-LTE radiative transfer in the 4.7 and 2.3 µm bands of CO: vibration-rotational non-LTE and its effects on limb radiance. JQSRT **57**, 317 (1997)
118. Shved GM, Kutepov AA, Ogibalov VP. Non-local thermodynamic equilibrium in CO_2 in the middle atmosphere. I. Input data and populations of the v_3 mode manifold states. J Atm Solar-Terrestrial Phys **60**, 289 (1998)
119. López-Valverde MA, López-Puertas M, López-Moreno JJ, Formisano V, *et al*. Analysis of CO_2 non-LTE emissions at 4.3 µm in the Martian atmosphere as observed by PFS/Mars Express and SWS/ISO. Planet Space Sci **53**, 1079 (2005)
120. López-Valverde MA, Drossart P, Carlson R, Mehlman R, *et al*. Non-LTE infrared observations at Venus: From NIMS/Galileo to VIRTIS/Venus Express. Planet Space Sci **55**, 1757 (2007)

121. Marsault-Herail F, Echargui M, Levi G, Marsault JP, et al. Collisional effects on the rotational and rotation–vibration Raman spectra of HD compressed by argon. J Chem Phys **77**, 2715 (1982)
122. Ozanne L, Ma Q, Nguyen-Van-Thanh, Brodbeck C, et al. Line mixing, finite duration of collision, vibrational shift, and non-linear density effects in the v_3 and the $3v_3$ bands of CO_2 perturbed by Ar up to 1000 bar. JQSRT **58**, 261 (1997)
123. Bulgakov YI, Storozhev AV, Strekalov ML. Comparison and analysis of rotationally inelastic collision models describing the Q-branch collapse at high density. Chem Phys **177**, 145 (1993)
124. Delalande C, Gale GM. Vibrational energy relaxation in fluid mixtures: Hydrogen in argon. J Chem Phys **73**, 1918 (1980)
125. Anderson PW, Talman JD. Bell Telephone System Tech. Publication no. 3117, University of Pittsburgh (1955)
126. Szudy J, Baylis WE. Unified Franck-Condon treatment of pressure broadening of spectral lines. JQSRT **15**, 641 (1975)
127. Szudy J, Baylis WE. Asymmetry in pressure-broadened spectral lines. JQSRT **17**, 681 (1977)
128. Royer A. Shift, width, and asymmetry of pressure-broadened spectral lines at intermediate densities. Phys Rev **A 22**, 1625 (1980)
129. Ciurylo R, Szudy J, Trawinski RS. Non-adiabatic approach to the asymmetry of pressure broadened spectral lines. JQSRT **57**, 551 (1997)
130. Marteau P, Boulet C, Robert D. Finite duration of collisions and vibrational dephasing effects on the Ar broadened HF infrared line shapes: asymmetric profiles. J Chem Phys **80**, 3632 (1984)
131. Boulet C, Flaud PM, Hartmann JM. Infrared line collisional parameters of HCl in argon, beyond the impact approximation: measurements and classical path calculations. J Chem Phys **120**, 11053 (2004)
132. Ciurylo R, Szudy J. Line mixing and collision-time asymmetry of spectral line shapes. Phys Rev **A 63**, 042714 (2001)
133. Merzbacher E. Quantum Mechanics. (2nd Ed., Willey, NY (1970)
134. Hutson JM. Vibrational dependence of the anisotropic intermolecular potential of Ar–HCl. J Phys Chem **96**, 4237 (1992)
135. Boulet C, Flaud PM, Hartmann JM. Infrared spectral shapes of HCl in Argon, beyond the impact approximation: experimental and theoretical studies. In Proc of the 17th International Conference on Spectral Line Shapes, p. 343, E. Dalimier Ed., Frontier group, (2004)
136. Bonamy J, Bonamy L, Robert D. Overlapping effects and motional narrowing in molecular band shapes: application to the Q branch of HD. J Chem Phys **67**, 4441 (1977)
137. Ben-Reuven A. Spectral line shapes in gases in the binary-collision approximation. Adv Chem Phys **33**, 235 (1975)
138. Fiutak J, Horodecki R. On the pressure broadening of the molecular spectra. I. The Liouville space methods. Acta Phys Pol **A 51**, 301 (1977)
139. Roney PL. Theory of spectral line shape. I. Formulation and line coupling. J Chem Phys **101**, 1037 (1994)
140. Ma Q, Tipping RH, Boulet C. A far wing line shape theory which satifies the detailed balance principle. JQSRT **59**, 245 (1998)
141. Royer A. Cumulant expansions and pressure broadening as an example of relaxation. Phys Rev **A 6**, 1741 (1972)
142. Royer A. Density expansion of the memory operator in pressure broadening theory. Phys Rev **A 7**, 1078 (1973)
143. Roney PL. Theory of spectral line shape. III. The Fano operator from near to the far wing. J Chem Phys **102**, 4757 (1995)

144. Shafer R, Gordon RG. Quantum scattering theory of rotationnal relaxation and spectral line shapes in H_2–He gas mixtures. J Chem Phys **58**, 5422 (1973)
145. Dillon TA, Smith EW, Cooper J, Mizushima M. Semiclassical treatment of strong collisions in pressure broadening. Phys Rev **A 2**, 1839 (1970)
146. Monchick L. The high energy asymptotic behavior of line shape cross sections and detailed balance. J Chem Phys **95**, 5047 (1991)
147. Neilsen WB, Gordon RG. On a semiclassical study of molecular collisions. I. General method. J Chem Phys **58**, 41318 (1973)
148. Boissoles J, Boulet C, Robert D, Green S. State-to-state rotational phase coherence effect on the vibration-rotation band shape: an accurate quantum calculation for CO–He. J Chem Phys **90**, 5392 (1989)
149. Schmidt J, Berman PR, Brewer RG. Coherent transient study of velocity changing collisions. Phys Rev Lett **31**, 1103 (1973)
150. Berman PR, Levy JM, Brewer RG. Coherent optical transient of molecular collisions: theory and observations. Phys Rev **A 11**, 1668 (1975)
151. Grossman SB, Schenzle A, Brewer RG. Pulse Fourier-transform optical spectroscopy. Phys Rev Lett **38**, 275 (1977)
152. Dicke RH. The effect of collisions upon the Doppler width of spectral lines. Phys Rev **89**, 472 (1953)
153. Dolbeau S, Berman R, Drummond JR, May AD. Dicke narrowing as an example of line mixing. Phys Rev **A 59**, 3506 (1999)
154. Pine AS, Fried A. Self-broadening in the fundamental bands of HF and HCl. J Mol Spectrosc **114**, 148 (1985)
155. Lavorel B, Millot G, Saint-Loup R, Wenger C, *et al*. Rotational collisional line broadening at high temperatures in the N_2 fundamental Q-branch studied with stimulated Raman spectroscopy. J de Physique **47**, 417 (1986)
156. Lavorel B, Guillot L, Bonamy J, Robert D. Collisional Raman linewidths of nitrogen at high temperature (1700–2400 K). Opt Lett **20**, 1189 (1995)
157. Birnbaum G. Microwave pressure broadening and its application to intermolecular forces. Adv Chem Phys **12**, 487 (1967)
158. Hartmann JM, Taine J, Bonamy J, Labani B, *et al*. Collisional broadening of rotation-vibration lines for asymmetric-top molecules. II. H_2O diode laser measurements in the 400–900 K range; calculations in the 300–2000 K range. J Chem Phys **86**, 144 (1987)
159. Wagner G, Birk M, Gamache RR, Hartmann JM. Collisional parameters of H_2O lines: effect of temperature. JQSRT **92**, 211 (2005)
160. Bouanich JP, Boulet C, Predoi-Cross A, Sharpe SW, *et al*. A multispectrum analysis of the ν_2 band of $H^{12}C^{14}N$: part II. Theoretical calculations of self-broadening, self-induced shifts and their temperature dependences. J Mol Spectrosc **231**, 85 (2005)
161. Murray JR, Javan A. Effects of collisions on Raman line profiles of hydrogen and deuterium gas. J Mol Spectrosc **42**, 1 (1972)
162. Schmücker N, Trojan Ch, Giesen T, Schieder R, *et al*. Pressure broadening and shift of some H_2O lines in the ν_2 band: revisited. J Mol Spectrosc **184**, 250 (1997)
163. Forsman JW, Sinclair PM, May AD, Drummond JR. Departures from the soft collision model for Dicke narrowing: Raman measurements in the Q branch of D_2. J Chem Phys **97**, 5355 (1992)
164. Wehr R. Dicke-narrowed spectral lines in carbon monoxide buffered by argon. Phd Thesis, University of Toronto, Canada (2005)
165. Wehr R, Ciurylo R, Vitcu A, Thibault F, *et al*. Dicke-narrowed spectral line shapes of CO in Ar: Experimental results and a revised interpretation. J Mol Spectrosc **235**, 54 (2006)

166. Berger JP, Saint-Loup R, Berger H, Bonamy J, et al. Measurement of vibrational line-profiles in H_2-rare gas mixtures. Determination of the speed dependence of the line shift. Phys Rev A **49**, 3396 (1994)
167. Henry A, Hurtmans D, Margottin-Maclou M, Valentin A. Confinement narrowing and absorber speed dependent broadening effects on CO lines in the fundamental band perturbed by Xe, Ar, Ne, He and N_2. JQSRT **56**, 647 (1996)
168. Thibault F, Boissoles J, Le Doucen R, Bouanich JP, et al. Pressure induced shifts of CO_2 lines. Measurements in the 00^00–00^03 band and theoretical analysis. J Chem Phys **96**, 4945 (1992)
169. Voigt W. Über das gesetz intensitätsverteilung innerhalb der linien eines gasspektrams. Sitzber. Bayr Akad. München Ber, 603 (1912)
170. Boissoles J, Thibault F, Boulet C, Bouanich JP, et al. Spectral lineshape parameters revisited for HF in a bath of Argon. II Asymmetries in the 0–0, 0–1, and 0–2 bands. J Mol Spectrosc **198**, 257 (1999)
171. Pine AS. Line shape asymmetries in Ar-broadened HF($v = 1$–0) in the Dicke-narrowing regime. J Chem Phys **101**, 3444 (1994)
172. Pine AS, Ciurylo R. Multispectrum fits of Ar-broadened HF with a generalized asymmetric line shape: Effects of correlation, hardness, speed dependence and collision duration. J Mol Spectrosc **208**, 180 (2001)
173. Daussy C, Guinet M, Amy-Klein A, Djerroud K, et al. Direct determination of the Boltzmann constant by an optical method. Phys Rev Lett **98**, 250801 (2007)
174. Wittke JP, Dicke RH. Redetermination of the hyperfine splitting in the ground state of atomic hydrogen. Phys Rev **103**, 620 (1956)
175. Looney JP. Comprehensive theory for the broadening, shifting and narrowing of HF and HCl fundamental band absorption profiles. Phd Thesis, Pennsylvania State Univ. USA (1987) and references herein.
176. Chandrasehkar S. Stochastic problems in physics and astronomy. Rev Mod Phys **15**, 1 (1943)
177. Lorentz HA. Absorption and Emission Lines of Gases. Proc Roy Acad Sci Amsterdam. **8**, 591 (1906)
178. Weisskopf V. Die Streuung des Lichts an angeregten Atomen. Z Physik **85**, 451 (1933)
179. Van Vleck JH, Weisskopf VF. On the shape of collision-broadened lines. Rev Mod Phys **17**, 227 (1945)
180. Herbert F. Spectrum line profiles: a generalized Voigt function including collisional narrowing. JQSRT **14**, 943 (1974)
181. Varghese PL, Hanson RK. Collisional narrowing effects on spectral line shapes measured at high resolution. Appl Opt **23**, 2376 (1984)
182. Abramowitz M, Stegun IA. Handbook of Mathematical Functions. M. Abramowitz and I.A. Stegun Eds., Dover Publications Inc., NY (1972)
183. Galatry L. Simultaneous effect of Doppler and foreign gas broadening on spectral lines. Phys Rev **122**, 1218 (1961)
184. Corey GC, McCourt FR. Dicke narrowing and collisional broadening of spectral lines in dilute molecular gases. J Chem Phys **81**, 2318 (1984)
185. Duggan P, Sinclair PM, Le Flohic MP, Forsman JW, et al. Testing the validity of the optical diffusion coefficient: line-shape measurements of CO perturbed by N_2. Phys Rev A **48**, 2077 (1993)
186. Chapman S, Cowling TG. Mathematical Theory of Non Uniform Gases. 2nd ed. Cambridge Univiversity Press, NY (1958)
187. Keilson J, Storer JE. On Brownian motion, Boltzmann's equation, and the Fokker-Planck equation. Quart Appl Math **10**, 243 (1952)

188. Bohm D, Gross EP. Theory of plasma oscillations. A. Origin of medium-like behavior. Phys Rev **75**, 1851 (1949)
189. Nelkin M, Ghatak A. Simple binary collision model for Van Hove's $G_s(r, t)$. Phys Rev **135**, 4 (1964)
190. Joubert P, Bonamy J, Robert D, Domenech JL, et al. A partially correlated strong collision model for velocity and state changing collisions. Application to Ar-broadened HF rovibrational line shape. JQSRT **61**, 519 (1999)
191. Bird GR. Interdependence of spectral line breadth, sound velocity dispersion, and viscosity of small molecules in the gas phase. J Chem Phys **38**, 2678 (1963)
192. Rank DH, Wiggins TA. Collision narrowing of spectral lines. H_2 quadrupole spectrum. J Chem Phys **39**, 1348 (1963)
193. Fink U, Wiggins TA, Rank DH. Frequency and intensity measurements on the quadrupole spectrum of molecular hydrogen. J Mol Spectrosc **18**, 384 (1965)
194. Rank DH, Fink U, Wiggins TA. Measurements on spectra of gases of planetary interest. II. H_2, CO_2, NH_3, and CH_4. Astrophys J **143**, 980 (1966)
195. Lallemand P, Simova P, Bret G. Pressure-induced line shift and collisional narrowing in hydrogen gas determined by stimulated Raman emission. Phys Rev Lett **17**, 1239 (1966)
196. Buijs HK, Gush HP. Static field induced spectrum of hydrogen. Can J Phys **49**, 2366 (1971)
197. Pine AS. Doppler-limited molecular spectroscopy by difference-frequency mixing. J Opt Soc Am **64**, 1683 (1974)
198. Pine AS. High-resolution methane ν_3-band spectra using stabalized tunable difference-frequency laser system. J Opt Soc Am **66**, 97 (1976)
199. Valentin A, Nicoles C, Henry L, Mantz AW. Tunable diode laser control by a stepping Michelson interferometer. Appl Opt **26**, 41 (1987)
200. Henry A, Valentin A, Margottin-Maclou M, Rachet F. Tunable diode laser spectrometer with controlled phase frequency emission. J Mol Spectrosc **166**, 41 (1994)
201. Henesian MA, Kulevskii L, Byer RL, Herbst RL. CW high-resolution CARS spectroscopy of H_2, D_2 and CH_4. Opt Comm **18**, 225 (1976)
202. Krynetsky BB, Kulevsky LA, Mishin VA, Prokhorov AM, et al. High resolution cw CARS spectroscopy in D_2 gas. Opt Comm **21**, 225 (1977)
203. Colmont JM, Nguyen L, Rohart F, Wlodarczak G. Lineshape analysis of the $J=3\leftarrow 2$ and $J=5\leftarrow 4$ rotational transitions of room temperature CO broadened by N_2, O_2, CO_2 and noble gases. J Mol Spectrosc **246**, 86 (2007)
204. Rodgers CD. Collisional narrowing: its effect on the equivalent widths of spectral lines. Appl Opt **15**, 714 (1976)
205. Gupta BK, May AD. Dicke narrowing and collision broadening of the depolarized Rayleigh and $S_0(1)$ Raman line in the hydrogen isotopes and H_2–He, H_2–Ne mixtures. Can J Phys **50**, 1747 (1972)
206. Humlicek J. An efficient method for evaluation of the complex probability function: the Voigt function and its derivatives. JQSRT **21**, 309 (1979)
207. Humlicek J. Optimized computation of the Voigt and complex probability functions. JQSRT **27**, 437 (1982)
208. Ouyang X, Varghese PL. Reliable and efficient program for fitting Galatry and Voigt profiles to spectral data on multiple lines. Appl Opt **28**, 1538 (1989)
209. Rosasco GJ, Hurst WS, Smyth KC, Petway LB, et al. A new feature in the Raman Q-branch of gaseous D_2. In Spectral Line Shapes vol. **4**, RJ Exton Ed., Deepak, Hampton, p. 575 (1987)
210. Smyth KC, Rosasco GJ, Hurst WS. Measurement and rate law analysis of D_2 Q-branch line broadening coefficients for collisions with D_2, He, Ar, H_2, and CH_4. J Chem Phys **87**, 1001 (1987)

211. Rosasco GJ, Bowers WJ, Hurst WS, Looney JP, et al. Simultaneous forward–backward Raman scattering studies of D_2 broadened by D_2, He, and Ar. J Chem Phys **94**, 7625 (1991)
212. Pine AS. Collisional narrowing of HF fundamental band spectral lines by neon and argon. J Mol Spectrosc **82**, 435 (1980)
213. Pine AS, Looney JP. N_2 and air broadening in the fundamental bands of HF and HCl. J Mol Spectrosc **122**, 41 (1987)
214. Pine AS. Asymmetries and correlations in speed-dependent Dicke-narrowed line shapes of argon-broadened HF. JQSRT **62**, 397 (1999)
215. Ciurylo R, Pine AS, Szudy J. A generalized speed-dependent line profile combining soft and hard partially correlated Dicke-narrowing collisions. JQSRT **68**, 257 (2001)
216. Mattick AT, Sanchez A, Kurnit NA, Javan A. Velocity dependence of collision-broadening cross section observed in an infrared transition of NH_3 gas at room temperature. Appl Phys Lett **23**, 675 (1973)
217. Mattick AT, Kurnit NA, Javan A. Velocity dependence of collision broadening cross section in NH_3. Chem Phys Lett **38**, 176 (1976)
218. Coy SL. Speed dependence of microwave rotational relaxation rates. J Chem Phys **73**, 5531 (1980)
219. Ritter KJ, Wilkerson TD. High-resolution spectroscopy of the oxygen A band. J Mol Spectrosc **121**, 1 (1987)
220. Luijendijk SCM. On the shape of pressure-broadened absorption lines in the microwave region. I. Deviations from the Lorentzian line shape. J Phys **B 10**, 1735 (1977)
221. Farrow RL, Rahn LA, Sitz GO, Rosasco RJ. Observation of a speed-dependent collisional inhomogeneity in H_2 vibrational line profiles. Phys Rev Lett **63**, 746 (1989)
222. May AD. Molecular dynamics and a simplified master equation for spectral line shapes. Phys Rev **A 59**, 3495 (1999)
223. Berger H. Etude des mécanisme de relaxation rovibrationnelle pour les mélanges H_2-gas rares, H_2–N_2 et H_2–H_2O par spectroscopie « Raman stimulé » en vue de la mesure de la température dans un moteur fusée. Thesis, Université de Bourgogne, Dijon, France, (1994)
224. Sinclair PM, Berger JP, Michaut X, Saint-Loup R, et al. Collisional broadening and shifting parameters of the Raman Q branch of H_2 perturbed by N_2 determined from speed-dependent line profiles at high temperatures. Phys Rev **A 54**, 402 (1996)
225. Robert D, Thuet JM, Bonamy J, Temkin S. Effect of speed-changing collisions on spectral line shape. Phys Rev **A 47**, R771 (1993)
226. Joubert P. Inhomogénéités dues à la dépendance en vitesse de la largeur et du déplacement collisionnels de H_2 et HF. Calculs *ab initio* et prédictions des profils spectraux à haute température. Thesis, Université de Besançon, Besançon, France, (1997)
227. Joubert P, Bruet X, Bonamy J, Robert D, et al. H_2 vibrational spectral signatures in binary and ternary mixtures: theoretical model, simulation and application to CARS thermometry in high pressure flames. CR Phys Acad Sci Paris **2**, 989 (2001)
228. Hoang PNM, Joubert P, Robert D. Speed-dependent line-shape models analysis from molecular dynamics simulations: the collision regime. Phys Rev **A 65**, 012507 (2002)
229. Duggan P, Sinclair PM, May AD, Drummond JR. Line-shape analysis of speed-dependent collisional width inhomogeneities in CO broadened by Xe, N_2, and He. Phys Rev **A 51**, 218 (1995)
230. Duggan P, Sinclair PM, Berman R, May AD, et al. Testing line-shape models: measurements for $v = 1$–0 CO broadened by He and Ar. J Mol Spectrosc **186**, 90 (1997)
231. Ciurylo R, Szudy J. Speed-dependent pressure broadening and shift in the soft collision approximation. JQSRT **57**, 411 (1997)

232. Burns MJ, Coy SL. Rotational relaxation rates for the OCS $J = 0–1$ pure rotational transition broadened by argon and helium. J Chem Phys **80**, 3544 (1984)
233. Haekel J, Mäder H. Speed-dependent T_2-relaxation rates of microwave emission signals. JQSRT **46**, 21 (1991)
234. Rohart F, Mäder H, Nicolaisen HW. Speed dependence of rotational relaxation induced by foreign gas collisions: studies on CH_3F by millimeter coherent transients. J Chem Phys **101**, 6475 (1994)
235. Kaghat F. Profils des transitions millimétriques: analyse par spectroscopie résolue en temps du rétrécissement et de l'assymétrie liés à la distribution des vitesses moléculaires. Thesis, Université de Lille, Lille, France, (1995)
236. Priem D, Rohart F, Colmont JM, Wlodarczak G, *et al.* Line-shape study of the $J = 3 \leftarrow 2$ rotational transition of CO perturbed by N_2 and O_2. J Mol Struct **517**, 435 (2000)
237. Shapiro DA, Ciurylo R, Jaworski R, May AD. Modeling the spectral line shapes with speed-dependent broadening and Dicke narrowing. Can J Phys **79**, 1209 (2001)
238. Rohart F, Ellendt A, Kaghat F, Mäder H. Self and polar foreign gas line broadening and frequency shifting of CH_3F: effect of the speed dependence observed by millimeter-wave coherent transients. J Mol Spectrosc **185**, 222 (1997)
239. Mizushima M. Velocity distribution in spectral line shape. JQSRT **7**, 505 (1967)
240. Mizushima M. Velocity distribution effect in pressure broadened spectral lines. JQSRT **11**, 471 (1971)
241. Edmonds FN. Line absorption coefficient profiles for velocity-dependent broadening. JQSRT **8**, 1447 (1968)
242. Berman PR. Speed-dependent collisional width and shift parameters in spectral profiles. JQSRT **12**, 1331 (1972)
243. Ward J, Cooper J, Smith EW. Correlation effects in the theory of combined Doppler and pressure broadening-I. Classical theory. JQSRT **14**, 555 (1974)
244. Lance B, Blanquet G, Walrand J, Bouanich JP. On the speed-dependent hard collision lineshape models: application to C_2H_2 perturbed by Xe. J Mol Spectrosc **185**, 262 (1997)
245. Lance B. Profils de raies d'absorption infrarouge, du régime Doppler au régime collisionnel. Expériences et modélisations. Thesis, Namur, Belgium (1998)
246. Ciurylo R. Shapes of pressure- and Doppler-broadened spectral lines in the core and near wings. Phys Rev **A 58**, 1029 (1998)
247. Robert D, Joubert P, Lance B. A velocity-memory model for the spectral line shape from the Doppler to the collision regime. J Mol Struct **517–518**, 393 (2000)
248. Ciurylo R, Jaworski R, Jurkowski J, Pine AS, *et al.* Spectral line shapes modeled by a quadratic speed-dependent Galatry profile. Phys Rev **A 63**, 032507 (2001)
249. Ciurylo R, Lisak D, Szudy J. Role of velocity- and speed-changing collisions on speed-dependent line shapes of H_2. Phys Rev **A 66**, 032701 (2002)
250. Pickett HM. Effects of velocity averaging on the shapes of absorption lines. J Chem Phys **73**, 6090 (1980)
251. Buffa G, Carocci S, Di Lieto A, Minguzzi P, *et al.* Speed dependence of pressure broadening in molecular rotational spectra using a novel technique. Phys Rev Lett **74**, 3356 (1995)
252. Lance B, Robert D. An analytical model for collisional effects on spectral line shape from the Doppler to the collision regime. J Chem Phys **109**, 8283 (1998)
253. Lance B, Robert D. Correlation effects on spectral line shape from the Doppler to the collision regime. J Chem Phys **111**, 789 (1999)
254. Chaussard F, Michaut X, Saint-Loup R, Berger H, *et al.* Collisional effects on spectral line shape from the Doppler to the collision regime: a rigorous test of a unified model. J Chem Phys **112**, 158 (2000)

255. Chaussard F. Effets collisionnels homogènes et inhomogènes dans les spectres Raman rovibrationnels de H_2 du régime Doppler au régime collisionnel. Application au diagnostic optique de la température dans des milieux en combustion. Thesis, Université de Bourgogne, Dijon, France, (2001)
256. Joubert P, Bruet X, Bonamy J, Robert D, et al. Inhomogeneous speed effects on H_2 vibrational line profiles in ternary mixtures. J Chem Phys **113**, 10056 (2000)
257. Joubert P, Hoang PNM, Bonamy L, Robert D. Speed-dependent line-shape model analysis from molecular-dynamics simulations: the collisional confinement narrowing regime. Phys Rev **A 66**, 042508 (2002)
258. Fraser GT, Pine AS. Van der Waals potentials from the infrared spectra of rare gas–HF complexes. J Chem Phys **85**, 2502 (1986)
259. Shapiro DA, May AD. Dicke narrowing for rigid spheres of arbitrary mass ratio. Phys Rev **A 63**, 012701 (2000)
260. Liao PF, Bjorkholm JE, Berman PR. Effects of velocity-changing collisions on two-photon and stepwise-absorption spectroscopic line shapes. Phys Rev **A 31**, 1927 (1980)
261. Parkhomenko AI, Shalagin AM. Effects of velocity dependence of the collision frequency on Dicke line narrowing. J Exp and Theo Phys **93**, 487500 (2001)
262. Brissaud A, Frisch U. Solving linear stochastic differential equations. J Math Phys **15**, 524 (1974)
263. Privalov T, Shalagin A. Exact solution of the one- and three-dimensional quantum kinetic equations with velocity-dependent collision rates: comparative analysis. Phys Rev **A 59**, 4331 (1999)
264. Berman PR. Quantum-mechanical transport equation for atomic systems. II. Inelastic collisions and general line-shape considerations. Phys Rev **A 6**, 2157 (1972)
265. Berman PR. Theory of collision effects on atomic and molecular line shapes. Appl Phys **A 6**, 283 (1975)
266. Berman PR. Collisions in atomic vapors. In New Trends in Atomic Physics, edited by Grynberg and Stora, Les Houches, North Holland, Amsterdam, vol. 38, p. 451 (1984)
267. Nienhuis G. Effects of the radiator motion in the classical and quantum-mechanical theories of collisional spectral-line broadening. JQSRT **20**, 275 (1978)
268. Lisak D, Rusciano G, Sasso A. An accurate comparison of lineshape models on H_2O lines in the spectral region around 3 µm. J Mol Spectrosc **227**, 162 (2004)
269. Lisak D, Rusciano G, Sasso A. Speed-dependent and correlation effects on the line shape of acetylene. Phys Rev **A 72**, 012503 (2005)
270. Robert D, Bonamy L. Memory effects in speed-changing collisions and their consequences for spectral lineshape: I. Collision regime. Eur Phys J **D 2**, 245 (1998)
271. Burshtein AI, Temkin SI. Spectroscopy of Molecular Rotation in Gases and Liquids. Cambridge University Press, (1994)
272. Shapiro DA. Spectral line narrowing in the Keilson-Storer model. J Phys **A 33**, L43 (2000)
273. Bonamy L, Tran Thi Ngoc H, Joubert P, Robert D. Memory effects in speed-changing collisions and their consequences for spectral line shape – II. From the collision to the Doppler regime. Eur Phys J **D 31**, 459 (2004)
274. McLennan JA. Introduction to Non-Equilibrium Statistical Mechanics. Prentice Hall, Englewood Cliffs (1989)
275. Tran H, Bermejo D, Domenech JL, Joubert P, et al. Collisional parameters of H_2O lines: velocity effects on the line-shape. JQSRT **108**, 126 (2007)
276. Skenderović H, Buckup T, Wohlleben W, Motzkus M. Determination of collisional line broadening coefficients with femtosecond time-resolved CARS. J Raman Spectrosc **33**, 866 (2002)

277. Monchick L, Hunter LW. Diatomic–diatomic molecular collision integrals for pressure broadening and Dicke narrowing: a generalization of Hess's theory. J Chem Phys **85**, 713 (1986), and *erratum* **86**, 7251 (1987)
278. Blackmore R. A modified Boltzmann kinetic equation for line shape functions. J Chem Phys **87**, 791 (1987)
279. Waldmann L. Die Boltzmann-Gleichung fur Gase mit rotierenden. Z. Naturforsch **A 12**, 660 (1957); Die Boltzmann-Gleichung fur Gase aus Spin-Teilchen. *ibid.* **A 13**, 609 (1958)
280. Snider RF. Quantum-mechanical modified Boltzmann equation for degenerate internal states. J Chem Phys **32**, 1051 (1960)
281. Tip A. Transport equations for dilute gases with internal degrees of freedom. Physica **52**, 493 (1971)
282. Snider RF, Sanctuary BC. Generalized Boltzmann equation for molecules with internal states. J Chem Phys **55**, 1555 (1971)
283. Hess S. Kinetic theory of spectral line shapes. The transition between Doppler broadening and collisional broadening. Physica **61**, 80 (1972)
284. Cercignani C. The Boltzmann Equation and its Applications. Springer-Verlag, NY, (1988)
285. Blackmore R, Green S, Monchick L. Dicke narrowing of the polarized Stokes–Raman Q branch of the $v = 0 \to 1$ transition of D_2 in He. J Chem Phys **91**, 3846 (1989)
286. Green S, Blackmore R, Monchick L. Comment on linewidths and shifts in the Stokes–Raman Q branch of D_2 in He. J Chem Phys **91**, 52 (1989)
287. Demeio L, Green S, Monchick L. Effects of velocity changing collisions on line shapes of HF in Ar. J Chem Phys **102**, 9160 (1995)
288. Schaefer J, Monchick L. Line broadening of HD immersed in He and H_2 gas. Astron Astrophys **265**, 859 (1992)
289. Meyer W, Hariharan PC, Kutzelnigg W. Refined *ab initio* calculation of the potential energy surface of the He–H_2 interaction with special emphasis to the region of the van der Waals minimum. J Chem Phys **73**, 1880 (1980)
290. Monchick L. Quantum kinetic equations incorporating the Fano collision operator: The generalized Hess method of describing line shapes. J Chem Phys **101**, 5566 (1994)
291. Blackmore R. Collision kernels for the Waldmann–Snider equation. J Chem Phys **86**, 4188 (1987)
292. Chen FM, Moraal H, Snider RF. On the evaluation of kinetic theory collision integrals: diamagnetic diatomic molecules. J Chem Phys **57**, 542 (1972)
293. Lindenfeld MJ, Shizgal B. Matrix elements of the Boltzmann collision operator for gas mixtures. Chem Phys **41**, 81 (1979)
294. Demeio L, Monchick L. Collision kernels for the Waldmann-Snider equation: generalization to gas mixture. Physica **A 214**, 95 (1995)
295. Canuto C, Hussaini MY, Quarteroni A, Zang TA. Spectral Methods in Fluid Dynamics. Springer-Verlag, NY, (1988)
296. Duderstadt JJ, Martin WR. Transport Theory. Wiley-Interscience, NY (1979)
297. Shizgal B, Blackmore R. Eigenvalues of the Boltzmann collision operator for binary gases: relaxation of anisotropic distributions. Chem Phys **77**, 417 (1983)
298. Shapiro DA, Ciurylo R, Drummond JR, May AD. Solving the line shape problem with speed-dependent broadening and shifting, and Dicke narrowing – Part I: formalism. Phys Rev **A 65**, 12501 (2002)
299. Ciurylo R, Shapiro DA, Drummond JR, May AD. Solving the line shape problem with speed-dependent broadening and shifting, and Dicke narrowing – Part II: application. Phys Rev **A 65**, 012502 (2002)

300. Wang-Chang CS, Uhlenbeck GE, de Boer J. The heat conductivity and viscosity of polyatomic gases. In Studies in Statistical Mechanics. North Holland, Amsterdam, vol. II, (1964)
301. Lindenfeld MJ. Self-structure factor of hard-sphere gases for arbitrary ratio of bath to test particle masses. J Chem Phys **73**, 5817 (1980)
302. Rautian SG. On the absorption line profile of a molecular gas. Opt & Spectrosc (English translation) **86**, 334 (1999)
303. Wehr R, Vitcu A, Ciurylo R, Thibault F, et al. Spectral line shape of the P(2) transition in CO–Ar: uncorrelated *ab initio* calculation. Phys Rev **A 66**, 62502 (2002)
304. Wehr R, Vitcu A, Thibault F, Drummond JR, et al. Collisional line shifting and broadening in the fundamental P-branch of CO in Ar between 214 and 324 K. J Mol Spectrosc **235**, 69 (2006)
305. Berman R, Sinclair PM, May AD, Drummond JR. Spectral profiles for atmospheric absorption by isolated lines: a comparison of model spectra with P- and R-branch lines of CO in N_2 and Ar. J Mol Spectrosc **198**, 283 (1999)
306. Toczylowski RR, Cybulski SM. An *ab initio* study of the potential energy surface and spectrum of Ar–CO. J Chem Phys **112**, 4604 (2000)
307. Rautian SG, Shalagin AM. Kinetic Problems of Non-Linear Spectroscopy. Elsevier Science, NY, (1991)
308. Hutson JM, Green S. Molscat version 14. Collaborative Computational Project no. 6 (CCP6), UK Science and Engineering Research Council, 1994
309. Flower DR, Bourhis G, Launay JM. MOLCOL: A program for solving atomic and molecular collision problems. Comp Phys Comm **131**, 187 (2000)
310. D'Eu JF, Lemoine B, Rohart F. Infrared HCN lineshapes as a test of Galatry and speed dependent Voigt profiles. J Mol Spectrosc **212**, 96 (2002)
311. Rohart F, Nguyen L, Buldyreva J, Colmont JM, et al. Lineshapes of the 172 and 602 GHz rotational transitions of $HC^{15}N$. J Mol Spectrosc **246**, 213 (2007)
312. Armstrong BH. Spectrum line profiles: the Voigt function. JQSRT **7**, 61 (1967)
313. Schreier F. The Voigt and complex error function. A comparison of computational methods. JQSRT **48**, 743 (1992)
314. Wells RJ. Rapid approximation to the Voigt/Faddeeva function and its derivative. JQSRT **62**, 29 (1999)
315. Hui AK, Armstrong BH, Wray AA. Rapid computation of the Voigt and complex error functions. JQSRT **19**, 509 (1978)
316. Karp AH. Efficient computation of spectral line shapes. JQSRT **20**, 379 (1978)
317. Cope D, Lovett RJ. A general expression for the Voigt profile. JQSRT **37**, 377 (1987)
318. Cope D, Khoury R, Lovett RJ. Efficient calculation of general Voigt profiles. JQSRT **39**, 163 (1988)
319. Cope D, Lovett RJ. Asymptotic coefficients for one-interacting-level Voigt profiles. JQSRT **39**, 173 (1988)
320. Luijendijk SCM. On the shape of pressure-broadened absorption lines in the microwave region. II. Collision-induced width and shift of some rotational absorption lines as a function of temperature. J Phys **B 10**, 1741 (1977)
321. Bouanich JP. Site-site lennard-Jones potential parameters for N_2, O_2, H_2, CO and CO_2. JQSRT **47**, 243 (1992)
322. Robert D, Bonamy J. Short range force effects in semiclassical molecular line broadening calculations. J de Physique **40**, 923 (1979)
323. Lance B, Blanquet G, Walrand J, Populaire JC, et al. Inhomogeneous lineshape profiles of C_2H_2 perturbed by Xe. J Mol Spectrosc **197**, 32 (1999)
324. Dore L. Using Fast Fourier Transform to compute the line shape of frequency-modulated spectral profiles. J Mol Spectrosc **221**, 93 (2003)

325. Berman R, Sinclair PM, May AD, Drummond JR. An algorithm for calculating a speed-dependent Lorentzian profile. J Mol Spectrosc **198**, 278 (1999)
326. Rothman LS, Jacquinet-Husson N, Boulet C, Perrin A. History and future of the molecular spectroscopic databases. CR Phys Acad Sci Paris **6**, 897 (2005)
327. Lévy A, Lacome N, Chackerian Jr C. Collisional line mixing. In Spectroscopy of the Earth Atmosphere and Interstellar Medium, Edited by K. N Rao and A Weber pp. 261–337, Academic Press, NY (1992)
328. Hadded S, Aroui H, Orphal J, Bouanich JP, et al. Line broadening and mixing in NH_3 inversion doublets perturbed by NH_3, He, Ar, and H_2. J Mol Spectrosc **210**, 275 (2001)
329. Brown LR, Benner DC, Malathy Devi V, Smith MAH, et al. Line mixing in self- and foreign-broadening water vapor at 6 µm. J Mol Struct **742**, 111 (2005)
330. Pine AS. N_2 and Ar broadening and line mixing in the P and R branches of the v_3 band of methane. JQSRT **57**, 157 (1997)
331. Cacciani P, Cosléou J, Herlemont F, Khelkahal M, et al. The role of relaxation in the nuclear spin conversion process. J Mol Struct **780–781**, 277 (2006)
332. Lavorel B, Oksengorn B, Fabre D, Saint-Loup R, et al. Stimulated Raman spectroscopy of the Q-branch of nitrogen at high pressure: collisional narrowing and shifting in the 150-680 bar range at room temperature. Mol Phys **75**, 397 (1992)
333. Tonkov MV, Khalil B, Boissoles J, Le Doucen R, et al. Measurements of Q-branch shapes of CO_2 near 15 µm. JQSRT **55**, 321 (1996)
334. Hartmann JM, Bouanich JP, Jucks KW, Blanquet G, et al. Line-mixing in N_2O Q branches. Model, laboratory, and atmospheric spectra. J Chem Phys **110**, 1959 (1999)
335. Bouanich JP, Rodrigues R, Hartmann JM, Domenech JL, et al. Line-mixing effects in Q branches of CO_2. II. The particular case of parallel bands. J Mol Spectrosc **186**, 269 (1997)
336. Pieroni D, Nguyen-Van-Thanh, Brodbeck C, Hartmann JM, et al. Experimental and theoretical study of line mixing in methane spectra. II. Influence of the collision partner (He and Ar) in the v_3 IR band. J Chem Phys **111**, 6850 (1999)
337. Ozanne L, Hartmann JM, Bouanich JP, Boulet C. Collisional narrowing of infrared bands of CO_2 at high density. Chem Phys Lett **261**, 353 (1996)
338. Bulanin MO, Dokuchaev AB, Tonkov MV, Filippov NN. Influence of line interference on the vibration-rotation band shapes. JQSRT **31**, 521 (1984)
339. Boissoles J, Menoux V, Le Doucen R, Boulet C, et al. Collisionally induced population transfer in IR absorption spectra II: the wing of the Ar-broadened v_3 band of CO_2. J Chem Phys **91**, 2163 (1989)
340. Knopp G, Radi P, Tulej M, Gerber T, et al. Collision induced rotational energy transfer probed by time-resolved coherent anti-Stokes Raman scattering. J Chem Phys **118**, 8223 (2003)
341. Lavorel B, Millot G, Bonamy J, Robert D. Study of rotational relaxation fitting laws from calculation of SRS N_2 Q-branch. Chem Phys **115**, 69 (1987)
342. Rosenkranz PW. Shape of the 5 mm oxygen band in the atmosphere. IEEE Trans Ant and Prop **AP-23**, 498 (1975)
343. Rodrigues R, Jucks KW, Lacome N, Blanquet G, et al. Model, software, and database for computation of line-mixing effects in infrared Q-branches of atmospheric CO_2. I. Symmetric isotopomers. JQSRT **61**, 153 (1999)
344. Strow LL, Reuter D. Effect of line mixing on atmospheric brightness temperatures near 15 µm. Appl Opt **27**, 872 (1988)
345. Rodrigues R, Hartmann JM. Simple modeling of the temperature dependence of first-order line-mixing coefficients Y_1. JQSRT **57**, 63 (1997)
346. Strow LL, Tobin DC, Hannon SE. A compilation of first-order line-mixing coefficients for CO_2 Q-branches. JQSRT **52**, 281 (1994)

347. Smith EW. Absorption and dispersion in the O_2 microwave spectrum at atmospheric pressures. J Chem Phys **74**, 6658 (1981)
348. Strekalov ML, Burshtein AI. Quantum theory of isotropic Raman spectra-changes with gas density. Chem Phys **60**, 133 (1981)
349. Hartmann JM, L'Haridon F. Simple modeling of line-mixing effects in IR bands. I. Linear molecules – Application to CO_2. J Chem Phys **103**, 6467 (1995)
350. Boulet C, Bouanich JP, Hartmann JM, Lavorel B, et al. Line mixing in the v_1 and $2v_2$ isotropic Raman Q-branch of CO_2 perturbed by argon and helium. J Chem Phys **111**, 9315 (1999)
351. Filippov NN, Tonkov MV. Semiclassical analysis of line mixing in the infrared bands of CO and CO_2. JQSRT **50**, 111 (1993)
352. Ozanne L, Nguyen-Van-Thanh, Brodbeck C, Bouanich JP, et al. Line-mixing and non linear density effects in the v_3 and $3v_3$ bands of CO_2 infrared bands perturbed by He up to 1000 bars. J Chem Phys **102**, 7306 (1995)
353. Niro F, Boulet C, Hartmann JM. Spectra calculations in central and wing regions of CO_2 IR bands between 10 and 20 µm. I: model and laboratory measurements. JQSRT **88**, 483 (2004)
354. Rothman LS, Hawkins RL, Wattson RB, Gamache RR. Energy levels, intensities, and linewidths of atmospheric carbon dioxide bands. JQSRT **48**, 537 (1992)
355. Malathy Devi V, Benner DC, Smith MAH, Rinsland CP. Nitrogen broadening and shift coefficients in the 4.2–4.5-µm bands of CO_2. JQSRT **76**, 289 (2003)
356. Hartmann JM, Boulet C. Line mixing and finite duration of collision effects in CO_2 infrared spectra. Fitting and scaling analysis. J Chem Phys **94**, 6406 (1991)
357. Le Doucen R, Cousin C, Boulet C, Henry A. Temperature dependence of the absorption in the region beyond the 4.3 µm band head of CO_2. I-Pure CO_2 case. Appl Opt **24**, 897 (1985)
358. Hartmann JM, Perrin MY. Measurements of pure CO_2 absorption beyond the v_3 bandhead at high temperature. Appl Opt **28**, 2550 (1989)
359. Pieroni D, Nguyen-Van-Thanh, Brodbeck C, Claveau C, et al. Experimental and theoretical study of line mixing in methane spectra. I. The N_2-broadened v_3 band at room temperature. J Chem Phys **110**, 7717 (1999)
360. Cousin C, Le Doucen R, Boulet C, Henry A, et al. Line coupling in the temperature and frequency dependences of absorption in the microwindows of the 4.3 µm CO_2 band. JQSRT **36**, 521 (1986)
361. Predoi-Cross AD, May AD, Vitcu A, Drummond JR, et al. Broadening and line mixing in the $20^00 \leftarrow 01^10$, $11^10 \leftarrow 00^00$ and $12^20 \leftarrow 01^10$ Q branches of carbon dioxide: experimental results and energy-corrected sudden modeling. J Chem Phys **120**, 10520 (2004)
362. Gordon RG, McGinnis RP. Lineshapes in molecular spectra. J Chem Phys **49**, 2455 (1968)
363. Green S. Rotational excitation of symmetric top molecules by collisions with atoms: close coupling, coupled states, and effective potential calculations for NH_3–He. J Chem Phys **64**, 3463 (1976)
364. Green S, Boissoles J, Boulet C. Accurate collision induced line coupling parameters for the fundamental band of CO in He: close coupling and coupled state scattering calculations. JQSRT **39**, 33 (1988)
365. Green S. Pressure broadening and line coupling in bending bands of CO_2. J Chem Phys **90**, 3603 (1989)
366. Temkin SI, Burshtein AI. On the shape of the Q-branch of Raman scattering spectra in dense media. Theory. Chem Phys Lett **66**, 52 (1979)
367. Hall RJ, Greenhalgh DA. Application of the rotational diffusion model to gaseous N_2 CARS spectra. Opt Comm **40**, 417 (1982)

368. Sala JP, Bonamy J, Robert D, Lavorel B, et al. A rotational thermalisation model for the calculation of collisionally narrowed isotropic Raman scattering spectra. Application to the SRS N_2 Q-Branch. Chem Phys **106**, 427 (1986)
369. Tonlov MV, Filippov NN, Timofeyev Yu M, Polyakov AV. A simple model of the line mixing effect for atmospheric applications: theoretical background and comparison with experimental profiles. JQSRT **56**, 783 (1996)
370. Domanskaya AV, Filippov NN, Grigorovich NM, Tonkov MV. Modeling of the rotational relaxation matrix in line-mixing effect calculations. Mol Phys **102**, 1843 (2004)
371. Hartmann JM, Boulet C, Margottin-Maclou M, Rachet F, et al. Simple modeling of Q-branch absorption. I. Theoretical model and application to CO_2 and N_2O. JQSRT **54**, 705 (1995)
372. Blanquet G, Walrand J, Hartmann JM, Bouanich JP. Simple modeling of Q-branch absorption– III. Pressure, temperature, and perturber dependences in the $2\nu_6$ Q-branch of $^{12}CH^{35}ClF_2$ (HCFC-22). JQSRT **55**, 289 (1996)
373. Hartmann JM, Bouanich JP, Boulet C, Blanquet G, et al. Simple modeling of Q-branch absorption. II. Application to molecules of atmospheric interest (CFC-22 and CH_3Cl). JQSRT **54**, 723 (1995)
374. Gillespie WD, Meinrenken CJ, Lempert WR, Miles RB. Interbranch line-mixing in CO_2 (10^01) and (02^01) combination bands. J Chem Phys **107**, 5995 (1997)
375. Hartmann JM, Nguyen-Van-Thanh, Brodbeck C, Benidar A, et al. Simple modeling of line-mixing effects in IR bands. II. Non linear molecules – Applications to O_3 and $CHClF_2$. J Chem Phys **104**, 2185 (1996)
376. Ballard J, Knight RJ, Newnham DA, Vander Auwera J, et al. An intercomparison of laboratory measurements of absorption cross-sections and integrated absorption intensities for HCFC-22. JQSRT **66**, 109 (2000)
377. Massie ST, Goldman A. The infrared absorption cross-section and refractive-index data in HITRAN. JQSRT **82**, 413 (2003)
378. Filippov NN, Tonkov MV. Line mixing in the infrared spectra of simple gases at moderate and high density. Spectrochim Acta A **52**, 901 (1996)
379. Filippov NN, Tonkov MV, Bouanich JP. Semi-classical line mixing analysis in the first overtone band of CO compressed by N_2. Infrared Phys Techn **35**, 897 (1994)
380. Brunner TA, Pritchard D. Fitting laws for rotationally inelastic collisions. In Dynamics of the Excited State, K.P. Lawley Ed., p. 589, Wiley, NY (1982)
381. Clary DC. Rotational and vbrational-rotational relaxation in collisions of CO_2 (01^10) with He atoms. J Chem Phys **78**, 4915 (1983)
382. Alexander MH, Clary DC. Propensity rules in rotationally inelastic collisions of CO_2. Chem Phys Lett **98**, 318 (1983)
383. Alexander MH, Dradigian PJ. Collision-induced transitions between molecular hyperfine levels: quantum formalism, propensity rules, and experimental study of $CaBr(X^2\Sigma^+) + Ar$. J Chem Phys **83**, 2191 (1985)
384. Flaud PM, Orphal J, Boulet C, Hartmann JM. Measurements and analysis of collisional line-mixing within hyperfine components of helium broadened HI lines. J Mol Spectrosc **235**, 157 (2006)
385. Smith EW, Giraud M, Cooper J. A semiclassical theory for spectral line broadening in molecules. J Chem Phys **65**, 1256 (1976)
386. Fanjoux G, Millot G, Saint-Loup R, Chaux R, et al. CARS study of collisional effects in the O_2–H_2O Q-branch for the cryogenic combustion in rocket engine. J Chem Phys **101**, 1061 (1994)
387. Strow LL, Gentry B. Rotationnal collisionnal narrowing in an infrared CO_2 Q branch studied with a tunable diode laser. J Chem Phys **84**, 1149 (1986)

388. Pine AS. Self-, N_2- and Ar-broadening and line mixing in HCN and C_2H_2. JQSRT **50**, 149 (1993)
389. Margottin-Maclou M, Henry A, Valentin A. Line mixing in the Q branches of the $v_1 + v_2$ band of nitrous oxide and of the $(11^10_I) \leftarrow (02^20)$ band of carbon dioxide. J Chem Phys **96**, 1715 (1992)
390. Rachet F, Margottin-Maclou M, Henry A, Valentin A. Q-Branch line mixing effects in the $(20^00)_I$–01^10 and $(12^00)_I$–01^10 bands of carbon dioxide perturbed by N_2, O_2 and Ar and in the 13^10–00^00 and 13^10–01^10 bands of pure nitrous oxide. J Mol Spectrosc **175**, 315 (1996)
391. Strow LL, Pine AS. Q-branch line mixing in N_2O: effects of l-type doubling. J Chem Phys **89**, 1427 (1988)
392. Parker J. Practical inclusion of CO_2 line mixing effects in a line-by-line atmospheric retrieval system. JQSRT **69**, 327 (2001)
393. Rodrigues R, Blanquet G, Walrand J, Khalil B, et al. Line-mixing effects in Q branches of CO_2. I. Influence of parity in Δ–Π bands. J Mol Spectrosc **186**, 256 (1997)
394. Rodrigues R, Khalil B, Le Doucen R, L Bonamy, et al. Temperature, pressure, and perturber dependencies of line-mixing effects in CO_2 infrared spectra. I. Π–Σ Q-branches. J Chem Phys **107**, 4118 (1997)
395. Bouanich JP, Hartmann JM, Blanquet G, Walrand J, et al. Line-mixing effects in He- and N_2- broadened $\Sigma \leftrightarrow \Pi$ infrared Q branches of N_2O. J Chem Phys **109**, 6684 (1998)
396. DeSouza-Machado S, Strow LL, Tobin D, Hannon S. Improved atmospheric radiance calculations using CO_2 P/R branch line-mixing. Proc SPIE Satellite remote sensing of clouds and the atmosphere **3867**,188 (1999)
397. Meinrenken CJ, Gillespie WD, Macheret S, Lempert WR, et al. Time domain modeling of spectral collapse in high density molecular gases. J Chem Phys **106**, 8299 (1997)
398. Rodrigues R, De Natale P, Di Lonardo G, Hartmann JM. Line-mixing effects in the rotational ^rQ- branches of $^{16}O_3$ perturbed by O_2 and N_2. J Mol Spectrosc **175**, 429 (1996)
399. Chackerian C, Brown LR, Lacome N, Tarrago G. Methyl chloride v_5 region lineshape parameters and rotational constants for the v_2, v_5, and $2v_3$ vibrational bands. J Mol Spectrosc **191**, 148 (1998)
400. Tran H, Jacquemart D, Lacome N, Mandin JY. Line mixing in the v_6 Q branches of methyl bromide self- and nitrogen-broadened: experiment and modelling. JQSRT **109**, 119 (2008)
401. Goldflam R, Green S, Kouri DJ. Infinite order sudden approximation for rotational energy transfer in gaseous mixtures. J Chem Phys **67**, 4165 (1977)
402. Goldflam R, Kouri DJ, Green S. On the factorization and fitting of molecular scattering information. J Chem Phys **67**, 5661 (1977)
403. DePristo AE, Augustin ST, Ramaswamy R, Rabitz H. Quantum number and energy scaling for nonreactive collisions. J Chem Phys **71**, 850 (1979)
404. DePristo AE. Collisional influence on vibration-rotation spectral line shapes: A scaling theoretical analysis and simplifications. J Chem Phys **73**, 2145 (1980)
405. Green S. Raman linewidths and rotationally inelastic collision rates in nitrogen. J Chem Phys **98**, 257 (1993)
406. Huo WM, Green S. Quantum calculations for rotational energy transfer in nitrogen collisions. J Chem Phys **104**, 7552 (1996)
407. Bonamy L, Emond F. Rotational-angular-momentum relaxation mechanisms in the energy-corrected-sudden scaling theory. Phys Rev **A 51**, 1235 (1995)
408. Hadded S, Thibault F, Flaud PM, Aroui H, et al. Experimental and theoretical study of line mixing in NH_3 spectra. I. Scaling analysis of parallel bands perturbed by He. J Chem Phys **116**, 7544 (2002)

409. Rodrigues R. Interférences entre raies et relaxation du moment angulaire de rotation dans les spectres infrarouges de CO_2. Etude des branches Q et applications atmosphériques. Thesis, Université d'Orsay (France), 1998
410. Green S. Rotational excitation of symmetric top molecules by collisions with atoms. II. Infinite sudden approximation. J Chem Phys **70**, 816 (1979)
411. Thibault F, Boissoles J, Boulet C, Ozanne L, et al. Energy corrected sudden calculations of linewidths and lineshapes based on coupled states cross sections: the test case of CO_2–Ar. J Chem Phys **109**, 6338 (1998)
412. Green S. Effect of nuclear hyperfine structure on microwave spectral pressure broadening. J Chem Phys **88**, 7331 (1988)
413. Martinez RZ, Domenech JL, Bermejo D, Thibault F, et al. Close coupling calculations for rotational relaxation of CO in argon: accuracy of energy corrected scaling procedures and comparison with experimental data. J Chem Phys **119**, 10563 (2003)
414. Boissoles J, Thibault F, Le Doucen R, Menoux V, et al. Line mixing effects in the $3\nu_3$ band of CO_2 in helium. III: E.C.S. simultaneous fit of linewidths and near wing profile. J Chem Phys **101**, 6552 (1994)
415. Tran H, Hartmann JM. An improved O_2 A-band absorption model and its consequences for retrievals of photon paths and surface pressures. J Geophys Res, submitted
416. Niro F. Etudes théoriques et expérimentales des profils collisionnels dans les centres et ailes des bandes infrarouges de CO_2. Applications à la simulation et à l'inversion de spectres atmosphériques. Thesis, Université d'Orsay, France, 2003
417. Jucks KW, Rodrigues R, Le Doucen R, Claveaux C, et al. Model, software, and database for computation of line-mixing effects in infrared Q branches of atmospheric CO_2. II. Minor and asymmetric isotopomers. JQSRT **63**, 31 (1999)
418. Millot G, Saint-Loup R, Santos J, Chaux R, et al. Collisional effects in the stimulated Raman Q-branch of O_2 and O_2–N_2. J Chem Phys **96**, 961 (1992)
419. Temkin S, Bonamy L, Bonamy J, Robert D. Angular momentum coupling in spectroscopic relaxation cross-sections. Consequences for line coupling in bending bands. Phys Rev A **47**, 1543 (1993)
420. Emond F. Modélisation des spectres infrarouge et de diffusion Raman cohérente: généralisation des lois d'échelle dynamiques à des modes vibrationnels excités avec durée finie des collisions. Thesis, Université de Franche-Comté, Besançon, France (1995)
421. Bonamy L, Huet JM, Bonamy J, Robert D. Local scaling analysis of state-to-state rotational energy-transfer rates in N_2 from direct measurements. J Chem Phys **95**, 3361 (1991)
422. Beaud P, Knopp G. Scaling rotationally inelastic collisions with an effective angular momentum parameter. Chem Phys Lett **371**, 194 (2003)
423. Wong CCK, Hanson EE, McCourt FRW. Classical trajectory calculation of transport and relaxation properties for CO_2–He mixtures. Mol Phys **74**, 497 (1991)
424. Ter Horst MA, Jameson CJ. A comparative study of CO_2–Ar potential surfaces. J Chem Phys **105**, 6787 (1996)
425. Burshtein AI, Storozhev AV, Strekalov ML. Rotational relaxation in gases and its spectral manifestations. Chem Phys **131**, 145 (1989)
426. Rodrigues R, Hartmann JM, L Bonamy Boulet C. Temperature, pressure, and perturber dependencies of line-mixing effects in CO_2 infrared spectra. II. Angular momemtum relaxation and spectral shift in Σ–Σ bands. J Chem Phys **109**, 3037 (1998)
427. Lavorel B, Fanjoux G, Millot G, Bonamy L, et al. Line coupling effects in anisotropic Raman Q-branches of the $\nu_1/2\nu_2$ Fermi dyad of CO_2. J Chem Phys **103**, 9903 (1995)

428. Corey GC, McCourt FR, Liu WK. Pressure-broadening cross sections of multiplet-Σ molecules: O_2-noble gas mixtures. J Phys Chem **88**, 2031 (1984)
429. Tran H, Boulet C, Hartmann JM. Line-mixing and collision induced absorption by oxygen in the A-band. Laboratory measurements, model, and tools for atmospheric spectra computations. J Geophys Res **D 111**, 6869 (2006). *See model changes in Ref. 415*
430. Richard AM, Depristo AE. Further development and application of the ECS scaling theory: non-linear molecules. Chem Phys **69**, 273 (1982)
431. Hadded S, Thibault F, Flaud PM, H Aroui, *et al*. Experimental and theoretical study of line mixing in NH_3 spectra. II. Effect of the perturber in infrared parallel bands. J Chem Phys **120**, 217 (2004)
432. Thibault F, Boissoles J, Grigoriev I, Filippov NN, *et al*. Line mixing effects in the v_3 band of CH_3F in helium: experimental bandshapes and ECS. analysis. Eur Phys J **D 6**, 343 (1999)
433. Green S, Huo W M. Quantum calculations for line shapes in Raman spectra of molecular nitrogen. J Chem Phys **104**, 7590 (1996)
434. Gomez L, Martinez RZ, Bermejo D, Thibault F, *et al*. Q-branch linewidths of N_2 perturbed by H_2: experiments and quantum calculations from an *ab initio* potential. J Chem Phys **126**, 20432 (2007)
435. Boissoles J, Boulet C, Bruet X. *Ab initio* lineshape cross-sections: on the need of off-the-energy shell calculations. J Chem Phys **116**, 7537 (2002)
436. Davis L, Boggs JE. Rate constants for rotational excitation in NH_3–He collisions. J Chem Phys **69**, 2355 (1978)
437. Hutson JM, Howard BJ. The intermolecular potential energy surface of Ar–HCl. Mol Phys **43**, 493 (1981)
438. Tsao CJ, Curnutte B. Line-widths of pressure-broadened spectral lines. JQSRT **2**, 41 (1962)
439. Robert D, Giraud M, Galatry L. Intermolecular potentials and width of pressure-broadened spectral lines. I. Theoretical formulation. J Chem Phys **51**, 2192 (1969)
440. Benedict WS, Herman R. The calculation of self-broadened line widths in linear molecules. JQSRT **3**, 265 (1963)
441. Benedict WS, Kaplan LD. Calculation of line widths in H_2O–H_2O and H_2O–O_2 collisions. JQSRT **4**, 453 (1964)
442. Murphy JS, Boggs JE. Collision broadening of rotational absorption lines. I. Theoretical formulation. J Chem Phys **47**, 691 (1967)
443. Mehrotra SC, Boggs JE. Effect of collision-induced phase shifts on the linewidths and line shifts of rotational spectral lines. J Chem Phys **67**, 5306 (1977)
444. Gamache RR, Davies RW. Theoretical N_2-, O_2-, and air-broadened halfwidths of $^{16}O_3$ calculated by quantum Fourier transform theory with realistic collision dynamics. J Mol Spectrosc **109**, 283 (1985)
445. Rosenmann L, Hartmann JM, Perrin MY, Taine J. Collisional broadening of CO_2 IR lines. II. Calculations. J Chem Phys **88**, 2999 (1988)
446. Hartmann JM, Camy-Peyret C, Flaud JM, Bonamy J, *et al*. New accurate calculations of ozone line-broadening by O_2 and N_2. JQSRT **40**, 489 (1988)
447. Looney JP, Herman RM. Air broadening of the hydrogen halides-I. N_2 broadening and shifting in the HCl fundamental. JQSRT **37**, 547 (1987)
448. Labani B, Bonamy J, Robert D, Hartmann JM, *et al*. Collisional broadening of rotation-vibration lines for asymmetric-top molecules. I. Theoretical model for both distant and close collisions. J Chem Phys **84**, 4256 (1986)
449. Labani B, Bonamy J, Robert D, Hartmann JM. Collisional broadening of rotation-vibration lines for asymmetric-top molecules. III. Self-broadening case; application to H_2O. J Chem Phys **87**, 2781 (1987)

450. Neshyba SP, Gamache RR. Improved line broadening coefficients for asymmetric rotor molecules: application to ozone perturbed by nitrogen. JQSRT **50**, 443 (1993)
451. Lynch R, Gamache RR, Neshyba SP. Fully complex implementation of the Robert-Bonamy formalism: halfwidths and line shifts of H_2O broadened by N_2. J Chem Phys **105**, 5711 (1996)
452. Neshyba SP, Lynch R, Gamache RR, Gabard T, et al. Pressure induced widths and shifts for the v_3 band of methane. J Chem Phys **101**, 9412 (1994)
453. Gabard T. Argon-broadened line parameters in the v_3 band of $^{12}CH_4$. JQSRT **57**, 177 (1997)
454. Gamache RR, Lacome N, Pierre G, Gabard T. Nitrogen broadening of SF_6 transitions in the v_3 band. J Mol Struct **599**, 279 (2001)
455. Joubert P, Bonamy J, Robert D. Exact trajectory in semiclassical line broadening and line shifting calculation. Test for H_2–He Q(1) line. JQSRT **61**, 19 (1999)
456. Buldyreva JV, Bonamy J, Robert D. Semiclassical calculations with exact trajectories for N_2 rovibrational Raman linewidths at temperatures below 300 K. JQSRT **62**, 321 (1999)
457. Bykov AD, Lavrent'eva NN, Sinitsa LN. Influence of the collisional bending of trajectories on shifts of the molecular spectral lines in the visible region. Atm Ocean Opt **5**, 587 (1992)
458. Domenech JL, Thibault F, Bermejo D, Bouanich JP. Ar-broadening of isotropic Raman lines in the v_2 band of acetylene. J Mol Spectrosc **225**, 48 (2004)
459. Nguyen L, Ivanov S, Buzykin OG, Buldyreva J. Comparative analysis of purely classical and semiclassical approaches to collisional line broadening of polyatomic molecules: II. C_2H_2–He case. J Mol Spectrosc **239**, 101 (2006)
460. Lerot C, Blanquet G, Bouanich JP, Walrand J, et al. Xe-broadening coefficients of $^{12}CH_3D$: a test of theoretical lineshapes. Mol Phys **103**, 1213 (2005)
461. Salem J, Bouanich JP, Walrand J, Aroui H, et al. Helium- and argon-broadening coefficients of phosphine lines in the v_2 and v_4 bands. J Mol Spectrosc **232**, 247 (2005)
462. Gamache RR, Lynch R. Argon-induced halfwidths and line shifts of water vapor transitions. JQSRT **64**, 439 (2000)
463. Boulet C, Rosenberg A. Pressure induced lineshifts in HCl/HF and HF/HCl systems. J Phys **42**, 203 (1981)
464. Buldyreva J, Benec'h S, and Chrysos M. Infrared nitrogen-perturbed NO linewidths in a temperature range of atmospheric interest: an extension of the exact trajectory model. Phys Rev A **63**, 012708 (2001)
465. Rohart F, Colmont JM, Wlodarczak G, Bouanich JP. N_2- and O_2-broadening coefficients and profiles for millimeter lines of $^{14}N_2O$. J Mol Spectrosc **222**, 159 (2003)
466. Afzelius M, Buldyreva J, Bonamy J. Exact treatment of classical trajectories governed by an isotropic potential for linewidths computations. Mol Phys **102**, 1759 (2004)
467. Buldyreva J, Bonamy J, Weikl MC, Beyrau F, et al. Linewidth modeling of C_2H_2–N_2 mixtures tested by rotational CARS measurements. J Raman Spectrosc **37**, 647 (2006)
468. Predoi-Cross A, Hambrook K, Brawley-Tremblay S, Bouanich JP, et al. Room-temperature broadening and pressure-shift coefficients in the v_2 band of CH_3D–O_2: measurements and semi-classical calculations. J Mol Spectrosc **236**, 75 (2006)
469. Colmont JM, Rohart F, Wlodarczak G, Bouanich JP. K-dependence and temperature dependence of N_2-, H_2-, and He-broadening coefficients for the $J = 12$–11 transition of acetonitrile CH_3CN located near 220.7 GHz. J Mol Spectrosc **238**, 98 (2006)
470. Gamache RR, Fisher J. Half-widths of $H_2^{16}O$, $H_2^{18}O$, $H_2^{17}O$, $HD^{16}O$, and $D_2^{16}O$: I Comparison between isotopomers. JQSRT **78**, 289 (2003)
471. Drouin BJ, Fischer J, Gamache RR. Temperature dependent pressure induced lineshape of the O_3 rotational transitions in air. JQSRT **83**, 63 (2004)
472. Gamache RR, Hartmann JM. Collisional parameters of H_2O lines: effects of vibration. JQSRT **83**, 119 (2004)

473. Blanquet G, Bouanich JP, Walrand J, Lepère M. Diode-laser measurements and calculations of N_2-broadening coefficients in the v_7 band of ethylene. J Mol Spectrosc **229**, 198 (2005)
474. Salem J, Aroui H, Bouanich JP, Walrand J, et al. Collisional broadening and line intensities in the v_2 and v_4 bands of PH_3. J Mol Spectrosc **225**, 174 (2004)
475. Predoi-Cross A, Hambrook K, Brawley-Tremblay S, Bouanich JP, et al. Measurements and theoretical calculations of self-broadening and self-shift coefficients in the v_2 band of CH_3D. J Mol Spectrosc **234**, 53 (2005)
476. Soufiani A, Hartmann JM. Measurements and calculations of $CO-H_2O$ line-widths at high temperature. JQSRT **37**, 205 (1987)
477. Blanquet G, Bouanich JP, Walrand J, Lepère M. Self-broadening coefficients in the v_7 band of ethylene at room and low temperatures. J Mol Spectrosc **222**, 284 (2003)
478. Antony BK, Neshyba S, Gamache RR. Self-broadening of water vapor transitions via the complex Robert–Bonamy theory. JQSRT **105**, 148 (2007)
479. Godon M, Bauer A, Gamache RR. The continuum of water vapor mixed with methane: absolute absorption at 239 GHz and linewidth calculations. J Mol Spectrosc **202**, 293 (2000)
480. Rabitz HA, Gordon RG. Semiclassical perturbation theory of molecular collisions. II. The calculation of collision cross sections. J Chem Phys **53**, 1831 (1970)
481. Boursier C, Ménard-Boursin F, Boulet C. Calculation of rotational state-to-state ozone relaxation rates for O_3-N_2 and O_3-O_3 collisions. J Chem Phys **101**, 9589 (1994)
482. Boissoles J, Boulet C, Robert D. Toward the calculation of line mixing effect: a semi-classical calculation of rotationally inelastic cross-sections. Chem Phys Lett **122**, 237 (1985)
483. Lam KS. Application of pressure broadening theory to the calculation of atmospheric oxygen and water vapor microwave absorption. JQSRT **17**, 351 (1977)
484. Buffa G, Di Lieto A, Minguzzi P, Tarrini O, et al. Nuclear-quadrupole effects in the pressure broadening of molecular lines. Phys Rev A **37**, 3790 (1988)
485. Belli S, Buffa G, Tarrini O. Collisional coupling between hyperfine and Stark components of molecular spectra. Phys Rev A **55**, 183 (1997)
486. Lemaire V, Dore L, Cazzoli G, Buffa G, et al. Broadening of CH_3F in presence of Stark fields. I. Self-broadening and self-shifting of isolated components. J Chem Phys **106**, 8995 (1997)
487. Cazzoli G, Cludi L, Cotti G, Degli Esposi C, et al. Self-collisional coupling and broadening in the asymmetric rotor CHF_2Cl. J Chem Phys **102**, 1149 (1995)
488. Lemaire V, Dore L, Cazzoli G, Buffa G, et al. Broadening of CH_3F in presence of Stark fields. II. Collisional coupling between the Stark components. J Chem Phys **110**, 9418 (1999)
489. Gabard T, Champion JP. Calculation of collision induced ernergy transfer rates in tetrahedral molecules. Application to $^{12}CH_4$ perturbed by argon. JQSRT **52**, 303 (1994)
490. Pine AS, Gabard T. Speed-dependent broadening and line mixing in CH_4 perturbed by Ar and N_2 from multispectrum fits. JQSRT **66**, 69 (2000)
491. Pieroni D, Hartmann JM, Brodbeck C, Hartmann JM, et al. Experimental and theoretical study of line mixing in methane spectra. IV. Influence of the vibrational transition and of temperature. J Chem Phys **113**, 5766 (2000)
492. Tran H, Flaud PM, Gabard T, Hase F, et al. Model, software and database for line-mixing effects in the v_3 and v_4 bands of CH_4 and tests using laboratory and planetary measurements—I: N_2 (and air) broadenings and the earth atmosphere. JQSRT **101**, 284 (2006)
493. Chapman WB, Schiffman A, Hutson JM, Nesbitt DJ. Rotationnaly inelastic scattering in $CH_4 + He$, Ne, and Ar: State-to-state cross sections via direct infrared laser absorption in crossed supersonic jets. J Chem Phys **105**, 3497 (1996)
494. Pieroni D, Hartmann JM, Camy-Peyret C, Jeseck P, et al. Influence of line mixing on absorption by CH_4 in atmospheric balloon-borne spectra near 3.3 μm. JQSRT **68**, 117 (2001)

495. Tran H, Flaud PM, Fouchet T, Gabard T, et al. Model, software and database for line-mixing effects in the ν_3 and ν_4 bands of CH_4 and tests using laboratory and planetary measurements-II: H_2 (and He) broadening and the atmospheres of Jupiter and Saturn. JQSRT **101**, 306 (2006)
496. Pieroni D, Hartmann JM, F Chaussard, X Michaut, et al. Experimental and theoretical study of line mixing in methane spectra. III. The Q branch of the Raman ν_1 band. J Chem Phys **112**, 1335 (2000)
497. Grisch F, Bertseva E, Habiballah M, Jourdanneau E, et al. CARS spectroscopy of CH_4 for implication of temperature measurements in supercritical LOX/CH_4 combustion. Aerosp Sci Technol **11**, 48 (2007)
498. Dhib M, Echargui MA, Aroui H, Orphal J. Shifting and line mixing parameters in the ν_4 band of NH_3 perturbed by CO_2 and He: Experimental results and theoretical calculations. J Mol Spectrosc **238**, 168 (2006)
499. Atom-Molecule Collision Theory (a guide for the experimentalist). R.B. Bernstein editor. Springer, NY, (1979)
500. Collision Theory for Atoms and Molecules. F. A. Gianturco editor. Springer, NY, (1989)
501. G. Delgado-Barrio editor. Dynamical Processes in Molecular Physics. Institute of Physics Publishing, Bristol and Philadelphia (1993)
502. Child MS. Molecular Collision Theory. Dover Publications Inc.; New edition (1997)
503. Flower D. Molecular Collisions in the Interstellar Medium (2nd Ed.). Cambridge University press (2007)
504. McGuire P, Kouri DJ. Quantum mechanical close coupling approach to molecular collisions. j_z-conserving coupled states approximation. J Chem Phys **60**, 2488 (1974)
505. Pack RT. Space-fixed vs body-fixed axes in atom-diatomic molecule scattering. Sudden approximations. J Chem Phys **60**, 633 (1974)
506. Ross KA, Willey DR. Low temperature pressure broadening of OCS by He. J Chem Phys **122**, 204308 (2005)
507. Lique F, Spielfieldel A, Dubernet ML, Feautrier N. Rotational excitation of sulfur monoxide by collisions with helium at low temperature. J Chem Phys **123**, 134316 (2005)
508. Cappelletti D, Bartolomei M, Carmona-Novillo E, Pirani F, et al. Intermolecular interaction potentials for the $Ar-C_2H_2$, $Kr-C_2H_2$, and $Xe-C_2H_2$ weakly bound complexes: Information from molecular beam scattering, pressure broadening coefficients, and rovibrational spectroscopy. J Chem Phys **126**, 064311 (2007)
509. Green S. Energy transfer in NH_3–He collisions. J Chem Phys **73**, 2740 (1980)
510. Bussery-Honvault B, Moszynski R, Boissoles J. Ab initio potential energy surface and pressure broadening coefficients for the $He-CH_3F$ complex. J Mol Spectrosc **232**, 73 (2005)
511. Green S, DeFrees SDJ, McLean AD. Calculations of water microwave line broadening in collisions with He atoms: sensitivity to potential energy surfaces. J Chem Phys **94**, 1346 (1991)
512. Smith LN, Secrest D. Close-coupling and coupled state calculations of argon scattering from normal methane. J Chem Phys **74**, 3882 (1981)
513. Wernli M, Valiron P, Faure A, Wiesenfeld L, et al. Improved low-temperature rate constants for rotational excitation of CO by H_2. Astron Astrophys **446**, 367 (2006)
514. Montero S, Thibault F, Tejeda G, Fernandez JM. Rotranslational state-to-state rates and spectral representation of inelastic collisions in low-temperature molecular hydrogen. J Chem Phys **125**, 124301 (2006)
515. Offer A, Flower DR. Propensity rules in NH_3–H_2 collisions. J Phys B **22**, 439 (1989)
516. Wiley DR, Timlin RE, Merlin JM, Sowa MM, et al. An experimental investigation of collisions of NH_3 with para-H_2 at the temperatures of cold molecular clouds. Astron J Suppl Series **139**, 191 (2002)

517. Phillips TR, Maluendes S, Green S. Collision dynamics for an asymmetric top rotor and a linear rotor: coupled channel formalism and application to H_2O–H_2. J Chem Phys **102**, 6024 (1995)
518. Dubernet ML, Daniel F, Grosjean A, Faure A, et al. Influence of a new potential energy surface on the rotational (de)excitation of H_2O by H_2 at low temperature. Astron Astrophys **460**, 323 (2006)
519. Arthur AM, Dalgarno A. The theory of scattering by a rigid rotator. Proc R Soc London **A 256**, 540 (1960)
520. Johnson BR. The renormalized Numerov method applied to calculating bound states of the coupled-channel Schroedinger equation. J Chem Phys **69**, 4678 (1978)
521. Alexander MH, Manolopoulos DE. A stable linear reference potential algorithm for solution of the quantum close-coupled equations in molecular scattering theory. J Chem Phys **86**, 2044 (1987)
522. Beaky MM, Goyette TM, De Lucia FC. Pressure broadening and line shift measurements of carbon monoxide in collision with helium from 1 to 600 K. J Chem Phys **105**, 3994 (1996)
523. Ball CD, Mengel M, De Lucia FC, Woon DE. Quantum scattering calculations for H_2S–He between 1–600 K in comparison with pressure broadening, shift, and time resolved double resonance experiments. J Chem Phys **111**, 8893 (1999)
524. Mengel M, Flatin DC, De Lucia FC. Theoretical and experimental investigation of pressure broadening and line shift of carbon monoxide in collision with hydrogen between 8 and 600 K. J Chem Phys **112**, 4069 (2000)
525. Thachuk M, Chuaqui CE, Le Roy RJ. Linewidths and shifts of very low temperature CO in He: a challenge for theory or experiment?. J Chem Phys **105**, 4005 (1996)
526. Thibault F, Mantz AW, Claveau C, Henry A, et al. Broadening of the R(0) and P(2) lines in the ^{13}CO fundamental by helium atoms from 300 K down to 12 K: Measurements and comparison with close-coupling calculations. J Mol Spectrosc **246**, 118 (2007)
527. Goldflam R, Kouri DJ. On accurate quantum mechanical approximation for molecular relaxation phenomena. Averaged J_z-conserving coupled states approximation. J Chem Phys **66**, 542 (1977)
528. Heil TG, Green S, Kouri DJ. The coupled states approximation for scattering of two diatoms. J Chem Phys **68**, 2562 (1978)
529. Thibault F, Martinez RZ, Domenech JL, Bermejo D, et al. Raman and infrared linewidths of CO in Ar. J Chem Phys **117**, 2523 (2002)
530. Roche CF, Dickinson AS, Hutson JM. A failing of coupled-states calculations for inelastic pressure-broadening cross sections: calculations on of CO_2–Ar. J Chem Phys **111**, 5824 (1999)
531. Thibault F, B. Calil, J. Buldyreva, M. Chrysos, et al. Experimental and theoretical CO_2–Ar pressure-broadening cross sections and their temperature dependence. Phys Chem Chem Phys **3**, 3924 (2001)
532. Boissoles J, Thibault F, Domenech JL, Bermejo D, et al. Temperature dependence of line mixing effects in the stimulated Raman Q-branch of CO in He: a further test of close coupling calculations. J Chem Phys **115**, 7420 (2001)
533. Brezina R, Liu WK, Green S. Close-coupling calculation of line mixing in the isotropic Raman Q branch of D_2 in He. Phys Rev **A 51**, 3645 (1995)
534. Brezina R, Lin SH, Liu WK. Quantum mechanical calculation of line shape parameters for the depolarized Raman Q branch of D_2 in He. Chem Phys Lett **262**, 437 (1996)
535. Sinclair PM, Forsman JW, Drummond JR, May AD. Line mixing and state-to-state rotational relaxation rates in D_2 determined from the Raman Q branch. Phys Rev **A 48**, 3030 (1993)
536. Crisp D, Atlas RM, Bréon FM, Brown LR, et al. The Orbiting Carbon Observatory (OCO) mission. Adv Space Res **34**, 700 (2004)

537. Miller CE, Brown LR, Toth RA, Benner DC, et al. Spectroscopic challenges for high accuracy retrievals of atmospheric CO_2 and the Orbiting Carbon Observatory (OCO) experiment. CR Phys Acad Sci Paris **6**, 876 (2005)
538. Malathy Devi V, Benner DC, Brown LR, Miller CE, et al. Line mixing and speed dependence in CO_2 at $6348\,cm^{-1}$: positions, intensities and air- and self-broadening derived with constrained multispectrum analysis. J Mol Spectrosc **242**, 90 (2007)
539. Thibault F, Boissoles J, Le Doucen R, Boulet C. Measurement of line interference parameters for the CO–He system. Eur Phys Lett **12**, 319 (1990)
540. Thibault F, Boissoles J, Le Doucen R, Farrenq R, et al. Line by line measurements of interference parameters for the 0–1 and 0–2 bands of CO in He, and comparison with coupled-states calculations. J Chem Phys **97**, 4623 (1992)
541. Benner DC, Rinsland CP, Malathy Devi V, Smith MAH, et al. A multispectrum nonlinear least squares fitting technique. JQSRT **53**, 705 (1995)
542. Pine AS, Gabard T. Multispectrum fits for line mixing in the v_3 band Q branch of methane. J Mol Spectrosc **217**, 105 (2003)
543. Malathy Devi V, Benner DC, Brown LR, Miller CE, et al. Line mixing and speed dependence in CO_2 at $6227.9\,cm^{-1}$: Constrained multispectrum analysis of intensities and line shapes in the 30013←00001 band. J Mol Spectrosc **245**, 52 (2007)
544. Predoi-Cross A, Unni AV, Heung H, Malathy Devi V, et al. Line mixing effects in the $v_2 + v_3$ band of methane. J Mol Spectrosc **246**, 65 (2007)
545. Kochanov VP. Collisional line narrowing and mixing of multiplet spectra. JQSRT **66**, 313 (2000)
546. Malathy Devi V, Benner DC, Smith MAH, Rinsland CP. Measurements of air-broadened width and air-induced shift coefficients and line mixing in the v_6 band of $^{12}CH_3D$. JQSRT **68**, 1 (2001)
547. Boyd R, Tak-San Ho, Rabitz H, Romanini D, et al. Inversion of absorption spectral data for relaxation matrix determination. I. Application to line mixing in the 106←000 overtone transition of HCN. J Chem Phys **108**, 392 (1998)
548. Boyd R, Ho TS, Rabitz H. Inversion of absorption spectral data for relaxation matrix determination. II. Application to Q-branch line mixing in HCN, C_2H_2, and N_2O. J Chem Phys **108**, 1780 (1998)
549. Tretyakov MYu, Koshelev MA, Dorovskikh VV, Makarov DS, et al. 60-GHz oxygen band: precise broadening and central frequencies of fine structure lines, absolute absorption profile at atmospheric pressure, revision of mixing coefficients. J Mol Spectrosc **231**, 1 (2005)
550. Rosenkranz PW. Interference coefficients for overlapping oxygen lines in air. JQSRT **39**, 287 (1988)
551. Beaud P, Gerber T, Radi PP, Tulej M, et al. Rotationally inelastic collisions between N_2 and rare gases: an extension of the angular momentum scaling law. Chem Phys Lett **373**, 251 (2003)
552. Gamache RR, Arié E, Boursier C, Hartmann JM. Pressure-broadening and pressure-shifting of spectral lines of ozone. Spectrochim Acta A **54**, 35 (1998)
553. Gamache RR, Hartmann JM. An intercomparison of measured pressure-broadening and pressure-shifting parameters of water vapor. Can J Chem **82**, 1013 (2004)
554. Malathy Devi V, Benner DC, Smith MAH, Rinsland CP, et al. Self- and N_2-broadening, pressure induced shift and line mixing in the v_5 band of $^{12}CH_3D$ using a multispectrum fitting technique. JQSRT **74**, 1 (2002)
555. Benner DC, Malathy Devi V, Smith MAH, Rinsland CP, et al. Temperature dependence of line mixing in the v_3 band of methane. paper WH02, 54th Int. Symp. on Mol. Spectrosc., Colombus (OH), June 14–18 1999

556. Duffour G, Hurtmans D, Henry A, Valentin A, et al. Line profile study from diode laser spectroscopy in the $^{12}CH_4$ $2\nu_3$ band perturbed by N_2, O_2, Ar, and He. J Mol Spectrosc **221**, 80 (2003)
557. Mondelain D, Payan S, Deng W, Camy-Peyret C, et al. Measurement of the temperature dependence of line mixing and pressure broadening parameters between 296 and 90 K in the ν_3 band of $^{12}CH_4$ and their influence on atmospheric methane retrievals. J Mol Spectrosc **244**, 130 (2007)
558. Millot G, Lavorel B, Steinfeld JI. Collisional broadening, line shifting, and line mixing in the stimulated Raman $2\nu_2$ Q-branch of CH_4. J Chem Phys **95**, 7938 (1991)
559. Sinclair PM, Duggan P, Berman R, May AD, et al. Line broadening, shifting, and mixing in the fundamental band of CO perturbed by N_2 at 301 K. J Mol Spectrosc **181**, 41 (1997)
560. Deroussiaux A, Lavorel B. Vibrational and rotational collisional relaxation in CO_2–Ar and CO_2–He mixtures studied by stimulated Raman-infrared double resonance. J Chem Phys **111**, 1875 (1999)
561. Predoi-Cross A, Liu W, Holladay C, Unni AV, et al. Line profile study of transitions in the 30012←00001 and 30013←00001 bands of carbon dioxide perturbed by air. J Mol Spectrosc **246**, 98 (2007)
562. Sheldon GD, Sinclair PM, Le Flohic MP, Drummond JR, et al. Line mixing and broadening in the Raman Q branch of HD at 304.6 K. J Mol Spectrosc **192**, 406 (1998)
563. Pine AS, Markov VN. Self- and foreign-gas-broadened lineshapes in the ν_1 band of NH_3. J Mol Spectrosc **228**, 121 (2004)
564. Millot G, Lavorel B, Fanjoux G. Pressure broadening, shift, and interference effect for a multiplet line in the rovibrational anisotropic stimulated Raman spectrum of molecular oxygen. J Mol Spectrosc **176**, 211 (1996)
565. Predoi-Cross A, Hambrook K, Keller R, Povey C, et al. Spectroscopic lineshape study of the self-perturbed oxygen A-band. J Mol Spectrosc **248**, 85 (2008)
566. Blanquet G, Walrand J, Bouanich JP, Boulet C. Line mixing effects in Ar-broadened doublets of a hot band of OCS. J Chem Phys **93**, 6962 (1990)
567. Niro F, KW Jucks, Hartmann JM. Spectra calculations in central and wing regions of CO_2 IR bands between 10 and 20 µm. IV: software and database for the computation of atmospheric spectra. JQSRT **95**, 469 (2005)
568. Frichot F, N Lacome, Hartmann JM. Pressure and temperature dependences of absorption in the ν_5 $^RQ_0$ branch of CH_3Cl in N_2. Measurements and modeling. J Mol Spectrosc **178**, 52 (1996)
569. Grigoriev IM, Le Doucen R, Benidar A, Filippov NN, et al. Line-mixing effects in the ν_3 parallel absorption band of CH_3F perturbed by rare gases. JQSRT **58**, 287 (1997)
570. Filippov NN, Ogibalov VP, Tonlov MV. Line mixing effect on the pure CO_2 absorption in the 15 µm region. JQSRT **72**, 315 (2002)
571. Predoi-Cross A, Baranov Y. Line-shape modeling of inter-branch intensity transfer in the $20^00\leftarrow01^10$, $11^10\leftarrow00^00$ and $12^20\leftarrow01^10$ Q-branches of carbon dioxide. J Mol Struct **742**, 77 (2005)
572. Lavorel B, Pykhov R, Millot G. Line mixing in the stimulated Raman spectrum of the ν_1 band of SiH4 at 0.4–1.0 bar. JQSRT **49**, 579 (1993)
573. Gentry B, Strow LL. Line-mixing in a N_2-broadened CO_2 Q branch observed with a tunable diode laser. J Chem Phys **86**, 5722 (1987)
574. Huet T, Lacome N, Lévy A. Line mixing effects in the Q branch of the $10^00\leftarrow01^10$ transition of CO_2. J Mol Spectrosc **138**, 141 (1989)
575. Margottin-Maclou M, Rachet F, Boulet C, Henry A, et al. Q-branch line mixing effects in the $(20^00)_I \leftarrow 01^01$ and $(12^20)_I \leftarrow 01^10$ bands of carbon dioxide. J Mol Spectrosc **172**, 1 (1995)

576. Lavorel B, Millot G, Saint-Loup R, Berger H, et al. Study of collisional effects on band shapes of the $v_1/2v_2$ Fermi dyad in CO_2 gas with stimulated Raman spectroscopy: I: rotational and vibrational relaxation in the $2v_2$ band. J Chem Phys **93**, 2176 (1990)
577. Roblin A, Bonamy J, Robert D, Lefèbvre M, et al. Rotational relaxation model for CO–N_2. Prediction of CARS profiles and comparison with experiment. J de Physique **II 2**, 285 (1992)
578. Romanini D, Lehmann KK. Line-mixing in the 106←000 overtone transition of HCN. J Chem Phys **105**, 81 (1996)
579. Rosasco GJ, May AD, Hurst WS, Smyth KC. Broadening and shifting of the Raman Q branch of HD. J Chem Phys **90**, 2115 (1989)
580. Bonamy J, Robert D, Hartmann JM, Gonze ML, et al. Line broadening, line shifting, line coupling effects on N_2–H_2O stimulated Raman spectra. J Chem Phys **91**, 5916 (1989)
581. Vitcu A, Ciurylo R, Wehr R, Drummond JR, et al. Broadening, shifting, and line mixing in the $03^10 \leftarrow 01^10$ parallel Q branch of N_2O. J Mol Spectrosc **226**, 71 (2004)
582. Hirono M, Ichikawa K. Line shapes in the R-branch of the 5.3 μm NO band. Opt Rev **4**, 362 (1997)
583. Lempert W, Rosasco GJ, Hurst WS. Rotational collisional narrowing in the NO fundamental Q branch studied with cw stimulated Raman spectroscopy. J Chem Phys **81**, 4241 (1984)
584. Rosenmann L, Hartmann JM, Plateaux JJ, Barbe A. Absorption in the Q-branch of the $v_1 + v_2 + v_3$ band of O_3. JQSRT **50**, 233 (1993)
585. Boissoles J, Boulet C, Robert D, Green S. IOS and ECS line coupling calculation for the CO–He system: influence on the vibration–rotation band shapes. J Chem Phys **87**, 3336 (1987)
586. Bouanich JP, Rodrigues R, Boulet C. Line-mixing effects in the 0-1 and 0-2 CO bands perturbed by CO and N_2 from low to high densities. JQSRT **54**, 683 (1995)
587. Hirono M. Line-mixing calculations for the 5.3 μm NO band. Opt Rev **9**, 45 (2002)
588. Millot G, Fanjoux G, Lavorel B. Fitting and scaling laws for high temperature Q branch collapse in the O_2 stimulated Raman spectra in O_2–H_2O mixtures. J Chem Phys **104**, 5347 (1996)
589. Blanquet G, Walrand J, Bouanich JP. Line-mixing effects in He- and N_2-broadened Q branches of C_2H_2 at low temperatures. J Mol Spectrosc **210**, 1 (2001)
590. Boissoles J, Thibault F and Boulet C. Line mixing effects in the 15 μm Q-branches of CO_2 in helium: theoretical analysis. JQSRT **56**, 835 (1996)
591. Hartmann JM, Rodrigues R, Nguyen-Van-Thanh, Brodbeck C, et al. Temperature, pressure, and perturber dependences of line-mixing effects in CO_2 infrared spectra. III. Second order rotational angular momentum relaxation and Coriolis effects in Π–Σ bands. J Chem Phys **110**, 7733 (1999)
592. Lavorel B, Millot G, Fanjoux G, Saint-Loup R. Study of collisional effects on band shapes of $v_1/2v_2$ Fermi dyad in CO_2 gas with stimulated Raman Spectroscopy. III. Modeling of collisional narrowing and study of vibrational shifting and broadening at high temperature. J Chem Phys **101**, 174 (1994)
593. Bonamy L, Bonamy J, Robert D, Deroussiaux A, et al. A direct study of the vibrational bending effect in line mixing: the hot degenerate $11^10 \leftarrow 01^10$ transition of CO_2. JQSRT **57**, 341 (1997)
594. Bonamy L, Bonamy J, Robert D, Lavorel B, et al. Rotationally inelastic rates for N_2–N_2 system from a scaling theoretical analysis of the SRS Q-branch. J Chem Phys **89**, 5568 (1988)
595. Sitz GO, Farrow RL. Pump–probe measurements of state-to-state rotational energy transfer rates in N_2 ($v = 1$). J Chem Phys **93**, 7883 (1990)
596. Bonamy L, Bonamy J, Robert D, Temkin SI, et al. Line coupling in anisotropic Raman branches. J Chem Phys **101** 7350 (1994)
597. Bruet X, Bonamy L, Bonamy J. Extension of the energy-corrected sudden moded to anisotropic Raman lines: application to pure N_2. Phys Rev **A 62**, 062702 (2000)

598. Birnbaum G, Buechele A, Thomas ME, Banta M, et al. Experimental and theoretical studies of absorption in microwindows of the v_4 band of methane and methane-hydrogen. JQSRT **72**, 637 (2002)
599. Grigoriev IM, Filippov NN, Tonkov MV, JP Champion, et al. Line parameters and shapes of high clusters: R branch of the v_3 band of CH_4 in He mixtures. JQSRT **74**, 431 (2002)
600. Filippov NN, Grigoriev IM, Grigorovich NM, Tonkov MV. Line mixing in v_3 and forbidden v_2 bands of CH_4 in gaseous helium. Mol Phys **104**, 2711 (2006)
601. Rinsland CP, Strow LL. Line mixing effects in solar occultation spectra of the lower atmosphere: measurements and comparisons with the calculations for the 1932-cm^{-1} CO_2 Q branch. Appl Opt **28**, 457 (1989)
602. Niro F, F Hase, Camy-Peyret C, Payan S, et al. Spectra calculations in central and wing regions of CO_2 IR bands between 10 and 20 μm. II: Atmospheric solar occultation spectra. JQSRT **90**, 43 (2005)
603. Niro F, von Clarmann T, Jucks KW, Hartmann JM. Spectra calculations in central and wing regions of CO_2 IR bands between 10 and 20 μm. III: Atmospheric emission spectra. JQSRT **90**, 61 (2005)
604. Niro F, Boulet C, Hartmann JM, Lellouch E. On the effect of line-mixing among CO_2 transitions on the emission of Mars atmosphere near 15 μm. JQSRT **95**, 483 (2005)
605. Kozlov DN, Smirnov VV, Volkov SYu. Study of processes of translational, rotational, and vibrational relaxation based on CARS spectra line shapes. Appl Phys **B 48**, 273 (1989)
606. Grigoriev IM, Le Doucen R, Boissoles J, Chalil B, et al. Spectral lineshape parameters revisited for HF in a bath of argon. I. Widths and shifts in the 0–0, 0–1, and 0–2 bands. J Mol Spectrosc **198**, 249 (1999)
607. Arcas P, Arié E, Valentin A, Henry A. Intensity of CO_2 bands in the 4.8- to 5.3-μm region. The $(11^10, 03^10)_I$–00^00 band. J Mol Spectrosc **96**, 288 (1982)
608. Boissoles J, Thibault F, Rachet F, Valentin A, et al. Line mixing effects in the Q-branch of CO_2 in helium near 4.7 μm: a further test of the E.C.S. formalism. JQSRT **57**, 519 (1997)
609. Ben-Aryeh Y, Sorgen A. Self-broadening of molecular spectral lines. I. General theory. Phys Rev **A 4**, 21708 (1971)
610. Pasmanter RA, Ben-Reuven A. Resonance-transfer contributions to resonance line broadening in the impact limit. JQSRT **13**, 57 (1973)
611. Roney PL. Theory of spectral line shape. II. Collision time theory and the line wing. J Chem Phys **101**, 1050 (1994)
612. Kouzov AP. Rotational relaxation matrix for fast non-Markovian collisions. Phys Rev **A 60**, 2921 (1999)
613. Filippov NN, Tonkov MV. Kinetic theory of band shapes in molecular spectra of gases: application to band wings. J Chem Phys **108**, 3608 (1998)
614. Menoux V, Le Doucen R, Boissoles J, Boulet C. Lineshape in the low frequency wing of self and N_2 broadened v_3 CO_2 lines: temperature dependence of the asymmetry. Appl Opt **30**, 281 (1991)
615. Perrin MY, Hartmann JM. Temperature-dependent measurements and modeling of absorption by CO_2–N_2 mixtures in the far line-wings of the 4.3 μm CO_2 band. JQSRT **42**, 311 (1989)
616. Clough SA, Kneizys FX, Davies RW. Line shape and the water vapor continuum. Atmos Res **23**, 229 (1989)
617. Winters BH, Silverman S, Benedict WS. Line shape in the wing beyond the band head of the 4.3 μm band of CO_2. JQSRT **4**, 527 (1964)
618. Burch DE, Gryvnak DA, Patty RR, Bartky CE. Shapes of collision-broadened CO_2 lines. J Opt Soc Am **59**, 267 (1969)

619. Cousin C, Le Doucen R, Boulet C, Henry A. Temperature dependence of the absorption in the region beyond the 4.3 µm band head of CO_2. II-N_2 and O_2 broadening. Appl Opt **24**, 3899 (1985)
620. Menoux V, Le Doucen R, Boulet C. Line shape in the low frequency wing of self-broadened CO_2 lines. Appl Opt **26**, 554 (1987)
621. Menoux V, Le Doucen R, Boulet C. Line shape in the low frequency wing of N_2 and O_2-broadened CO_2 lines. Appl Opt **26**, 5183 (1987)
622. Cann MWP, Nicholls RW, Ronay PL, Blanchard A, et al. Spectral line shapes for carbon dioxide in the 4.3 µm band. Appl Opt **24**, 1374 (1985)
623. Hartmann JM. Measurements and calculations of CO_2 room-temperature high-pressure spectra in the 4.3 µm region. J Chem Phys **90**, 2944 (1989)
624. Tsuboi T, Arimitsu N, Hartmann JM. High temperature absorption by pure CO_2 far line wings in the 4 µm region. Jap J Appl Phys **32**, L1778 (1993)
625. Tonkov MV, Filippov NN, Bcrtscv VV, Bouanich JP, et al. Measurements and empirical modeling of pure CO_2 absorption in the 2.3 µm region at room temperature: far wings, allowed and collision induced transitions. Appl Opt **35**, 4863 (1996)
626. Brodbeck C, Bouanich JP, Nguyen-Van-Thanh, Hartmann JM, et al. Absorption of radiation by gases from low to high pressures. II. Measurements and calculations of CO infrared spectra. J de Physique **II 4**, 2101 (1994)
627. Hartmann JM, Perrin MY, Ma Q, Tipping RH. The infrared continuum of pure water vapor: calculations and high-temperature measurements. JQSRT **49**, 675 (1993)
628. Hartmann JM, Boulet C, Fouchet T, Drossart P. A far wing line shape for H_2 broadened CH_4 infrared transitions. JQSRT **72**, 117 (2002)
629. Bailly D, Birnbaum G, Buechele A, Flaud PM, et al. Absorption by pure NH_3 and NH_3–H_2 mixtures n the 5 µm window region. JQSRT **83**, 1 (2004)
630. Houdeau JP, Boulet C, Robert D. A theoretical and experimental study of the infrared line shape from resonance to the wings for uncoupled lines. J Chem Phys **82**, 1661 (1985)
631. Strow LL, Tobin DC, McMillan WW, Hannon SE, et al. Impact of a new water vapor continuum and line shape model on observed high resolution infrared radiances. JQSRT **59**, 303 (1998)
632. Strow L, Hannon S, De Souza-Machado S, Motteler H, et al. An overview of the AIRS radiative transfer model. IEEE Trans on Geosci Remote and Sensing. **41**, 303 (2003)
633. Burch DE, Gryvnak DA. Continuum absorption by H_2O vapor in the infrared and millimeter regions. In Atmospheric Water Vapor, p. 47, A. Deepak Ed., Academic Press (1980)
634. Clough SA, Shephard MW, Mlawer EJ, Delamere JS, et al. Atmospheric radiative transfer modeling: a summary of the AER codes. JQSRT **91**, 233 (2005). The up-to-date version of the CKD continuum and of the LBLRTM radiative transfer code by T. Clough are available at the address http://www.rtweb.aer.com/lblrtm_code.html
635. Tobin DC, Best FA, Brown PD, Clough SA, et al. Downwelling spectral radiance observations at the SHEBA ice station: Water vapor continuum measurements from 17 to 26 microns. J Geophys Res **D 104**, 2081 (1999)
636. Mlawer EJ, Tobin DC, Clough SA. A new water vapor continuum Model MT_CKD_1.0. Proc. of the 13th Atmospheric Radiation Measurement (ARM) Program Science Team Meeting CONF-2003, Ed. by D.Carrothers, U.S. Department of Energy, Richland, WA
637. Burch DE. Absorption by H_2O in narrow windows between 3000 and 4200 cm^{-1}. Report No AFGL-TR-85-0036, Air Force Geophysics Laboratory, Hanscom AFB (1985)
638. Read WG, Waters JW, Wu DL, Stone EM, et al. UARS microwave limb sounder upper tropospheric humidity measurements: method and validation. J Geophys Res **D 106**, 32207 (2001)
639. Liebe H J, Layton D H. Millimeter wave properties of the atmosphere: laboratory studies and propagation modeling. NTIA Report 87-224, Natl Telecommun and Inf Admin, Boulder, CO, (1987)

640. Bauer A, Godon M. Temperature dependence of water vapor absorption in linewings at 190 GHz. JQSRT **46**, 211, (1991)
641. Kuhn T, Bauer A, Godon M, Bühler S, et al. Water vapor continuum: absorption measurements at 350 GHz and model calculations. JQSRT **74**, 545 (2002)
642. Liebe HJ. MPM-an atmospheric millimeterwave propagation model. Int J Infrared Millim Waves **10**, 631 (1989)
643. Liebe HJ, Hufford GH, Cotton MG. Propagation modeling of moist air and suspended water/ice particles at frequencies below 1000 GHz. AGARD Conference proceedings 542, 3–10, (1993)
644. Rosenkranz PW. Water vapor microwave continuum absorption: a comparison of measurements and models. Radio Sci **33**, 919 (1998) and *erratum* in **34**, 1025 (1999)
645. Ma Q, Tipping RH. A simple analytical parameterization for the water vapor millimeter wave foreign continuum. JQSRT **82**, 517 (2003)
646. Gautier D, Bézard B, Marten A, Baluteau JP, et al. The C/H ratio in Jupiter from the Voyager infrared investigation. Astrophys J **257**, 901 (1982)
647. Rosenkranz PW. Pressure broadening of rotational bands I. A statistical theory. J Chem Phys **83**, 6139 (1985)
648. Rosenkranz PW. Pressure broadening of rotational bands. II. Water vapor from 300 to 1100 cm^{-1}. J Chem Phys **87**, 163 (1987)
649. Boulet C, Boissoles J, Robert D. Collisionally induced population transfer in I.R. absorption spectra. I-A line-by-line coupling theory from resonances to the far wings. J Chem Phys **89**, 625 (1988)
650. Ma Q, Tipping RH. The atmospheric water continuum in the infrared: extension of the statistical theory of Rosenkranz. J Chem Phys **93**, 7066 (1990)
651. Ma Q, Tipping RH. A far wing lineshape theory and its application to the foreign-broadened water continuum absorption. J Chem Phys **97**, 818 (1992)
652. Ma Q, Tipping RH, Boulet C. The frequency detuning and band-average approximations in a far wing line shape theory satisfying detailed balance. J Chem Phys **104**, 9678 (1996)
653. Ma Q, Tipping RH. The distribution of density matrices over potential energy surfaces: application to the calculation of the far wing line shapes for CO_2. J Chem Phys **108**, 3386 (1998)
654. Ma Q, Tipping RH, Boulet C, Bouanich JP. Theoretical far wing line shape and absorption for high temperature CO_2. App Opt **38**, 599 (1999)
655. Ma Q, Tipping RH. The density matrix of H_2O–N_2 in the coordinate representation: a Monte Carlo calculation of the far wing line shape. J Chem Phys **112**, 574 (2000)
656. Ma Q, Tipping RH. The frequency detuning correction and the asymmetry of line shapes: the far wings of H_2O–H_2O. J Chem Phys **116**, 4102 (2002)
657. Ma Q, Tipping RH. Water vapor millimeter wave foreign continuum: a Lanczos calculation in the coordinate representation. J Chem Phys **117**, 10581 (2002)
658. Burch DE, Alt RL. Continuum absorption by H_2O in the 700–1200 cm^{-1} and 2400–2600 cm^{-1} windows. Report No AFGL-TR-84-0128, Air Force Geophysics Laboratory, Hanscom AFB (1984)
659. Cormier JG, Ciurylo R, Drummond JR. Cavity ringdown spectroscopy measurements of the infrared water vapor continuum. J Chem Phys **116**, 1030 (2002)
660. Podobedov VB, Plusquellic DF, Siegrist KE, Fraser GT, et al. New measurements of the water vapor continuum in the region from 0.3 to 2.7 THz. JQSRT **109**, 458 (2008)
661. Borysow A, Moraldi M. Rototranslational Raman scattering in hydrogen. Phys Rev A **40**, 1251 (1989)
662. Bafile U, Ulivi L, Zoppi M, Barocchi F, et al. Depolarized-light-scattering spectrum from gaseous hydrogen at 50 K: the density squared component. Phys Rev A **42**, 6916 (1990)
663. Borysow A, Moraldi M. Effects of the intermolecular interactions on the depolarized rototranslational Raman spectra of hydrogen. Phys Rev A **48**, 3036 (1993)

664. Yi Fu, Borysow A, Moraldi M. High frequency wings of rototranslational Raman spectra of gaseous nitrogen. Phys Rev A **53**, 201 (1996)
665. Bonamy L. Private communication.
666. Bonamy L, Buldyreva JV. Non-Markovian far-wing rotational Raman spectrum from translational modeling. Phys Rev A **63**, 012715 (2001)
667. Bancewicz T, Glaz W, Kielich S. The dipole-octupole contribution to the Rayleigh wings of gaseous nitrogen. Phys Lett A **148**, 78 (1990)
668. Bancewicz T, Teboul V, Le Duff Y. High-frequency interaction-induced rototranslational scattering from gaseous nitrogen. Phys Rev A **46**, 1349 (1992)
669. Van der Avoird A, Worner PES, Jansen APJ. An improved intermolecular potential for nitrogen. J Chem Phys **84**, 1629 (1986)
670. Camy-Peyret C, Vigasin AA (Eds.). Weakly Interacting Molecular Pairs: Unconventional Absorbers of Radiation in the Atmosphere. NATO Science Series: IV: Earth and Environmental Sciences, Vol. 27, Kluwer, Dordrecht, (2003)
671. Le Duff Y, Teboul V. Collision induced light scattering in the far Rayleigh wing of gaseous nitrogen. Phys Lett A **157**, 44 (1991)
672. Buldyreva JV, Bonamy L. Non-Markovian energy corrected sudden model for the rototranslational spectrum of N_2. Phys Rev A **60**, 370 (1999)
673. Kouzov AP, Buldyreva JV. Orthogonal transformations in the line space and modelling of rotational relaxation in the Raman spectra of linear tops. Chem Phys **221**, 103 (1997)
674. Birnbaum G, Cohen ER. Theory of line shape in pressure induced absorption. Can J Phys **54**, 593 (1976)
675. Benec'h S, Rachet F, Chrysos M, Buldyreva J, *et al*. On far-wing Raman profiles by CO_2. J Raman Spectrosc **33**, 934 (2002)
676. Teboul V, Le Duff Y, Bancewicz T. Collision induced scattering in CO_2 gas. J Chem Phys **103**, 1384 (1995)
677. Atmospheric Water Vapor. Eds. Deepak A, Wilkerson TD, Ruhnke LH; Academic Press, NY, (1980)
678. Thermal Microwave Radiation – Application for Remote Sensing. Eds Mätzler C, Rosenkranz PW, Battaglia A, Wigneron JP; IEE Electromagnetic Waves Series, London UK (2006)
679. Aumann HH, Pagano RJ. Atmospheric sounder on the Earth observing system. Opt Eng **33**, 776 (1994)
680. Pardo JR, Cernicharo J, Serabyn E. Atmospheric Transmission at Microwaves (ATM): an improved model for millimeter/submillimeter applications. IEEE Trans Antennas Propag **49**, 1683 (2001)
681. Prigent C, Pardo JR, Rossow WB. Comparisons of the millimeter and submillimeter bands for atmospheric temperature and water vapor soundings for clear and cloudy skies. J Appl Meteor **45**, 1622 (2006)
682. Roberts RE, Selby JE, Biberman LM. Infrared continuum absorption by atmospheric water vapor in the 8–12 μm window. Appl Opt **15**, 2085 (1976)
683. Tobin DC, Strow LL, Lafferty WJ, Olson WB. Experimental investigation of the self and N_2-broadened continuum within the ν_2 band of water vapor. Appl Opt **35**, 4724 (1996)
684. Grant WD. Water vapor absorption coefficient in the 8–13 μm spectral region: a critical review. Appl Opt **29**, 451 (1990)
685. Thomas ME. Infrared and millimeter wavelength continuum absorption in the atmospheric windows: measurements and models. Infrared Phys **30**, 161 (1990)
686. Sierk B, Solomon S, Daniel JS, Portmann RW, *et al*. Field measurements of water vapor continuum absorption in the visible and near infrared. J Geophys Res D **109**, 08307 (2004)

687. Cormier JG, Hodges JT, Drummond JR. Infrared water vapor continuum absorption at atmospheric temperatures. J Chem Phys **122**, 114309 (2005)
688. Baranov YI, Lafferty WJ, Ma Q, Tipping RH. Water vapor continuum absorption in the 800 cm^{-1} to 1250 cm^{-1} spectral region at temperatures from 311 to 363 K. JQSRT, in press (2008).
689. Han Y, Shaw JA, Churnside JH, Brown PD, et al. Infrared spectral radiance measurements in the tropical Pacific atmosphere. J Geophys Res **D 102**, 4353 (1997)
690. Mlawer EJ, Tobin DC, Clough SA. A revised perspective on the water vapor continuum: the MT_CKD model. JQSRT, paper in preparation.
691. Fomin BA, Udalova TA, Zhitnitskii EA. Evolution of spectroscopic information over the last decade and its effect on line-by-line calculations for validation of radiation codes for climate models. JQSRT **86**, 73 (2004)
692. Ptashnik IV. Evaluation of suitable spectral intervals for near infrared laboratory detection of water vapor continuum absorption. JQSRT **108**, 146 (2007)
693. Tipping RH, Ma Q. Theory of the water vapor continuum and validations. Atmos Res **36**, 69(1995)
694. Hettner G. Uber das ultrarote absorptionsspektrum des wasserdampfes. Ann Phys **55**, 476 (1918)
695. Elsasser WM. Mean absorption and equivalent absorption coefficient of a band spectrum. Phys Rev **54**, 126 (1938)
696. Varanasi P, Chou S, Penner SS. Absorption coefficient for water vapor in the 600–1000 cm^{-1} region. JQSRT **8**, 1537 (1968)
697. Suck SH, Wetmore AE, Chen TS, Kassner JL Jr. Role of various water clusters in infrared absorption in the 8–14 μm window region. Appl Opt **21**, 1610 (1982)
698. Mlawaer EJ, Clough SA, Brown PD, Tobin DC. Collision induced effects and the water vapor continuum. Proc. of the 8th atmospheric radiation measurement (ARM) science team meeting DOE/ER-0738, p. 503; US Department of Energy, Washington DC (1998)
699. Bauer A, Godon M. Continuum for H_2O-X mixtures in the H_2O spectral window at 239 GHz; X = C_2H_4, C_2H_6. Are collision-induced absorption processes involved?. JQSRT **69**, 277 (2001)
700. Vigasin AA. Water vapor continuous absorption in various mixtures: a possible role of weakly bound complexes. JQSRT **64**, 25 (2000)
701. Scribano Y, Leforestier C. Contribution of water dimer absorption to the millimeter and far infrared atmospheric water continuum. J Chem Phys **126**, 234301 (2007)
702. Hinderling J, Sigrist MW, Kneubuhl FK. Laser-photoacoustic spectroscopy of water-vapor continuum and line absorption in the 8 to 14 μm atmospheric window. Infrared Phys **27**, 63 (1987)
703. Loper GL, O'Neill MA, Gelbwachs JA. Water-vapor continuum CO_2 laser absorption spectra between 27 C and −10 C. Appl Opt **22**, 3701 (1983)
704. Ma Q, Tipping RH, Leforestier C. Temperature dependence of mechanisms responsible for the water continuum absorption. I. Far wings of allowed lines. J Chem Phys, **128**, 124313 (2008)
705. Tipping RH, Ma Q. Far wing lineshapes: an application to the water vapor continuum. In Ref. 670, p. 137
706. Fulghum SF, Tilleman MM. Interferometric calorimeter for the measurement of water vapor absorption. J Opt Soc Am **B 8**, 2401 (1991)
707. Ptashnik IV, Smith KM, Shine KP, Newnham DA. Laboratory measurements of water vapor continuum absorption in spectral region 5000–5600 cm^{-1}: evidence for water dimers. Quart J Royal Met Soc **130**, 2391 (2004)
708. Schofield DP, Kjaergaard HG. Calculated OH-stretching and HOH-bending vibrational transitions in the water dimer. Phys Chem Chem Phys **5**, 3100 (2003)

709. Brown A, Tipping RH. Collision induced absorption in dipolar molecule- homonuclear diatomic pairs. In Ref. 670, p. 93
710. Daniel JS, Solomon S, Kjaergaard HG, Schofield DP. Atmospheric water vapor complexes and the continuum. Geophys Res Lett **31**, L06118 (2004)
711. Tikhominov AB, Ptashnik IV, Tikhominov BA. Measurements of the continuum absorption coefficient of water vapor near $14400\,\text{cm}^{-1}$ ($0.694\,\mu\text{m}$). Optics and Spectrosc **101**, 80 (2006)
712. Reichert L, Andres-Hernandez MD, Burrows JP, Tikhominov AB, *et al*. First CRDS measurements of water vapor continuum in the 940 nm absorption band. JQSRT **105**, 303 (2007)
713. Jenouvrier A, Daumont L, Régalia-Jarlot L, Tyuterev VG, *et al*. Fourier transform measurements of water vapor line parameters in the 4200–$6600\,\text{cm}^{-1}$ region. JQSRT **105**, 326 (2007)
714. Crawford MF, Welsh HL, Locke JL. Infra-red absorption of oxygen and nitrogen induced by intermolecular forces. Phys Rev **75**, 1607 (1949)
715. Phenomena Induced by Intermolecular Interaction. Edited by G. Birnbaum, Plenum, NY, (1985)
716. Hunt JL, Poll JD. A second bibliography on collision-induced absorption. Mol Phys **59**, 163 (1986)
717. Borysow A, Frommhold L, Dore P. Far infrared absorption by pairs of nonpolar molecules. Intern J Infrared Millim Waves **8**, 381 (1987)
718. Borysow A, Frommhold L. Collision induced light scattering: a bibliography. Adv Chem Phys **75**, 439 (1989)
719. Collision- and Interaction-Induced Spectroscopy. Edited by G.C. Tabisz and M.N. Neuman, Kluwer, Dordrecht, (1995)
720. Li X, Hunt KLC. Nonadditive three-body dipoles of inert gas trimers and H_2–H_2–H_2: Long-range effects in far infrared absorption and triple vibrational transitions. J Chem Phys **107**, 4133 (1997)
721. Bachet G, Cohen ER, Dore P, Birnbaum G. The translational–rotational absorption spectrum of hydrogen. Can J Phys **61**, 591 (1983)
722. Brodbeck C, Nguyen-Van-Thanh, Jean-Louis A, Bouanich JP, *et al*. Collision-induced absorption by H_2 pairs in the fundamental band at 78 and 298 K. Phys Rev **A 50**, 484 (1994)
723. Le Duff Y, Ouillon R. Depolarized scattering from para and normal hydrogen. J Chem Phys **82**, 1 (1985)
724. Maté B, Lugez C, Fraser GT, Lafferty WJ. Absolute intensities for the O_2 $1.27\,\mu\text{m}$ continuum absorption. J Geophys Res **D 104**, 30585 (1999)
725. Maté B, Lugez CL, Solodov AM, Fraser GT, *et al*. Investigation of the collision induced absorption by O_2 near $6.4\,\mu\text{m}$ in pure O_2 and O_2/N_2 mixtures. J Geophys Res **D 105**, 22225 (2000)
726. Smith KM, Newnham DA. Near-infrared absorption cross sections and integrated absorption intensities of molecular oxygen. J Geophys Res **D 105**, 7383 (2000)
727. Moreau G, Boissoles J, Le Doucen R, Boulet C, *et al*. Experimental and theoretical study of the collision-induced fundamental absorption spectra of N_2–O_2 and O_2–N_2 pairs. JQSRT **69**, 245 (2001)
728. Sneep M, Ubachs W. Cavity ring-down measurement of the O_2–O_2 collision-induced absorption resonance at 477 nm at sub-atmospheric pressures. JQSRT **78**, 171 (2003)
729. Baranov YI, Lafferty WJ, Fraser GT. Infrared spectrum of the continuum and dimer absorption in the vicinity of the O_2 vibrational fundamental in O_2/CO_2 mixtures. J Mol Spectrosc **228**, 432 (2004)
730. Birnbaum G, Borysow A, Buechele A. Collision-induced absorption in mixtures of symmetrical linear and tetrahedral molecules: Methane–nitrogen. J Chem Phys **99**, 3234 (1993)

731. Hartmann JM, Flaud PM, Tipping RH, Ma Q. Collision induced absorption in the v_2 fundamental band of CH_4. II. Dependence on the perturber. J Chem Phys **116**, 123 (2002)
732. Buckingham AD. Permanent and induced molecular moments and long-range intermolecular forces. Adv Chem Phys **12**, 107 (1967)
733. Buckingham AD, Tabisz GC. Collision-induced rotational Raman scattering by tetrahedral and octahedral molecules. Mol Phys **36**, 583 (1978)
734. Hunt KLC, Li X. Collision induced dipoles and polarizabilities for S state atoms or diatomic molecules. In Ref. 719, p. 61
735. Tabisz GC, Meinander N, Penner AR. Interaction induced rotational light scattering in molecular gases. In Ref. 715, p. 345
736. Meyer W. *Ab initio* calculations of collision induced dipole moments. In Ref. 715, p. 29
737. Meyer W, Frommhold F, Birnbaum G. Rototranslational absorption spectra of H_2–H_2 pairs in the far infrared. Phys Rev A **39**, 2434 (1989)
738. Meyer W, Frommhold L. *Ab initio* interaction-induced dipoles and related spectra. In Ref. 719, p. 441
739. Li X, Ahuja C, Harrison JF, Hunt KLC. The collision-induced polarizability of a pair of hydrogen molecules. J Chem Phys **126**, 214302 (2007)
740. Poll JD, Hunt JL. Analysis of the far infrared spectrum of gaseous N_2. Can J Phys **59**, 1448 (1981)
741. Hunt JL, Poll JD. Lineshape analysis of collision induced spectra of gases. Can J Phys **56**, 950 (1978)
742. Trafton LM. Induced spectra in planetary atmospheres. In Ref. 719, p. 517
743. Borysow A. Collision induced molecular absorption in stellar atmospheres. In Ref. 719, p. 529
744. Meyer W, Borysow A, Frommhold L. Absorption spectra of H_2–H_2 pairs in the fundamental band. Phys Rev A **40**, 6931 (1989)
745. Meyer W, Borysow A, Frommhold L. Collision-induced first overtone band of gaseous hydrogen from first principles. Phys Rev A **47**, 4065 (1993)
746. Bouanich JP, Brodbeck C, Drossart P, Lellouch E. Collision-induced absorption for H_2–H_2 and H_2–He interactions at 5 µm. JQSRT **42**, 141 (1989)
746b. Borysow J, Frommhold L, Birnbaum G. Collision induced rototranslational absorption spectra of H_2–He pairs at temperatures from 40 to 3000 K. Astrophys J **326**, 509 (1988)
747. Gautier D, Marten A, Baluteau JP, Bachet G. About unidentified features in the Voyager far infrared spectra of Jupiter and Saturn. Can J Phys **61**, 1455 (1983)
748. Tipping RH. Collision induced effects in planetary atmospheres. In Ref. 715, p. 727
749. Borysow A, Trafton J, Frommhold L, Birnbaum G. Modeling of pressure-induced far-infrared absorption spectra. Molecular hydrogen pairs. Astrophys J **296**, 644 (1985)
750. Borysow A, Frommhold L. A new computation of the infrared absorption by H_2 pairs in the fundamental band at temperatures from 600 to 5000 K. Astrophys J Lett **348**, L41 (1990)
751. Ling MSH, Rigby M. Toward an intermolecular potential for nitrogen. Mol Phys **51**, 855 (1984)
752. Buontempo U, Cunsolo S, Jacucci G, Weis JJ. The far infrared absorption spectrum of N_2 in the gas and liquid phases. J Chem Phys **63**, 2570 (1975)
753. Dagg IR, Anderson A, Yan S, Smith W, *et al.* Collision-induced absorption in nitrogen at low temperatures. Can J Phys **63**, 625 (1985)
754. Borysow A, Frommhold L. Theoretical collision induced rototranslational absorption spectra for the outer planets: H_2–CH_4 pairs. Astrophys J **304**, 849 (1986)
755. Borysow A, Frommhold L. Collision induced rototranslational absorption spectra of binary methane complexes (CH_4–CH_4). J Mol Spectrosc **123**, 293 (1987)
756. Buser M, Frommhold L, Gustafsson M, Moraldi M, *et al.* Far-infrared absorption by collisionally interacting nitrogen and methane molecules. J Chem Phys **121**, 2617 (2004)

757. Buser M, Frommhold L. Infrared absorption by collisional $CH_4 + X$ pairs, with $X = He$, H_2, or N_2. J Chem Phys **122**, 024301 (2005)
758. Li X, Champagne MH, Hunt KLC. Long-range, collision-induced dipoles of T_d–$D_{\infty h}$ molecule pairs: theory and numerical results for CH_4 or CF_4 interacting with H_2, N_2, CO_2, or CS_2. J Chem Phys **109**, 8416 (1998)
759. Borysow J, Frommhold L. The infrared and Raman lineshapes of pairs of interacting molecules. In Ref. 715, p. 67
760. Borysow A, Frommhold L. Theoretical collision induced rototranslational absorption spectra for modeling Titan's atmosphere: H_2–N_2 pairs. Astrophys J **303**, 495 (1986)
761. Zheng C, Borysow A. Rototranslational CIA spectra of H_2–H_2 at temperatures between 600 and 7000 K. Astrophys J **441**, 960 (1995)
762. Vigasin AA. Bound, metastable and free states of bimolecular complexes. Infrared Phys **32**, 461 (1991)
763. Kouzov AP, Tokhadze KG, Utkina SS. Buffer-gas effect on the rotovibrational line intensity distribution: analysis of possible mechanisms. Eur Phys J **D 12**, 153 (2000)
764. Tipping RH, Brown A, Ma Q, Boulet C. Collision-induced absorption in the $a^1\Delta_g(v'=0) \leftarrow X^3\Sigma_g^-(v=0)$ transition in O_2–CO_2, O_2–N_2, and O_2–H_2O Mixtures. J Mol Spectrosc **209**, 88 (2001)
765. Brown A, Tipping RH. Theoretical study of the collision-induced double transitions in $CO_2 - X$ ($X = H_2$, N_2, and O_2) pairs. J Mol Spectrosc **205**, 319 (2001)
766. Maté B, Fraser GT, Lafferty WJ. Intensity of the simultaneous vibrational absorption $CO_2(v_3=1) + N_2(v=1) \leftarrow CO_2(v_3=0) + N_2(v=0)$ at $4680\,cm^{-1}$. J Mol Spectrosc **201**, 175 (2000)
767. Borysow A, Moraldi M. Effects of anisotropic interaction on collision induced absorption by pairs of linear molecules. Phys Rev Lett **68**, 3686 (1992)
768. Borysow A, Moraldi M. On the role of the anisotropic interaction on collision induced absorption of systems containing linear molecules: The CO_2–Ar case. J Chem Phys **99**, 8424 (1993)
769. Gruszka M, Borysow A. Spectral moments of collision induced absorption of CO_2 pairs: The role of an interaction potential. J Chem Phys **101**, 3573 (1994)
770. Glaz W, Tabisz GC. Collisional propagation effects in collision-induced rotational spectra. Phys Rev **A 54**, 3903 (1996)
771. Birnbaum G, Chu SI, Frommhold L, Wright EL. Theory of collision-induced translation-rotation spectra: H_2–He. Phys Rev **A 29**, 595 (1984)
772. Moraldi M, Borysow A, Borysow J, Frommhold L. Collision induced rototranslational spectra of H_2–He: accounting for the anisotropic interaction. Phys Rev **A 34**, 632 (1986)
773. Schaefer J, McKellar ARW. Faint features of the rotational $S_0(0)$ and $S_0(1)$ transitions of H_2 A comparison of calculations and measurements at 77 K. Z Phys **D 15**, 51 (1990)
774. McKellar ARW, Schaefer J. Far-infrared spectra of hydrogen dimers: comparisons of experiment and theory for $(H_2)_2$ and $(D_2)_2$ at 20 K. J Chem Phys **95**, 3081 (1991)
775. Gustafsson M, Frommhold L, Meyer W. Infrared absorption spectra by H_2–He collisional complexes: the effect of the anisotropy of the interaction potential. J Chem Phys **113**, 3641 (2000)
776. Gustafsson M, Frommhold L. The HD–He complex: interaction-induced dipole surface and infrared absorption spectra. J Chem Phys **115**, 5427 (2001)
777. Gustafsson M, Frommhold L, Meyer W. The H_2–H complex: interaction-induced dipole surface and infrared absorption spectra. J Chem Phys **118**, 1667 (2003)
778. Gustafsson M, Frommhold L, Bailly D, Bouanich JP, et al. Collision-induced absorption in the rototranslational band of dense hydrogen gas. J Chem Phys **119**, 12264 (2003)
779. Gustafsson M, Frommhold L. Infrared absorption by H_2–Ar collisional complexes and the anisotropy of the intermolecular interaction potential. Phys Rev **A 74**, 054703 (2006)

780. Birnbaum G. Far infrared absorption in H_2 and H_2–He mixtures. JQSRT **19**, 51 (1978)
781. Birnbaum G, Bachet G, Frommhold L. Experimental and theoretical investigation of the far-infrared spectrum of H_2–He mixtures. Phys Rev **A 36**, 3729 (1987)
782. Ho W, Birnbaum G, Rosenberg A. Far-infrared collision-induced absorption in CO_2. I. Temperature dependence. J Chem Phys **55**, 1028 (1971)
783. Wishnow EH, Gush HP, Ozier I. Far-infrared spectrum of N_2 and N_2-noble gas mixtures near 80 K. J Chem Phys **104**, 3511 (1996)
784. Brocks G. Bound and rotational resonance states and the infrared spectrum of N_2Ar. J Chem Phys **88**, 578 (1988)
785. Wang F, McCourt FRW, Le Roy RJ. Dipole moment surfaces and the mid- and far-IR spectra of N_2–Ar. J Chem Phys **113**, 98 (2000)
786. Orlando JJ, Tyndall GS, Nickerson KE, Calvert JG. The temperature dependence of collision induced absorption by oxygen near 6 μm. J Geophys Res **D 96**, 20755 (1991)
787. Moreau G, Boissoles J, Le Doucen R, Boulet C, et al. Metastable dimer contributions to the collision-induced fundamental absorption spectra of N_2 and O_2 pairs. JQSRT **70**, 99 (2001)
788. McKellar ARW. Infrared spectra of weakly bound complexes and collision induced effects involving atmospheric molecules. In Ref. 670, p. 223
789. McKellar ARW. Infrared studies of van der Waals complexes: the low temperature limit of collision induced spectra. In Ref. 719, p. 467
790. Bafile U, Zoppi M, Barocchi F, Brown MS, et al. Depolarized light scattering of parahydrogen gas at low temperature. Phys Rev **A 40**, 1654 (1989)
791. Poll JD. The infrared spectrum of HD. In Ref. 715, p. 677
792. Lu Z, Tabisz GC, Ulivi L. Temperature dependence of the pure rotational band of HD: interference, widths, and shifts. Phys Rev **A 47**, 1159 (1993)
793. Ulivi L, Lu Z, Tabisz GC. Interference of allowed and collision induced transitions in HD: experiment. In Ref. 719, p. 407
794. McKellar ARW, Rich NH. Interference effects in the spectrum of HD: II. The fundamental band for HD-rare gas mixtures. Can J Phys **62**, 1665 (1984)
795. Fano U. Effects of configuration interaction on intensities and phase shifts. Phys Rev **124**, 1866 (1961)
796. Poll JD, Tipping RH, Prasad RDG, Reddy SP. Intracollisional interference in the spectrum of HD mixed with rare gases. Phys Rev Lett **36**, 248 (1976)
797. Tipping RH, Poll JD, McKellar ARW. The influence of intracollisional interference on the dipole spectrum of HD. Can J Phys **56**, 75 (1978)
798. Herman RM. Analysis of the $R_1(J)$- and $P_1(J)$-branch absorption spectrum of HD-rare-gas mixtures: An example of positive intercollisional interference. Phys Rev Lett **42**, 1206 (1979)
799. Herman RM, Tipping RH, Poll JD. Shape of the R and P lines in the fundamental band of gaseous HD. Phys Rev **A 20**, 2006 (1979)
800. Tabisz GC, Nelson JB. Rotational-level mixing and intracollisional interference in the pure rotational spectrum of HD gas. Phys Rev **A 31**, 1160 (1985)
801. Ma Q, Tipping RH, Poll JD. Mixing of rotational levels and intracollisional interference in the pure rotational $R_0(J)$ transitions of gaseous HD. Phys Rev **A 38**, 6185 (1988)
802. Gao B, Tabisz GC, Trippenbach M, Cooper J. Spectral line shape arising from collisional interferences between electric-dipole-allowed and collision-induced transitions. Phys Rev **A 44**, 7379 (1991)
803. Gao B, Cooper J, Tabisz GC. Rotational spectrum of HD perturbed by He or Ar gases: the effects of rotationally inelastic collisions on the interference between allowed and collisionally induced components. Phys Rev **A 46**, 5781 (1992)

804. Tabisz GC. Interference effects in the infrared spectrum of HD. In Ref. 670, p. 83
805. Gustafsson M, Frommhold L. Intracollisional interference of R lines of HD in mixtures of deuterium hydride and helium gas. Phys Rev **A 63**, 052514 (2001)
806. McKellar ARW, MacTaggart JW, Welsh HL. Studies in molecular dynamics by collision-induced infrared absorption in H_2-rare gas mixtures. III. H_2–He mixtures at low temperatures and densities. Can J Phys **53**, 2060 (1975)
807. Kelley JD, Bragg SL. Asymmetry of the intercollisional interference dips in the collision-induced absorption spectrum of molecular hydrogen. Phys Rev **A 29**, 1168 (1984)
808. Welsh HL. The pressure induced infrared spectrum of hydrogen and its application to the study of planetary atmospheres. J of Atm Sci **26**, 835 (1969)
809. Van Kranendonk J. Intercollisional interference effects in pressure-induced infrared spectra. Can J Phys **46**, 1173 (1968)
810. Lewis JC. Intercollisional interference effects. In Intermolecular Spectroscopy and Dynamical Properties of Dense Systems. J Van Kranendonk Ed., Elsevier science (1980)
811. Lewis JC. Intercollisional interference; theory and experiment. In Ref. 715, p. 215
812. Moraldi M, Frommhold L. Three body induced dipole moments and infrared absorption : the H_2 fundamental band. Phys Rev **A 49**, 4508 (1994)
813. Moraldi M, Frommhold L. Triple transition $Q_1(J_1) + Q_1(J_2) + Q_1(J_3)$ near $12466 \, cm^{-1}$ in compressed hydrogen. Phys Rev Lett **74**, 363 (1995)
814. Gustafsson M, Frommhold L. Spectra of two and three-body van der Waals complexes. In Ref. 670, p. 3
815. Moraldi M, Celli M, Barocchi F. Theory of virial expansion of correlation functions and spectra: application to interaction-induced spectroscopy. Phys Rev **A 40**, 1116 (1989)
816. Dore P, Filabozzi A, Birnbaum G. Rototranslational absorption in gaseous H_2–Ar mixtures at intermediate densities. Can J Phys **67**, 599 (1989)
817. Bafile U, Ulivi L, Zoppi M, Moraldi M, *et al*. Third virial coefficients of collision induced depolarized light scattering of hydrogen. Phys Rev **A 44**, 4450 (1991)
818. Moraldi M, Frommhold L. Three-body components of collision-induced absorption. Phys Rev **A 40**, 6260 (1989)
819. PaddyReddy S, Xiang F, Varghese G. Observation of the new triple transitions $Q_1(J_1) + Q_1(J_2) + Q_1(J_3)$ in molecular hydrogen in its second overtone region. Phys Rev Lett **74**, 367 (1995)
820. Tabisz GC, Allin EJ, Welsh HL. Interpretation of the visible and near infrared absorption spectra of compressed oxygen as collision induced electronic transitions. Can J Phys **47**, 2859 (1969)
821. Hermans C, Vandaele AC, Fally S, Carleer M, *et al*. Absorption cross-sections of the collision induced bands of oxygen from the UV to the NIR. In Ref. 670, p. 193
822. Sneep M, Ubachs W. Cavity ring down spectroscopy of O_2–O_2 collisional induced absorption. In Ref. 670, p. 203
823. Moraldi M, Frommhold L. Dipole moments induced in three interacting molecules. J Molec Liquids **70**, 143 (1996)
824. Penner SS. Quantitative Molecular Spectroscopy and Gas Emissivities. Addison-Wesley, Reading, MA, 1959
825. Siegel R, Howell JR. Thermal Radiation Heat Transfer (4th edition). Taylor & Francis, NY, (2001)
826. Modest M. Radiative Heat Transfer. Elsevier, (2003)
827. Goody RM, Yung YL. Atmospheric Radiation: Theoretical Basis. Oxford Univ. Press (1989)
828. Hartmann JM, Bouanich JP, Boulet C, Sergent M. Absorption of radiation by gases from low to high pressures. I. Empirical line-by-line and narrow-band statistical models. J de Physique **II 1**, 739 (1991)

829. Coelho PJ. Detailed numerical simulation of radiative transfer in a nonluminous turbulent jet diffusion flame. Combust Flame **136**, 481 (2004)
830. Chen Y, Liou KN. A Monte Carlo method for 3D thermal infrared radiative transfer. JQSRT **101**, 166 (2006)
831. Kneizys FX, Shettle EP, Gallery WO, Chetwynd JH, *et al*. Atmospheric transmittance/radiance: computer code LOWTRAN 6. Air Force Geophysics Laboratroy, Report AFGL-TR-83-0187, Hanscom AFB, MA. 1983
832. Berk A, Bernstein LW, Robertson DC. MODTRAN: A moderate resolution model for LOWTRAN 7. Philips Laboratroy, Report AFGL-TR-89-0122, Hanscom AFB, MA. (1989)
833. Houghton JT, Taylor FW, Rodgers CD. Remote Sounding of Atmospheres. Cambridge Planetary Science Series 5, Cambridge University Press, 1984
834. Rodgers CD. Inverse Methods for Atmospheric Sounding: Theory and Practice. World Scientific Publ, Singapore (2000)
835. Hanel R, Conrath B, Jennings D, Samuelson R. Exploration of the Solar System by Infrared Remote Sensing. 2nd edition, Cambridge University Publishers (2003)
836. Jay Barker (Edt) Jeffries, Kohse-Hoinghaus. Applied Combustion Diagnostics. Taylor & Francis (2002)
837. Steck T. Methods for determining regularization for atmospheric retrieval problems. Appl Opt **41**, 1788 (2002)
838. Rodgers CD, Connor BJ. Intercomparison of remote sounding instruments. J Geophys Res **D 108**, 4116 (2003)
839. Koner PK, Drummond JR. A comparison of regularization techniques for atmospheric trace gases retrievals. JQSRT **109**, 514 (2008)
840. Goldman A. On simple approximations to the equivalent width of a Lorentz line. JQSRT **8**, 829 (1968)
841. Durry G, Mégie G. Atmospheric CH_4 and H_2O monitoring with near-infrared InGaAs diodes by the SDLA, a balloon borne spectrometer for tropospheric and stratospheric in situ measurements. Appl Opt **38**, 7342 (1999)
842. Stiller GP, Gunson MR, Lowes LL, Abrams MC, *et al*. Stratospheric and mesospheric pressure-temperature profiles from rotational analysis of CO_2 lines in atmospheric trace molecule spectroscopy/ATLAS 1 infrared solar occultation spectra. J Geophys Res **D 100**, 3107 (1995)
843. Teichert H, Fernholz T, Ebert V. Simultaneous in situ measurement of CO, H_2O, and gas temperatures in a full-sized coal-fired power plant by near-infrared diode lasers. Appl Opt **42**, 2043 (2003)
844. Zhou X, Jeffries JB, Hanson RK. Development of a fast temperature sensor for combustion gases using a single tunable diode laser. Appl Phys **B 81**, 711 (2005)
845. Liljegren JC, Boukabara SA, Cady-Pereira K, Clough SA. The effect of the half-width of the 22-GHz water vapor line on retrievals of temperature and water vapor profiles with a 12-channel microwave radiometer. IEEE Trans on Geosci Remote Sensing **43**, 1102 (2005)
846. Hussong J, Stricker W, Bruet X, Joubert P, *et al*. Hydrogen CARS thermometry in H_2–N_2 mixtures at high pressure and medium temperatures. Influence of the linewidth model. Appl Phys **B 70**, 447 (2000)
847. Zuev VV, Ponomarev YN, Solodov AM, Tikhomirov BA, *et al*. Influence of the shift H_2O absorption lines with air pressure on the accuracy of the atmospheric humidity profiles measured by the differential-absorption method. Opt Lett **10**, 318 (1985)
848. Pumphrey HC, Bühler S. Instrumental and spectral parameters: their effect on and measurement by microwave limb sounding of the atmosphere. JQSRT **64**, 421 (2000)
849. Sokolov A, Khomenko G, Dubuisson P. Sensitivity of atmospheric-surface parameters retrieval to the spectral stability of channels in thermal IR. JQSRT, in press, (2008)

850. Ortland DA, Skinner WR, Hays PB, Burrage MD, et al. Measurements of stratospheric winds by high resolution Doppler imager. J. Geophys Res. **101**, 10351 (1996)
851. Soufiani A, Taine J. High-resolution spectroscopy temperature measurements in laminar channel flows. Appl Opt **27**, 3754 (1988)
852. Durry G, Zéninari V, Parvitte B, Le Barbu T, et al. Pressure broadening coefficients and line strengths of H_2O near 1.39 µm: application to the in situ sensing of the middle atmosphere with balloon-borne diode lasers. JQSRT **94**, 387 (2005)
853. Barret B, Hurtmans D, Carleer MR, De Mazière M, et al. Line narrowing effect on the retrieval of HF and HCl vertical profiles from ground-based FTIR measurements. JQSRT **95**, 499 (2005)
854. Hancock RD, Bertagnolli KE, Lucht RP. Nitrogen and hydrogen CARS temperature measurements in a hydrogen/air flame using a near-adiabatic flat-flame burner. Combust Flame **109**, 323 (1997)
855. Boone CD, Walker KA, Bernath PF. Speed-dependent Voigt profile for water vapor in infrared remote sensing applications. JQSRT **105**, 525 (2007)
856. Kratz DP. The sensitivity of radiative transfer calculations to the changes in the HITRAN database from 1982 to 2004. JQSRT, in press (2008)
857. Millot G, Lavorel B, Fanjoux G, Wenger C. Determination of temperature by stimulated Raman scattering of molecular nitrogen, oxygen and carbon dioxide. Appl Phys **B 56**, 287 (1993)
858. Jourdanneau E, Gabard T, Chaussard F, Saint-Loup R, et al. CARS methane spectra: simulations for temperature diagnostic purposes. J Mol Spectrosc **246**, 167 (2007)
859. Armstrong RL. Line mixing in the ν_2 band of CO_2. Appl Opt **21**, 2141 (1982)
860. Camy-Peyret C. Balloon-borne infrared Fourier transform spectroscopy for measurements of atmospheric trace species. Spectrochim Acta **A 51**, 1143 (1995)
861. Smith WL, Revercomb HE, Howell HB, Woolf HM. HIS – A satellite instrument to observe temperature and moisture profiles with high vertical resolution. Preprints of the Fifth Conference on Atmospheric Radiation. Amer Meteor Soc, 1–9 ((1983) See also http://deluge.ssec.wisc.edu/~his/his/his.html
862. Funke B, Stiller GP, von Clarmann T, Eschle G, et al. CO_2 line-mixing in MIPAS limb emission spectra and its influence on retrieval of atmospheric parameters. JQSRT **59**, 215 (1998)
863. Endemann M. MIPAS instrument concept and performances. Proc of the European symposium on atmospheric measurements from space **1**, 29. Noordwijk, Netherland, January 1999
864. Kopp G, Berg H, Blumenstock Th, Fischer H, et al. Evolution of ozone and ozone related species over Kiruna during the THESEO 2000 – SOLVE campaign retrieved from ground-based millimeter wave and infrared observations. J Geophys Res **D 108**, 8308 (2003)
865. Stiller GP, von Clarmann T, Funke B, Glatthor N, et al. Sensitivity of trace gas abundances retrievals from infrared limb emission spectra to simplifying approximations in radiative transfer modeling. JQSRT **72**, 249 (2002)
866. Sherlock VJ, Jones NB, Matthews WA, Murcray FJ, et al. Increase in the vertical column abundance of HCFC-22 ($CHClF_2$) above Lauder, New Zealand, between 1985 and 1994. J Geophys Res **D 102**, 8861 (1997)
867. Flaud JM, Brizzi G, Carlotti M, Perrin A, et al. MIPAS database: Validation of HNO_3 line parameters using MIPAS satellite measurements. Atmos Chem Phys **6**, 5037 (2006)
868. Vander Auvera J, Moazzen-Ahmadi N, Flaud JM. Toward an accurate database for the 12 µm region of the ethane spectrum. Astrophys J **662**, 750 (2007)
869. Niro F, Brizzi G, Carlotti M, Papandrea E, et al. Precision improvements in the geo-fit retrieval of pressure and temperature from MIPAS limb observations by modeling CO_2 line-mixing. JQSRT **103**, 14 (2007)

870. Rinsland CP, Gunson MR, Zander R, Lopez-Puertas M. Middle and upper atmosphere pressure-temperature profiles and the abundances of CO_2 and CO in the upper atmosphere from ATMOS/Spacelab 3 observations. J Geophys Res **D 97**, 20479 (1992)
871. Carlotti M, Dinelli BM, Raspolini P, Ridolfi M. Geo-Fit approach to the analysis of limb-scanning satellite measurement. Appl Opt **40**, 1872 (2001)
872. Shephard M, Payne V, Clough T. Line-By-Line Radiative Transfer Model (LBLRTM). Improvements and validations. Paper in preparation
873. Clough SA, Iacono MJ. Line-by-line calculations of atmospheric fluxes and cooling rates: Application to carbon dioxide, ozone, methane, and the halocarbons. J Geophys Res **D 100**, 16519 (1995)
874. Liljegren J, Cadeddu MP. Retrievals of atmospheric temperature and water vapor in the arctic. 16th ARM science team meeting proceedings, Albukerque, NM, March (2006). See also IEEE Microrad 2006, p. 241
875. Cadeddu MP, Cady-Pereira K, Clough SA, Liljegren J. Improving the modeling of oxygen-band absorption: A model-measurement comparison. IEEE MicroRad 2006, 259 (2006)
876. Boukabara S, Clough SA, Moncet JL, Krupnov AF, et al. Uncertainties in the temperature dependence of the line-coupling parameters of the microwave oxygen band: impact study. IEEE Trans on Geosci Remote and Sensing **43**, 1109 (2005)
877. Rosenkranz PW. Comment on "Uncertainties in the temperature dependence of the line-coupling parameters of the microwave oxygen band: impact study". IEEE Trans on Geosci Remote Sensing **43**, 2160 (2005) and reply to comment in the following two pages
878. Funk O, Pfeilsticker K. Photon path lengths distributions for cloudy skies – Oxygen A-band measurements and model calculations. Annal Geophysicae **21**, 615 (2003)
879. Corradini S, Cervino M. Aerosol extinction coefficient profile retrieval in the oxygen A-band considering multiple scattering atmosphere. Test case: SCIAMACHY nadir simulated measurements. JQSRT **97**, 354 (2006)
880. Pitts MMC, Thomason LW. Remote sensing of temperature and pressure by the Strastospheric Aerosol and Gas Experiment III. In Satellite Remote Sensing of Clouds and the Atmosphere IV, proceedings of SPIE, edited by J. E. Russell, vol **3867**, 206 (1999)
881. Sugita T, Yokota T, Nakajima T, Nakajima H, et al. Temperature and pressure retrievals from O_2 A-band absorption measurements made by ILAS: retrieval algorithm and error analyses. In Optical Remote Sensing of the atmosphere and clouds II, Proceedings of SPIE. Y. Sasano, J. Wang, and T. Hayasaka Eds., vol 4150, 94 (2001)
882. Van Diedenhoven B, Hasekamp OP, Aben I. Surface pressure retrieval from SCIAMACHY measurements in the O_2 A band: validation of the measurements and sensitivity on aerosols. Atmos Chem Phys **5**, 2109 (2005)
883. Bösch H, Toon GC, Sen B, Washenfelder RA, et al. Space-based near-infrared CO_2 measurements: testing the Orbiting Carbon Observatory retrieval algorithm and validation concept using SCIAMACHY observations over Park Falls, Wisconsin. J Geophys Res **D 111**, 7080 (2006)
884. Yang Z, Wennberg PO, Cageao RP, Pongetti TJ, et al. Ground-based photon path measurements from solar absorption spectra of the O_2 A-band. JQSRT **90**, 309 (2005)
885. Crisp D, Titov D. The thermal balance of the Venus atmosphere. In Venus II, Edited by Bougher SW, Hunten DM, Philips RJ. Tucson, AZ: University of Arizona Press, (1997)
886. Afanasenko TS, Rodin AV. The effect of collisional line broadening on the spectrum and fluxes of thermal radiation in the lower atmosphere of Venus. Solar System Research **39**, 187 (2005)
887. Ibgui L, Valentin A, Mérienne MF, Jenouvrier A, et al. An optimized line by line code for plume signature calculations. II. Comparisons with measurements. JQSRT **74**, 401 (2002)
888. Clough SA, Iacono MJ, Moncet JL. Line-by-line calculations of atmospheric fluxes and cooling rates: application to water vapor. J Geophys Res **D 97**, 15761 (1992)

889. Vogelmann AM, Ramanathan V, Conant WC, Hunter WE. Observational constraints on non-Lorentzian continuum effects in the near-infrared solar spectrum using ARM ARESE data. JQSRT **60**, 231 (1998)
890. Iacono MJ, Mlawer EJ, Clough SA, Morcrette JJ. Impact of an improved longwave radiation model, RRTM, on the energy budget and thermodynamic properties of the NCAR communityclimate model. J Geophys Res **D 105**, 14873 (2000)
891. Buehler SA, von Engeln A, Brocard E, John VO, et al. Recent developments in the line-by-line modeling of outgoing longwave radiation. JQSRT **98**, 446 (2006)
892. De Berg C, Bézard B, Crisp D, Maillard JP, et al. Water in the deep atmosphere of Venus from high-resolution spectra of the night side. Adv Space Res **15**, 79 (1995)
893. Marcq E, Encrenaz T, Bézard B, Birlan M. Remote sensing of Venus' lower atmosphere from ground-based IR spectroscopy: latitudinal and vertical distribution of minor species. Planet Space Sci **54**, 1360 (2006)
894. Clough SA. The Water vapor continuum and its role in remote sensing. In Optical Remote Sensing of the Atmosphere, Vol. 2, 1995, OSA Technical Digest Series, (Optical Society of America, Washington, DC, 1995), p. 76
895. Pollack JB, Dalton JB, Grinspoon D, Wattson RB, et al. Near-infrared light from Venus' nightside: a spectroscopic analysis. Icarus **103**, 1 (1993)
896. French AN, Norman JM, Anderson MC. A simple and fast atmospheric correction for spaceborne remote sensing of surface temperature. Remote Sens and Env **87**, 326 (2003)
897. Sobrino JA, Jiménez-Muñoz JC, Zarco-Tejada PJ, Sepulcre-Cantó G, et al. Land surface temperature derived from airborne hyperspectral scanner thermal infrared data. Remote Sens and Envir **102**, 99 (2006)
898. Negrao A, Coustenis A, Lellouch E, Maillard JP, et al. Titan's surface albedo from near-infrared CFHT/FTS spectra: modeling dependence on the methane absorption. Planet Space Sci **54**, 1225 (2006)
899. Greenhough J, Remedios JJ, Sembhi H, Kramer LJ. Toward cloud detection and cloud frequency distributions from MIPAS infra-red observations. Adv Space Res **36**, 800 (2005)
900. Coustenis A, Negrao A, Salama A, Schulz B, et al. Titan's 3-micron spectral region from ISO high-resolution spectroscopy. Icarus **180**, 176 (2006)
901. Griffith CA, Penteado P, Rannou P, Brown R, et al. Evidence for ethane clouds on Titan from Cassini VIMS observations. Science **313**, 1620 (2006)
902. Strow LL, Hannon SE, De-Souza Machado S, Motteler HE, et al. Validation of the Atmospheric Infrared Sounder radiative transfer algorithm. J Geophy Res **D 111**, 09S06 (2006)
903. Roos-Serote M, P Drossart P, Encrenaz Th, Lellouch E, et al. The thermal structure and dynamics of the atmosphere of Venus between 70 and 90 km from the Galileo-NIMS spectra. Icarus **114**, 300 (1995)
904. Zasova LV, Moroz VI, Formisano V, Ignatiev NI, et al. Infrared spectrometry of Venus: IR Fourier spectrometer on Venera 15 as a precursor of PFS for Venus express. Adv Space Res **34**, 1655 (2004)
905. Rinsland CP, McHugh MJ, Irion FW. Lower stratospheric densities from solar occultation measurements of continuum absorption near $2400\,cm^{-1}$. J Geophys Res **D 109**, 1301 (2003)
906. Solomon S, Portmann RW, Sanders RW, Daniel JS. Absorption of solar radiation by water vapor, oxygen, and related collision pairs in the Earth's atmosphere. J Geophys Res **D 103**, 3847 (1998)
907. Chagas JCS, Newnham DA, Smith KM, Shine KP. Impact of new measurements of oxygen collision-induced absorption on estimates of short-wave atmospheric absorption. Quart J Royal Met Soc **128**, 2377 (2002)

908. Tolchenov R, Tennyson J. Water line parameters from refitted spectra constrained by empirical upper state levels: study of the 9500–14,500 cm^{-1} region. JQSRT **109**, 559 (2008)
909. Burgdorf M, Orton GS, van Cleve J, Meadows V, et al. Detection of new hydrocarbons in Uranus' atmosphere by infrared spectroscopy. Icarus **184**, 634 (2006)
910. Orton GS, Gustafsson M, Burgdorf M, Meadows V. Revised *ab initio* models for H$_2$–H$_2$ collision-induced absorption at low temperatures. Icarus **189**, 544 (2007)
911. Courtin R, Gautier D, McKay CP. Titan's thermal emission spectrum: Reanalysis of the Voyager infrared measurements. Icarus **114**, 144 (1995)
912. Samuelson RE, Nath NR, Borysow A. Gaseous abundances and methane supersaturation in Titan's troposphere. Planet Space Sci **45**, 959 (1997)
913. Orton GS, Aitken DK, Smith C, Roche PF, et al. The spectra of Uranus and Neptune at 8–14 and 17–23 µm. Icarus **70**, 1 (1987)
914. Oertel O, Spänkuch D, Jahn H, Becker-Ross H, et al. Infrared spectrometry of Venus from "Venera-15" and "Venera-16". Adv Space Res **5**, 25 (1985)
915. Macko P, Romanini D, Mikhailenko SN, Naumenko OV, et al. High sensitivity CW-cavity ring down spectroscopy of water in the region of the 1.5 µm atmospheric window. J Mol Spectrosc **227**, 90 (2004)
916. Flaud JM, Perrin A, Orphal J, Kou Q, et al. New analysis of the $v_5 + v_9 - v_9$ hot band of HNO$_3$. JQSRT **77**, 355 (2003)
917. Grisch F, Bouchardy P, Clauss W. CARS thermometry in high pressure rocket combustors. Aerosp Sci Technol **7**, 317 (2003)
918. Chaussard F, Michaut X, Saint-Loup R, Berger H, et al. Optical diagnostic of temperature in rocket engines by coherent Raman techniques. CR Phys Acad Sci Paris **5**, 249 (2004)
919. Brown LR, Benner DC, Champion JP, Devi VM, et al. Methane line parameters in HITRAN. JQSRT **82**, 218 (2003)
920. Pine AS. Line mixing sum rules for the analysis of multiplet spectra. JQSRT **57**, 145 (1997)
921. Gabard T. Etude des effets collisionnels dans les molécules tétraédriques. Applications au méthane perturbé par l'argon. Thesis, Université de Bourgogne, Dijon, France, (1996)
922. Pine AS. Self-, N$_2$-, O$_2$-, H$_2$-, Ar- and He-broadening in the v_3 band Q branch of CH$_4$. J Chem Phys **97**, 773 (1992)
923. Brault JW, Brown LR, Chackerian C, Freedman R, et al. Self-broadened ^{12}C^{16}O line shapes in the $v = 2 \leftarrow 0$ band. J Mol Spectrosc **222**, 220 (2003)
924. Varanasi P, Sarangi SK, Tejwani GDT. Line shape parameters for HCl and HF in a CO$_2$ atmosphere. JQSRT **12**, 857 (1972)
925. Tokhaze KG, Mielke Z. About origin of Q component on the vibration-rotation band of HF in simple solvent. J Chem Phys **99**, 5071 (1993)
926. Herman RM. Theory of the rare-gas-induced Q branch in HCl vibration–rotation spectra at high pressures. J Chem Phys **52**, 2040 (1970)
927. Piollet-Mariel E, Boulet C, Lévy A. Theoretical calculation of the intensity modification of HCl vibration-rotation lines by rare gases at moderate densities. J Chem Phys **76**, 787 (1982)
928. Flaud JM, Piccolo C, Carli B. A spectroscopic database for MIPAS. Proc. of Envisat Validation Workshop, Frascati, Italy, 9–13 December 2002. ESA (August 2003) SP-531
929. Jacquemart D, Gamache RE, Rothman LS. Semi-empirical calculation of air-broadened half-widths and air pressure-induced frequency shifts of water-vapor absorption lines. JQSRT **96**, 205 (2005)
930. Baldacchini G, Buffa G, D'Amato F, Tarrini O, et al. New results for the temperature dependence of self-broadening and shift in the v_2 ammonia band. JQSRT **67**, 365 (2000)
931. Fisher J, Gamache RR, Goldman A, Rothman LS, et al. Total internal partition sums for molecular species in the 2000 edition of the HITRAN database. JQSRT **82**, 401 (2003)

932. Birnbaum G, Borysow A, Orton GS. Collision induced absorption of H_2–H_2 and H_2–He in the rotational and fundamental bands (0–6000 cm^{-1}) for planetary applications. Icarus **123**, 4 (1996). Data are available at http://www.astro.ku.dk/~aborysow/
933. Taylor BN. Guide for the Use of the International System of Units (SI). NIST Special Publication 811, available at http://physics.nist.gov/Pubs/SP811/contents.html
934. Taylor BN. Guide for the Use of the International System of Units (SI): The metric system (2nd Ed.). Diane Pub Co, (1995)

SUBJECT INDEX

NOTE In order to keep the subject index of reasonable size, the information given in the *Table of contents* and in the *Acronyms list* (page 357) is not duplicated here. The reader is kindly invited to refer to these parts of the book if this index does not provide the information searched for. Similarly, the information concerning *line-mixing* effects for various molecular species provided in Sec. IV.5 (page 228) is not repeated here.

Absorption coefficient:
    CIA, 30, 282
    continuum, 247, 268
    general, 12, 249
    quadrupolar, 35
    susceptibility, 33
    with χ factors, 244
Adiabaticity corrections, 193, 263

CFC's:
    optical soundings, 322
    Q branch, 176
CH_3F:
    band, 198
    manifolds (Stark), 209
    non Voigt lines, 95
    Q branch, 198
CH_4:
    CIA, 275, 281, 332
    manifolds, 152, 168, 210, 225, 340
    non Voigt lines, 340
    optical soundings, 313, 319, 320, 332
    Q branch, 210, 235, 319
    radiative transfer, 331
    wings, 210, 244
CO:
    band, 217
    broadening and shifting, 69, 213, 214
    non Voigt lines, 93, 136
    Q branch, 217
    wings, 217, 244
CO_2:
    band, 153, 196, 226
    broadening and shifting, 69, 190
    CIA/CILS, 275, 283, 289
    dimers, 293

    optical soundings, 318, 321, 323, 328, 330
    Q branch, 161, 164, 186, 196
    radiative transfer, 325
    wings, 48, 153, 167, 196, 244, 256, 264
Collisions:
    binary, 17, 47
    duration, 47, 49, 51
    finite duration effects, 52, 72, 110, 167, 242, 345
    hard, 81, 138, 174, 335
    soft, 80, 138
    speed changing, 70, 122, 138
    ternary, 301
    velocity changing, 26, 75, 122

Dimers and bound states, 32, 282f, 290, 347

H_2 and D_2:
    broadening and shifting, 69, 81, 115
    CIA/CILS, 260, 275, 278, 283, 287, 294
    dimers, 280, 289
    non Voigt lines, 68, 87, 90, 107, 115, 125, 129, 133
    optical soundings, 313, 316, 318, 333
    radiative transfer, 331
    wings, 260, 287
H_2O:
    bands, 266
    broadening and shifting, 207
    dimers, 270
    manifolds, 223
    optical soundings, 315, 317, 329
    radiative transfer, 326
    wings, 244, 246, 254, 266
HCl:
    bands, 347
    broadening and shifting, 66, 213, 239, 346

HCl: (*continued*)
  optical soundings, 315
  Q branch, 347
  wings, 347
HD:
  broadening and shifting, 71
  CIA, 297
  intracollisional, 297
HF:
  broadening and shifting, 48, 72, 144, 202, 213, 236, 239
  non Voigt lines, 88, 111, 130
  optical sounding, 315
Higher order density corrections:
  excluded volume, 48
  intercollisional, 29, 298, 300, 301
  spectral example, 287

Interaction potential:
  atom-atom, 205
  diatom-atom, 201, 212, 285
  electrostatic, 204
  molecule-molecule, 251, 259
  radiative coupling, 286

Line asymmetry:
  collision duration, 50, 72
  intracollisional, 287
  line mixing, 158
  parameter, 63, 123
  speed dependence, 92, 122
  velocity changes, 86, 88, 122
Line broadening:
  Dicke, 78
  diffusional, 76
  Doppler, 74
  Lorentz:
    calculations, 199, 211
    concentration effect, 91, 115
    equivalent width, 311
    general, 64
    in optical soundings, 312, 315
    speed dependence, 96, 101, 350
    T-dependence, 103, 207, 349
  natural, 73
  resonant, 238
  vibrational, 163, 174, 235

Line coupling:
  with Doppler effect, 160, 172
  equivalent lines, 163, 165, 168, 171, 337
  first-order approximation, 158, 169, 224, 338
  general, 174
  narrowing/broadening of structures, 163, 185
  second-order approximation, 161
  strong overlapping, 162
  sum rules, 157, 160, 173, 225, 339
Line intensity:
  allowed lines, 157, 219
  equivalent lines, 172
  induced lines, 282
Line shifting:
  due to line mixing, 221
  Lorentz:
    calculations, 199, 211
    general, 67
    in optical soundings, 313
    speed dependence, 96, 101, 108, 350
    T-dependence, 103, 350
  vibrational, 174, 235

N_2:
  broadening and shifting, 67, 183
  CIA/CILS, 260, 264, 275, 280, 291, 294
  dimers, 292
  optical soundings, 316, 318, 333
  Q branch, 48, 151, 184, 197
  radiative transfer, 332
  wings, 260, 264
NH_3:
  manifolds, 199, 210, 221
  Q branch, 199
  wings, 244

O_2:
  bands, 198
  CIA/CILS, 7, 275
  Dimers, 293
  non Voigt lines, 90
  optical soundings, 318, 324
  radiative transfer, 331

Relaxation matrix:
  broadening and shifting, 173, 174
  detailed balance, 61, 173, 261

Relaxation matrix: (*continued*)
 diagonalization procedure, 171, 336
 on/of energy shell, 54, 61, 243
 impact, 173
 non impact, 257, 261, 343
 renormalization procedure, 192
 in spectral density, 21, 24
 speed dependence, 336
 sum rule, 157, 166, 173, 235, 237, 262
 symmetrized, 157, 261

Spectral shapes:
 CIA (Birnbaum and Cohen), 281
 Dicke, 78
 diffusional, 76
 dispersive:
  finite collision duration, 50, 72
  intracollisional ((Fano), 298
  line mixing (Rosenkranz), 160, 338
 Doppler, 74
 Galatry, 80, 140
 Lorentz, 22, 78, 156
 models for isolated lines, 138

 Super Lorentz, 245, 268
 Van Vleck-Weisskopf, 157
 Voigt, 79, 140
 χ factors, 243
Spectrum fits:
 multispectrum, 221
 for optical soundings, 307
 spectrum-by-spectrum, 219

Time domain experiments:
 general, 42
 line mixing, 197, 227
 optical soundings, 316
 speed dependence, 95, 125
Times:
 collision duration, 47, 54, 263
 free flight, 47, 53
 of momentum relaxation, 194
 of velocity changes, 122, 124
Transition moment:
 general, 29
 induced, 30, 276, 299
 reduced, 18, 189, 193, 261